Biology of the Sialic Acids

Biology of the Sialic Acids

Edited by

Abraham Rosenberg

University of California, Los Angeles
Los Angeles, California

SPRINGER SCIENCE+BUSINESS MEDIA, LLC

Library of Congress Cataloging-in-Publication Data

Biology of the sialic acids / edited by Abraham Rosenberg.
p. cm.
Includes bibliographical references and index.
ISBN 978-1-4757-9506-6 ISBN 978-1-4757-9504-2 (eBook)
DOI 10.1007/978-1-4757-9504-2
1. Sialic acids--Physiological effect. I. Rosenberg, Abraham, 1924- .
QP801.S47B566 1995
574.19'24--dc20 95-19224
CIP

ISBN 978-1-4757-9506-6

Originally published by Plenum Press, New York in 1995
Softcover reprint of the hardcover 1st edition 1995

10 9 8 7 6 5 4 3 2 1

Contributors

Manju Basu Department of Chemistry and Biochemistry, University of Notre Dame, Notre Dame, Indiana 46556

Shib Sankar Basu Department of Chemistry and Biochemistry, University of Notre Dame, Notre Dame, Indiana 46556

Subhash Basu Department of Chemistry and Biochemistry, University of Notre Dame, Notre Dame, Indiana 46556

Veer P. Bhavanandan Department of Biochemistry and Molecular Biology, The Milton S. Hershey Medical Center, Pennsylvania State University, Hershey, Pennsylvania 17033

Kiyoshi Furukawa Department of Biochemistry, Institute of Medical Science, Univeristy of Tokyo, Tokyo 108, Japan; *present address:* Department of Biosignal Research, Tokyo Metropolitan Institute of Gerontology, Tokyo 173, Japan

Sen-itiroh Hakomori The Biomembrane Institute, and Departments of Pathobiology and Microbiology, University of Washington, Seattle, Washington 98195

Jürgen Hausmann Virology Institute, University of Marburg, D-35037 Marburg, Germany

Georg Herrler Virology Institute, University of Marburg, D-35037 Marburg, Germany

Masao Iwamori Department of Biochemistry, Faculty of Medicine, The University of Tokyo, Tokyo 113, Japan

Sörge Kelm Biochemistry Institute, University of Kiel, D-24098 Kiel, Germany

Hans-Dieter Klenk Virology Institute, University of Marburg, D-35037 Marburg, Germany

Yoshitaka Nagai Mitsubishi Kasei Institute of Life Sciences; Machida City, Tokyo, and the Glycobiology Research Group, Frontier Research Program, The Institute of Physical and Chemical Research (RIKEN), Wako City, Saitama, Japan

Gerd Reuter Biochemistry Institute, University of Kiel, D-24098 Kiel, Germany

Peter Roggentin Biochemistry Institute, University of Kiel, D-24098 Kiel, Germany

Abraham Rosenberg Neuropsychiatric Institute and Brain Research Institute, University of California, Los Angeles, Los Angeles, California 90024-1759

Megumi Saito Department of Biochemistry and Molecular Biophysics, Medical College of Virginia, Virginia Commonwealth University, Richmond, Virginia 23298

Roland Schauer Biochemistry Institute, University of Kiel, D-24098 Kiel, Germany

Lee Shaw Biochemistry Institute, University of Kiel, D-24098 Kiel, Germany

Kunihiko Suzuki Brain and Development Research Center, Departments of Neurology and Psychiatry, University of North Carolina School of Medicine, Chapel Hill, North Carolina 27599

Frederic A. Troy II Department of Biological Chemistry, University of California, School of Medicine, Davis, California 95616

Robert K. Yu Department of Biochemistry and Molecular Biophysics, Medical College of Virginia, Virginia Commonwealth University, Richmond, Virginia 23298

Preface

Breakthroughs worldwide have placed the cellular and molecular biology of the sialic acids in the forefront of current scientific inquiry. The accumulating wealth of new information in this field of study is presented in this volume for use by both students and experts. The contributing authors are highly ranked among those presently building up the store of knowledge gained from this very lively international research effort. It is with wonderment that I have come to realize that their gifts of knowledge are made at the time of my seventieth birthday. This book is dedicated with great affection to my wife, Estelle, to my children, Ruth Ann and Jonathan, to their mates, Stephen and Anne-Marie, and to grandson Turner.

Abraham Rosenberg

Contents

Chapter 8
Biochemistry and Function of Sialidases

Megumi Saito and Robert K. Yu

Chapter 9
Sialic Acid as Receptor Determinant of Ortho- and Paramyxoviruses

Georg Herrler, Jürgen Hausmann, and Hans-Dieter Klenk

Chapter 10
Sialic Acid in Biochemical Pathology

Kunihiko Suzuki

Chapter 11
Neurite Loss Caused by Ganglioside Undersialylation

Abraham Rosenberg

Chapter 1

The Beginnings of Sialic Acid

Abraham Rosenberg

1. INTRODUCTION

The very early history of sialic acid research has been reviewed by pioneers who were intimately engaged in creating it. The interested reader is referred to those accounts for the flavor and excitement of discovery in breaking open a field of research that now has come to fruition. Of particular historical interest are the summations published by Blix working in Sweden (Blix *et al.*, 1955), Zilliken working in the United States (Zilliken and Whitehouse, 1958), and Gottschalk working in Germany (Gottschalk, 1960).

As seen by a very long distance runner glancing back, some past highlights still glitter along that now distant origin of the path beaten to current sialobiology. This chapter will scan, briefly and selectively, those earlier times.

2. DISCOVERY OF SIALIC ACID

Landsteiner and Levene (1927) in the United States and Walz (1927) in Germany independently noticed that certain animal lipid preparations possessed a sugarlike component that gave an unusual purple color reaction with Bial's reagent (orcinol and ferric ion in mineral acid) rather than the green color given by then known hexoses. Klenk (1935) working in Cologne some years later

Abraham Rosenberg Neuropsychiatric Institute and Brain Research Institute, University of California, Los Angeles, California 90024-1759.

Biology of the Sialic Acids, edited by Abraham Rosenberg. Plenum Press, New York, 1995.

observed that a new brain glycolipid fraction, which he was the first to isolate and give the name "ganglioside," also gave a purple color with Bial's reagent, and he removed the chromogen by methanolysis, purifying it to crystalline form. He called it "neuraminic acid." It is now known to have been the de-*N*-acylated methoxy derivative of the compound sialic acid. In the next year, Blix (1936) working in Uppsala used the milder process of autohydrolysis to isolate the purple Bial's chromogen from submaxillary mucin. He purified it to crystalline form, presuming it to be an acidic hexosamine-containing disaccharide. He gave it the universally accepted name "sialic acid" (Blix *et al.*, 1957). The compound that Blix had produced is now known to represent the intact unaltered sialic acid molecule.

It is apparent in retrospect that between them Blix and Klenk had discovered sialic acid and, respectively, the mucin sialoglycoproteins and the ganglioside sialoglycosphingolipids which are among the major sialoglycoconjugate components in the vertebrate organism. Although sometimes appearing to be at odds, Gunnar Blix and Ernst Klenk in a true sense started up the pursuit of sialobiology. Subsequently, mild hydrolytic conditions as practiced by Blix (1936) were used successfully over the next 20 years to isolate intact crystallizable sialic acids once more from bovine submaxillary mucin (Klenk and Faillard, 1954), from bovine colostrum (Klenk *et al.*, 1954), and from "mucolipids" or gangliosides (Blix and Odin, 1955; Rosenberg and Chargaff, 1956). Thus, widespread natural occurrence of sialic acid in biologically important glycoconjugate classes was established without knowledge of the actual construction of the sialic acid molecule.

3. SCALING THE SIALIC ACID STRUCTURE

In this author's opinion, certain key findings cited here stand out as having been essential among the numerous quite useful early efforts leading to the determination of the unique chemical structure of sialic acid. Klenk and Faillard (1954) discovered that strong base treatment of sialic acid produces a pyrrole-2-carboxylic acid, indicating that the sialic acid molecule should have a terminal carboxyl group and an adjacent carbonyl group four carbons removed from an amino group, a separation permitting these two latter functional groups to condense and give rise to the pyrrole ring. Gottschalk (1955) intelligently divined that sialic acid could be the product of condensation of pyruvic acid and an *N*-acyl hexosamine. Zilliken and Glick (1956) as well as Kuhn and Brossmer (1956) then found that more regulated organic base treatment of sialic acid broke it down into pyruvic acid and an *N*-acetyl hexosamine determined to be *N*-acetyl glucosamine, not realizing that *N*-acetyl glucosamine was an artifact produced by base epimerization of the true component found later actually to be *N*-acetyl mannosamine. In that same year, Blix *et al.* (1956) determined that elemental analysis of several sialic acids gave carbon, nitrogen, hydrogen, and oxygen compositions consistent with the chemical structure of a nine-carbon acylamino-sugar carboxylic acid in which the *N*-acyl group is often acetyl but sometimes *N*-glycoloyl instead, and that one or more of the sugar hydroxyls also may be acetylated.

Following the early lead of Gottschalk and Lind (1949), who first found that the receptor-destroying enzymes of the eubacterium *vibrio cholerae* and the myxoviruses of influenza removed sialic acid from ovomucin, Kuhn and Brossmer (1956) used influenza virus preparations to cleave sialic acid from cow colostrum which contains a considerable amount of sialyl lactose (Zilliken *et al.*, 1956). Then Rosenberg *et al.* (1956) working in New York discovered that the receptor-destroying enzyme of myxoviruses behaves as a sialidase that removes sialic acid from gangliosides. At the same time and working in close association in the same building of the College of Physicians and Surgeons of Columbia University in New York, Heimer and Meyer (1956) showed that sialidase activity toward submaxillary mucin was contained in an extract of pneumococci. Also, a biochemical breakthrough occurred when Heimer and Meyer found as well that an enzyme produced by *V. cholerae* could split sialic acid into pyruvic acid and an *N*-acetyl hexosamine which understandably but erroneously they identified as *N*-acetyl glucosamine because ninhydrin reaction produced xylose, the expected breakdown product for that aminosugar, not suspecting that the real aminosugar moiety of sialic acid is *N*-acetyl mannosamine, which likewise yields xylose. Then Comb and Roseman (1958a,b) succeeded in working out the enzymatic pathway for biosynthesis of *N*-acetyl mannosamine and the enzymatic synthesis of sialic acid itself, making the elegant discovery that a sialic acid aldolase purified from *Clostridium perfringens* cleaved sialic acid to produce pyruvic acid and *N*-acetyl mannosamine, which for identification is chromatographically separable from *N*-acetyl glucosamine.

In the same year as the latter Comb and Roseman discoveries, Brug and Paerels (1958) chemically synthesized sialic acid by condensing *N*-acetyl mannosamine and oxaloacetic acid whereupon the latter decarboxylates yielding the other obligate precursor pyruvate, and they showed meanwhile that Cornforth *et al.* (1958) who used *N*-acetyl glucosamine in a like synthesis had succeeded because the latter had base-epimerized to *N*-acetyl mannosamine in their reaction system. Finally, Roseman *et al.* (1961) as well as Warren and Felsenfeld (1961a,b) succeeded in biochemically synthesizing sialic acid as the 9-phosphate by condensing phosphoenolpyruvate with *N*-acetyl mannosamine-6-phosphate through the agency of a ubiquitous vertebrate enzyme aptly known as sialic acid-9-phosphate synthetase. Therewith *N*-acetyl mannosamine, which hitherto had been a biological unknown, made its entry as an obligate metabolic precursor of sialic acid, introducing its unique stereochemical form into carbon atoms 5–9 of the sialic acid structure. Extrapolating from the structure of the obligate precursors, the precise features of sialic acid began to emerge and, as covered by this brief history, the first segment of the now well traveled path toward modern research in sialobiology was laid out. Ancillary studies in the chemistry of the sialic acids have been reviewed in detail by Ledeen and Yu (1976) in an earlier treatment of this subject.

4. HOW SIALOBIOLOGY BEGAN

CMP-sialic acid synthase, the pyrophosphorylase that synthesizes the unique nucleotide-activated sialyltransferase substrate cytosine-5′-monophospho-sialic acid from free sialic acid and cytosine triphosphate, was found simultaneously by Warren and Blacklow (1962) in bacteria and importantly in animal tissues by Roseman (1962). Thereby there opened up the possibility of study of the specific incorporation of sialic acid into physiologically vital sialoglycoconjugates, and the road toward elucidation of the activity of the wide range of sialyltransferases whose genetic expression regulates this important phenomenon began. Over the period following the early 1960s, there occurred a huge explosion of information gathered worldwide on the chemistry, natural occurrence, metabolism, molecular and cell surface properties, and pathological associations of sialic acid for which detailed documentation may be found in a prior overview of the field (Rosenberg and Schengrund, 1976). Further elucidation of the history of sialobiology will be provided in appropriate context in each of the subsequent chapters in this book.

REFERENCES

Blix, G., 1936, Über die Kohlenhydratgruppen des Submaxillarismucins, *Hoppe-Seylers Z.* **240:**43–54.

Blix, G., and Odin, L., 1955, Isolation of sialic acids from gangliosides, *Acta Chem. Scand.* **9:**1541–1543.

Blix, G., Lindberg, E., Odin, L., and Werner, J., 1955, Sialic acids, *Nature*, **175:**340.

Blix, G., Lindberg, E., Odin, L., and Werner, I., 1956, Studies on sialic acids, *Acta Soc. Med. Ups.* **61:**1–26.

Blix, G., Gottschalk, A., and Klenk, E., 1957, Proposed nomenclature in the field of neuraminic and sialic acids, *Nature* **175:**340–341.

Brug, J., and Paerels, G. B., 1958, Configuration of N-acetyl neuraminic acid, *Nature* **182:**1159–1160.

Comb, D. G., and Roseman, S., 1958a, Enzymatic synthesis of N-acetyl-D-mannosamine, *Biochim. Biophys. Acta* **29:**653–654.

Comb, D. G., and Roseman, S., 1958b, Composition and enzymatic synthesis of N-acetylneuraminic (sialic) acid, *J. Am. Chem. Soc.* **80:**497–499.

Cornforth, J. W., Firth, M. E., and Gottschalk, A., 1958, The synthesis of N-acetylneuraminic acid, *Biochem. J.* **68:**57.

Gottschalk, A., 1955, 2-Carboxypyrrole: Its preparation from its precursors in muco-proteins, *Biochem. J.* **61:**298–307.

Gottschalk, A., 1960, *The Chemistry and Biology of Sialic Acid and Related Substances,* University Press, Cambridge, England.

Gottschalk, A., and Lind, P. E., 1949, Product of interaction between influenza virus enzyme and ovomucin, *Nature* **160:**232.

Heimer, R., and Meyer, K., 1956, Studies on sialic acid of submaxillary mucoid, *Proc. Natl. Acad. Sci. USA* **42:**531–534.

Klenk, E., 1935, Über die Natur der Phosphatide und anderer Lipoide des Gehirns und der Leber in Niemann-Pickscher Krankheit, *Hoppe-Seylers Z.* **235:**24–36.

Klenk, E., and Faillard, H., 1954, Zur Kenntnis der Kohlenhydratgruppen der Mucoproteide, *Hoppe-Seylers Z.* **298:**232.

Klenk, R., Brossmer, R., and Schulz, W., 1954, Über die Prostetische Gruppe der Mucoproteine des Kuh-Colostrums, *Chem. Ber.* **87:**123.

Kuhn, R., and Brossmer, R., 1956, Über O-acetyl-lactosaminsäurelactose aus Kuh-Colostrum und Ihre Spaltbarkeit durch Influenza-Virus, *Chem. Ber.* **89:**2013.

Landsteiner, K., and Levene, P. A., 1927, On some new lipoids, *J. Biol. Chem.* **75:**607–612.

Ledeen, R. W., and Yu, R. K., 1976, Chemistry and analysis of sialic acid, in: *Biological Roles of Sialic Acid* (A. Rosenberg and C.-L. Schengrund, eds.), Plenum Press, New York, pp. 1–57.

Roseman, S., 1962, Enzymatic synthesis of cytidine-5′-monophosphosialic acid, *Proc. Natl. Acad. Sci. USA* **48:**437–441.

Roseman, S., Jourdian, G. W., Watson, D., and Rood, R., 1961, Enzymatic synthesis of sialic acid-9-phosphate, *Proc. Natl. Acad. Sci. USA* **47:**958–961.

Rosenberg, A., and Chargaff, E., 1956, Nitrogenous constituents of an ox brain mucolipid, *Biochim. Biophys. Acta* **21:**588–590.

Rosenberg, A., and Schengrund, C.-L. (eds.), 1976, *Biological Roles of Sialic Acid,* Plenum Press, New York.

Rosenberg, A., Howe, C., and Chargaff, E., 1956, Inhibition of influenza virus haemagglutination by a brain lipid fraction, *Nature* **177:**234–235.

Walz, E., 1927, Uber das Vorkommen von Kerasin in normaler Ochsenmilz, Hoppe-Segler's, *Z. Physiol. Chem.* **166:**210.

Warren, L., and Blacklow, R. S., 1962, The biosynthesis of cytidine-5′-monophospho-N-acetylneuraminic acid by an enzyme from *Neisseria meningitidis, J. Biol. Chem.* **237:**3527–3534.

Warren, L., and Felsenfeld, H., 1961a, The biosynthesis of N-acetylneuraminic acid, *Biochem. Biophys. Res. Commun.* **4:**232–235.

Warren, L., and Felsenfeld, H., 1961b, N-Acetyl-mannosamine-6-phosphate and N-acetylneuraminic acid-9-phosphate as intermediates in sialic acid biosynthesis, *Biochem. Biophys. Res. Commun.* **5:**185–190.

Zilliken, F., and Glick, M. C., 1956, Alkalischer Abbau von Gynaminsäure zu Brenztraubensäure und N-acetyl-D-glucosamine, *Naturwissenschaften* **43:**536.

Zilliken, F., and Whitehouse, M. W., 1958, The nonulosaminic acids. Neuraminic acids and related compounds (sialic acids), *Adv. Carbohydr. Chem.* **13:**237–263.

Zilliken, F., Bran, G. A., and Gyorgi, P., 1956, "Gynaminic acid" and other naturally occurring forms of N-acetylneuraminic acid, *Arch. Biochem. Biophys.* **63:**394–402.

Chapter 2

Biochemistry and Role of Sialic Acids

Roland Schauer, Sörge Kelm, Gerd Reuter,
Peter Roggentin, and Lee Shaw

1. INTRODUCTION

Sialic acids mainly occur as terminal components of cell surface glycoproteins and glycolipids, playing as such a major role in the chemical and biological diversity of glycoconjugates. Cell-type-specific expression of glycosyltransferases, particularly of sialyltransferases (Paulson and Colley, 1989; van den Eijnden and Joziasse, 1993), leads to specific sialylation patterns of oligosaccharides which can be considered as key determinants in the makeup of cells. Striking differences have been found in the sialoglycosylation patterns of cells during development, activation, aging, and oncogenesis. Research on the structures, metabolism, and molecular biology, as well as on the biological and clinical importance of sialic acids as components of these glycoconjugates, has therefore intensified during the past several years.

Structurally, sialic acids comprise a family of 36 naturally occurring derivatives of neuraminic acid that are all *N*-acylated to form *N*-acetylneuraminic acid (Neu5Ac) or *N*-glycoloylneuraminic acid (Neu5Gc) as fundamental molecules (Figure 1). Most additional modifications result from *O*-acetylation at one or several of the hydroxyl functions, at C-4, -7, -8, or -9, or to the introduction of a double bond between C-2 and C-3 (Table I) (Rosenberg and Schengrund, 1976a;

Roland Schauer, Sörge Kelm, Gerd Reuter, Peter Roggentin, and Lee Shaw Biochemistry Institute, University of Kiel, D-24098 Kiel, Germany.

Biology of the Sialic Acids, edited by Abraham Rosenberg. Plenum Press, New York, 1995.

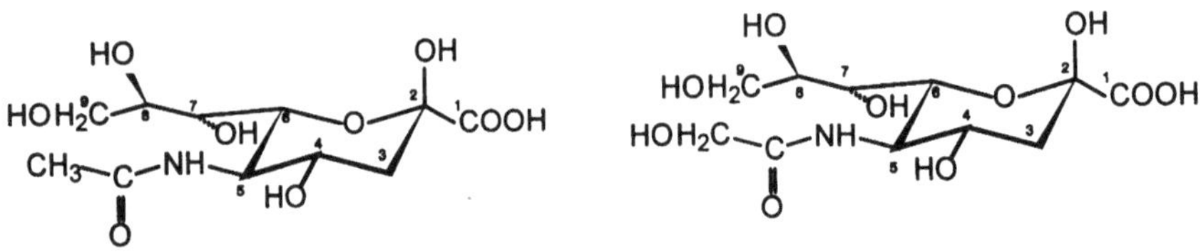

FIGURE 1. $^{2}C_{5}$-conformation of *N*-acetyl-neuraminic acid (left) and *N*-glycoloylneuraminic acid (right).

Corfield and Schauer, 1982a; Reuter *et al.*, 1983; Schauer *et al.*, 1984a,b; Reuter and Schauer, 1988, 1994; Varki, 1992a). Sialic acids in these various structural forms are found on some viruses, microorganisms such as bacteria and protozoa, in a large number of higher animals, and on cell membranes and in body fluids of all mammals. The saturated sialic acids usually occupy terminal, nonreducing positions of oligosaccharide chains of complex carbohydrates on outer and inner (e.g., lysosomal) membrane surfaces in various linkages mainly to galactose, *N*-acetylgalactosamine, and to sialic acid itself. Sialic acids are among the first molecules encountered by other cells or by compounds coming into contact with the cell, a feature that is important in the expression of their biological role (Schauer, 1982a,b, 1991). The 2,3-didehydro-sialic acids and 2,7-anhydro-*N*-acetylneuraminic acid have been found only as free sugars in body fluids and secretions, since they lack the linkage-forming glycosidic hydroxyl group. In Sections 2 and 3, some new aspects of the occurrence and analysis of sialic acids will be discussed.

In the past decade, little progress has been made in elucidating the enzymatic and regulatory mechanisms involved in the long pathway of Neu5Ac biosynthesis starting from glucose (Corfield and Schauer, 1982b). However, further insight has been gained into the biosynthesis and role of modified sialic acids, especially enzymatic *N*-acetyl-hydroxylation, *O*-acetylation, and *O*-methylation. This will be discussed in Section 4. Numerous studies have been carried out on the transfer of sialic acids onto oligosaccharides, polysaccharides, glycoproteins, and glycolipids mediated by sialyltransferases, which in most cases result in the completion of their oligosaccharide chain extensions (see Chapter 3). Such studies have led to the elucidation of the acceptor specificity of at least ten sialyltransferases (Paulson and Colley, 1989; van den Eijnden and Joziasse, 1993). By now, eight cDNA clones of sialyltransferases have been obtained (Weinstein *et al.*, 1987; Grundmann *et al.*, 1990; Gillespie *et al.*, 1992; Wen *et al.*, 1992; Kitagawa and Paulson, 1993, 1994; Lee *et al.*, 1993; Kurosawa *et al.*, 1994). All of these cDNA clones contain a stretch of about 50 amino acids displaying high homology, the sialyl motif possibly involved in substrate binding (Wen *et al.*, 1992). Using polymerase chain reactions with degenerate primers deduced from this motif, additional cDNA clones were obtained (Livingston and Paulson, 1993; Kurosawa *et al.*, 1993) for which no sialyltransferase activity could be assigned as yet.

Table I
Naturally Occurring Sialic Acids and Suggested Abbreviations[a,b]

N-acetylneuraminic acid	Neu5Ac
N-acetyl-4-*O*-acetylneuraminic acid	Neu4,5Ac_2
N-acetyl-7-*O*-acetylneuraminic acid	Neu5,7Ac_2
N-acetyl-8-*O*-acetylneuraminic acid	Neu5,8Ac_2
N-acetyl-9-*O*-acetylneuraminic acid	Neu5,9Ac_2
N-acetyl-4,9-di-*O*-acetylneuraminic acid	Neu4,5,9Ac_3
N-acetyl-7,9-di-*O*-acetylneuraminic acid	Neu5,7,9Ac_3
N-acetyl-8,9-di-*O*-acetylneuraminic acid	Neu5,8,9Ac_3
N-acetyl-7,8,9-tri-*O*-acetylneuraminic acid	Neu5,7,8,9Ac_4
N-acetyl-9-*O*-acetyl-8-*O*-methylneuraminic acid*	Neu5,9$Ac_2$8Me
N-acetyl-9-*O*-lactoylneuraminic acid	Neu5Ac9Lt
N-acetyl-4-*O*-acetyl-9-*O*-lactoylneuraminic acid	Neu4,5$Ac_2$9Lt
N-acetyl-8-*O*-methylneuraminic acid	Neu5Ac8Me
N-acetylneuraminic acid 9-phosphate	Neu5Ac9P
N-acetylneuraminic acid 8-sulfate	Neu5Ac8S
2-deoxy-2,3-didehydro-*N*-acetylneuraminic acid	Neu5Ac2en
2-deoxy-2,3-didehydro-*N*-acetyl-9-*O*-acetylneuraminic acid*	Neu5,9$Ac_2$2en
2-deoxy-2,3-didehydro-*N*-acetyl-9-*O*-lactoylneuraminic acid*	Neu5Ac2en9Lt
2,7-anhydro-*N*-acetylneuraminic acid*	2,7-anhydro-Neu5Ac
N-glycoloylneuraminic acid	Neu5Gc
N-glycoloyl-4-*O*-acetylneuraminic acid	Neu4Ac5Gc
N-glycoloyl-7-*O*-acetylneuraminic acid	Neu7Ac5Gc
N-glycoloyl-9-*O*-acetylneuraminic acid	Neu9Ac5Gc
N-glycoloyl-7,9-di-*O*-acetylneuraminic acid	Neu7,9$Ac_2$5Gc
N-glycoloyl-8,9-di-*O*-acetylneuraminic acid	Neu8,9$Ac_2$5Gc
N-glycoloyl-7,8,9-tri-*O*-acetylneuraminic acid	Neu7,8,9$Ac_3$5Gc
N-glycoloyl-9-*O*-acetyl-8-*O*-methylneuraminic acid*	Neu9Ac5Gc8Me
N-(*O*-acetyl)glycoloylneuraminic acid*	Neu5GcAc
N-glycoloyl-9-*O*-lactoylneuraminic acid*	Neu5Gc9Lt
N-glycoloyl-8-*O*-methylneuraminic acid*	Neu5Gc8Me
N-glycoloylneuraminic acid 8-sulfate	Neu5Gc8S
2-deoxy-2,3-didehydro-*N*-glycoloylneuraminic acid*	Neu2en5Gc
2-deoxy-2,3-didehydro-*N*-glycoloyl-9-*O*-acetylneuraminic acid*	Neu9Ac2en5Gc
2-deoxy-2,3-didehydro-*N*-glycoloyl-9-*O*-lactoylneuraminic acid*	Neu2en5Gc9Lt
2-deoxy-2,3-didehydro-*N*-glycoloyl-8-*O*-methylneuraminic acid*	Neu2en5Gc8Me
Ketodeoxynonulosonic acid*	Kdn

[a]Reuter and Schauer (1988).
[b]Asterisks identify those sialic acids discovered since the publication of the last book review (Schauer, 1982a).

Research has also been focused on the enzymes of sialic acid catabolism, mainly the sialidases, which may initiate the degradation of sialoglycoconjugates and thus are involved in regulation of the turnover of these substances (Corfield *et al.*, 1981a; Corfield and Schauer, 1982b) (see Chapter 8). Sialidases are important enzymes, since they may destroy the sialic acid-mediated biological functions of many glycoconjugates (Rosenberg and Schengrund, 1976a; Corfield, 1992; Corfield *et al.*, 1992). This will be further elucidated in Section 6, where sialic acid-specific cellular recognition and adhesion are described. Furthermore, desialylation can lead by unmasking to the formation of new recognition sites, frequently galactose or novel antigens (see below).

In addition to the common exo-α-sialidases (EC 3.2.1.18; Cabezas, 1991), a rare enzyme has been found, so far only in the leech (Li *et al.*, 1993), yielding 2,7-anhydro-Neu5Ac as its reaction product. This compound may also be a product of an as yet unknown microbial sialidase, since it was found in human cerumen (Suzuki *et al.*, 1985).

An extraordinary sialidase is expressed in several trypanosomal species, *Trypanosoma cruzi* (Zingales *et al.*, 1987; Schenkman *et al.*, 1991), *T. brucei* (Engstler *et al.*, 1993), and *T. congolense* (Engstler *et al.*, 1994). This enzyme transfers sialic acids (Neu5Ac or Neu5Gc) from oligosaccharide chains either onto water, which constitutes a normal sialohydrolase reaction, or with greater preference, onto nonsialylated glycan chains with terminal galactose residues, leading to the formation of α2,3 sialic acid linkages. The existence of this trans-sialidase explains the occurrence of sialic acids on the surface of *T. cruzi,* since this parasite is not able to synthesize sialic acid (Schauer *et al.*, 1983). The biological significance of this unusual sialylation mechanism is to protect these pathogens from the host's immune or proteolytic defense systems (reviewed by Engstler and Schauer, 1993).

One particular exosialidase, from the loach *Misgurnus fossilis,* is specialized for cleaving 2-keto-3-deoxy-D-*glycero*-D-*galacto*-nononic acid (Kdn) residues (Section 2), and can therefore be described as a Kdn'ase (Li *et al.*, 1993). It also acts slowly on sialic acid glycosidic linkages.

An endo-α-sialidase (EC 3.2.1.129; Cabezas, 1991) exists in bacteriophages, which hydrolyzes α2,8 bonds within the polysialyl chains of bacterial colominic acid and mammalian "cell adhesion" molecules (Long *et al.*, 1993).

Since sialidases are extremely important in the pathogenesis of microbial diseases, much research on their diagnosis and inhibition has been carried out. Influenza A and B viruses contain sialidases as receptor-destroying enzymes, which are necessary for the propagation of the viruses, possibly for the release of virus particles from inhibitory mucins or during budding of new viruses (see Chapter 9). Based on earlier observations that an *O*-acetyl group at C-4 of Neu5Ac completely prevents release of the corresponding sialic acid by bacterial and mammalian sialidases, but not by the influenza virus enzyme (Kleineidam *et al.*, 1990), inhibitors for viral sialidases were designed. The rationale for this was the observation of a pocket in the active center of influenza virus sialidases, elucidated by X-ray crystallography, into which the 4-*O*-acetyl group of sialic acid fits and thus allows the (slow) hydrolysis of this substrate. (Nonviral sialidases do not seem to have this pocket.) Correspondingly, Neu5Ac2en, known as an inhibitor of most sialidases including the viral ones, could be transformed into a much more potent inhibitor (inhibitor constants up to 10^{-9} M) by the addition of substituents at C-4, as in 4-amino- and 4-guanidino-Neu5Ac2en (Corfield, 1993; von Itzstein *et al.*, 1993).

In *Clostridium* infections, the bacterial sialidases can be identified in wound fluids or blood serum by immunological methods (T. Roggentin *et al.*, 1993). In this way, infecting bacterial species can be recognized at an early stage of the disease, thus enabling a prompt and specific therapy of lethal diseases such as gas edema, which is most frequently caused by *Clostridium perfringens*. Studies on these bacterial enzymes led us to the investigation of their gene structures, which revealed a sialidase gene family, thus giving support to theories about the molecular relationship between microbial and animal sialidases. Section 5 is devoted to these studies.

Investigations on the biological functions of sialic acids are rapidly growing fields of research. Several aspects have already been mentioned and more aspects, such as the significance of sialic acid *O*-acetylation, *N*-acetyl-hydroxylation, and sialic acid-binding proteins, will be discussed in the following sections. Here, only an introduction to the main aspects of sialobiology can be given, based on numerous experiments performed in many laboratories. (For reviews, see Rosenberg and Schengrund, 1976a; Schauer, 1982a,b; and Varki, 1992a,b, 1993.)

The functions of sialic acids may be grouped into three sections: one is a less specific, more general role, related to the fact that sialic acids are relatively large, hydrophilic and acidic molecules that exert physicochemical effects on the glycoconjugates to which they are bound, and on the environmental molecules *in situ*, e.g., in cell membranes. Therefore, sialyl groups influence and stabilize the conformation of both the glycan and the protein parts of glycoconjugates, resulting in modified properties, e.g., higher thermal and proteolytic stability of glycoproteins (Schauer, 1982a,b). These oligosaccharide moieties have therefore been considered to "mimic the effect of a molecular chaperone" (Jaenicke, 1993). Mucin glycoproteins of vertebrates are highly sialylated (the name sialic acid is derived from the Greek "sialos," meaning saliva) which gives them a high viscosity in solution because of their negative charges. The protective and lubricative effects of mucins are well known (Schauer, 1992).

Two other functions of the sialic acids deal with molecular and cellular recognition. Here, sialic acids have dual and opposite effects. On the one hand, sialic acids act as masks to prevent biological recognition, e.g., of subterminal galactose residues, and on the other they serve as recognition sites (Schauer, 1982a,b, 1985; Schauer *et al.*, 1984b). The best studied example of the masking function is the binding and uptake of desialylated serum glycoproteins by hepatocytes (Harford *et al.*, 1984).

In a similar way, sialidase-treated erythrocytes, lymphocytes, and thrombocytes are bound and partly phagocytized by macrophages, mediated by a galactose-specific lectin (Müller *et al.*, 1983; Fischer *et al.*, 1991; Kluge *et al.*,

1992). It was observed during the programmed cell death (apoptosis) of rodent thymocytes that these cells are phagocytized after the loss of cell surface sialic acids during this process (Savill *et al.*, 1993). These studies show the role of sialic acids in maintaining the life span of molecules and cells which, together with demasked galactose residues, regulate these and many more biological processes, as well as pathological events like the spreading of cancer.

In contrast, it has long been known that influenza viruses recognize sialic acids and bind to them on cell surfaces via hemagglutinin (see Chapter 9). Similar receptors are also involved in the infection mechanism of many bacteria and protozoa (see Table V in Section 6). Many physiological processes also are regulated by sialic acid recognition, and this field of research is rapidly expanding. Section 6 is devoted to reviewing characteristic features of sialic acid-dependent receptors focusing on those mammalian proteins that are well defined as molecules, in terms of specificity and possible functions. Further aspects discussed are the nature of ligands for these receptors and the potential modulatory role of sialic acid modifications.

The intention of this chapter is to give the reader a feeling for the great chemical and biological diversity of sialic acids, which is unique in glycobiology. It may also contribute to an understanding of the vulnerability of sialobiology to misuse by pathogens and transformed cells, and to the possibility that microbial sialidases from infectious agents can severely disturb normal sialobiological function.

2. OCCURRENCE OF SIALIC ACIDS

Since publication of the last book dealing with various aspects of sialic acids (Schauer, 1982a), 12 new derivatives have been structurally identified, the 9-*O*-acetylated and 9-*O*-lactoylated derivatives of the unsaturated 2-deoxy-2,3-didehydro-*N*-acetylneuraminic acid, 2-deoxy-2,3-didehydro-*N*-glycoloylneuraminic acid together with the corresponding 9-*O*-acetylated, 9-*O*-lactoylated, and 8-*O*-methylated derivatives, and the 9-*O*-lactoylated derivative of *N*-glycoloylneuraminic acid (Schauer *et al.*, 1984a,b; Shukla *et al.*, 1987) (Table I). In the starfish *Asterias rubens*, in which Neu2en5Gc8Me was detected, Neu5,9$Ac_2$8Me and Neu9Ac5Gc8Me were also found (Bergwerff *et al.*, 1992), thus extending the series of 8-*O*-methylated sialic acids discovered in echinoderms. Finally, a derivative of *N*-glycoloylneuraminic acid was isolated from rat thrombocytes, in which the OH group of the *N*-glycoloyl function is acetylated. This sialic acid is very labile and seems not to occur on other rat blood cells (Kluge *et al.*, 1992). A substitution of the *N*-glycoloyl OH function in Neu5Gc8Me by another Neu5Gc8Me has been characterized from *A. rubens* as sialo-di- and trisaccharide (Bergwerff *et al.*, 1992, 1993) (Figure 2).

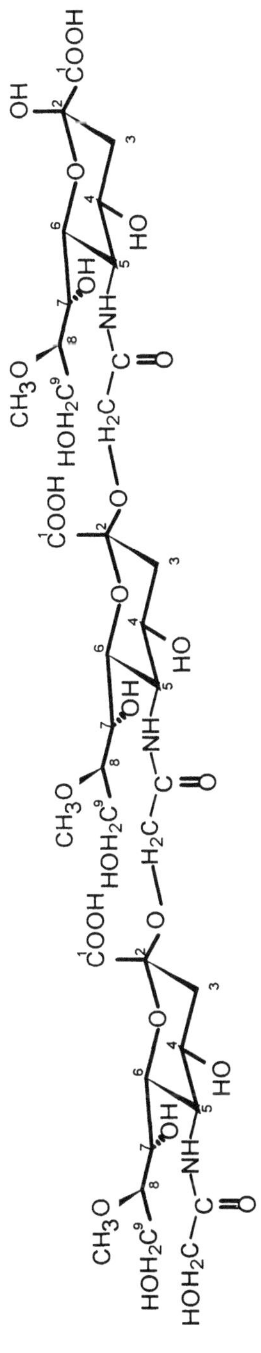

FIGURE 2. Structure of the Neu5Gc8Me-trisaccharide identified from *Asterias rubens*.

With the 5-deaminated derivative of neuraminic acid, 3-deoxy-D-*glycero*-D-*galacto*-nonulosonic acid (ketodeoxynonulosonic acid, Kdn) (Figure 3), an unusual modification was found in the eggs of rainbow trout (Inoue, 1993) and the egg jelly coat from the newt *Pleurodeles waltlii* (Strecker *et al.*, 1992a), from *Axolotl mexicanum* (Strecker *et al.*, 1992b), and from *Xenopus laevis* (Strecker *et al.*, 1993). In trout, it is a component of gangliosides (Song *et al.*, 1991) and glycoproteins, where it can terminate long sialic acid chains consisting of Neu5Gc (Nadano *et al.*, 1986) or form poly-Kdn chains (Kitazume *et al.*, 1992; Inoue et al., 1992). In some cases its 9-*O*-acetylated derivative was also found (Strecker *et al.*, 1993). The metabolism of Kdn is beginning to be elucidated (Terada *et al.*, 1993). Recently, an enzyme releasing Kdn from its glycosidic linkage has been found in the loach *Misgurnus fossilis* (Li *et al.*, 1993). Although Kdn is not structurally a neuraminic acid, it is denominated as a sialic acid.

In *Shigella boydii* and *Pseudomonas aeruginosa*, other rare neuraminic acid derivatives ("pseudaminic acids") were identified with a different stereochemistry at C-5, -7, and -8 and a 3-hydroxybutyramido or formamido substituent at C-7 (Knirel *et al.*, 1985, 1986, 1987a,b). A further unusual derivative was found in the *O*-specific chain of *Legionella pneumophila* serogroup 1 lipopolysaccharide having an α2,4 homopolymer of 5-acetamidino-7-acetamido-8-*O*-acetyl-3,5,7,9-tetradeoxy-D-*glycero*-L-*galacto*-nonulosonic acid (Knirel *et al.*, 1994).

More and more bacteria have been found to contain sialic acids. Only Neu5Ac (Moran *et al.*, 1991; Krauss *et al.*, 1988, 1992) and in few cases Neu5,9Ac$_2$ (Varki and Higa, 1993) were identified. The position of sialic acids within the oligosaccharide chain of bacterial lipopolysaccharides is unusual since nonterminal linkages were found such as that in Galβ1,3GalNAcβ1,4Gal-α1,4Neu5,9Ac$_2$ (Gamian *et al.*, 1991) or Galα1,6Glcβ1,7Neu5Ac (Krauss *et al.*, 1992) in addition to the polysialic acid chains of *Escherichia coli, Neisseria meningitidis, Pasteurella haemolytica,* and *Moraxella nonliquefaciens* (Vann *et al.*, 1993).

Recently, the first proof for the occurrence of sialic acids in insects has been published (Roth *et al.*, 1992). Neu5Ac was identified by mass spectrometry only

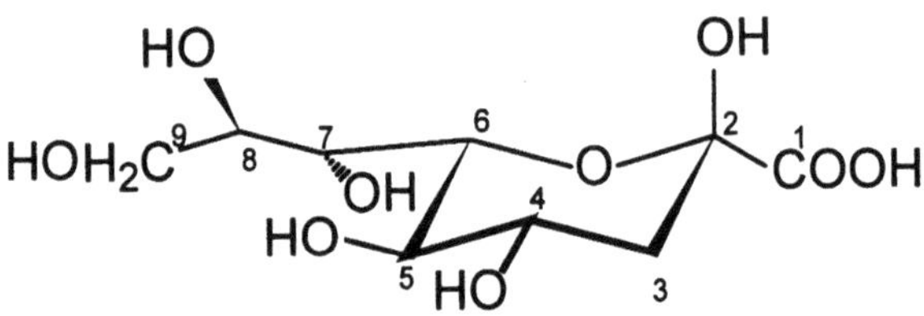

FIGURE 3. Structure of 3-deoxy-D-*glycero*-D-*galacto*-nonulosonic acid (ketodeoxynonulosonic acid, Kdn).

Table II
Sialic Acid Composition of Various Mouse Erythrocytes

Strain	%Neu5Ac	%Neu5Gc	%Neu5,9Ac$_2$	μg sialic acid per 10^{10} cells
DBA2	86	7	7	58
C57BL6	73	13	14	60
Swiss	57	7	36	6
Balb c	82	0	18	61
C3H	50	45	5	56
CBA	81	7	12	71

in certain developmental stages of *Drosophila* larvae, where it exists most probably in the form of a polysialic acid chain, as detected with corresponding antibodies (see Chapter 4).

In man, the spectrum of identified sialic acids is still restricted to Neu5Ac, Neu5,9Ac$_2$, Neu5Ac9Lt, and Neu5Ac2en as previously described (Corfield and Schauer, 1982a). In tumors such as colon and mammary carcinoma, gastric and liver cancer, malignant lymphoma and teratoma, Neu5Gc has been detected in low amounts (e.g., Higashi *et al.*, 1985; Devine *et al.*, 1991; Kawai *et al.*, 1991; Hanisch *et al.*, 1992). Only in the latter two reports was a chemical identification of this sialic acid type carried out; rather, most authors use antibodies raised against certain Neu5Gc-containing structures for detection.

In human melanoma, a tumor-associated expression of the gangliosides GD3 and GD2, each with one 9-*O*-acetylated sialic acid residue, was identified (Thurin *et al.*, 1985; Sjoberg *et al.*, 1992), so adding to other reports about modified sialic acid *O*-acetylation in some tumors (Muchmore *et al.*, 1987; Hutchins *et al.*, 1988). The finding of Neu5,9Ac$_2$ on human T lymphocytes only in cancer patients (Stickl *et al.*, 1991) was not confirmed by Suguri *et al.* (1993) who reported approximately 5% Neu5,9Ac$_2$ in the total sialic acid pool isolated from normal human B as well as T cells.

Many more recent findings not only provide further evidence for the long-established species-specific pattern of sialic acids (Corfield and Schauer, 1982a; Varki, 1992a), but also indicate a strain, tissue, or cell specificity within a given species and a developmental regulation of the expression of certain sialic acids. The sialic acid composition of mouse erythrocytes, for example, shows great variation among the strains analyzed (Klotz *et al.*, 1992; Table II).

A comparative analysis of 13 different adult and fetal bovine tissues (Schauer *et al.*, 1991) revealed a concentration of 0.1–3.1 mg total sialic acids/g wet weight in adult bovine tissues, while in the corresponding fetal tissues the sialic acid concentration was higher and ranged from 1.1 to 12.5 mg/g wet tissue. The percentage of Neu5Gc in the total sialic acid pool was always greater in the adult than in the fetus, ranging from 9 to 66% and from 0 to 49%, respectively. Chickens were found to acquire the full, i.e., adult, level of $Neu5,9Ac_2$ on their erythrocytes only on day 21 after hatching, whereas freshly hatched chickens do not have detectable amounts of this sialic acid on their red blood cells (Herrler *et al.*, 1987). Similar examples of developmentally regulated expression of Neu5Gc or of sialic acid *O*-acetylation were reported for rat small intestine (Bouhours and Bouhours, 1983, 1988), human and rat colon (Muchmore *et al.*, 1987), transgenic mice (Varki *et al.*, 1991), and for hibernating animals (Rahmann *et al.*, 1987).

3. SIALIC ACID ANALYSIS

Since a number of biological phenomena depend not only on the general presence of sialic acids, but also on a specific sialic acid type, the exact analysis of these sugars within a biological system is of great importance. The tools formerly described for sialic acid analysis (Schauer, 1978a, 1982b, 1987a,b; Schauer and Corfield, 1982; Vliegenthart *et al.*, 1982; Kamerling and Vliegenthart, 1982; Reuter *et al.*, 1983; Reuter and Schauer, 1986) are still successfully used for this purpose. Over the past 10 years, however, a number of improvements have been made in the various techniques (Reuter and Schauer, 1994).

One of the major problems in sialic acid analysis is still the liberation of this sugar from its glycosidic linkage. Although complete release requires milder conditions than for hexoses, there is no method available that can prevent the simultaneous cleavage of sialic acid ester substituents. The two-step acid hydrolysis using formic acid (pH 2, 1 hr, 70°C) followed by hydrochloric acid (pH 1, 1 hr, 80°C) completely releases sialic acids from sialoglycoconjugates but will result in a loss of 20–50% of *O*-acyl groups together with some degradation of sialic acid itself (Schauer, 1978a, 1987a,b; Reuter and Schauer, 1994). Other conditions used for acid hydrolysis (Varki and Diaz, 1984) did not improve the recovery, especially of *O*-acetylated sialic acids, in our hands. The use of sialidases for liberation of sialic acids does not result in significant losses of these ester substituents. However, only a portion of all sialic acids present is often released (Corfield and Schauer, 1982b; Reuter and Schauer, 1994), though this depends on the sialidase type used for hydrolysis.

These problems in the analysis of free sialic acids demonstrate the need to investigate intact sialoglycoconjugates that can often be isolated from biological material without significant modification of sialic acids. By ^{1}H-NMR spectroscopy, which has become a routine application for some laboratories using machines with high magnetic fields, and various two-dimensional techniques, even complex sialoglycoconjugates can be analyzed. The information obtained includes structural details like the type of monosaccharides, including sialic acid substitution, sequence, anomeric configuration, and the type of linkage as well as conformational data and dynamic behavior of the whole molecule (Vliegenthart *et al.*, 1982, 1983; Dabrowski, 1989).

In addition to the more conventional electron impact and chemical ionization mass spectrometric techniques that have been applied to the analysis of free sialic acids (Kamerling and Vliegenthart, 1982; Reuter *et al.*, 1983; Reuter and Schauer, 1986), fast-atom bombardment mass spectrometry (FAB-MS) allows the analysis of whole glycoconjugates (Egge *et al.*, 1985; Dell, 1987). FAB-MS requires about 5 μg of substance and gives information about the molecular weight, type of monosaccharide unit in terms of hexose, *N*-acetylhexosamine, deoxyhexose, etc., and about branching of an oligosaccharide chain. However, it does not allow the localization of the position of a substituent within a given residue; thus, a discrimination between, e.g., 7- and 9-*O*-acetylated sialic acids is not possible using this technique.

Both methods, NMR and FAB-MS, require preparation and purification of individual sialylated oligosaccharides, glycopeptides, or gangliosides before analysis, making this approach less attractive if only low amounts of the corresponding materials are available. Therefore, sensitive methods for isolation and analysis of sialic acids and sialoglycoconjugates are required. Several HPLC methods have been developed that fulfill this need.

Derivatization of free sialic acids with 1,2-diamino-4,5-methylenedioxybenzene (DMB) yields fluorigenic derivatives that can be separated on reversed phase columns with a detection limit in the femtomole range (Hara *et al.*, 1986, 1987, 1989). Since the derivatization reaction is specific for α-keto acids, Kdo and Kdn will also be detected by this method (Reuter and Schauer, 1994). Even with this specificity, a structural identification by HPLC is not possible. To ascertain the preliminary assignment made on the basis of retention times, treatment with specific reagents or enzymes is very helpful (Shukla and Schauer, 1986; Schauer, 1987b; Reuter and Schauer, 1994). Incubation with sialate 9-*O*-acetylesterase (EC 3.1.1.53), e.g., from influenza C virus (Schauer *et al.*, 1988a,c), will result in a decrease of the peak for 9-*O*-acetylated sialic acids with a corresponding increase of unmodified Neu5Ac or Neu5Gc. Alkali treatment (Reuter and Schauer, 1994) hydrolyzes all esters, not only 9-*O*-acetyl groups, again accompanied by an increase of Neu5Ac and Neu5Gc. These latter sialic acids are cleaved by incubation with *N*-acylneuraminate-pyruvate lyase (EC 4.1.3.3, aldolase) that also acts on *O*-acetylated derivatives but significantly slower. Thus, a combination of various treatments of the sialic acid pool followed by fluorimetric HPLC analysis indicates the presence of this sugar even in very low quantities (Figure 4). In addition, the use of sialic acid-specific enzymes in this assay allows identification of sialic acids, which is not possible by HPLC alone.

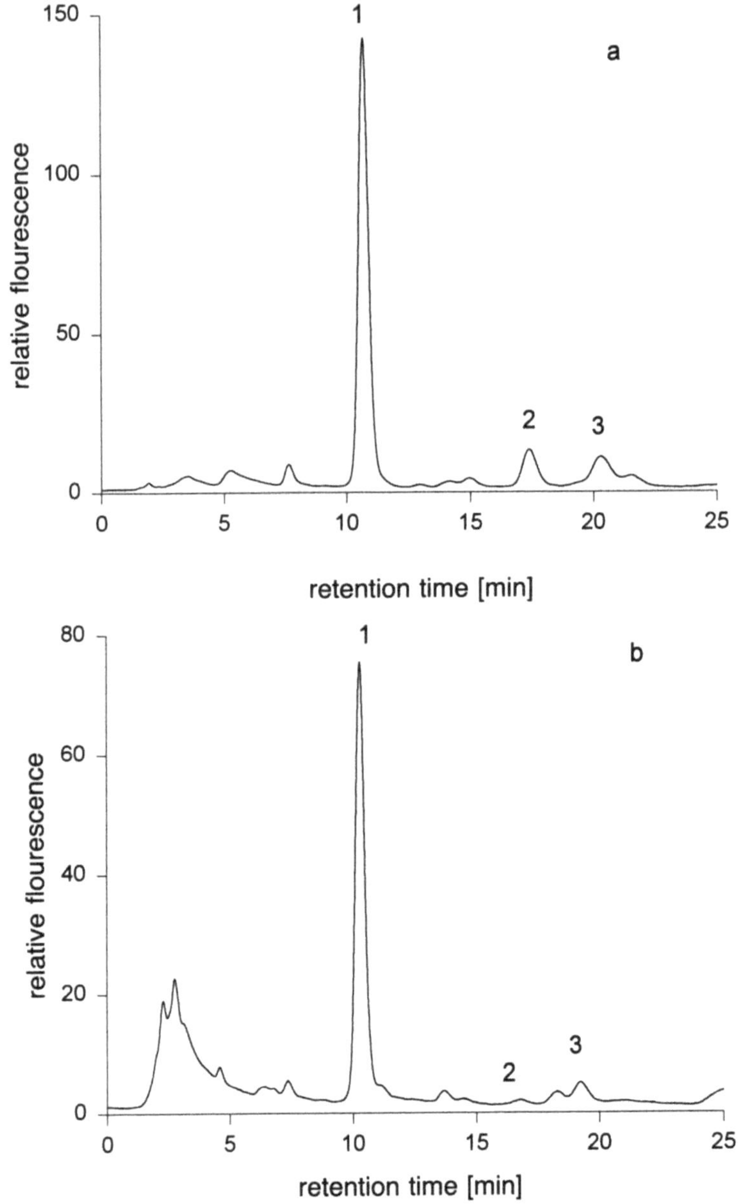
a
b
1
2
3
relative flourescence
retention time [min]
0
50
100
150
20
40
60
80
5
10
15
25

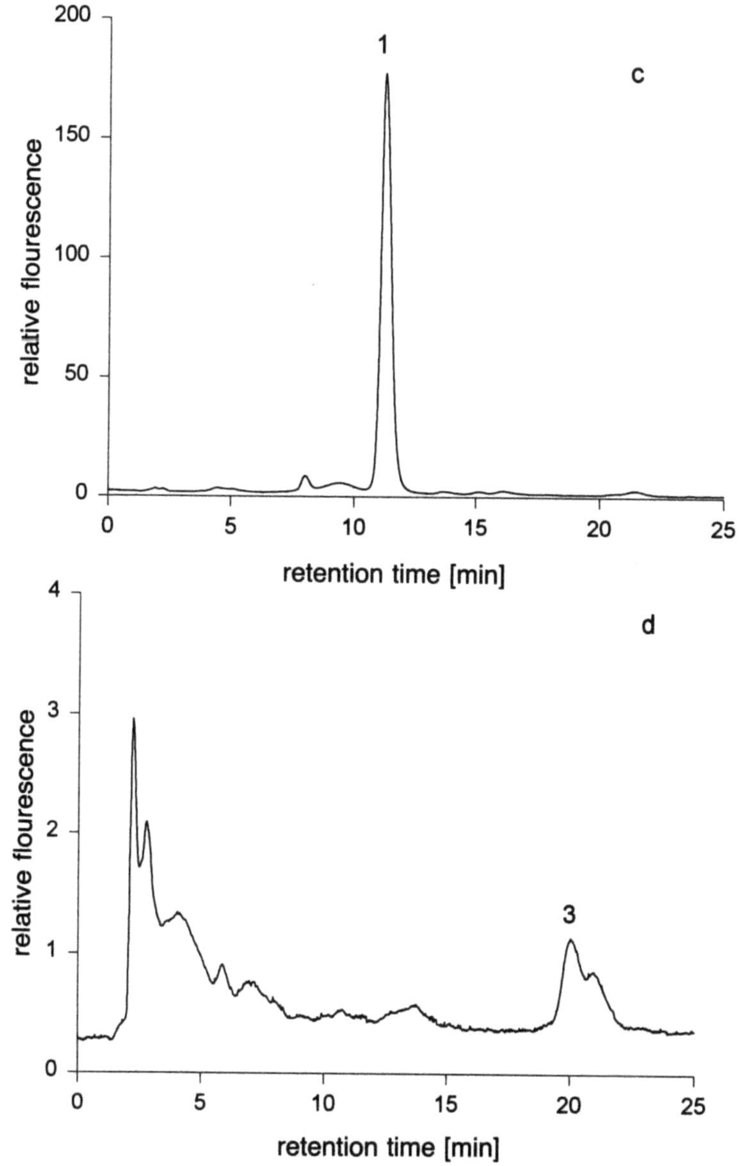

FIGURE 4. Analysis of sialic acids from human colon as DMB derivatives by fluorimetric HPLC on RP-18. (a) Sialic acids after direct derivatization with DMB; (b) after treatment with influenza C virus; (c) after treatment with 0.1 M NaOH; (d) after treatment with aldolase. The following sialic acid peaks were identified by comparison with known standards: 1, Neu5Ac; 2, Neu5,9Ac$_2$; 3, higher *O*-acetylated sialic acids.

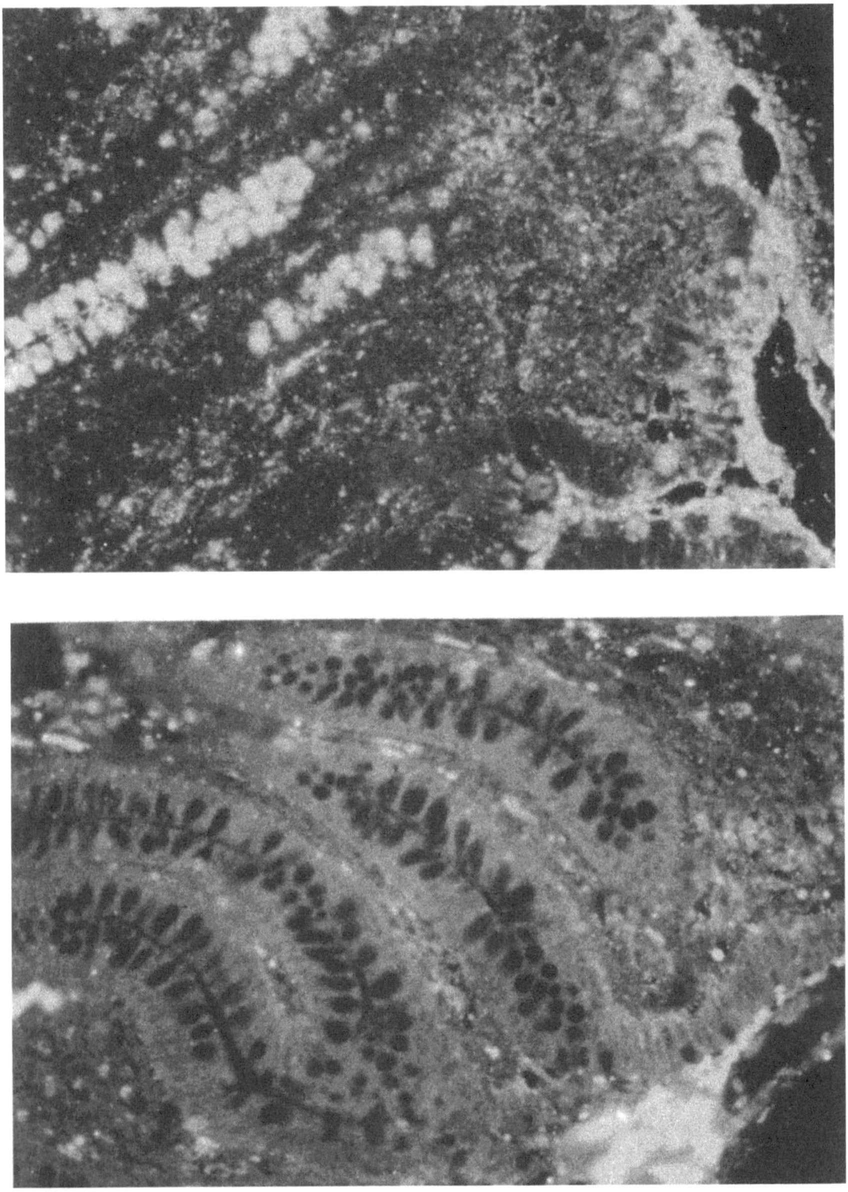

FIGURE 5. Thin section of human colon overlaid with influenza C virus. Bound virus was visualized by an immunoassay with FITC-conjugated anti-influenza antibodies. (Top) Normal staining; (Bottom) same as a after sialidase treatment.

A separation of neutral or sialylated oligosaccharides or glycopeptides can be achieved by HPLC on anion-exchange resins using eluents with strong alkali followed by pulsed amperometric detection (Townsend *et al.*, 1988), allowing analysis of these compounds at the picomole level. Because of the high pH of the eluent, analysis of *O*-acetylated sialic acids is not possible. For separation of unmodified gangliosides, a new HPLC method has been established using anion-exchange resins (Unland and Müthing, 1993).

In addition to these methods, histochemical techniques may be used. Various lectins, antibodies, and viruses are known to specifically detect certain sialylated epitopes in glycoconjugates (summarized by Reuter and Schauer, 1994, and see Section 6.1). Recently, influenza C virus, which is well suited for detection of glycosidically linked terminal Neu5,9Ac_2 residues via its hemagglutinin (Zimmer *et al.*, 1992; see Chapter 9), has been used for histochemical analysis of thin sections (Harms *et al.*, 1993) (Figure 5).

4. ENZYMATIC MODIFICATIONS OF SIALIC ACIDS

4.1. *O*-Acylation of Sialic Acids

Over the past several years, the modification of sialic acids by *O*-acetylation has been found to be of great importance (Reuter and Schauer, 1987; Schauer, 1991; Varki, 1992a). Probably the best established example is the binding of influenza viruses to sialic acids. Whereas influenza A and B viruses exclusively bind to non-*O*-acetylated sialic acids, influenza C virus only recognizes Neu5,9Ac_2 (see Chapter 9). Furthermore, sialic acid *O*-acetylation reduces or even prevents the action of sialidases and *N*-acylneuraminate-pyruvate lyase (Schauer, 1982b; Corfield and Schauer, 1982b). The biological significance of this influence is illustrated by the observation that sialic acid de-*O*-acetylation followed by sialidase treatment of rat erythrocytes significantly increases the binding rate of these cells to peritoneal macrophages when compared to sialidase treatment alone (Kiehne and Schauer, 1992). An increased level of sialic acid *O*-acetylation was discovered in human melanoma gangliosides that are considered to be tumor-associated antigens (Thurin *et al.*, 1985; Sjoberg *et al.*, 1992). For human colon cancer, however, a reduction of sialic acid *O*-acetylation was found (Hutchins *et al.*, 1988). The effect of Neu5,9Ac_2 on complement activation has also been described (Varki and Kornfeld, 1980). Thus, this type of molecular modification seems to be a very important tool for the organism to regulate recognition and cellular interaction.

Although 9-*O*-acetylation of sialic acids is the most frequent modification found in many species from bacteria to humans (Corfield and Schauer, 1982a; Schauer, 1982b, 1991), the mechanism of the acetylation reaction is not clear.

The data available up to now indicate an enzymatic transfer of the acetyl function from acetyl-CoA onto glycoconjugate-bound sialic acid and not onto free sialic acids or their CMP-glycosides (Schauer, 1982b, 1987b, 1991; Varki, 1992a). The complete pattern of sialic acid side chain *O*-acetylation may require several sialate *O*-acetyltransferases with a specificity for only one position. However, the primary insertion site for the *O*-acetyl function may be the 7-OH alone from where the ester group can migrate even under physiological conditions to the 9-position, presumably via C-8 (Kamerling *et al.*, 1987; Schauer, 1987b), leaving the 7-OH ready for a new transfer. Thus, for complete side chain *O*-acetylation, only one enzyme, acetyl-CoA:sialate 7-*O*-acetyltransferase (EC 2.3.1.45), may be necessary together with nonenzymatic migration of this substituent. In rat liver Golgi vesicles 7- and 9-*O*-acetylation of glycoconjugates were found (Diaz *et al.*, 1989; Higa *et al.*, 1989a,b), 7-*O*-acetylation primarily in lysosomal membranes and 9-*O*-acetylation in plasma membranes (Butor *et al.*, 1993a). In bovine submandibular glands, sialate *O*-acetyltransferase was detected in a cytosolic fraction as well as associated with smooth and mitochondrial membranes (Schauer *et al.*, 1988a) using de-*O*-acetylated bovine submandibular gland mucin or immobilized fetuin as substrates. Sialic acid *O*-acetyltransferase has been reported to be specific for Neu5Ac in *N*-glycans in rat liver (Butor *et al.*, 1993a) and for Neu5Ac in gangliosides in human melanoma cells (Sjoberg *et al.*, 1992), suggesting that *O*-acetyltransferase may also exhibit substrate specificity for the oligosaccharide to which the sialic acid is bound. So far, however, sialate *O*-acetyltransferases have not been purified to homogeneity, which may be a prerequisite for a detailed characterization of these enzymes.

4-*O*-Acetylation has been found in horse, donkey, guinea pig, and *Echidna* (Corfield and Schauer, 1982a) and requires a different *O*-acetyltransferase, acetyl-CoA:sialate 4-*O*-acetyltransferase (EC 2.3.1.44), which has been described for equine submandibular gland as a membrane-associated enzyme (Schauer, 1978b, 1987b).

Little is known about sialic acid *O*-lactoylation, which has been found as Neu5Ac9Lt in human, cow, horse, and trout (Corfield and Schauer, 1982a; Schmelter *et al.*, 1993; Corfield *et al.*, 1993). Recent investigations with a particulate fraction from horse liver suggested that this modification occurs enzymatically, although the exact mechanism and type of lactoyl donor are unknown (Kleineidam *et al.*, 1993).

4.2. De-O-Acetylation of Sialic Acids

Sialate 9-*O*-acetylesterases (EC 3.1.1.53) have been isolated and characterized from bovine brain, horse and rat liver, and influenza C virus (Schauer *et al.*, 1988b,c, 1989; Butor *et al.*, 1993b). These enzymes specifically release 9-*O*-acetyl groups from free and glycosidically bound sialic acids; 9-*O*-lactoyl groups, Neu5,7Ac_2, or sialic acid methyl esters are not substrates. An additional specificity for 4-*O*-acetyl groups was discovered for the esterase from horse liver (Schauer *et al.*, 1988b), which may render this type of sialic acid susceptible to sialidases, since a 4-*O*-acetyl function in sialic acids prevents their release from glycoconjugates by mammalian and bacterial sialidases (Schauer, 1982b; Corfield and Schauer, 1982b), including as well the sialidase from horse liver. Deesterification of side-chain *O*-acetylated sialic acids also increases the catabolism of these sugars (Schauer, 1987b). These esterases, which, however, also attack *O*-acetyl groups of nonphysiological substances, e.g., naphthylacetate, belong to the few esterases that have been recognized to have a physiological function (Schauer *et al.*, 1988c).

Influenza C viruses express a surface glycoprotein (HEF) consisting of three different domains, two of which have been found to recognize *O*-acetylated sialic acids, the receptor-destroying enzyme that was characterized as sialate 9-*O*-acetylesterase (Herrler *et al.*, 1985), and the hemagglutinin that binds to 9-*O*-acetylated sialic acids of glycoproteins and glycolipids (Zimmer *et al.*, 1992), as well as to Neu5,7Ac_2-GD3 (Harms *et al.*, 1993). Whereas the enzymatic activity is optimal at 37°C and hardly detectable at 4°C, the hemagglutinin is fully capable of binding at the lower temperature. On the basis of this temperature difference in the two activities, a test was developed for the detection of glycoconjugate-bound 9-*O*-acetylated sialic acids on thin-layer chromatograms, SDS gels after blotting, microtiter plates, and tissue sections (Zimmer *et al.*, 1992; 1994; Harms *et al.*, 1993). The third domain of the HEF is involved in fusion of the virus with the cell membrane. The primary sequence of HEF has been elucidated by molecular cloning (Vlasak *et al.*, 1987).

4.3. *O*-Methylation of Sialic Acids

Sialic acid *O*-methylation is a modification that seems to be restricted to *Echinodermata* (Corfield and Schauer, 1982a). *O*-Methyl groups have only been found at the 8-position of Neu5Gc and in much lower quantities on Neu5Ac (Bergwerff *et al.*, 1992, 1993). As outlined in Section 2, 9-*O*-acetylated derivatives of 8-*O*-methylated sialic acids were also detected. The existence of a sialate 8-*O*-methyl transferase needed for the biosynthesis of this sialic acid was first demonstrated to occur in the starfish *Asterias rubens* (Schauer and Wember, 1985; Bergwerff *et al.*, 1992). The enzyme has now been further characterized from a crude preparation of starfish gonads as a membrane-associated protein that transfers methyl groups from *S*-adenosyl methionine preferably onto glycosidically linked Neu5Gc residues. Neither free Neu5Gc nor CMP-Neu5Gc are substrates for this enzyme. Horse erythrocytes containing almost exclusively Neu5Gc residues are excellent substrates for the sialate *O*-methyltransferase (de Freese *et al.*, 1993).

4.4. *N*-Acetyl-Hydroxylation of Sialic Acids

4.4.1. Occurrence of *N*-Glycoloylneuraminic Acid

N-Glycoloylneuraminic acid (Neu5Gc) is formally derived by the addition of a hydroxyl group onto the *N*-acetyl function of Neu5Ac. Neu5Gc is present in essentially all animal groups possessing sialylated glycoconjugates, i.e., spanning the deuterostomate lineage from the echinoderms up to mammals (Warren, 1963; Corfield and Schauer, 1982a; Varki, 1992a). The extent of glycoconjugate sialylation with Neu5Gc is very much dependent on the species (Warren, 1963; Corfield and Schauer, 1982a), tissue (Reuter *et al.*, 1988), stage in development (Muchmore *et al.*, 1987; Sherblom *et al.*, 1988; Schauer *et al.*, 1991; Budd *et al.*, 1992), and the presence of certain pathogenic conditions (Higashi *et al.*, 1985). Humans and chickens are, however, notable since they lack Neu5Gc in healthy tissues but do express small amounts in certain tumors (Gottschalk, 1960; Haverkamp *et al.*, 1976; Ledeen and Yu, 1976; Corfield and Schauer, 1982a; Higashi *et al.*, 1985; Kawai *et al.*, 1991). The significance of this is discussed later in this section. Although several bacterial species incorporate Neu5Ac and even Neu5,9Ac$_2$ into surface lipopolysaccharides (Dutton *et al.*, 1987; Gibson *et al.*, 1993) and capsular polysialic acid (Higa and Varki, 1988; Troy, 1992), the existence of Neu5Gc in bacteria has never been reported. The pathogenic protozoan *Trypanosoma cruzi* has been shown to contain Neu5Gc (Schauer *et al.*, 1983). This, however, most probably originates from sialoglycoconjugates in the culture medium. The transfer reaction may be catalyzed by a trans-sialidase (Schenkman *et al.*, 1991), though activity with a Neu5Gc-containing sialyl donor has only been demonstrated for the trans-sialidase from *Trypanosoma brucei* (Engstler and Schauer, 1993).

Neu5Gc is present in gangliosides and glycoproteins, exhibiting no pronounced preference for a particular oligosaccharide type. However, in porcine submaxillary gland mucin Neu5Gc was found to occur preferentially in α2,6 linkage to GalNAc (Savage *et al.*, 1986). Furthermore, in *N*-glycans of porcine vitronectin, Neu5Gc residues are located predominantly on the Man α1,6 arm (Yoneda *et al.*, 1993). Like Neu5Ac, Neu5Gc is also derivatized in a variety of positions, including *O*-acetylation in positions 4, 7, 8, and 9 (Corfield and Schauer, 1982b), *O*-lactoylation in position 9 (Reuter *et al.*, 1988), and a rather unusual 8-*O*-methylation occurring in certain starfish (Warren, 1964; Corfield and Schauer, 1982a; Bergwerff *et al.*, 1992). The additional C-5 hydroxyl group in Neu5Gc may also be acetylated to yield Neu5GcAc (Kluge *et al.*, 1992) present in rat thrombocytes, or may be glycosylated, as demonstrated in certain starfish gangliosides (Smirnova *et al.*, 1987) and oligosaccharides (Bergwerff *et al.*, 1992, 1993). While oligomers of Neu5Gc have been detected in the polysialic acid on the vitelline coat glycoproteins of several fish eggs (see Section 2;

Kitajima *et al.*, 1988; Kanamori *et al.*, 1990), the presence of Neu5Gc in mammalian NCAM-associated polysialic acid has yet to be demonstrated (Troy, 1992). (See Chapter 4.)

4.4.2. Biological Role of Neu5Gc

Since the sialic acid residues of glycoconjugates play a variety of roles, as discussed elsewhere in this chapter, it is unlikely that Neu5Gc has one specific function. However, it seems to modulate the functions exerted by sialic acids. For example, cell surface glycoconjugates may help to protect tissues from attack by certain pathogen-derived toxic enzymes. In order to aid the digestion of cell surface glycoconjugates, several microorganisms secrete a sialidase, which may facilitate spreading such as that of bacteria during infection (Ezepchuk *et al.*, 1973; Corfield, 1992). Since the cleavage rate of Neu5Gc by bacterial and viral sialidases is generally lower than that of Neu5Ac, the expression of Neu5Gc may be an adaptation to retard the effects of sialidases (Corfield *et al.*, 1981b). The further degradation of Neu5Gc by the acylneuraminate-pyruvate lyase also occurs at a slower rate than for Neu5Ac (Corfield and Schauer, 1982b). It is therefore conceivable that Neu5Gc may mask subterminal galactose residues of oligosaccharide chains thus preventing their recognition and phagocytosis by macrophages more potently than does Neu5Ac (Schauer, 1985).

Over the last few years, certain sialic acid-containing oligosaccharide structures have been found to function as ligands for a number of receptors involved in important cell–cell recognition phenomena (Varki, 1992b). Such receptors include the selectins (Cummings and Smith, 1992), the B-lymphocyte-associated glycoprotein CD22 (Sgroi *et al.*, 1993), and the sialoadhesin associated with mouse stromal macrophages (Crocker *et al.*, 1991), as will be discussed in detail in Section 6. In addition, several sialic acid-binding lectins have been described, which exhibit specificity for either Neu5Ac or Neu5Gc (Section 6, Table IV).

Although several bacterial (Karlsson, 1989; Liukkonen *et al.*, 1992) and viral (Higa *et al.*, 1985) pathogens exploit glycoconjugate-bound sialic acid residues for adherence to a cell surface prior to infection (Table V), only the enterotoxic *E. coli* strain K99, which infects young pigs, is specific for Neu5Gc (Kyogashima *et al.*, 1989).

4.4.3. Biosynthesis of Neu5Gc

4.4.3.a. *N*-Acylation with Activated Glycolic Acid. The mechanism of the biosynthesis of Neu5Gc has been the subject of several investigations. The possibility that the *N*-glycoloyl group might arise by hexosamine acylation with glycoloyl-CoA at an early step in sialic acid biosynthesis was advanced by Jourdian and Roseman (1962). However, later studies on glycoprotein biosynthesis in sheep colon largely excluded this pathway (Allen and Kent, 1968). Nevertheless, in a recent publication it was put forward that glycoloyl-CoA derived from fatty acid by ω-oxidation with subsequent β-oxidation might give rise to Neu5Gc (Vamecq *et al.*, 1992).

4.4.3.b. Oxidative Hydroxylation of Neu5Ac: Specificity of a Sialic Acid Hydroxylase. Several studies on the production of mucin-bound Neu5Gc in pig submandibular glands indicated that the *N*-glycoloyl group results from the hydroxylation of the amino-sugar-bound *N*-acetyl moiety at some point during Neu5Ac biosynthesis (Schauer *et al.*, 1968). Subsequent experiments with cell-free extracts of this tissue suggested that free (Schauer, 1970) and glycoconjugate-bound Neu5Ac (Buscher *et al.*, 1977) are the substrates for an NAD(P)H-dependent Neu5Ac hydroxylase, which could exist in soluble and membrane-bound forms.

However, a number of later investigations using fractionated homogenates of pig submandibular gland and mouse liver unequivocally established that Neu5Gc is synthesized exclusively by the hydroxylation of the sugar nucleotide sialic acid metabolite CMP-Neu5Ac, giving CMP-Neu5Gc as the product (Shaw and Schauer, 1988, 1989). The isolation of CMP-Neu5Gc from pig submandibular glands is consistent with the suggested substrate specificity (Buscher *et al.*, 1977). Moreover, pulse–chase experiments using mouse myeloma cells incubated in culture with [6-^{3}H]mannosamine revealed that labeled Neu5Gc was first detected in the CMP-sialic acid pool, confirming that the enzyme exhibits the same substrate specificity in living cells (Muchmore *et al.*, 1989).

Interestingly, the existence of CMP-Neu5Ac hydroxylase activity was also demonstrated in the starfish *Asterias rubens* (Bergwerff *et al.*, 1992; Schlenzka *et al.*, 1993a). Since this is one of the evolutionarily least advanced organisms known to possess Neu5Gc-containing glycoconjugates, the pathway leading to Neu5Gc is evidently highly conserved throughout evolution.

4.4.3.c. Properties of CMP-Neu5Ac Hydroxylase. In all mammalian tissues and cell lines so far tested, CMP-Neu5Ac hydroxylase is extracted in a particle-free fraction, suggesting that this enzyme is cytosolic (Shaw and Schauer, 1988, 1989; Bouhours and Bouhours, 1989; Shaw *et al.*, 1991). Similarly, the CMP-Neu5Ac hydroxylase from gonads of *A. rubens* was also predominantly found in a high-speed supernatant fraction, though some membrane-associated activity was also detected (Schlenzka *et al.*, 1993a).

The hydroxylase in supernatants of various tissues exhibits an apparent K_M for CMP-Neu5Ac ranging from 18 μM for the enzyme from *A. rubens* (Schlenzka *et al.*, 1993a) through to 2.5 μM for the hydroxylase in pig submandibular glands (Muchmore *et al.*, 1989), 1.3 μM for the mouse liver enzyme (Shaw and Schauer, 1989), and 0.6 μM for the hydroxylase from rat small intestine (Bouhours and Bouhours, 1989). Although the complex substrate and cofactor requirements of this enzyme (see later) only permit the measurement of an apparent K_M, the values cited above indicate that the hydroxylase has a very high affinity for its CMP-Neu5Ac substrate. The fact that the K_M of the Golgi membrane-associated CMP-sialic acid antiporter from mouse and rat also lies in the low micromolar range, is consistent with both the hydroxylase and the transporter being exposed to the same cytosolic pool of CMP-Neu5Ac (Carey *et al.*, 1980; Lepers *et al.*, 1989, 1990).

CMP-Neu5Ac hydroxylase exhibits a high level of substrate specificity. Thus, neither glycoprotein-bound Neu5Ac nor free Neu5Ac is hydroxylated by this enzyme (Shaw and Schauer, 1988, 1989). Moreover, relatively high concentrations of free Neu5Ac and a variety of cytidine nucleotides are unable to inhibit the hydroxylase from mouse (Shaw *et al.*, 1992) and starfish (Schlenzka *et al.*, 1993a) to any significant degree. Even CMP-Neu5Gc, the product of the hydroxylase reaction, is only a weak inhibitor (Shaw *et al.*, 1992). A number of sugar-nucleotides (i.e., GDP-Man, UDP-Glc, UDP-GlcNAc, UDP-Gal, and UDP-GalNAc), relevant to glycoconjugate biosynthesis, also have little influence on the activity of the hydroxylase from mouse (Shaw *et al.*, 1992). These results show that the hydroxylase has stringent structural requirements for its CMP-sialic acid substrate and also suggest that this enzyme is unlikely to be regulated by any of the above-mentioned substances.

Experiments with the hydroxylase in high-speed supernatants of several tissues revealed that this enzyme requires a number of substrates and coenzymes for activity. The removal of oxygen from reaction media causes severe inhibition of the hydroxylase, suggesting that the enzyme is a monooxygenase (Shaw and Schauer, 1989; Schlenzka *et al.*, 1993a). This conclusion, however, requires confirmation using the purified enzyme. A reducing cofactor is also essential for the turnover of CMP-Neu5Ac hydroxylase. Although several reducing agents such as ascorbate, dithiothreitol, and tetrahydrobiopterin support a low level of activity, reduced pyridine nucleotides are by far the most effective cofactors, NADH generally being the preferred coenzyme (Shaw and Schauer, 1988, 1989; Kozutsumi *et al.*, 1990; Schlenzka *et al.*, 1993a).

Most NAD(P)H-dependent monooxygenases usually consist of two or three loosely associated protein components. These form an electron transport chain where an NAD(P)H oxidoreductase passes electrons to a terminal acceptor which catalyzes the reductive activation of oxygen and the subsequent hydroxylation of the substrate (Hayaishi, 1974). The inhibitory effect of increased ionic strength and dilution on the activity of the hydroxylase from mammalian sources points to the involvement of protein–protein interactions during catalysis (Shaw *et al.*, 1992). This has been confirmed in a number of studies in which the participation of cytochrome b_5 and cytochrome b_5 reductase in the hydroxylase reaction of mammalian tissues was demonstrated (Kozutsumi *et al.*, 1990; Shaw *et al.*,

1992; Kawano *et al.*, 1993a). The postulated scheme for the interaction between the various components of this system is depicted in Figure 6. On the basis of these properties, the EC number 1.14.13.45 has been suggested.

In mammalian cells, cytochrome b_5 and its reductase occur in membrane-bound forms associated with the endoplasmic reticulum, where they serve as a source of reducing equivalents for several membrane-bound redox enzymes, such as cytochrome P450 (Chiang, 1981) and stearoyl-CoA desaturase (Strittmatter *et al.*, 1974). The majority of cytochrome b_5-dependent enzymes are integral membrane proteins bound to the endoplasmic reticulum, functioning in conjunction with the cytochrome b_5 system associated with this membrane compartment (Arinç, 1991). CMP-Neu5Ac hydroxylase, however, is a soluble protein and only draws electrons from the microsomal cytochrome b_5 system at a very low rate *in vitro*. The source of reducing equivalents *in vivo* is still unclear, though some evidence points to the involvement of an amphiphilic cytochrome b_5 system present in the cytosol (Shaw *et al.*, 1994). Although cytochrome b_5 is present in starfish (den Besten *et al.*, 1990), it is not known whether it participates in the activity of CMP-Neu5Ac hydroxylase in this organism. Indeed, the insensitivity of the enzyme from starfish to high ionic strength and dilution suggests that this enzyme may consist of more firmly associated components (Schlenzka *et al.*, 1993a), contrasting with the mouse enzyme system.

All cytochrome b_5-dependent enzymes so far described possess an iron-containing prosthetic group and several lines of evidence point to the participation of an iron cofactor in the reaction of the hydroxylase. For example, several iron-binding compounds are potent inhibitors of this enzyme (Kozutsumi *et al.*, 1991; Shaw *et al.*, 1992). Furthermore, the addition of iron salts can stimulate the hydroxylase (Shaw and Schauer, 1988, 1989; Schlenzka *et al.*, 1993a). Although no firm conclusions can yet be made, the above observations rule out the presence of a heme group, a suggestion that is supported by preliminary spectroscopic measurements on the purified enzyme (Kawano *et al.*, 1993b).

The characterization of CMP-Neu5Ac hydroxylase described above was, for the most part, carried out using unfractionated supernatants and many of the investigations need refining using the purified hydroxylase as well as purified

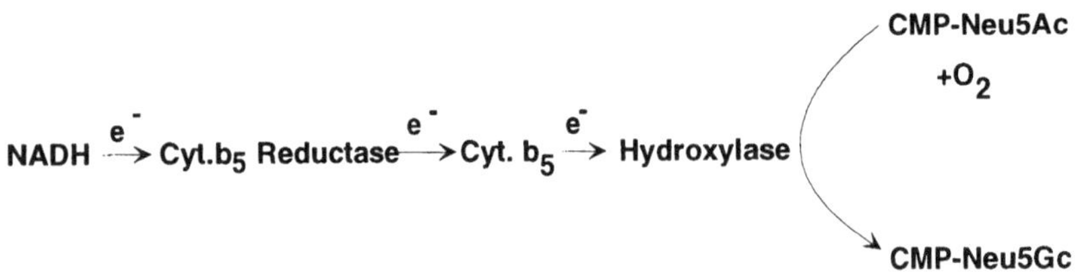

FIGURE 6. Redox protein components involved in the hydroxylation of CMP-Neu5Ac.

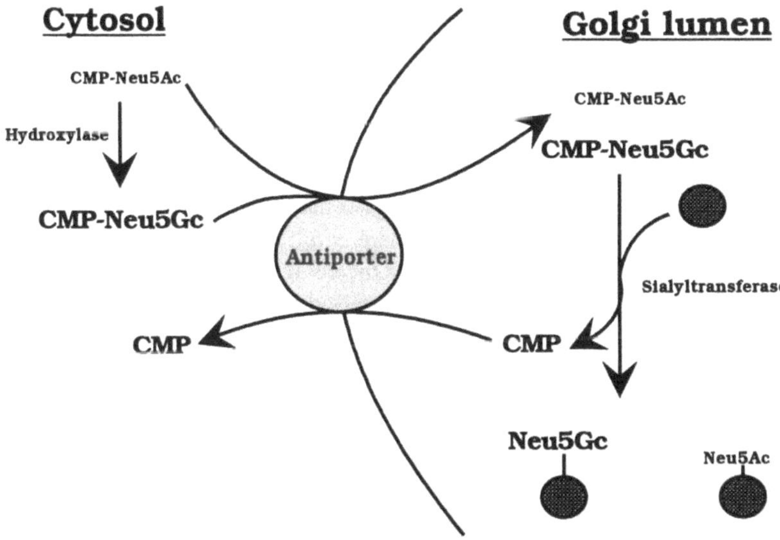

FIGURE 7. Cellular compartmentalization and regulation of Neu5Gc biosynthesis and incorporation into glycoconjugates.

cytochrome b_5 and reductase. The purification of the hydroxylase from mouse liver (Kawano *et al.*, 1993b; Schneckenburger *et al.*, 1993) and pig submandibular glands (Schlenzka *et al.*, 1993b) has been reported recently. The enzyme from both sources was found to be a monomer with a molecular mass of 65 kDa.

4.4.3.d. Role of CMP-Neu5Ac Hydroxylase in Regulating the Incorporation of Neu5Gc into Glycoconjugates. The activity of CMP-Neu5Ac hydroxylase is probably the main factor influencing the level of sialylation with Neu5Gc. Using rat small intestine mucosal cells, Bouhours and Bouhours (1989) showed that the increase in Neu5Gc-GM3 content occurring in the period 15–20 days after birth correlates with an increase in the activity of the hydroxylase, though other tissues did not exhibit such a clear correlation (Muchmore, 1992). Similarly, the relative level of Neu5Gc and Neu5Ac in rat and mouse liver (Lepers *et al.*, 1990), as well as in several related mouse lymphoma cell lines, correlates with the activity of CMP-Neu5Ac hydroxylase (Shaw *et al.*, 1991). Neither the Golgi CMP-sialic acid antiporter nor the sialyltransferases exhibit a pronounced preference for CMP-Neu5Ac or CMP-Neu5Gc, and thus have little or no role in regulating the relative incorporation of Neu5Gc or Neu5Ac into glycoconjugates (Figure 7) (Lepers *et al.*, 1990). The activity of the hydroxylase may thus be tuned so that the ratio of Neu5Gc and Neu5Ac required in the resulting glycoconjugates is generated in the form of CMP-glycosides in the cytosol. Since the cytochrome b_5 system has a multiplicity of functions and the hydroxylase is not apparently influenced by any metabolites, the rate of production of CMP-Neu5Gc is presumably regulated at the level of expression of the 65-kDa monooxygenase component. The tissue-specific and developmental factors affecting the expression of this protein still remain to be elucidated.

Although *O*-acetyl modifications of sialic acids can be enzymatically removed (see Section 4.2), no mechanism for the dehydroxylation of Neu5Gc is known. Neu5Gc from recycled glycoconjugates may thus augment the hydroxylase-mediated Neu5Gc biosynthesis (Muchmore *et al.*, 1989).

4.4.3.e. Neu5Gc as a Human Oncofetal Antigen. Glycoconjugates sialylated with Neu5Gc have not been detected in normal human and chicken tissues and are in fact antigenic in both species, inducing the formation of heterophile antibodies historically referred to as Hanganutziu–Deicher antibodies (Fujii *et al.*, 1982; Schauer, 1988; Higashi, 1990). Antigens reacting with Hanganutziu–Deicher antibodies have been detected in a variety of human tumor types, including colon cancers (Higashi *et al.*, 1985; Hirabayashi *et al.*, 1987a), retinoblastoma (Ohashi *et al.*, 1983; Higashi *et al.*, 1988), melanoma (Hirabayashi *et al.*, 1987b; Saida *et al.*, 1990), and breast cancer (Hanisch *et al.*, 1992). Cancerous tissue from chickens has also been shown to contain Neu5Gc (Kawai *et al.*, 1991). The majority of these antigens have been identified as gangliosides sialylated with Neu5Gc. Tumor-associated glycoproteins sialylated with Neu5Gc have also been reported in a human gastric cell line grown in serum-free medium (Fukui *et al.*, 1989) and in a mucin from breast cancer (Devine *et al.*, 1991). The frequency of these antigens is variable and depends on the type of tumor, but generally varies from 30 to 50% of all cancers (Higashi *et al.*, 1984, 1985). In all human tumors so far tested, the amount of Neu5Gc, as a proportion of total sialic acid, is extremely low, usually less than 1% and frequently in the range 0.01–0.1% (Higashi *et al.*, 1985; Kawai *et al.*, 1991; Hanisch *et al.*, 1992). Such low levels of Neu5Gc are generally detected with immunological methods rather than with chemical techniques. Some authors have, however, identified Neu5Gc in tumors by gas chromatography and mass spectrometry and the amounts of Neu5Gc determined are in the same range as those estimated by immunological methods (Kawai *et al.*, 1991; Hanisch *et al.*, 1992).

To date one can only speculate on the origin of Neu5Gc in human tumor tissues. Neu5Gc observed in HeLa cells grown in the presence of fetal calf serum could have arisen by incorporation of Neu5Gc present in the serum (Carubelli and Griffin, 1968). However, in the majority of investigations, pathological tissue removed directly from patients was analyzed. The abnormal uptake or incorporation of dietary Neu5Gc or *N*-glycoloylated amino-sugar metabolites is a possible source, though in animal experiments, the main part of orally administered Neu5Gc is excreted (Nöhle *et al.*, 1982). The possibility that abnormal glycoloyl-CoA metabolism could give rise to Neu5Gc must also be considered (Vamecq *et al.*, 1992). The simplest mechanism leading to Neu5Gc formation in human tumors is the anomalous expression of an otherwise dormant or repressed CMP-Neu5Ac hydroxylase gene. However, no evidence has been obtained for this assumption.

5. RELATIONSHIP AND EVOLUTIONARY DISTRIBUTION OF MICROBIAL SIALIDASES

Sialidases (EC 3.2.1.18) are essential tools in sialic acid catabolism (see Chapter 8). These enzymes cleave the *O*-glycosidic linkages between the terminal sialic acids and the subterminal sugars of free and glycoconjugate-bound oligosaccharides as one of the first steps in sialoglycoconjugate degradation. Sialidase, as well as its substrate, is common in metazoan animals of the deuterostomate lineage from echinoderms to mammals (Rosenberg and Schengrund, 1976b; Corfield *et al.*, 1981a; Corfield and Schauer, 1982b). Diverse viruses and microorganisms, like fungi, protozoa, and bacteria, also produce sialidases (Müller, 1974; Corfield *et al.*, 1981a; Corfield, 1992; Engstler *et al.*, 1993), although they mostly lack sialic acids. A common property of these organisms is their close contact to animals as commensals or pathogens (Schauer and Vliegenthart, 1982), whereby sialidase is used primarily for nutritional purposes (Corfield, 1992; Müller, 1992). However, this enzyme may additionally be employed for adhesion to host cells (Gabriel *et al.*, 1984) and as a spreading or virulence factor in invasive infections (Ezepchuk *et al.*, 1973; Godoy *et al.*, 1993). Close interactions between animals and microorganisms may have led to the exchange of sialidase genes between these groups and also between the microorganisms involved.

It is remarkable that the occurrence of sialidase in microorganisms is frequently not in accordance with the phylogenetic relationship of bacterial species or strains. Highly related species, e.g., *Clostridium sordellii* and *C. bifermentans* (Roggentin *et al.*, 1985), or even strains of one species, e.g., *C. butyricum* (Popoff and Dodin, 1985) or *Salmonella typhimurium* (Hoyer *et al.*, 1992), differ with respect to sialidase production. This irregular distribution of the enzyme indicates that sialidases evolved by other mechanisms than their producers in the bacterial kingdom. It is therefore of interest to gain insight into the relatedness and the directions of propagation of the factor sialidase, the possession of which possibly is the result of an adaptation of the microorganisms to a "new" sugar presented by their animal hosts.

A comparison of the properties of the sialidases from a variety of microbial sources revealed that these enzymes are highly diverse with respect to molecular mass, number of subunits, isoelectric point, temperature optimum, influence of Ca^{2+} on activity, substrate specificity, and specific activity (Nees *et al.*, 1975; Uchida *et al.*, 1979; Reuter *et al.*, 1987; Roggentin *et al.*, 1987; Teufel *et al.*,

Table III
Summary of Cloned and Sequenced Sialidases from Eukarya and Bacteria

Organism	Type of sialidase	Reference
Rattus rattus	Cytosolic	Miyagi *et al.* (1993)
Trypanosoma cruzi	Cell surface, trans-sialidase	Pereira *et al.* (1991)
Trypanosoma rangeli	Secreted (sialidase-like sequence)	Buschiazzo *et al.* (1993)
Actinomyces viscosus	Cell surface	Henningsen *et al.* (1991), Yeung (1993)
Bacteriodes fragilis	Secreted (partial sequence)	Russo *et al.* (1990)
Clostridium perfringens	"Large" secreted isoenzyme	Traving *et al.* (1993)
	"Small" cytosolic isoenzyme	Roggentin *et al.* (1988)
Clostridium septicum	Secreted	Rothe *et al.* (1991)
Clostridium sordellii	Secreted	Rothe *et al.* (1989)
Micromonospora viridifaciens	Secreted	Sakurada *et al.* (1992)
Salmonella typhimurium	Cytosolic	Hoyer *et al.* (1992)
Vibrio cholerae	Secreted	Galen *et al.* (1992)

1989; Aisaka *et al.*, 1991; Heuermann *et al.*, 1991; Hoyer *et al.*, 1991; Tanaka *et al.*, 1992; Engstler *et al.*, 1992, 1994; Zenz *et al.*, 1993). The only common property is the acidic pH optimum (pH 5.0–6.1), though the trypanosomal trans-sialidases exhibit maximum activity at a neutral pH (Engstler *et al.*, 1993). As these properties did not reveal any relationship between microbial sialidases, the enzymes were further investigated on the DNA level, from which interesting information became available.

In the first primary structures of microbial sialidases, obtained by cloning and sequencing of the respective genes from *Clostridium perfringens* (Roggentin *et al.*, 1988), *Vibrio cholerae* (Galen *et al.*, 1992), *Clostridium sordellii* (Rothe *et al.*, 1989), and *Salmonella typhimurium* (Hoyer *et al.*, 1992), an amino acid sequence motif was detected, which is repeated fourfold in each protein: S-X-D-X-G-X-T-W (Roggentin *et al.*, 1989). This motif, named Asp-box, was found in all sialidases of animals and bacteria that have been sequenced (Table III). From the alignment of more and more sialidase sequences, it became obvious that the Asp-boxes II and IV are more degenerate than I and III, whereby box III appears to be virtually complete in each protein (Figure 8). In viral sialidases, however, the motif was rarely detectable (e.g., only an N9 influenza A virus strain exhibits the complete motif) (Air *et al.*, 1987), and has probably undergone mutational alterations.

The function of this conserved and repeated motif is as yet unknown. In the N9 influenza A virus sialidase (Air *et al.*, 1987), the single Asp-box is located as part of a β-pleated sheet polypeptide at the connections between the four protein

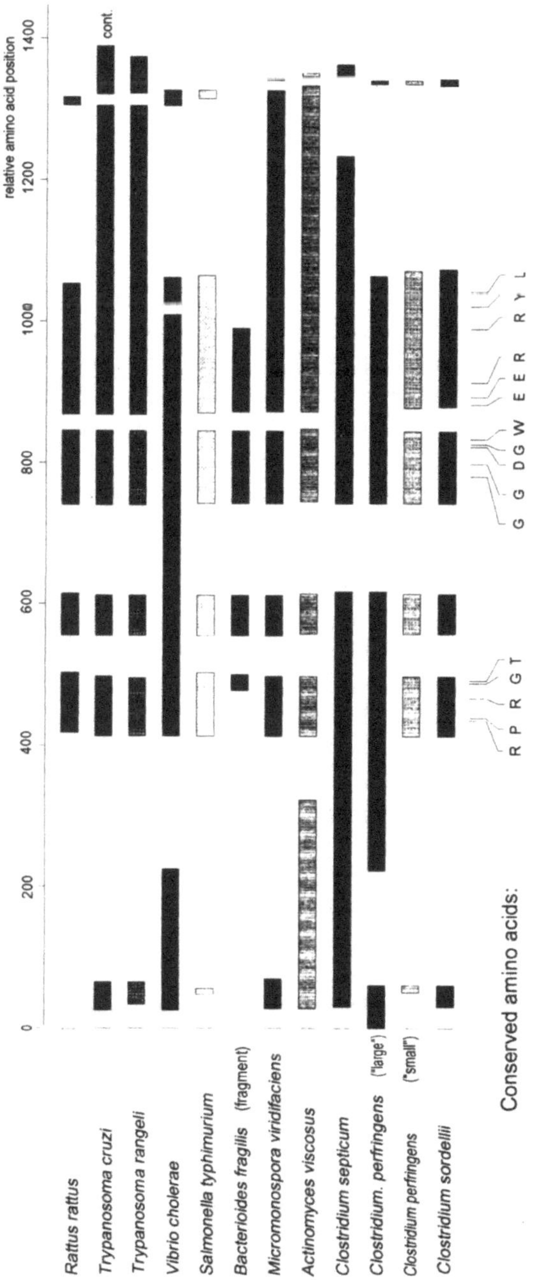

FIGURE 8. Schematic survey of an alignment of 12 sialidase primary structures and the location of 16 conserved amino acids.

subunits. Immunological studies revealed that the Asp-box I of *Trypanosoma cruzi* sialidase is inaccessible to antibodies, which has been taken as an indication that it is not part of the external catalytic domain (Prioli *et al.*, 1992). Site-directed mutagenesis experiments of the "small" sialidase isoenzyme of *Clostridium perfringens* resulted in only small alterations of enzyme activity by changing some of the Asp-box amino acids (Roggentin *et al.*, 1992), while the exchange of other highly conserved amino acids drastically reduced enzyme activity and increased the K_M value, e.g., by replacement of the N-terminal conserved arginine (Figure 8) with lysine.

In addition to the Asp-boxes, further identical motifs, or single amino acids, became evident by an alignment of sialidase protein sequences, which is schematically demonstrated in Figure 8. Gaps had to be introduced as a consequence of the differences in protein sizes. The central regions of these proteins are especially homologous and exhibit most of the conserved amino acids. A further motif, the FRIP-region, which is located N-terminally from the first Asp-box, is highly conserved in clostridial sialidases, but was found to be degenerated to X-R-X-P, when further bacterial and animal sialidase sequences were included in the alignment. Nevertheless, 16 amino acids have been found to be conserved. The presence of conserved motifs and single amino acids indicates that the enzymes are interrelated and originate from one source.

By a pairwise comparison of the sequences, further amino acids were found to be identical at certain positions, which allows a calculation of similarity values. This gives the percentage of identical amino acids (excluding the conserved amino acids mentioned in Figure 8) from the total number of amino acids situated at the same points after alignment as 100%. A dendrogram (Figure 9) based on the average linkage method (Anderberg, 1973) combines the values obtained from the pairwise comparisons. It shows that some of the sialidases are related in accordance with the phylogenetic distances of their producers, e.g., *Micromonospora viridifaciens* and *Actinomyces viscosus* sialidase, or the "large" (72 kDa) isoenzyme of *Clostridium perfringens* and the *Clostridium septicum* sialidase. On the other hand, the three "small" sialidases of lower molecular mass (42–44 kDa) produced by *Salmonella typhimurium, Clostridium perfringens,* and *Clostridium sordellii* exhibit a higher similarity than is expected from the relationship of the bacterial species. These gram-positive (clostridia) or gram-negative (salmonella) bacteria are quite distinct from an evolutionary point of view, but are found at the same ecological location, e.g., the intestine of vertebrates. Here, an exchange of genes even between unrelated microorganisms might be possible via phages, transposons, or plasmids by mechanisms of transfection or conjugation. The participation of phages is indicated by typical sequence motifs up- and downstream from sialidase genes of *Salmonella typhimurium* and *Micromonospora viridifaciens* (Hoyer *et al.*, 1992; Sakurada *et al.*, 1992), and from the observation that the "small" sialidase gene is located

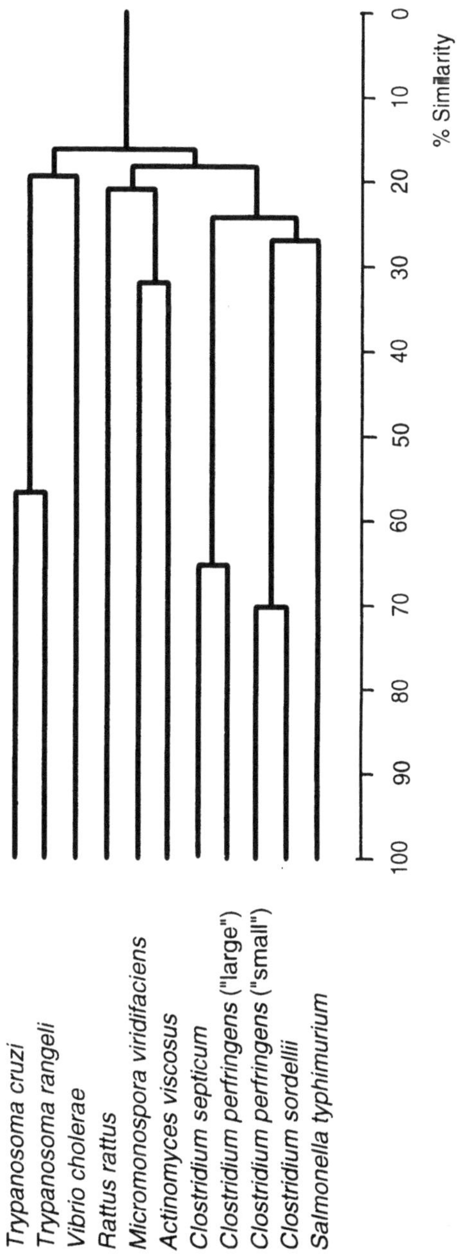

FIGURE 9. Dendrogram of similarities among sialidase primary structures based on identical amino acids.

near a phage attachment site on the chromosome of *Clostridium perfringens* (Canard and Cole, 1990).

Such a horizontal gene transfer would explain the cytoplasmic location (Table III) of active sialidases, thus hindering them from contact with their extracellular substrate, because of the absence of signal peptides. If a gene is acquired from a foreign source, it usually has to be adapted for bacterial purposes, or it is eliminated. The presence of unadapted but active gene products, such as the cytoplasmic sialidases in *Clostridium perfringens* and *Salmonella typhimurium,* therefore indicates that the gene transfer may have occurred recently. The sialidase of *Clostridium sordellii,* which is highly related to these sialidases, probably originated from the same source. In the meantime, it has been modified for secretion and exhibits a broader substrate specificity than the cytosolic enzymes, which are highly selective for α(2-3)-linked sialic acids of oligosaccharides, a structure that is regularly found in higher animals. In contrast, the well-adapted microbial sialidases cited in Table III, e.g., of *Actinomyces viscosus, Bacteroides fragilis, Clostridium perfringens* ("large" isoenzyme), *Clostridium septicum, Micromonospora viridifaciens,* or *Vibrio cholerae,* must have been acquired earlier in evolution, which is also indicated by a relatively poor similarity between the amino acid sequences of sialidase proteins produced by these different genera.

Investigations on the structural sialidase gene composition often revealed a percentage of the bases G + C which is atypical for the chromosomal DNA of the respective bacterial species (e.g., *Salmonella typhimurium:* 40.9/50–53 mol% G + C, or *Actinomyces viscosus:* 70.8/56–58 mol% G + C for the sialidase and the chromosome, respectively; P. Roggentin *et al.*, 1993). Summing up the mol% G + C data available, the origin of sialidase is expected in an organism exhibiting genes that contain about 45 mol% G + C. This value is not uncommon in higher animals.

In conclusion, the hypothesis put forward by Schauer and Vliegenthart (1982) is supported, that sialidase as well as its substrate originated in deuterostomate animals, and that its gene was possibly captured by cell-lysing microorganisms as a reaction to the expression of a "new" sugar.

6. SIALIC ACID-DEPENDENT RECEPTORS AND THEIR LIGANDS

6.1. Occurrence

Many functional roles for the carbohydrate chains of glycoconjugates have been proposed in cellular recognition events, as extensively reviewed by Varki (1993). Whereas for a long time the main function of terminal sialic acid residues was considered to be that of a mask for recognition sites on cell surfaces, e.g., galactose residues or other antigens (Schauer, 1985), only recently have receptors been described that seem to be necessary to connect the structural diversity of sialylated glycoconjugates to specific functions in cellular interaction.

Many microorganisms, plants, and animals express proteins that bind to sialic acids occurring mostly as components of glycoconjugates. A list of such sialic acid-recognizing proteins or lectins from plants and invertebrates is given in Table IV. They show more or less specific binding to different sialic acids or even specific sialic acid linkages and oligosaccharide sequences, as reviewed by Zeng and Gabius (1992b). Since these organisms do not express sialic acids themselves, it is unlikely that these lectins play roles as sialic acid-binding proteins in their own cellular functions. However, they may function in the defense of sialic acid-containing microorganisms. Some of the sialic acid-binding lectins have proved to be useful tools in the analysis and histochemistry of glycoconjugates. Most frequently used are the agglutinins from wheat germ

Table IV
Sialic Acid-Binding Lectins from Plants and Invertebrates[a]

Source	Specificity
Plants	
Wheat germ *Triticum vulgare*	Neu5Ac < GlcNAc
Elderberry *Sambucus nigra*	Neu5Acα2,6Gal/GalNAc
Maackia amurensis	Neu5Acα2,3Galβ1,4GlcNAc
Invertebrates	
Snail *Dolabella*	Neu5Ac
Slug *Limax flavus*	Neu5Ac > Neu5Gc
Snail *Cepaea hortensis*	Neu5Ac > Neu5Gc
Snail *Achatina fulica*	NeuAcα2,3Gal > Neu5Acα2,6Gal
Snail *Pila globosa*	Neu5Gc
Oyster *Crassostrea gigas*	Neu5Ac
Horseshoe crab *Limulus polyphemus*	Neu5Ac
Lobster *Homarus americanus*	Neu5Ac, Neu5Gc
Horseshoe crab *Tachypleus tridentatus*	Neu5Ac, Neu5Gc
Scorpion *Androctonus australis*	Sialyllactose
Horseshoe crab *Carcinoscorpius rotunda*	Neu5Acα2,6Gal > Neu5Acα2,3Gal
Scorpion *Centruroides sculpturatus*	Neu5Ac, Neu5Gc
Prawn *Macrobrachium rosenbergii*	Neu5Ac
Scorpion *Masticoproctus giganteus*	Neu5Ac
Spider *Aphonopelma cepaeahortensis*	Sialoglycoproteins
Scorpion *Heterometrus granulomanus*	Neu5Acα2,3Lac
Prawn *Peneaus monodon*	Neu5Ac
Scorpion *Paruroctonus mesaenis*	Sialoglycoproteins

[a]Zeng and Gabius (1992b, and references therein).

Table V
Pathogenic Microorganisms and Toxins Binding to Sialic Acids on Host Cells

Pathogen	Specificity	Reference
	Viruses	
Influenza A and B	Neu5Ac (some strains prefer Neu5Acα2,3Gal or Neu5Acα2,6Gal, dependent on host specificity)	Paulson (1985)
Influenza C	Neu5,9Ac$_2$	Rogers *et al.* (1986)
Corona virus	Neu5,9Ac$_2$	Vlasak *et al.* (1988)
Sendai virus	Neu5Ac	Markwell and Paulson (1980)
Polyoma virus	Neu5Acα2,3Galβ1,3GalNAc	Cahan and Paulson (1980)
Rotavirus group C	Neu5Ac	Svensson (1992)
	Mycoplasma	
Mycoplasma pneumoniae	Neu5Acα2,3Gal on polylactosamine chains	Loomes *et al.* (1984)
	Bacteria	
Streptococcus sanguis	*O*-linked sialylated tetrasaccharides	Murray *et al.* (1982)
Escherichia coli K99	Neu5Gc-containing glycolipids	Ouadia *et al.* (1992)
Escherichia coli, S-fimbriae	Neu5Acα2,3Galβ1,3GalNAc	Parkkinen *et al.* (1986)
(newborn human meningitis)	Neu5Gc-GM3, GD3, GD1b	Hanisch *et al.* (1993)
Bordetella bronchiseptica	Neu5Ac	Ishikawa and Isayama (1987)
Pseudomonas aeruginosa	Neu5Ac	Ko *et al.* (1987)
Helicobacter (*Campylobacter*) *pylori*	Neu5Acα2,3Lac > Neu5Acα2,6Lac	Evans *et al.* (1988)
Streptococcus suis	Neu5Acα2,3Galβ1,4GlcNAcβ1,3Gal	Liukkonen *et al.* (1992)
	Protozoa	
Malaria (MSA-1) *Plasmodium falciparum*	Neu5Ac	Perkins and Rocco (1988)
Chagas disease *Trypanosoma cruzi*	Neu5Ac	Schenkman and Eichinger (1993)
	Toxins	
Vibrio cholerae toxin	GM1	Schengrund and Ringler (1989)
Pertussis toxin	Neu5Ac	Brennan *et al.* (1988)
Tetanus toxin	Sialoglycolipids	Schiavo *et al.* (1991)

(WGA), *Limax flavus* (LFA), *Sambucus nigra* (SNA), and *Maackia amurensis* (MAA).

It has been known for many years that microbial pathogens, i.e., viruses, mycoplasma, bacteria, and protozoa, take advantage of cell surface sialic acids to adhere to their respective host cells (Table V). In Chapter 9 the role of sialic acids in infection by myxoviruses will be discussed. Sialic acid-specific adhesion

Table VI
Mammalian Sialic Acid-Binding Proteins (for Selectins See Table VII)

Source	Specificity	Reference
Frog egg	Sialylated glycoproteins	Titani *et al.* (1987)
Rat uterus	Neu5Ac	Chakraborty *et al.* (1993)
Rat brain	Neu5Ac, Neu5Gc	Popoli and Mengano (1988)
Rat brain myelin	Gangliosides, preferentially GT1b, GQ1b, GD1b	Tiemeyer *et al.* (1989)
Human placenta (IgG)	*O*-acetylated sialic acids	Ahmed and Gabius (1989)
Blood (factor H of alternative complement pathway)	Sialylated glycoconjugates, other polyanionic molecules	Meri and Pangburn (1990)
Murine macrophages (sialoadhesin)	Neu5Acα2→3Galβ1→3GalNAc on glycoproteins and glycolipids	Crocker *et al.* (1991)
Bovine heart (calcyclin)	Neu5Ac, Neu5Gc	Zeng and Gabius (1991)
Human placenta (sarcolectin)	Neu5Ac, Neu5Gc	Zeng and Gabius (1992a)
B lymphocytes (CD22)	Neu5Acα2→6Galβ1→4GlcNAc	Sgroi *et al.* (1993), Powell *et al.* (1993)

of bacteria is a phenomenon drawing an increasing level of interest, since it often is a critical step in infectious diseases. Examples are the inflammation of gastric mucosa by *Helicobacter pylori* after adhesion to sialoglycoproteins of the cell surface (Evans *et al.*, 1988), and meningitis of infants by *Escherichia coli* (Parkkinen *et al.*, 1986; Hanisch *et al.*, 1993). Several bacterial toxins are known that bind to gangliosides in a sialic acid-dependent manner, e.g., cholera or tetanus toxins (Schauer, 1982b; Table V). In addition, it should be mentioned that various antibodies have been described that recognize epitopes containing sialic acids (Schauer, 1988; Zeng and Gabius, 1992b; Varki, 1993).

Sialic acid-dependent receptors have been recognized to play an important role in the adhesion of mammalian cells. This line of research was initiated by the discovery of a family of cell adhesion molecules now generally called selectins (Bevilacqua *et al.*, 1991). Other well-defined sialic acid-dependent adhesion receptors are sialoadhesin (Crocker and Gordon, 1986; Crocker *et al.*, 1991) found on specific subsets of macrophages in bone marrow and lymphatic tissues (e.g., lymph nodes and spleen) as well as CD22 (Stamenkovic *et al.*, 1991, 1992), a B-cell-restricted protein belonging to the immunoglobulin superfamily. In addition, other sialic acid-binding activities from various mammalian tissues have been described and are listed in Table VI.

6.2. Selectins

All three members of this family (Bevilacqua *et al.*, 1991) show a similar primary structure (Bevilacqua *et al.*, 1989; Johnston *et al.*, 1989; Lasky *et al.*, 1989; Larsen *et al.*, 1989; Siegelman *et al.*, 1989; reviewed in Lasky, 1992;

Table VII
Distribution and Binding Specificity of Selectins

Selectin	Cell type	Ligand determinant
E-selectin	Activated endothelia	Sialyl-Le^x and sialyl-Le^a Neu5Acα2,3Galβ1,4(Fucα1,3)GlcNAc and Neu5Acα2,3Galβ1,3(Fucα1,4)GlcNAc
L-selectin	Leukocytes	Sialylated, sulfated, and fucosylated *O*-glycans
P-selectin	Activated platelets and endothelia	Sialyl-Le_x and sialyl-Le_a

McEver, 1992; Varki, 1992b; Bevilacqua, 1993; Bevilacqua and Nelson, 1993; Rosen, 1993). They represent type I transmembrane glycoproteins containing an amino-terminal carbohydrate recognition domain (CRD), a single epidermal growth factor (EGF)-like domain, a variable number of short consensus repeats (SCR) (two for L-selectin, six for E-selectin, and nine for P-selectin), and a relatively short carboxy-terminal cytoplasmic domain. Whereas the homology between the three selectins is 60–70% in the CRD and EGF-like domain, in the SCR only about 40% homology was found. The essential structural elements of the CRD as defined by Drickamer (1988) were deduced from homologies between the hepatic asialoglycoprotein receptor and other sugar-binding proteins. This motif consists of about 120 amino acids with 32 conserved residues—18 identical and 14 conservative. For E-selectin, the crystal structure of a recombinant protein containing the CRD and EGF-like domains has been obtained (Graves *et al.*, 1994).

Besides their structural homologies, the selectins also share common aspects in their function. They all play crucial roles in the initial event of white blood cell adhesion to specific endothelia, the so-called rolling (Lawrence and Springer, 1991). Before firm adhesion, cells flowing in the bloodstream start to slow down by rolling along the endothelial lining of the vessel. This is mediated by selectins interacting with sialic acid-containing ligands. The specificity of this interaction is accomplished by the expression pattern of the receptors and their appropriate ligands (Table VII). At least two of the selectins recognize the same carbohydrate structures sialyl-Lewis^x (sLe^x) or sialyl-Lewis^a (sLe^a), which are expressed at high levels on leukocytes and some tumor cells. Therefore, it has been proposed that selectins are also involved in metastatic events. The cooperation of the selectin family with ICAM-1 and -2, sLe^x on CD11/CD18 during the rolling and adhesion of granulocytes on endothelial cells under the influence of inflammatory cytokines is illustrated in Figure 10 (Kotovuori *et al.*, 1993).

6.2.1. L-Selectin

L-Selectin is found on lymphocytes and other leukocytes and was described under various names (MEL-14 Ag, gp90MEL, LAM-1, Leu8, Ly22, TQ1,

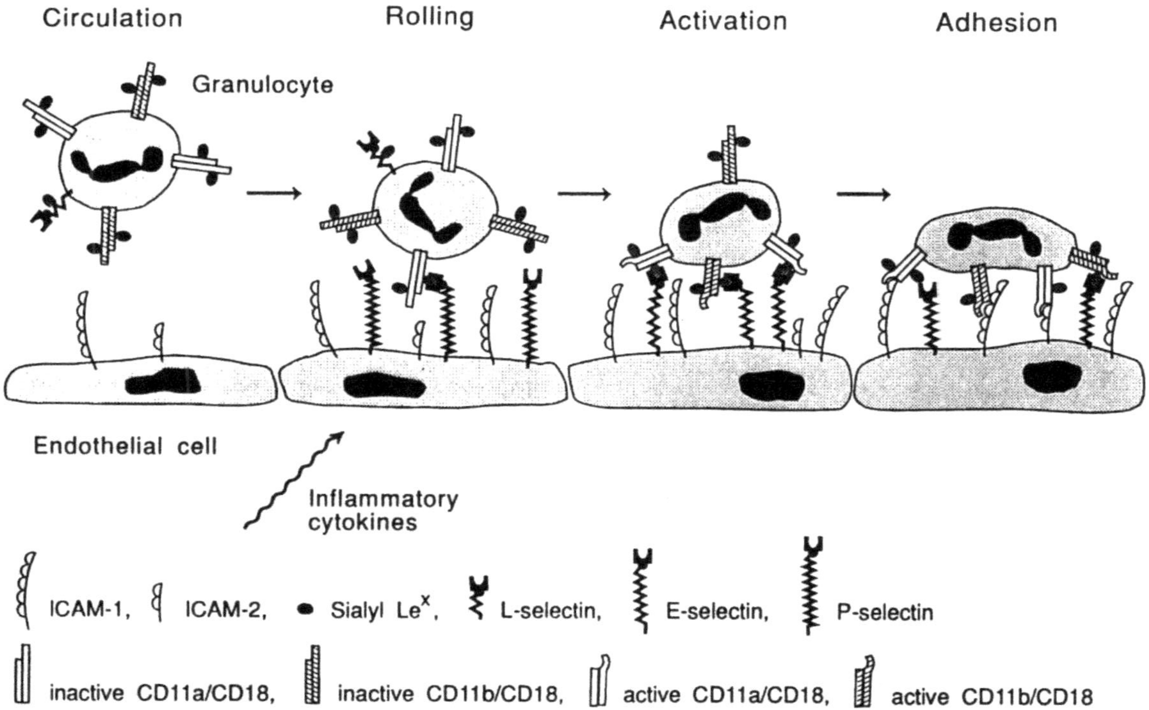

FIGURE 10. Model for granulocyte binding to activated endothelial cells. Circulating granulocytes express L-selctin and inactivated CD11/CD18. Unstimulated endothelial cells express only smalll amounts of ICAM-1 and ICAM-2. After induction by cytokines. ICAM-1 and ICAM-2 expression is upregulated and E- and P-selectin expression is induced. Binding of E- and P-selectin to sLe^x on CD11/cd18 triggers activation of these integrins leading to firm attachment through binding to ICAM molecules. (From Kotovuori *et al.*, 1993.)

DREG56). The functional aspects of L-selectin and recent progress in the characterization of its putative ligands have been reviewed in detail by Rosen (1993). It functions as an adhesion molecule in the homing of lymphocytes to high endothelial venules (HEV) in peripheral lymph nodes, but it is also involved in other leukocyte trafficking events, such as the recruitment of neutrophils to sites of acute inflammation (Lewinsohn *et al.*, 1987; Watson *et al.*, 1991). The requirement of sialic acids in this homing event was first described by Rosen *et al.* (1985). Until now, three mucinlike proteins (Imai *et al.*, 1991), GlyCAM-1 (Lasky *et al.*, 1992), CD34 (Baumhueter *et al.*, 1993), and MAdCAM-1 (Berg *et al.*, 1993) were characterized as potential ligands for L-selectin in HEVs of peripheral lymph nodes. In addition, it was shown that the *O*-linked oligosaccharides from GlyCAM-1 contain essential sialic acid and sulfate residues besides fucose and *N*-acetylglucosamine (Imai and Rosen, 1993; Imai *et al.*, 1993). Possible implications of the characteristic features of these ligands will be discussed in Section 6.6.

6.2.2. E-Selectin

E-Selectin was originally described under the name ELAM-1 as an adhesion molecule expressed on endothelia activated by cytokines (e.g., by interleukin-1 or tumor necrosis factor-α) where it mediates the rolling of leukocytes. Characteristic for the expression of E-selectin is the lag time of about 4 hr between activation of the endothelial cell and occurrence of the receptor on the cell surface due to the requirement of transcription and translation. This is in contrast to the rapid expression of P-selectin (Section 6.2.3). The discovery of a CRD homologous to the other selectins (Bevilacqua *et al.*, 1989) and the knowledge of specific carbohydrate structures found on granulocytes led to the discovery of sLe^x as oligosaccharide recognized by this receptor, in a number of laboratories (Lowe *et al.*, 1990; Phillips *et al.*, 1990; Walz *et al.*, 1990; Tiemeyer *et al.*, 1991). Later, other structures were shown also to bind E-selectin, like the structural analogues of sLe^a or sLe^x containing sulfate instead of sialic acid (Yuen *et al.*, 1992; see Section 6.6 for further discussion). A specific protein ligand for murine E-selectin was purified and partially characterized (Levinovitz *et al.*, 1993).

6.2.3. P-Selectin

P-Selectin also was originally described under different names (GMP-140, PADGEM, CD62), stemming from independent discoveries. It is found on activated platelets and endothelia where it is expressed on the cell surface briefly after activation by cytokines, because of its storage in vesicles under the cell surface. Like E-selectin, it is involved in rolling events of leukocytes on activated endothelia. The importance of P-selectin in rolling and leukocyte trafficking has been shown convincingly by targeted gene disruption in the mouse (knockout mouse). These mice appear normal and healthy, but their level of circulating leukocytes is elevated and the recruitment of these cells to sites of inflammation is impaired, although they express normal levels of E- and L-selectin and sLe^x (Mayadas *et al.*, 1993). As for E-selectin, shortly after the discovery of its CRD by molecular cloning (Johnston *et al.*, 1989; Larsen *et al.*, 1989), binding of P-selectin to sLe^x and sLe^a as the oligosaccharide structures recognized was demonstrated (Polley *et al.*, 1991; Zhou *et al.*, 1991; Handa *et al.*, 1991; Table VII). For P-selectin a high-affinity ligand protein also was characterized (Moore *et al.*, 1992; Norgard *et al.*, 1993b, Sako *et al.*, 1993; see Section 6.6).

6.3. CD22

This receptor is found on B cells. As a member of the superfamily of immunoglobulin (Ig)-like molecules, it is a type I transmembrane glycoprotein consisting of seven Ig-like domains, a transmembrane domain, and a C-terminal cytoplasmic domain (Stamenkovic and Seed, 1990; Stamenkovic *et al.*, 1991; Wilson *et al.*, 1991). Alternatively spliced transcripts lacking domain 3 and 4 were originally described as CD22α (Stamenkovic and Seed, 1990). However, later studies failed to demonstrate the presence of this protein form on the surface of human B cells (Engel *et al.*, 1993). The murine homologue has also been described (Torres *et al.*, 1992). In contrast to the high homology between murine and human selectins, only 62% of the amino acids are identical between human and murine CD22, with the highest homology in the extracellular domain 7, the transmembrane and the cytoplasmic domain (71, 68, and 67% identity, respectively). The overall genomic organizations of the human (Wilson *et al.*, 1993) and the murine (Law *et al.*, 1993) homologues were resolved, showing the same structure composed of 15 exons. Each Ig-like domain is contained in a separate exon (exon 4 to 10). Interestingly, the gene of CD22 maps to the same chromosome as the most closely related proteins, carcinoembryonic antigen (CEA) and myelin-associated glycoprotein (MAG), suggesting that the genes of these adhesion proteins may have arisen from the same ancestral gene (Wilson *et al.*, 1993; Law *et al.*, 1993).

Recombinant soluble CD22 constructs and CD22-transfected COS cells were used to analyze the specificity of the receptor for different blood cells (Stamenkovic *et al.*, 1992; Torres *et al.*, 1992; Engel *et al.*, 1993; Crocker *et al.*, 1995). These studies demonstrated that CD22 can mediate binding of B cells to B and T cells as well as to neutrophils, monocytes, and erythrocytes. The interaction with T cells is proposed to be involved in early B-cell activation. CD22 has also been implicated to modulate signaling through the surface IgM (sIgM)–cell receptor complex. Along this line, Leprince *et al.* (1993) showed that CD22 coprecipitates with the complex and that it is phosphorylated rapidly upon sIgM cross-linking. Interestingly, all six tyrosine residues in the cytoplasmic domain are conserved in the human and murine homologues (Torres *et al.*, 1992).

First evidence for sialic acid-dependent binding came from a study (Stamenkovic *et al.*, 1991) reporting that CD22 interacts with CD45RO on T cells and CD75 on B cells, an epitope that depends on the expression of α2,6 sialyltransferase in these cells (Bast *et al.*, 1992; Munro *et al.*, 1992; Stamenkovic *et al.*, 1992). In addition, proteins isolated as ligands for CD22 contain α2,6-linked sialic acids on branched, *N*-linked oligosaccharides which are determinants for binding (Sgroi *et al.*, 1993; Powell *et al.*, 1993; see Section 6.6). Interestingly, in the α2,6 sialyltransferase gene a B-cell-specific promoter leads to cell-type-specific regulation of this enzyme during B-cell development (Wang *et al.*, 1993).

6.4. Sialoadhesin

Sialoadhesin is a receptor found on specific macrophage subpopulations in murine bone marrow, spleen, and lymph nodes (Crocker and Gordon, 1986). High expression of sialoadhesin is restricted to resident bone marrow macrophages in hematopoietic clusters, to marginal zone macrophages in spleen, and to macrophages in the subcapsular sinuses and medullary cord of lymph nodes (Crocker and Gordon, 1989; Crocker *et al.*, 1992). A striking distribution of sialoadhesin was observed in bone marrow on the ultrastructural level, where the receptor is highly enriched at contact sites between macrophages and developing myeloid cells (Crocker *et al.*, 1990). In contrast, no staining was observed at contact sites of the same macrophages to erythroblasts. Whereas most studies on sialoadhesin were done in mouse, the existence of a homologous protein in rat was demonstrated in spleen and lymph node macrophages (van den Berg *et al.*, 1992). In addition, reports on the specificity of a ganglioside binding activity on rat alveolar macrophages (Riedl *et al.*, 1982; Boltz-Nitulescu *et al.*, 1984; Förster *et al.*, 1986) point to the possibility that sialoadhesin may also be present in these macrophages, although murine alveolar macrophages express only low amounts of the receptor (Crocker and Gordon, 1989). Using glycoproteins and glycolipids as well as erythrocytes that, after sialidase treatment, were resialylated with purified sialyltransferases to contain only specific sialylated glycoconjugates, it was shown that sialoadhesin recognizes the sequence Neu5Acα2→3Galβ1→3GalNAc on glycoproteins and glycolipids on cell surfaces (Crocker *et al.*, 1991).

Possible functions for sialoadhesin have been implicated in the development of myeloid cells in the bone marrow and the trafficking of leukocytes in lymphatic organs (Crocker *et al.*, 1991, 1992; van den Berg *et al.*, 1992). Evidence for this hypothesis comes from the distribution of the receptor in bone marrow (see above) and from cell binding experiments. In experiments with purified sialoadhesin and macrophages expressing the receptor, the receptor bound preferentially to inflammatory and circulating neutrophils (Crocker *et al.*, 1995). Whereas total bone marrow cells were still bound well, lymphocytes and monocytes showed intermediate binding, and binding to murine erythrocytes was barely detectable. In binding assays with frozen sections of spleen and lymph nodes, sialoadhesin can mediate the adhesion of lymphocytes and lymphoma cells (van den Berg *et al.*, 1992). In quantitative binding assays, activated T cells bind better than resting cells, and lowest binding in this cell lineage is to thymocytes. In contrast, activated B cells do not bind sialoadhesin better than resting B cells.

6.5. Other Mammalian Sialic Acid-Dependent Receptors

Besides the receptors described above, other proteins from mammalian sources capable of sialic acid recognition have been described in the literature (Table VI). However, their further characterization as sialic acid-dependent receptors is still needed, since their purification, specificity, or possible function has not as yet been described. Four examples should be mentioned here. (1) Factor H of the alternate complement pathway binds to sialic acid residues and other polyanionic molecules on "nonactivating" cell surfaces and has been inferred to facilitate access of the H protein to C3b on cell surfaces (Meri and Pangburn, 1990). However, to date no study on the specificity of this interaction has been described. (2) A ganglioside binding activity was characterized in membranes from myelin sheets with preferential binding to GD1b, GT1b, and GQ1b (Tiemeyer *et al.*, 1989, 1990), although no protein has yet been purified. (3) From human placenta, a protein was isolated with a specificity toward *O*-acetylated sialic acids (Ahmed and Gabius, 1989), which was shown to be an IgG (Zeng and Gabius, 1992b). (4) Finally, the potential role of a membrane-bound sialidase found in myelin, which binds GM1, should be mentioned here (Saito and Yu, 1993). This phenomenon is discussed in Chapter 8.

6.6. Ligands for Sialic Acid-Dependent Receptors

All receptors discussed above recognize specific carbohydrate structures containing terminal sialic acid residues. Whereas the restrictions toward these structures are more or less stringent depending on the receptor, different glycoconjugates can carry identical oligosaccharides and consequently are potentially bound by the same receptor. In addition, the same glycoprotein can carry different oligosaccharide structures ("microheterogeneity"), depending on the glycosylation machinery of the cell producing it. Since the oligosaccharide determines whether a glycoconjugate can serve as ligand ("counterreceptor"), we could call the oligosaccharide "ligand determinant." However, the oligosaccharide in question will only be recognized if presented appropriately, emphasizing the importance of the carrier molecule. Another feature of carbohydrate recognition is that only a few functional groups of the oligosaccharide structure are required for binding, as has been shown in many examples of sugar–protein interactions. On some of these functions relatively major changes are tolerated, whereas others cannot be modified at all. Because of the large variability in naturally occurring oligosaccharide structures, sometimes related structures are recognized as well. One example is the similar potential of the isomers sLe^x and sLe^a to serve as ligand determinants for E- and P-selectin (see Sections 6.2.2 and 6.2.3, Table VII). Furthermore, E- and L-selectin bind to glycolipids containing structural analogues of sLe^x or sLe^a containing sulfate instead of sialic acid (Yuen *et al.*, 1992; Green *et al.*, 1992). Considering these aspects, we cannot expect precise receptor–ligand pairs involving simple kinetics in sialic acid-dependent interactions. Studies dealing with the "specificity" of these receptors have to take this into account, especially since experimental approaches often involve artificial presentation of the molecules investigated. One example is the TLC overlay technique, where the oligosaccharides may be presented in an unnatural way including high clustering of single molecules which might not occur on cell surfaces. Also the commonly used expression of the receptor in transfected cells leads to high levels of these molecules on the cell surface not always existing in the natural environment. Although these conditions might lead to the adhesion of cells not found naturally with sufficient expression levels of receptors, the observed specificity reflects the flexibility of protein–carbohydrate interaction, which also allows binding to low-affinity ligands, if the density is high enough. One such example is the binding of soluble P-selectin to HL-60 cells compared to CHO cells transfected with fucosyltransferase expressing sLe^x (Zhou *et al.*, 1991). Whereas on HL-60 cells P-selectin bound to a low number of high-affinity sites, on CHO cells P-selectin bound to a high number of low-affinity sites. Despite the obvious artificial situation in the CHO cells, we cannot rule out that such high number of low-affinity interactions may have biologically relevant roles.

Besides the description of rather diverse molecules that can be bound by selectins, specific glycoproteins binding to these receptors were recently isolated and characterized. Their structural features supply evidence for the importance of the presentation of ligand determinants. Ligands for L-selectin are GlyCAM1 (Lasky *et al.*, 1992), CD34 (Baumhueter *et al.*, 1993), and MAdCAM (Berg *et al.*, 1993), which all are highly glycosylated proteins with mucinlike structures carrying dense clusters of *O*-linked oligosaccharides (Chapter 5). The tissue-specific functionality of MAdCAM is an example of how the cell-specific oligosaccharide structures can function as ligand determinants (Berg *et al.*, 1993).

Although the final structure(s) of the ligand determinant(s) on the native ligand isolated from HEVs of lymph nodes has not been resolved, sialic acid and sulfate are structural components necessary for their biological function (Imai and Rosen, 1993; Imai *et al.*, 1993). The importance of the carrier molecule of ligand determinants was also demonstrated for a P-selectin ligand isolated from neutrophils (Norgard *et al.*, 1993b). Although this glycoprotein carries only a minor fraction of the sLe^x found on these cells, it was the only molecule in cell extracts binding to P-selectin. The primary sequence of a ligand protein cloned from HL-60 cells contains mucinlike elements, like the ligands for L-selectin, since it contains clusters of *O*-linked oligosaccharides required for binding activity (Sako *et al.*, 1993). Also for E-selectin, a specific ligand glycoprotein could be isolated from murine neutrophils and a myeloid cell line (Levinovitz *et al.*, 1993). However, its carbohydrate structure(s) serving as ligand determinant(s) have not been studied. In conclusion, although specific high-affinity ligands for selectins have been described, it is not clear whether only these or also low-affinity ligands are involved in adhesion events.

Considerable progress has been made in the determination of ligands for CD22 (Sgroi *et al.*, 1993; Powell *et al.*, 1993). CD22 requires sialic acids α2,6-bound to branched *N*-linked oligosaccharides as ligands. From B and T cells as well as from lymphoma cell lines, a number of surface glycoproteins could be isolated on CD22 columns, which is in contrast to the very limited number of ligands for selectins. The number and size of these glycoproteins were dependent on the type of cell used as source for CD22 ligands (Sgroi *et al.*, 1993), supplying evidence that for CD22 different cell-type-specific carrier proteins ("counter-receptors") probably with similar ligand determinants can exist. The fraction of *N*-linked oligosaccharides, released from purified ligands, which bound to a CD22 affinity column contained more α2,6-linked sialic acid residues than the oligosaccharides which were only retarded or not bound at all. In addition, a portion of the retarded fraction could be converted to a binding fraction by sialylation with α2,6 sialyltransferase (Powell *et al.*, 1993).

For sialoadhesin, no specific protein or glycolipid ligand has been isolated so far. However, in TLC overlay assays with glycolipid extracts from inflammatory neutrophils or bone marrow cells, sialoadhesin bound to specific ganglioside bands, whereas similar extracts on Western blots did not reveal glycoprotein ligands for sialoadhesin (Kelm *et al.*, 1993), although in principle it is possible to detect glycoprotein ligands for sialoadhesin by Western blotting, as was shown for erythrocyte membrane extracts (Crocker *et al.*, 1991). In addition, treatment of bone marrow cells or inflammatory neutrophils with proteases including *O*-sialoglycoprotease (Abdullah *et al.*, 1992) had little if any effect. Although this is not proof that glycolipids are the ligands on these cells, it supports such a possibility.

6.7. Modifications of Sialic Acids

O-Acetylation and *N*-acetyl-hydroxylation are modifications of sialic acids that are known to have an antirecognition effect on the binding of influenza A and B viruses (Higa *et al.*, 1985; Section 4.1). This is in contrast to the specific recognition of 9-*O*-acetylated sialic acids by influenza C virus (Rogers *et al.*, 1986; see Section 4.1 and Chapter 9).

Only a few studies are known that deal with the effect of sialic acid modifications on mammalian adhesion molecules. Shortening of the glycerol side chain of sialic acids by periodate/borohydrate treatment revealed that this moiety is not important for the binding of E-selectin (Tyrrell *et al.*, 1991), of P-selectin (Norgard *et al.*, 1993b), or of L-selectin (Norgard *et al.*, 1993a). In contrast, for the ligand determinant for CD22, the glycerol side chain seems to be an essential structural element (Sgroi *et al.*, 1993; Powell *et al.*, 1993). However, in none of these studies were naturally occurring sialic acid modifications used.

Experiments with various glycoconjugates and cells demonstrated that sialoadhesin only recognizes Neu5Ac and not Neu5Gc or Neu5,9Ac_2 (Kelm *et al.*, 1994). This leads to the following possible modulatory situation for sialic acid modifications. Cells expressing the ligand determinant for sialoadhesin could mask their ligands by acetylation. However, this mask could be removed by appropriate extracellular esterases thus again allowing interactions of ligands with sialoadhesin. In contrast, Neu5Gc can only be removed by sialidases. However, this would lead to the asialooligosaccharides, which are not ligand determinants for sialoadhesin. A cell expressing Neu5Gc on the surface could only interact with sialoadhesin after consumption of the intracellular pool of CMP-Neu5Gc and exchange of the cell surface sialic acids with Neu5Ac. This model represents an example of how modifications of sialic acid can modulate cellular interactions and how this is affected by their different metabolic pathways.

6.8. Perspectives

Since the characterization of the first sialic acid-dependent receptors at the beginning of this decade, new aspects regarding the role of sialic acids in cell recognition and cell–cell interaction can be investigated. The rather broad specificity toward different ligands (see Section 6.6) potentially allows an almost continuous spectrum of ligands with different affinities to occur on cell surfaces. Theoretically, this adds a new potential in regulating and fine-tuning cellular interactions not possible with highly specific receptor–ligand systems. Therefore, an important aspect of future research will be to examine whether these low-affinity ligands have a biological function (and how this is accomplished), or whether they are just side products of a glycosylation machinery necessary to produce the high-affinity ligands. In addition, increasing knowledge on mechanisms like clustering of ligand determinants, e.g., on mucinlike molecules, leading to high-affinity ligands, will be helpful in the design of therapeutic drugs specific for certain receptor-mediated interactions. Examples are the increasing efforts to develop carbohydrate-based drugs as inhibitors for selectins for the treatment of undesired inflammatory reactions or for the prevention of cancer metastasis.

The discovery of the selectin family and their close localization on the genome in less than 300 kb raises the intriguing possibility that other gene families of sialic acid-dependent receptors also exist. For example, the gene for CD22 as a member of the Ig superfamily is also localized next to the most homologous genes. Future experiments are needed to clarify whether other members of this family of adhesion receptors recognize sialic acid-containing carbohydrates on their ligands.

REFERENCES

Abdullah, K. M., Udoh, E. A., Shewen, P. E., and Mellors, A., 1992, A neutral glycoprotease of *Pasteurella haemolytica* A1 specifically cleaves O-sialoglycoproteins, *Infect. Immun.* **60:**56–62.

Ahmed, H., and Gabius, H.-J., 1989, Purification and properties of a Ca^{2+}-independent sialic acid binding lectin from human placenta with preferential affinity to O-acetylsialic acids, *J. Biol. Chem.* **264:**18673–18678.

Air, G. M., Webster, R. G., Colman, P. M., and Laver, W. G., 1987, Distribution of sequence differences in influenza N9 neuraminidase of tern and whale viruses and crystallisation of the whale neuraminidase complexed with antibodies, *Virology* **160:**346–354.

Aisaka, K., Igarashi, A., and Uwajima, T., 1991, Purification, crystallization, and characterization of neuraminidase from *Micromonospora viridifaciens, Agric. Biol. Chem.* **55:**997–1004.

Allen, A., and Kent, P. W., 1968, Studies on the enzymic N-acylation of amino sugars in the sheep colonic mucosa, *Biochem. J.* **107:**589–598.

Anderberg, M. R. (ed.), 1973, *Cluster Analysis for Application,* Academic Press, New York.

Arinç, E., 1991, Essential features of NADH-dependent cytochrome b_5 reductase and cytochrome b_5 of liver and lung microsomes, in: *Molecular Aspects of Monooxygenases and Bioactivation of Toxic Compounds* (E. Arinç, J. B. Schenkman, and E. Hodgson, eds.), NATO ASI Series, Ser. A, Vol. 202, Plenum Press, New York, pp. 149–170.

Bast, B.J.E.G., Zhou, L. J., Freeman, G. J., Colley, K. J., Ernst, T. J., Munro, J. M., and Tedder, T. F., 1992, The HB-6, CDw75, and CD76 and differentiation antigens are unique cell-surface carbohydrate determinants generated by the β-galactoside α2,6-sialyltransferase, *J. Cell Biol.* **116:**423–435.

Baumhueter, S., Singer, M. S., Henzel, W., Hemmerich, S., Renz, M., Rosen, S. D., and Lasky, L. A., 1993, Binding of L-selectin to the vascular sialomucin CD34, *Science* **262:**436–438.

Berg, E. L., McEvoy, L. M., Berlin, C., Bargatze, R. F., and Butcher, E. C., 1993, L-Selectin-mediated lymphocyte rolling on MAdCAM-1, *Nature* **366:**695–698.

Bergwerff, A. A., Hulleman, S.H.D., Kamerling, J. P., Vliegenthart, J.F.G., Shaw, L., Reuter, G., and Schauer, R., 1992, Nature and biosynthesis of sialic acids in the starfish *Asterias rubens.* Identification of sialo-oligomers and detection of S-adenosyl-L-methionine:N-acylneuraminate

8-O-methyltransferase and CMP-N-acetylneuraminate monooxygenase activities, *Biochimie* **74:**25–37.

Bergwerff, A. A., Hulleman, S.H.D., Kamerling, J. P., Vliegenthart, J.F.G., Shaw, L., Reuter, G., and Schauer, R., 1993, Structural and biosynthetic aspects of N-glycoloyl-8-O-methylneuraminic acid-oligomers linked through their N-glycoloyl groups, isolated from the starfish *Asterias rubens,* in: *Polysialic Acids* (J. Roth, U. Rutishauser, and F. A. Troy II, eds.), Birkhäuser Verlag, Basel, pp. 201–212.

Bevilacqua, M. P., 1993, Endothelial–leukocyte adhesion molecules, *Annu. Rev. Immunol.* **11:**767–804.

Bevilacqua, M. P., and Nelson, R. M., 1993, Selectins, *J. Clin. Invest.* **91:**379–387.

Bevilacqua, M. P., Stengelin, S., Gimbrone, M. A., Jr., and Seed, B., 1989, Endothelial leukocyte adhesion molecule 1: An inducible receptor for neutrophils related to complement regulatory proteins and lectins, *Science* **243:**1160–1165.

Bevilacqua, M., Butcher, E., Furie, B., Furie, B., Gallatin, M., Gimbrone, M., Harlan, J., Kishimoto, K., Lasky, L., McEver, R., Paulson, J., Rosen, S., Seed, B., Siegelman, M., Springer, T., Stoolman, L., Tedder, T., Varki, A., Wagner, D., Weissman, I., and Zimmerman, G., 1991, Selectins—A family of adhesion receptors, *Cell* **67:**233.

Boltz-Nitulescu, G., Ortel, B., Riedl, M., and Förster, O., 1984, Ganglioside receptor of rat macrophages—Modulation by enzyme treatment and evidence for its protein nature, *Immunology* **51:**177–184.

Bouhours, D., and Bouhours, J. F., 1983, Developmental changes of hematoside of rat small intestine. Postnatal hydroxylation of fatty acids and sialic acid, *J. Biol. Chem.* **258:**299–304.

Bouhours, D., and Bouhours, J. F., 1988, Tissue-specific expression of GM3 (NeuGc) and GD3 (NeuGc) in epithelial cell of the small intestine of strains of inbred rats. Absence of NeuGc in intestine and presence in kidney gangliosides of brown Norway and spontaneously hypertensive rats, *J. Biol. Chem.* **263:**15540–15545.

Bouhours, J.-F., and Bouhours, D., 1989, Hydroxylation of CMP-Neu5Ac controls the expression of N-glycolylneuraminic acid in G_{M3} ganglioside of the small intestine of inbred rats, *J. Biol. Chem.* **264:**16992–16999.

Brennan, M. J., David, J. L., Kenimer, J. G., and Manclark, C. R., 1988, Lectin-like binding of pertussis toxin to a 165-kilodalton Chinese hamster ovary cell glycoprotein, *J. Biol. Chem.* **263:**4895–4899.

Budd, T., Dolman, C. D., Lawson, A. M., Chai, W., Saxton, J., and Hemming, F. W., 1992, Comparison of the N-glycoloylneuraminic acid and N-acetylneuraminic acid content of platelets and their precursors using high-performance anion exchange chromatography, *Glycoconjugate J.* **9:**274–278.

Buscher, H.-P., Casals-Stenzel, J., Mestres-Ventura, P., and Schauer, R., 1977, Biosynthesis of N-glycoloylneuraminic acid in porcine submandibular glands. Subcellular site of hydroxylation of N-acetylneuraminic acid in the course of glycoprotein biosynthesis, *Eur. J. Biochem.* **77:**297–310.

Buschiazzo, A., Cremona, M. L., Campetella, O., Frasch, A.C.C., and Sanchez, D. O., 1993, Sequence of a *Trypanosoma rangeli* gene closely related to *Trypanosoma cruzi* trans-sialidase, *Mol. Biochem. Parasitol.* **62:**115–116.

Butor, C., Diaz, S., and Varki, A., 1993a, High level O-acetylation of sialic acids on N-linked oligosaccharides of rat liver membranes, *J. Biol. Chem.* **268:**10197–10206.

Butor, C., Higa, H. H., and Varki, A., 1993b, Structural, immunological, and biosynthetic studies of a sialic acid-specific O-acetylesterase from rat liver, *J. Biol. Chem.* **268:**10207–10213.

Cabezas, J. A., 1991, Some questions and suggestions on the type references of the official nomenclature (IUB) for sialidase(s) and endosialidase, *Biochem. J.* **278:**311–312.

Cahan, L. D., and Paulson, J. C., 1980, Polyoma virus adsorbs to specific sialyloligosaccharide receptors on erythrocytes, *Virology* **103:**505–509.
Canard, B., and Cole, S. T., 1990, Lysogenic phages of *Clostridium perfringens:* Mapping of the chromosomal attachment sites, *FEMS Lett.* **66:**323–326.
Carey, D. J., Sommers, L. W., and Hirschberg, C. B., 1980, CMP-N-acetylneuraminic acid: Isolation from and penetration into mouse liver microsomes, *Cell* **19:**597–605.
Carubelli, R., and Griffin, M. J., 1968, On the presence of N-glycoloylneuraminic acid in HeLa cells, *Biochim. Biophys. Acta* **170:**446–448.
Chakraborty, I., Mandal, C., and Chowdhury, M., 1993, Modulation of sialic acid-binding proteins of rat uterus in response to changing hormonal milieu, *Mol. Cell. Biochem.* **126:**77–86.
Chiang, J.Y.L., 1981, Interaction of purified microsomal cytochrome P-450 with cytochrome b_5, *Arch. Biochem. Biophys.* **211:**662–673.
Corfield, A. P., and Schauer, R., 1982a, Occurrence of sialic acids, in: *Sialic Acids* (R. Schauer, ed.), Springer, New York, pp. 5–55.
Corfield, A. P., and Schauer, R., 1982b, Metabolism of sialic acids, in: *Sialic Acids* (R. Schauer, ed.), Springer, New York, pp. 195–261.
Corfield, A. P., Michalski, J.-C., and Schauer, R., 1981a, The substrate specificity of sialidases from microorganisms and mammals, in: *Sialidases and Sialidoses Perspectives in Inherited Metabolic Diseases,* Vol. 4 (G. Tettamanti, P. Durand, and S. Di Donato, eds.), Edi Ermes, Milan, pp. 3–70.
Corfield, A. P., Veh, R. W., Wember, M., Michalski, J.-C., and Schauer, R., 1981b, The release of N-acetyl- and N-glycollylneuraminic acid from soluble complex carbohydrates and erythrocytes by bacterial, viral and mammalian sialidases, *Biochem. J.* **197:**293–299.
Corfield, A. P., Lambré, C. R., Michalski, J.-C., and Schauer, R., 1992, Role of sialic acids and sialidases in molecular recognition phenomena, in: *Conférences Philippe Laudat 1991,* INSERM, Paris, pp. 111–175.
Corfield, A. P., Wagner, S. A., Safe, A., Mountford, R. A., Clamp, J. R., Kamerling, J. P., Vliegenthart, J.F.G., and Schauer, R., 1993, Sialic acids in human gastric aspirates. Detection of 9-O-lactyl- and 9-O-acetyl-N-acetylneuraminic acids in gastric aspirates and a decrease in total sialic acid concentration with age, *Clin. Sci.* **84:**573–579.
Corfield, T., 1992, Bacterial sialidases—Roles in pathogenicity and nutrition, *Glycobiology* **2:**509–521.
Corfield, T., 1993, Tailor made sialidase inhibitors home in on influenza virus, *Glycobiology* **3:**413–415.
Crocker, P. R., and Gordon, S., 1986, Properties and distribution of a lectin-like hemagglutinin differentially expressed by murine stromal tissue macrophages, *J. Exp. Med.* **164:**1862–1875.
Crocker, P. R., and Gordon, S., 1989, Mouse macrophage hemagglutinin (sheep erythrocyte receptor) with specificity for sialylated glycoconjugates characterized by a monoclonal antibody, *J. Exp. Med.* **169:**1333–1346.
Crocker, P. R., Freeman, S., Gordon, S., and Kelm, S., 1995, Sialoadhesion binds preferentially to cells of the granulocytic lineage, *J. Clin. Invest.,* **95:**635–643.
Crocker, P. R., Werb, Z., Gordon, S., and Bainton, D. F., 1990, Ultrastructural localisation of a macrophage-restricted sialic acid binding hemagglutinin, SER, in macrophage–hematopoietic cell clusters, *Blood* **76:**1131–1138.
Crocker, P. R., Kelm, S., Dubois, C., Martin, B., McWilliam, B. S., Shotten, D. M., Paulson, J. C., and Gordon, S., 1991, Purification and properties of sialoadhesin, a sialic acid-binding receptor of murine tissue macrophages, *EMBO J.* **10:**1661–1669.
Crocker, P. R., Kelm, S., Morris, L., Bainton, D. F., and Gordon, S., 1992, Cellular interactions

between stromal macrophages and haematopoietic cells, in: *Mononuclear Phagocytes* (R. v. Furth, ed.), Kluwer, Amsterdam, pp. 55–69.

Cummings, R. D., and Smith, D. F., 1992, The selectin family of carbohydrate-binding proteins: Structure and importance of carbohydrate ligands for cell adhesion, *BioEssays* **14:**849–856.

Dabrowski, J., 1989, Two-dimensional proton magnetic resonance spectroscopy, *Methods Enzymol.* **179:**122–156.

de Freese, A., Shaw, L., Reuter, G., and Schauer, R., 1993, Characterization of S-adenosylmethionine:sialate 8-O-methyltransferase from gonads of the starfish *Asterias rubens, Glycoconjugate J.* **10:**330.

Dell, A., 1987, F.A.B.-mass spectrometry of carbohydrates, *Adv. Carbohydr. Chem. Biochem.* **45:**19–72.

den Besten, P. J., Herwig, H. J., van Donselaar, E. G., and Livingstone, D. R., 1990, Cytochrome P-450 monooxygenase system and benzo(a)pyrene metabolism in echinoderms, *Mar. Biol.* **107:**171–177.

Devine, P. L., Clark, B. A., Birrell, G. W., Layton, G. T., Ward, B. G., Alewood, P. F., and McKenzie, I.F.C., 1991, The breast tumor-associated epitope defined by monoclonal antibody 3E1-2 is an O-linked mucin carbohydrate containing N-glycolylneuraminic acid, *Cancer Res.* **51:**5826–5836.

Diaz, S., Higa, H. H., Hayes, B. K., and Varki, A., 1989, O-Acetylation and de-O-acetylation of sialic acids. 7- and 9-O-acetylation of α2,6-linked sialic acids on endogenous N-linked glycans in rat liver Golgi vesicles, *J. Biol. Chem.* **264:**19416–19426.

Drickamer, K., 1988, Two distinct classes of carbohydrate-recognition domains in animal lectins, *J. Biol. Chem.* **263:**9557–9560.

Dutton, G.G.S., Parolis, H., and Parolis, L.A.S., 1987, The structure of the neuraminic acid-containing capsular polysaccharide of *Escherichia coli* serotype K9, *Carbohyd. Res.* **170:**193–206.

Egge, H., Peter-Katalinic, J., Reuter, G., Schauer, R., Ghidoni, R., Sonnino, S., and Tettamanti, G., 1985, Analysis of gangliosides using fast atom bombardment mass spectrometry, *Chem. Phys. Lipids* **37:**127–141.

Engel, P., Nojima, Y., Rothstein, D., Zhou, L. J., Wilson, G. L., Kehrl, J. H., and Tedder, T. F., 1993, The same epitope on CD22 of B-lymphocytes mediates the adhesion of erythrocytes, T-lymphocytes and B-lymphocytes, neutrophils, and monocytes, *J. Immunol.* **150:**4719–4732.

Engstler, M., and Schauer, R., 1993, Sialidases from African trypanosomes, *Parasitol. Today* **9:**222–225.

Engstler, M., Reuter, G., and Schauer, R., 1992, Purification and characterization of a novel sialidase found in procyclic culture forms of *Trypanosoma brucei, Mol. Biochem. Parasitol.* **54:**21–30.

Engstler, M., Reuter, G., and Schauer, R., 1993, The developmentally regulated trans-sialidase from *Trypanosoma brucei* sialylates the procyclic acidic repetitive protein, *Mol. Biochem. Parasitol.* **61:**1–14.

Engstler, M., Schauer, R., and Brun, R., 1994, Distribution of developmentally regulated trans-sialidase in kinetoplastida and identification of a procyclic *Trypanosoma congolense* shed trans-sialidase activity, *Acta Trop.* in press.

Evans, D. G., Evans, D. J., Moulds, J. J., and Graham, D. Y., 1988, N-acetylneuraminyllactose-binding fibrillar hemagglutinin of *Campylobacter pylori:* A putative colonization factor antigen, *Infect. Immun.* **56:**2896–2906.

Ezepchuk, Y. V., Vertiev, Y. V., and Kostyukova, N. N., 1973, Neuraminidase of *Corynebacterium diphtheriae* as a factor of pathogenicity with a spreading function, *Byull. Eksp. Biol. Med.* **75:**63–65.

Fischer, C., Kelm, S., Ruch, B., and Schauer, R., 1991, Reversible binding of sialidase-treated rat lymphocytes by homologous peritoneal macrophages, *Carbohydr. Res.* **213:**263–273.

Förster, O., Boltz-Nitulescu, G., Holzinger, C., Wiltschke, C., Riedl, M., Ortel, B., Fellinger, A., and Bernheimer, H., 1986, Specificity of ganglioside binding to rat macrophages, *Mol. Immunol.* **23:**1267–1273.

Fujii, Y., Higashi, H., Ikuta, K., Kato, S., and Naiki, M., 1982, Specificities of human heterophilic Hanganutziu and Deicher (H-D) and avian antisera against H-D antigen-active glycosphingolipids, *Mol. Immunol.* **19:**87–94.

Fukui, Y., Maru, M., Ohkawara, K., Miyake, T., Osada, Y., Wang, D., Ito, T., Higashi, H., Naiki, M., Wakamiya, N., and Kato, S., 1989, Detection of glycoproteins as tumor-associated Hanganutziu–Deicher antigen in human gastric cancer cell line, NUGC4, *Biochem. Biophys. Res. Commun.* **160:**1149–1154.

Gabriel, O., Heeb, M. J., and Hinrichs, M., 1984, Interaction of the surface adhesins of the oral *Actinomyces* ssp. with mammalian cells, in: *ASM Molecular Basis of Oral Microbial Adhesion*, Proceedings of a workshop held in Philadelphia, pp. 45–52.

Galen, J. E., Ketley, J. M., Fasano, A., Richardson, S. H., Wasserman, S. S., and Kaper, J. B., 1992, Role of *Vibrio cholerae* neuraminidase in the function of cholera toxin, *Infect. Immun.* **60:**406–415.

Gamian, A., Romanowska, E., Dabrowski, U., and Dabrowski, J., 1991, Structure of the O-specific, sialic acid-containing polysaccharide chain and its linkage to the core region in lipopolysaccharide from *Hafnia alvei* strain-2 as elucidated by chemical methods, gas–liquid chromatography/mass-spectrometry, and ^{1}H NMR spectroscopy, *Biochemistry* **30:**5032–5038.

Gibson, B. W., Melaugh, W., Phillips, N. J., Apicella, M. A., Campagnari, A. A., and Grifiss, J. M., 1993, Investigation of the structural heterogeneity of lipopolysaccharides from pathogenic *Haemophilus* and *Neisseria* species and of the R-type lipopolysaccharides from *Salmonella typhimurium* by electrospray mass spectrometry, *J. Bacteriol.* **175:**2702–2712.

Gillespie, W., Kelm, S., and Paulson, J. C., 1992, Cloning and expression of the Galβ1, 3GalNAcα2,3-sialyltransferase, *J. Biol. Chem.* **267:**21004–21010.

Godoy, V. G., Miller Dallas, M., Russo, T. A., and Malamy, M. H., 1993, A role for *Bacteroides fragilis* neuraminidase in bacterial growth in two model systems, *Infect. Immun.* **61:**4415–4426.

Gottschalk, A., 1960, *The Chemistry and Biology of Sialic Acids and Related Substances*, Cambridge University Press, London.

Graves, B. J., Crowther, R. L., Chandran, C., Rumberger, J. M., Li, S., Huang, K.-S., Presky, D. H., Familletti, P. C., Wolitzky, B. A., and Burns, D. K., 1994, Insight into E-selectin/ligand interaction from the crystal structure and mutagenesis of the Lec/EGF domains, *Nature* **367:**532–538.

Green, P. J., Tamatani, T., Watanabe, T., Miyasaka, M., Hasegawa, A., Kiso, M., Yuen, C. T., Stoll, M. S., and Feizi, T., 1992, High affinity binding of the leucocyte adhesion molecule L-selectin to 3′-sulphated-Le^a and 3′-sulphated-Le^x oligosaccharides and the predominance of sulphate in this interaction demonstrated by binding studies with a series of lipid-linked oligosaccharides, *Biochem. Biophys. Res. Commun.* **188:**244–251.

Grundmann, U., Nerlich, C., Rein, T., and Zettlmeissl, G. 1990, Complete cDNA sequence encoding human β-galactoside α-2,6-sialyltransferase, *Nucleic Acids Res.* **18:**667.

Handa, K., Nudelman, E. D., Stroud, M. R., Shiozawa, T., and Hakomori, S.-i., 1991, Selectin GMP-140 (CD62; PADGEM) binds to sialosyl-Le^a and sialosyl-Le^x, and sulfated glycans modulate this binding, *Biochem. Biophys. Res. Commun.* **181:**1223–1230.

Hanisch, F.-G., Witter, B., Crombach, G. A., Schänzer, W., and Uhlenbruck, G., 1992, N-Glycolylneuraminic acid is a chemical marker of gangliosides from human breast carcinoma, in: *Tumor-Associated Antigens, Oncogenes, Receptors, Cytokines in Tumor Diagnosis and Therapy at the Beginning of the Nineties. Cancer of the Breast—State and Trends in Diagnosis and Therapy* (R. Klapdor, ed.), Zuckschwerdt, Munich, pp. 367–370.

Hanisch, F.-G., Hacker, J., and Schroten, H., 1993, Specificity of S-fimbriae on recombinant *Escherichia coli*: Preferential binding to gangliosides expressing NeuGcα(2,3)Gal and NeuAcα(2,8)NeuAc, *Infect. Immun.* **61**:2108–2115.

Hara, S., Yamaguchi, M., Takemori, Y., and Nakamura, M., 1986, Highly sensitive determination of N-acetyl- and N-glycolyl-neuraminic acids in human serum and rat serum by reversed phase liquid chromatography with fluorescence detection, *J. Chromatogr.* **377**:111–119.

Hara, S., Takemori, Y., Yamaguchi, M., Nakamura, M., and Ohkura, Y., 1987, Fluorimetric high-performance liquid chromatography of N-acetyl- and N-glycolylneuraminic acids and its application to their microdetermination in human and animal sera, glycoproteins, and glycolipids, *Anal. Biochem.* **164**:138–145.

Hara, S., Yamaguchi, M., Takemori, Y., Furuhata, K., Ogura, H., and Nakamura, M., 1989, Determination of mono-O-acetylated N-acetylneuraminic acids in human and rat sera by fluorimetric high-performance liquid chromatography, *Anal. Biochem.* **179**:162–166.

Harford, J., Klausner, R. D., and Ashwell, G., 1984, Inhibition of the endocytic pathway for asialoglycoprotein catabolism, *Biol. Cell* **51**:173–179.

Harms, G., Corfield, A. P., Schauer, R., and Reuter, G., 1993, Recognition of O-acetylated sialic acids by influenza C virus, *Biol. Chem. Hoppe-Seyler* **374**:946–947.

Haverkamp, J., Schauer, R., Wember, M., Farriaux, J.-P., Kamerling, J. P., Versluis, C., and Vliegenthart, J.F.G., 1976, Neuraminic acid derivatives newly discovered in humans: N-Acetyl-9-O-L-lactoylneuraminic acid, N,9-O-diacetyl-neuraminic acid and N-acetyl-2,3-dehydro-2-deoxyneuraminic acid, *Hoppe-Seyler's Z. Physiol. Chem.* **357**:1699–1705.

Hayaishi, O. (ed.), 1974, *Molecular Mechanisms of Oxygen Activation*, Academic Press, New York.

Henningsen, M., Roggentin, P., and Schauer, R., 1991, Cloning, sequencing and expression of the sialidase gene from *Actinomyces viscosus* DSM 43798, *Biol. Chem. Hoppe-Seyler* **372**:1065–1072.

Herrler, G., Rott, R., Klenk, H.-D., Müller, H. P., Shukla, A. K., and Schauer, R., 1985, The receptor-destroying enzyme of influenza C virus is neuraminate O-acetylesterase, *EMBO J.* **4**:1503–1506.

Herrler, G., Reuter, G., Rott, R., Klenk, H.-D., and Schauer, R., 1987, N-Acetyl-9-O-acetylneuraminic acid, the receptor determinant for influenza C virus, is a differentiation marker on chicken erythrocytes, *Biol. Chem. Hoppe-Seyler* **368**:451–454.

Heuermann, D., Roggentin, P., Kleineidam, R. G., and Schauer, R., 1991, Purification and characterization of a sialidase from *Clostridium chauvoei* NC08596, *Glycoconj. J.* **8**:95–101.

Higa, H. H., and Varki, A., 1988, Acetyl-coenzyme A: polysialic acid O-acetyltransferase from K1 positive *Escherichia coli:* The enzyme responsible for the O-acetyl plus phenotype and for O-acetyl form variation, *J. Biol. Chem.* **263**:8872–8878.

Higa, H. H., Rogers, G. N., and Paulson, J. C., 1985, Influenza virus hemagglutinins differentiate between receptor determinants bearing N-acetyl, N-glycollyl and N,O-diacetylneuraminic acids, *Virology* **144**:279–282.

Higa, H. H., Butor, C., Diaz, S., and Varki, A., 1989a, O-Acetylation and de-O-acetylation of sialic acids. O-Acetylation of sialic acids in rat liver Golgi apparatus involves an acetyl intermediate and essential histidine and lysine residues—A transmembrane reaction? *J. Biol. Chem.* **264**:19427–19434.

Higa, H. H., Manzi, A., and Varki, A., 1989b, O-Acetylation and de-O-acetylation of sialic acids. Purification, characterization, and properties of a glycosylated rat liver esterase specific for 9-O-acetylated sialic acids, *J. Biol. Chem.* **264**:19435–19442.

Higashi, H., 1990, N-Glycolylneuraminic acid-containing glycoconjugate as tumor-associated antigen: Hanganutziu–Deicher antigen, *Trends Glycosci. Glycotechnol.* **2**:7–15.

Higashi, H., Nishi, Y., Fukui, Y., Kazuyoshi, I., Ueda, S., Kato, S., Fujita, M., Nakano, Y., Taguchi, T., Sakai, S., Sako, M., and Naiki, M., 1984, Tumor-associated expression of glycosphingolipid Hanganutziu–Deicher antigen in human cancers, *Jpn. J. Cancer Res. (Gann)* **75:**1025–1029.

Higashi, H., Hirabayashi, Y., Fukui, Y., Naiki, M., Matsumoto, M., Ueda, S., and Kato, S., 1985, Characterization of N-glycolylneuraminic acid-containing gangliosides as tumor-associated Hanganutziu–Deicher antigen in human colon cancer, *Cancer Res.* **45:**3796–3802.

Higashi, H., Sasabe, T., Fukui, Y., Maru, M., and Kato, S., 1988, Detection of gangliosides as N-glycolylneuraminic acid-specific tumor-associated Hanganutziu–Deicher antigen in human retinoblastoma cells, *Jpn. J. Cancer Res. (Gann)* **79:**952–956.

Hirabayashi, Y., Kasakura, H., Matsumoto, M., Higashi, H., Kato, S., Kasai, N., and Naiki, M., 1987a, Specific expression of unusual GM2 ganglioside with Hanganutziu–Deicher antigen activity on human colon cancers, *Jpn. J. Cancer Res. (Gann)* **78:**251–260.

Hirabayashi, Y., Higashi, H., Kato, S., Taniguchi, M., and Matsumoto, M., 1987b, Occurrence of tumor-associated ganglioside antigens with Hanganutziu–Deicher antigenic activity on human melanomas, *Jpn. J. Cancer Res. (Gann)* **78:**614–620.

Hoyer, L. L., Roggentin, P., Schauer, R., and Vimr, E. R., 1991, Purification and properties of cloned *Salmonella typhimurium* LT2 sialidase with virus-typical kinetic preference for sialyl α2-3 linkages, *J. Biochem.* **110:**462–467.

Hoyer, L. L., Hamilton, A. C., Steenbergen, S. M., and Vimr, E. R., 1992, Cloning, sequencing and distribution of the *Salmonella typhimurium* LT-2 sialidase gene, nanH, provides evidence for interspecies gene transfer, *Mol. Microbiol.* **6:**873–884.

Hutchins, J. T., Reading, C. L., Giavazzi, R., Hoaglund, J., and Jessup, J. M., 1988, Distribution of mono-O-acetylated, di-O-acetylated, and tri-O-acetylated sialic acids in normal and neoplastic colon, *Cancer Res.* **48:**483–489.

Imai, Y., and Rosen, S. D., 1993, Direct demonstration of heterogeneous, sulfated O-linked carbohydrate chains on an endothelial ligand for L-selectin, *Glycoconj. J.* **10:**34–39.

Imai, Y., Singer, M. S., Fennie, C., Lasky, L. A., and Rosen, S. D., 1991, Identification of a carbohydrate-based endothelial ligand for a lymphocyte homing receptor, *J. Cell Biol.* **113:**1213–1221.

Imai, Y., Lasky, L. A., and Rosen, S. D., 1993, Sulphation requirement for GlyCAM-1, an endothelial ligand for L-selectin, *Nature* **361:**555–557.

Inoue, S., Iwasaki, M., Kitajima, K., Kanamori, A., Kitazume, S., and Inoue, Y., 1993, Analytical methods for identifying and quantitating poly (α2→8Sia) structures containing Neu5Ac, Neu5Gc, and KDN, in: *Polysialic Acid* (J. Roth, U. Rutishauser, and F. A. Troy II, eds.), Birkhäuser, Basel, pp. 183–189.

Inoue, Y., 1993, Glycobiology of fish egg polysialoglycoproteins (PSGP) and deaminated neuraminic acid-rich glycoproteins (KDN-gp), in: *Polysialic Acid* (J. Roth, U. Rutishauser, and F. A. Troy II, eds.), Birkhäuser, Basel, pp. 171–181.

Ishikawa, H., and Isayama, Y., 1987, Evidence for sialyl glycoconjugates as receptors for *Bordetella bronchiseptica* on swine nasal mucosa, *Infect. Immun.* **55:**1607–1609.

Jaenicke, R., 1993, What does protein refolding *in vitro* tell us about protein folding in the cell? *Philos. Trans. R. Soc. London Ser. B* **339:**287–295.

Johnston, G. I., Cook. R. G., and McEver, R. P., 1989, Cloning of GMP-140, a granule membrane protein of platelets and endothelium: Sequence similarity to proteins involved in cell adhesion and inflammation, *Cell* **56:**1033–1044.

Jourdian, G. W., and Roseman, S., 1962, The sialic acids: II. Preparation of N-glycolylhexosamines, N-glycolylhexosamine-6-phosphates, glycolyl-CoA and glycolylglutathione, *J. Biol. Chem.* **237:**2442–2446.

Kamerling, J. P., and Vliegenthart, J.F.G., 1982, Gas–liquid chromatography and mass spectrometry of sialic acids, in: *Sialic Acids* (R. Schauer, ed.), Springer, New York, pp. 95–125.

Kamerling, J. P., Schauer, R., Shukla, A. K., Stoll, S., van Halbeek, H., and Vliegenthart, J.F.G., 1987, Migration of O-acetyl groups in N,O-acetylneuraminic acids, *Eur. J. Biochem.* **162:**601–607.

Kanamori, A., Inoue, S., Iwasaki, M., Kitajima, K., Kawai, G., Yokoyama, S., and Inoue, Y., 1990, Deaminated neuraminic acid rich glycoprotein of rainbow trout egg vitelline envelope: Occurrence of a novel α-2,8-linked oligo (deaminated neuraminic acid) structure in O-linked glycan chains, *J. Biol. Chem.* **265:**21811–21819.

Karlsson, K. A., 1989, Animal glycosphingolipids as membrane attachment sites for bacteria, *Annu. Rev. Biochem.* **58:**309–350.

Kawai, T., Kato, A., Higashi, H., Kato, S., and Naiki, M., 1991, Quantitative determination of N-glycolylneuraminic acid expression in human cancerous tissues and avian lymphoma cell lines as a tumor-associated sialic acid by gas chromatography–mass spectrometry, *Cancer Res.* **51:**1242–1246.

Kawano, T., Kozutsumi, Y., Takematsu, H., Kawasaki, T., and Suzuki, A., 1993a, Regulation of biosynthesis of N-glycolylneuraminic acid-containing glycoconjugates: Characterization of factors required for NADH-dependent cytidine 5′-monophosphate-N-acetylneuraminic acid hydroxylation, *Glycoconj. J.* **10:**109–115.

Kawano, T., Kozutsumi, Y., Kawasaki, T., and Suzuki, A., 1993b, The key enzyme which regulates CMP-N-acetylneuraminic acid hydroxylation is a novel protein, *Glycoconj. J.* **10:**331.

Kelm, S., Schauer, R., Manuguerra, J.-C., Gross, H.-J., and Crocker, P. R., 1994, Modifications of cell surface sialic acids modulate cell adhesion mediated by sialoadhesion and CD22, *Glycoconj. J.* **11:**576–585.

Kelm, S., Dubois, C., Müthing, J., Schauer, R., and Crocker, P. R., 1993, Gangliosides as ligands for sialoadhesin, a cell adhesion receptor on macrophages, *J. Cell Biochem.* Suppl. **17A:**372.

Kiehne, K., and Schauer, R., 1992, The influence of α- and β-galactose residues and sialic acid O-acetyl groups of rat erythrocytes on the interaction with peritoneal macrophages, *Biol. Chem. Hoppe-Seyler* **373:**1117–1123.

Kitagawa, H., and Paulson, J. C., 1993, Cloning and expression of human Galβ1,3(4)-GlcNAc α2,3-sialyltransferase, *Biochem. Biophys. Res. Commun.* **194:**375–382.

Kitagawa, H., and Paulson, J. C., 1994, Cloning of a novel α2,3-sialyltransferase that sialylates glycoprotein and glycolipid carbohydrate groups, *J. Biol. Chem.* **269:**1394–1401.

Kitajima, K., Inoue, S., Inoue, Y., and Troy, F. A., 1988, Use of a bacteriophage-derived endo-N-acetylneuraminidase and an equine antipolysialyl antibody to characterise the polysialyl residues in salmonid fish egg polysialoglycoproteins: Substrate and immunospecificity studies, *J. Biol. Chem.* **263:**18269–18276.

Kitazume, S., Kitajima, K., Inoue, S., and Inoue, Y., 1992, Detection, isolation, and characterization of oligo/poly(sialic acid) and oligo/poly(deaminoneuraminic acid) units in glycoconjugates, *Anal. Biochem.* **202:**25–34.

Kleineidam, R. G., Furuhata, K., Ogura, H., and Schauer, R., 1990, 4-Methylumbelliferyl-α-glycosides of partially O-acetylated N-acetylneuraminic acids as substrates of bacterial and viral sialidases, *Biol. Chem. Hoppe-Seyler* **371:**715–719.

Kleineidam, R. G., Hofmann, O., Reuter, G., and Schauer, R., 1993, Indications for the enzymic synthesis of 9-O-lactoyl-N-acetylneuraminic acid in equine liver, *Glycoconj. J.* **10:**116–119.

Klotz, F. W., Orlandi, P. A., Reuter, G., Cohen, S. J., Haynes, J. D., Schauer, R., Howard, R. J., Palese, P., and Miller, L. H., 1992, Binding of *Plasmodium falciparum* 175kilodalton erythrocyte binding antigen and invasion of murine erythrocyte requires N-acetylneuraminic acid but not its O-acetylated form, *Mol. Biochem. Parasitol.* **51:**49–54.

Kluge, A., Reuter, G., Lee, H., Ruch-Heeger, B., and Schauer, R., 1992, Interaction of rat peritoneal macrophages with sialidase-treated thrombocytes in vitro: Biochemical and morphological studies. Detection of N-(O-acetyl)-glycoloyl-neuraminic acid, *Eur. J. Cell Biol.* **59:**12–20.

Knirel, Y. A., Vinogradov, E. V., Shashkov, A. S., and Kochetkov, N. K., 1985, Identification of 5-acetamido-3,5,7,9-tetradeoxy-7-[(R)3-hydroxybutyramido]-L-glycero-L-manno-nonulosonic acid as a component of bacterial polysaccharide, *Carbohydr. Res.* **141:**C1–C3.

Knirel, Y. A., Kocharova, N. A., Shashkov, A. S., and Kochetkov, N. K., 1986, The structure of *Pseudomonas aeruginosa* immunotype 6 O-antigen: Isolation and identification of 5-acetamido 3,5,7,9-tetradeoxy 7-muramido L-glycero-L-manno nonulosonic acid, *Carbohydr. Res.* **145:** C1–C4.

Knirel, Y. A., Vinogradov, E. V., Shashkov, A. S., Dimitriev, B. A., Kochetkov, N. K., Stanislavsky, E. S., and Mashilova, G. M., 1987a, Somatic antigens of *Pseudomonas aeruginosa*. The structure of the O-specific polysaccharide chain of the lipopolysaccharide from *P. aeruginosa* O13 (Lanyi), *Eur. J. Biochem.* **163:**627–637

Knirel, Y. A., Kocharova, N. A., Shashkov, A. S., Dimitriev, B. A., Kochetkov, N. K., Stanislavsky, E. S., and Mashilova, G. M., 1987b, Somatic antigens of *Pseudomonas aeruginosa*. The structure of O-specific polysaccharide chains of the lipopolysaccharides from *P. aeruginosa* O5 (Lanyi) and immunotype 6 (Fisher), *Eur. J. Biochem.* **163:**639–652.

Knirel, Y. A., Rietschel, E. T., Marre, R., and Zähringer, U., 1994, The structure of the O-specific chain of *Legionella pneumophila* serogroup 1 lipopolysaccharide, *Eur. J. Biochem.* **221:**239–245.

Ko, H. L., Beuth, J., Sölter, J., Schroten, H., Uhlenbruck, G., and Pulverer, G., 1987, *In vitro* and *in vivo* inhibition of lectin mediated adhesion of *Pseudomonas aeruginosa* by receptor blocking carbohydrates, *Infection* **15:**237–240.

Kotovuori, P., Tontti, E., Pigott, R., Shepherd, M., Kiso, M., Hasegawa, A., Renkonen, R., Nortamo, P., Altieri, D. C., and Gahmberg, C. G., 1993, The vascular E-selectin binds to the leukocyte integrins CD11/CD18, *Glycobiology* **3:**131–136.

Kozutsumi, Y., Kawano, T., Yamakawa, T., and Suzuki, A., 1990, Participation of cytochrome b_5 in CMP-N-acetylneuraminic acid hydroxylation in mouse liver cytosol, *J. Biochem.* **108:**704–706.

Kozutsumi, Y., Kawano, T., Kawasaki, H., Suzuki, K., Yamakawa, T., and Suzuki, A., 1991, Reconstitution of CMP-N-acetylneuraminic acid hydroxylation activity using a mouse liver cytosol fraction and soluble cytochrome b_5 purified from horse erythrocytes, *J. Biochem.* **110:**429–435.

Krauss, J. H., Reuter, G., Schauer, R., Weckesser, J., and Mayer, H., 1988, Sialic acid-containing lipopolysaccharide of purple nonsulfur bacteria, *Arch. Microbiol.* **150:**584–589.

Krauss, J. H., Himmelspach, K., Reuter, G., Schauer, R., and Mayer, H., 1992, Structural analysis of a novel sialic acid-containing trisaccharide from *Rhodobacter capsulatus* 37b4 lipopolysaccharide, *Eur. J. Biochem.* **204:**217–223.

Kurosawa, N., Hamamoto, T., Lee, Y.-L., Nakaoka, T., and Tsuji, S., 1993, cDNA Cloning of three groups of sialyltransferase from chick embryo, *Glycoconj. J.* **10:**236.

Kurosawa, N., Hamamoto, T., Lee, Y.-L., Nakaoka, T., Kojima, N., and Tsuji, S., 1994, Molecular cloning and expression of GalNAc α2,6-sialyltransferase, *J. Biol. Chem.* **269:**1402–1409.

Kyogashima, M., Ginsburg, V., and Krivan, H., 1989, *Escherichia coli* K99 binds to N-glycolylsialoparagloboside and N-glycolyl-GM3 found in piglet small intestine, *Arch. Biochem. Biophys.* **270:**391–397.

Larsen, E., Celi, A., Gilbert, G. E., Furie, B. C., Erban, J. K., Bonfanti, R., Wagner, D. D., and Furie, B., 1989, PADGEM protein: A receptor that mediates the interaction of activated platelets with neutrophils and monocytes, *Cell* **59:**305–312.

Lasky, L. A., 1992, Selectins: Interpreters of cell-specific carbohydrate information during inflammation, *Science* **258:**964–969.

Lasky, L. A., Singer, M. S., Yednock, T. A., Dowbenko, D., Fennie, C., Rodriguez, H., Nguyen,

T., Stachel, S., and Rosen, S. D., 1989, Cloning of a lymphocyte homing receptor reveals a lectin domain, *Cell* **56:**1045–1055.

Lasky, L. A., Singer, M. S., Dowbenko, D., Imai, Y., Henzel, W. J., Grimley, C., Fennie, C., Gillett, N., Watson, S. R., and Rosen, S. D., 1992, An endothelial ligand for L-selectin is a novel mucin-like molecule, *Cell* **69:**927–938.

Law, C.-L., Torres, R. M., Sundberg, H. A., Parkhouse, R.M.E., Brannan, C. I., Copeland, N. G., Jenkins, N. A., and Clark, E. A., 1993, Organization of the murine *Cd22* locus—Mapping to chromosome 7 and characterization of two alleles, *J. Immunol.* **151:**175–187.

Lawrence, M. B., and Springer, T. A., 1991, Leukocytes roll on a selectin at physiological flow rates: Distinction from and prerequisite for adhesion through integrins, *Cell* **65:**859–873.

Ledeen, R. W., and Yu, R. W., 1976, Chemistry and analysis of sialic acids, in: *Biological Roles of Sialic Acid* (A. Rosenberg and C. L. Schengrund, eds.), Plenum Press, New York, pp. 1–48.

Lee, Y-C., Kurosawa, N., Hamamoto, T., Nakaoka, T., and Tsuji, S., 1993, Molecular cloning and expression of Galβ1,3GalNAcα2,3-sialyltransferase from mouse brain, *Eur. J. Biochem.* **216:** 377–385.

Lepers, A., Shaw, L., Cacan, R., Schauer, R., Montreuil, J., and Verbert, A., 1989, Transport of CMP-N-glycoloylneuraminic acid into mouse liver Golgi vesicles, *FEBS Lett.* **250:**245–250.

Lepers, A., Shaw, L., Schneckenburger, P., Cacan, R., Verbert, A., and Schauer, R., 1990, A study on the regulation of N-glycoloylneuraminic acid biosynthesis and utilization in rat and mouse liver, *Eur. J. Biochem.* **193:**715–723.

Leprince, C., Draves, K. E., Geahlen, R. L., Ledbetter, J. A., and Clark, E. A., 1993, CD22 associates with the human surface IgM B cell antigen receptor complex, *Proc. Natl. Acad. Sci. USA* **90:**3236–3240.

Levinovitz, A., Mühlhoff, J., Isenmann, S., and Vestweber, D., 1993, Identification of a glycoprotein ligand for E-selectin on mouse myeloid cells, *J. Cell Biol.* **121:**449–459.

Lewinsohn, D. M., Bargatze, R. F., and Butcher, E. C., 1987, Leukocyte–endothelial cell recognition: Evidence for a common molecular mechanism shared by neutrophils, lymphocytes and other leukocytes, *J. Immunol.* **138:**4313–4321.

Li, Y.-T., Yuziuk, J. A., Li, S.-C., Nematalla, A., Hasegawa, A., Tsutsumi, M., and Nakagawa, H., 1993, A novel sialidase capable of cleaving 2-keto-3-deoxy-D-glycero-D-galacto-nononic acid (KDN), *Glycobiology* **10:**525.

Liukkonen, J., Haataja, S., Tikkanen, K., Kelm, S., and Finne, J., 1992, Identification of N-acetylneuraminyl α-2,3 poly-N-acetyllactosamine glycans as the receptors of sialic acid binding *Streptococcus suis* strains, *J. Biol. Chem.* **267:**21105–21111.

Livingston, B. D., and Paulson, J. C., 1993, Polymerase chain reaction cloning of a developmentally regulated member of the sialyltransferase gene family, *J. Biol. Chem.* **268:**11504–11507.

Long, G. S., Taylor, P. W., and Luzio, J. P., 1993, Characterization of bacteriophage E endosialidase specific for alpha-2,8-linked polysialic acid, in: *Polysialic Acid* (J. Roth, U. Rutishauser, and F. A. Troy II, eds.), Birkhäuser Verlag, Basel, pp. 137–144.

Loomes, L. M., Uemura, K.-I., Childs, R. A., Paulson, G. N., Rogers, G. N., Scudder, P. R., Michalski, J.-C., Hounsell, E. F., Taylor-Robinson, D., and Feizi, T., 1984, Erythrocyte receptors for *Mycoplasma pneumoniae* are sialylated oligosaccharides of Ii antigen type, *Nature* **307:**560–563.

Lowe, J. B., Stoolman, L. M., Nair, R. P., Larsen, R. D., Berhend, T. L., and Marks, R. M., 1990, ELAM-1-dependent cell adhesion to vascular endothelium determined by a transfected human fucosyltransferase cDNA, *Cell* **63:**475–484.

McEver, P. R., 1992, Leukocyte–endothelial interactions, *Curr. Opin. Cell Biol.* **4:**840–849.

Markwell, M. A. K., and Paulson, J. C., 1980, Sendai virus utilizes specific sialyloligosaccharides as host cell receptor determinants, *Proc. Natl. Acad. Sci. USA* **77:**5693–5697.

Mayadas, T. N., Johnson, R. C., Rayburn, H., Hynes, R. O., and Wagner, D. D., 1993, Leukocyte rolling and extravasation are severely compromised in P-selectin-deficient mice, *Cell* **74:**541–554.

Meri, S., and Pangburn, M. K., 1990, Discrimination between activators and nonactivators of the alternative pathway of complement: Regulation via sialic acid/polyanion binding site of factor H, *Proc. Natl. Acad. Sci. USA* **87:**3982–3986.

Miyagi, T., Konno, K., Emori, Y., Kawasaki, H., Suzuki, K., Yasui, A., and Tsuiki, S., 1993, Molecular cloning and expression of cDNA encoding rat skeletal muscle cytosolic sialidase, *J. Biol. Chem.* **268:**26435–26440.

Moore, K. L., Stults, N. L., Diaz, S., Smith, D. F., Cummings, R. D., Varki, A., and McEver, R. P., 1992, Identification of a specific glycoprotein ligand for P-selectin (CD62) on myeloid cells, *J. Cell Biol.* **118:**445–456.

Moran, P., Raab, H., Kohr, W. J., and Caras, I. W., 1991, Glycophospholipid membrane anchor attachment—Molecular analysis of the cleavage/attachment site, *J. Biol. Chem.* **266:**1250–1257.

Muchmore, E. A., 1992, Developmental sialic acid modifications in rat organs, *Glycobiology* **2:**337–343.

Muchmore, E., Varki, N., Fukuda, M., and Varki, A., 1987, Developmental regulation of sialic acid modifications in rat and human colon, *FASEB J.* **1:**229–235.

Muchmore, E. A., Milewski, M., Varki, A., and Diaz, S., 1989, Biosynthesis of N-glycolylneuraminic acid: The primary site of hydroxylation of N-acetylneuraminic acid is the cytosolic sugar nucleotide pool, *J. Biol. Chem.* **264:**20216–20223.

Müller, E., Schröder, C., Sharon, N., and Schauer, R., 1983, Binding and phagocytosis of sialidase-treated rat erythrocytes by a mechanism independent of opsonins, *Hoppe-Seyler's Z. Physiol. Chem.* **364:**1410–1420.

Müller, H. E., 1974, Neuraminidases of bacteria and protozoa and their pathogenic role, *Behring Inst. Mitt.* **55:**34–56.

Müller, H. E., 1992, Relationships between ecology and pathogenicity of microorganisms, *BIOforum* **1/2:**16–22.

Munro, S., Bast, B.J.E.G., Colley, K. J., and Tedder, T. F., 1992, The lymphocyte-B surface antigen CD75 is not an α-2,6-sialyltransferase but is a carbohydrate antigen, the production of which requires the enzyme, *Cell* **68:**1003–1004.

Murray, P. A., Levine, M. J., Tabak, L. A., and Reddy, M. S., 1982, Specificity of salivary–bacterial interactions: II. Evidence for a lectin on *Streptococcus sanguis* with specificity for a Neu5Acα2,3Galβ1,3GalNAc sequence, *Biochem. Biophys. Res. Commun.* **106:**390–396.

Nadano, D., Iwasaki, M., Endo, S., Kitajima, K., Inoue, S., and Inoue, Y., 1986, A naturally occurring deaminated neuraminic acid, 3-deoxy-D-glycero-D-galacto-nonulosonic acid (KDN). Its unique occurrence at the nonreducing ends of oligosialyl chains in polysialoglycoprotein of rainbow trout eggs, *J. Biol. Chem.* **261:**11550–11557.

Nees, S., Veh, R. W., and Schauer, R., 1975, Purification and characterization of neuraminidase from *Clostridium perfringens*, *Hoppe-Seyler's Z. Physiol. Chem.* **356:**1027–1042.

Nöhle, U., Beau, J. M., and Schauer, R., 1982, Uptake metabolism and excretion of orally and intravenously administered double-labeled N-glycoloylneuraminic acid and single-labeled 2-deoxy-2,3-dehydro-N-acetylneuraminic acid in mouse and rat, *Eur. J. Biochem.* **126:**543–548.

Norgard, K. E., Han, H., Powell, L., Kriegler, M., Varki, A., and Varki, N. M., 1993a, Enhanced interaction of L-selectin with the high endothelial venule ligand via selectively oxidized sialic acids, *Proc. Natl. Acad. Sci. USA* **90:**1068–1072.

Norgard, K. E., Moore, K. L., Diaz, S., Stults, N. L., Ushiyama, S., McEver, R. P., Cummings, R. D., and Varki, A., 1993b, Characterization of a specific ligand for P-selectin on myeloid cells—A minor glycoprotein with sialylated O-linked oligosaccharides, *J. Biol. Chem.* **268:**12764–12774.

Ohashi, Y., Sasabe, T., Nishida, T., Nishi, Y., and Higashi, H., 1983, Hanganutziu–Deicher heterophile antigen in human retinoblastoma cells, *Am. J. Ophthalmol.* **96:**321–325.

Ouadia, A., Karamanos, Y., and Julien, R., 1992, Detection of the ganglioside N-glycolyl-neuraminyl-lactosyl-ceramide by biotinylated *Escherichia coli* K99 lectin, *Glycoconj. J.* **9:**21–26.

Parkkinen, A., Rogers, G. N., Korhonen, T., Dahr, W., and Finne, J., 1986, Identification of the O-linked sialyloligosaccharides of glycophorin A as the erythrocyte receptors for S-fimbriated *Escherichia coli, Infect. Immun.* **54:**37–42.

Paulson, J. C., 1985, Interactions of animal viruses with cell surface receptors, in: *The Receptors,* Vol. 2 (M. Conn, ed.), Academic Press, New York, pp. 131–219.

Paulson, J. C., and Colley, K. J., 1989, Glycosyltransferases: Structure, localization, and control of cell type-specific glycosylation, *J. Biol. Chem.* **264:**17615–17618.

Pereira, M.E.A., Mejia, J. S., Ortega-Barria, E., Matzilevich, D., and Prioli, R., 1991, The *Trypanosoma cruzi* neuraminidase contains sequences similar to bacterial neuraminidases, to YWTD repeats of the LDL receptor, and to type III modules of fibronectin, *J. Exp. Med.* **174:**179–191.

Perkins, M. E., and Rocco, L. J., 1988, Sialic acid-dependent binding of *Plasmodium falciparum* merozoite surface antigen, Pf200, to human erythrocytes, *J. Immunol.* **141:**3190–3196.

Phillips, M. L., Nudelman, E., Gaeta, F.C.A., Perez, M., Singhal, A. K., Hakomori, S.-i., and Paulson, J. C., 1990, ELAM-1 mediates cell adhesion by recognition of a carbohydrate ligand, sialyl-Le^x, *Science* **250:**1130–1132.

Polley, M. J., Phillips, M. L., Wayner, E., Nudelman, E., Singhal, A. K., Hakomori, S.-i., and Paulson, J. C., 1991, CD62 and endothelial cell leukocyte adhesion molecule-1 (ELAM-1) recognize the same carbohydrate ligand, sialyl-$Lewis^x$, *Proc. Natl. Acad. Sci. USA* **88:**6224–6228.

Popoff, N. R., and Dodin, A., 1985, Survey of neuraminidase production by *Clostridium butyricum, Clostridium beijerinckii* and *Clostridium difficile* strains from clinical and nonclinical sources, *J. Clin. Microbiol.* **22:**873–876.

Popoli, M., and Mengano, A., 1988, A hemagglutinin specific for sialic acids in a rat brain synaptic vesicle-enriched fraction, *Neurochem. Res.* **13:**63–67.

Powell, L. D., Sgroi, D., Sjoberg, E. R., Stamenkovic, I., and Varki, A., 1993, Natural ligands of the B-cell adhesion molecule CD22-β carry N-linked oligosaccharides with α-2,6-linked sialic acids that are required for recognition, *J. Biol. Chem.* **268:**7019–7027.

Prioli, R. P., Ortega-Barria, E., Mejia, J. S., and Pereira, M.E.A., 1992, Mapping of the B-cell epitope present in the neuraminidase of *Trypanosoma cruzi, Mol. Biochem. Parasitol.* **52:**85–96.

Rahmann, H., Hilbig, R., Marx, J., Beitinger, H., and Mehlfeld, R., 1987, Brain gangliosides and hibernation, *J. Therm. Biol.* **12:**81–85.

Reuter, G., and Schauer, R., 1986, Comparison of electron and chemical ionization mass spectrometry of sialic acids, *Anal. Biochem.* **157:**39–46.

Reuter, G., and Schauer, R., 1987, Isolation and analysis of gangliosides with O-acetylated sialic acids, in: *Gangliosides and Modulation of Neuronal Functions,* NATO ASI Series, Vol. H7 (H. Rahmann, ed.), Springer-Verlag, Berlin, pp. 155–165.

Reuter, G., and Schauer, R., 1988, Nomenclature of sialic acids, *Glycoconj. J.* **5:**133–135.

Reuter, G., and Schauer, R., 1994, Determination of sialic acids, *Methods Enzymol.* **230:**168–199.

Reuter, G., Pfeil, R., Stoll, S., Schauer, R., Kamerling, J. P., Versluis, C., and Vliegenthart, J.F.G., 1983, Identification of new sialic acids derived from glycoprotein of bovine submandibular gland, *Eur. J. Biochem.* **134:**139–143.

Reuter, G., Schauer, R., Prioli, R., and Pereira, M.E.A., 1987, Isolation and properties of a sialidase from *Trypanosoma rangeli, Glycoconj. J.* **4:**339–348.

Reuter, G., Stoll, S., Kamerling, J. P., Vliegenthart, J.F.G., and Schauer, R., 1988, Sialic acids on erythrocytes and in blood plasma of mammals, in: *Sialic Acids 1988, Proceedings of the German-Japanese Symposium on Sialic Acids* (R. Schauer and T. Yamakawa, eds.), Verlag Wissenschaft + Bildung, Kiel, pp. 88–89.

Riedl, M., Förster, O., Rumpold, H., and Bernheimer, H., 1982, A ganglioside-dependent cellular binding mechanism in rat macrophages, *J. Immunol.* **128:**1205–1210.

Rogers, G. N., Herrler, G., Paulson, J. C., and Klenk, H.-D., 1986, Influenza C virus uses 9-O-acetyl-N-acetylneuraminic acid as high affinity receptor determinant for attachment to cells, *J. Biol. Chem.* **261:**5947–5951.

Roggentin, P., Gutschker-Gdaniec, G., Schauer, R., and Hobrecht, R., 1985, Correlative properties for a differentiation of two *Clostridium sordellii* phenotypes and their distinction from *Clostridium bifermentans, Zbl. Bakt. Hyg. A* **260:**319–328.

Roggentin, P., Berg, W., and Schauer, R., 1987, Purification and characterization of sialidase from *Clostridium sordellii* G12, *Glycoconj. J.* **4:**349–359.

Roggentin, P., Rothe, B., Lottspeich, F., and Schauer, R., 1988, Cloning and sequencing of a *Clostridium perfringens* sialidase gene, *FEBS Lett.* **238:**31–34.

Roggentin, P., Rothe, B., Kaper, J. B., Galen, J., Lawrisuk, L., Vimr, E. R., and Schauer, R., 1989, Conserved sequences in bacterial and viral sialidases, *Glycoconj. J.* **6:**349–353.

Roggentin, P., Schauer, R., Hoyer, L. L., and Vimr, E. R., 1993, The sialidase superfamily and its spread by horizontal gene transfer, *Mol. Microbiol.* **9:**915–921.

Roggentin, T., Kleineidam, R. G., Schauer, R., and Roggentin, P., 1992, Effects of site-specific mutations on the enzymic properties of a sialidase from *Clostridium perfringens, Glycoconj. J.* **9:**235–240.

Roggentin, T., Kleineidam, R. G., Majewski, D. M., Tirpitz, D., Roggentin, P., and Schauer, R., 1993, An immunoassay for the rapid and specific detection of three sialidase-producing clostridia causing gas gangrene, *J. Immunol. Methods* **157:**125–133.

Rosen, S. D., 1993, L-selectin and its biological ligands, *Histochemistry* **100:**185–191.

Rosen, S. D. Singer, M. S., Yednock, T. A., and Stoolman, L. M., 1985, Involvement of sialic acid on endothelial cells in organ-specific lymphocyte recirculation, *Science* **228:**1005–1007.

Rosenberg, A., and Schengrund, C.-L. (eds.), 1976a, *Biological Roles of Sialic Acid,* Plenum Press, New York.

Rosenberg, A., and Schengrund, C.-L., 1976b, Sialidases, in: *Biological Roles of Sialic Acid* (A. Rosenberg and C.-L. Schengrund, eds.), Plenum Press, New York, pp. 295–359.

Roth, J., Kempf, A., Reuter, G., Schauer, R., and Gehring, W. J., 1992, Occurrence of sialic acids in *Drosophila melanogaster, Science* **256:**673–675.

Rothe, B., Roggentin, P., Blöcker, H., Frank, R., and Schauer, R., 1989, Cloning, sequencing and expression of a sialidase gene of *Clostridium sordellii, J. Gen. Microbiol.* **135:**3087–3096.

Rothe, B., Rothe, B., Roggentin, P., and Schauer, R., 1991, The sialidase gene from *Clostridium septicum:* Cloning, sequencing, expression in *Escherichia coli,* and identification of conserved sequences in sialidases and other proteins, *Mol. Gen. Genet.* **226:**190–197.

Russo, T. A., Thompson, J. S., Godoy, V. G., and Malamy, M. H., 1990, Cloning and expression of the *Bacteroides fragilis* TAL2480 neuraminidase gene, *nanH,* in *Escherichia coli, J. Bacteriol.* **172:**2594–2600.

Saida, T., Ikegawa, S., Takizawa, Y., and Kawachi, S., 1990, Immunohistochemical detection of heterophile Hanganutziu–Deicher antigen in human malignant melanoma, *Arch. Dermatol. Res.* **282:**179–182.

Saito, M., and Yu, R. K., 1993, Possible role of myelin-associated neuraminidase in membrane adhesion, *J. Neurosci. Res.* **36:**127–132.

Sako, D., Chang, X.-J., Barone, K. M., Vachino, G., White, H. M., Shaw, G., Veldman, G. M.,

Bean, K. M., Ahern, T. J., Furie, B., Cumming, D. A., and Larsen, G. R., 1993, Expression cloning of a functional glycoprotein ligand for P-selectin, *Cell* **75:**1179–1186.

Sakurada, K., Ohta, T., and Hasegawa, M., 1992, Cloning, expression, and characterization of the *Micromonospora viridifaciens* neuraminidase gene in *Streptomyces lividans, J. Bacteriol.* **174:**6896–6903.

Sasaki, K., Watanabe, E., Kawashima, K., Sekine, S., Dohi, T., Oshima, M., Hanai, N., Nishi, T., and Hasegawa, M., 1993, Expression cloning of a novel Galβ(1-3/1-4)GlcNAc α2,3-sialyltransferase using lectin resistance selection, *J. Biol. Chem.* **268:**22782–22787.

Savage, A. V., Koppen, P. L., Schiphorst, W. E. C. M., Trippelvitz, L. A. W., van Halbeek, H., Vliegenthart, J. F. G., and van den Eijnden, D. H., 1986, Porcine submaxillary mucin contains α2→3- and α2→6-linked N-acetyl- and N-glycolyl-neuraminic acid, *Eur. J. Biochem.* **160:**123–129.

Savill, J., Fadok, V., Henson, P., and Haslett, C., 1993, Phagocyte recognition of cells undergoing apoptosis, *Immunol. Today* **14:**131–136.

Schauer, R., 1970, Biosynthese der N-Glykoloylneuraminsäure durch eine von Ascorbinsäure bzw. NADPH abhängige N-Acetyl-hydroxylierende "N-Acetylneuraminat:O_2 Oxidoreduktase" in Homogenaten der Unterkieferspeicheldrüse vom Schwein, *Hoppe-Seyler's Z. Physiol. Chem.* **351:**783–791.

Schauer, R., 1978a, Characterization of sialic acids, *Methods Enzymol.* **50:**64–89.

Schauer, R., 1978b, Biosynthesis of sialic acids, *Methods Enzymol.* **50:**374–386.

Schauer, R. (ed.), 1982a, *Sialic Acids—Chemistry, Metabolism and Function,* Springer, New York.

Schauer, R., 1982b, Chemistry, metabolism and biological functions of sialic acids, *Adv. Carbohydr. Chem. Biochem.* **40:**131–234.

Schauer, R., 1985, Sialic acids and their role as biological masks, *Trends Biochem. Sci.* **10:**357–360.

Schauer, R., 1987a, Analysis of sialic acids, *Methods Enzymol* **138:**132–161.

Schauer, R., 1987b, Metabolism of O-acetyl groups of sialic acids, *Methods Enzymol.* **138:**611–626.

Schauer, R., 1988, Sialic acids as antigenic determinants of complex carbohydrates, *Adv. Exp. Med. Biol.* **228:**47–72.

Schauer, R., 1991, Biosynthesis and function of N- and O-substituted sialic acids, *Glycobiology* **1:**449–452.

Schauer, R., 1992, Sialinsäurereiche Schleime - bioaktive Schmierstoffe, *Nachr. Chem. Tech. Lab.* **40:**1227–1231.

Schauer, R., and Corfield, A. P., 1982, Colorimetry and thin-layer chromatography of sialic acids, in: *Sialic Acids—Chemistry, Metabolism and Function* (R. Schauer, ed.), Springer, New York, pp. 77–94.

Schauer, R., and Vliegenthart, J.F.G., 1982, Introduction, in: *Sialic Acids—Chemistry, Metabolism and Function* (R. Schauer, ed.), Springer, New York, pp. 1–3.

Schauer, R., and Wember, M., 1985, Sialidase and sialate-8-O-methyltransferase in the starfish *Asterias rubens,* in: *Glycoconjugates—Proc. VIIIth Int. Symp.*, Vol. 1 (E. A. Davidson, J. C. Williams, and N. M. Di Ferrante, eds.), Praeger, New York, pp. 266–267.

Schauer, R., Schoop, H. J., and Faillard, H., 1968, Zur Biosynthese der N-Glykolyl-Gruppe der N-Glykolylneuraminsäure. Die oxidative Umwandlung der N-Acetyl-Gruppe zur Glykolyl-Gruppe, *Hoppe-Seyler's Z. Physiol. Chem.* **349:**645–652.

Schauer, R., Reuter, G., Mühlpfordt, H., Andrade, A.F.B., and Pereira, M.E.A., 1983, The occurrence of N-acetyl and N-glycoloylneuraminic acid in *Trypanosoma cruzi, Hoppe-Seyler's Z. Physiol. Chem.* **364:**1053–1057.

Schauer, R., Schröder, C., and Shukla, A. K., 1984a, New techniques for the investigation of structure and metabolism of sialic acids, *Adv. Exp. Med. Biol.* **174:**75–86.

Schauer, R., Shukla, A. K., Schröder, C., and Müller, E., 1984b, The anti-recognition function of sialic acids: Studies with erythrocytes and macrophages, *Pure Appl. Chem.* **56:**907–921.

Schauer, R., Reuter, G., and Stoll, S., 1988a, Sialate O-acetylesterases, key enzymes in sialic acid catabolism, *Biochimie* **70:**1511–1519.

Schauer, R., Casals-Stenzel, J., Corfield, A. P., and Veh, R. W., 1988b, Subcellular site of biosynthesis of O-acetylated sialic acids in bovine submandibular gland, *Glycoconj. J.* **5**:257–270.
Schauer, R., Reuter, G., Stoll, S., Posadas del Rio, F., Herrler, G., and Klenk, H.-D., 1988c, Isolation and characterization of sialate 9(4)-O-acetylesterase from influenza C virus, *Biol. Chem. Hoppe-Seyler* **369**:1121–1130.
Schauer, R., Reuter, G., Stoll, S., and Shukla, A. K., 1989, Isolation and characterization of sialate 9(4)-O-acetylesterase from bovine brain, *J. Biochem.* **106**:143–150.
Schauer, R., Stoll, S., and Reuter, G., 1991, Differences in the amount of N-acetyl-and N-glycoloylneuraminic acid, as well as O-acylated sialic acids, of fetal and adult bovine tissues, *Carbohydr. Res.* **213**:353–359.
Schauer, R., Kelm, S., Reuter, G., and Shaw, L., 1993, New insights into the mode and role of enzymatic sialic acid modifications, *Glycoconj. J.* **10**:327.
Schengrund, C.-L., and Ringler, N. J., 1989, Binding of *Vibrio cholerae* toxin and the heat-labile enterotoxin of *Escherichia coli* to G_{M1}, derivatives of G_{M1}, and nonlipid oligosaccharide polyvalent ligands, *J. Biol. Chem.* **264**:13233–13237.
Schenkman, S., and Eichinger, D., 1993, *Trypanosoma cruzi* trans-sialidase and cell invasion, *Parasitol. Today* **9**:218–222.
Schenkman, S., Jiang, M. S., Hart, G. W., and Nussenzweig, V., 1991, A novel cell surface trans-sialidase of *Trypanosoma cruzi* generates a stage-specific epitope required for invasion of mammalian cells, *Cell* **65**:1117–1125.
Schiavo, G., Demel, R., and Montecucco, C., 1991, On the role of polysialoglycosphingolipids as tetanus toxin receptors, *Eur. J. Biochem.* **199**:705–711.
Schlenzka, W., Shaw, L., and Schauer, R., 1993a, Catalytic properties of the CMP-N-acetylneuraminic acid hydroxylase from the starfish *Asterias rubens:* Comparison with the mammalian enzyme, *Biochim. Biophys. Acta* **1161**:131–138.
Schlenzka, W., Shaw, L., and Schauer, R., 1993b, Purification of CMP-Neu5Ac hydroxylase from pig submandibular gland and inhibition of the enzyme by rabbit anti-(hydroxylase) antiserum, *Biol. Chem. Hoppe-Seyler* **374**:955.
Schmelter, T., Ivanov, S., Wember, M., Stagier, P., Thiem, J., and Schauer, R., 1993, Partial purification and characterization of cytidine-5′-monophosphosialate synthase from rainbow trout liver, *Biol. Chem. Hoppe-Seyler* **374**:337–342.
Schneckenburger, P., Shaw, L., and Schauer, R., 1993, Purification and kinetic properties of CMP-Neu5Ac hydroxylase from mouse liver, *Biol. Chem. Hoppe-Seyler* **374**:956.
Sgroi, D., Varki, A., Braesch-Andersen, S., and Stamenkovic, I., 1993, CD22, a B-cell specific immunoglobulin superfamily member, is a sialic acid binding lectin, *J. Biol. Chem.* **268**:7011–7018.
Shaw, L., and Schauer, R., 1988, The biosynthesis of N-glycoloylneuraminic acid occurs by the hydroxylation of the CMP-glycoside of N-acetylneuraminic acid, *Biol. Chem. Hoppe-Seyler* **369**:477–486.
Shaw, L., and Schauer, R., 1989, Detection of CMP-N-acetylneuraminic acid hydroxylase activity in fractionated mouse liver, *Biochem. J.* **263**:355–363.
Shaw, L., Yousefi, S., Dennis, J. W., and Schauer, R., 1991, CMP-N-acetylneuraminic acid hydroxylase activity determines the wheat germ agglutinin-binding phenotype in two mutants of the lymphoma cell line MDAYD2, *Glycoconj. J.* **8**:434–444.
Shaw, L., Schneckenburger, P., Carlsen, J., Christiansen, K., and Schauer, R., 1992, Mouse liver cytidine-5′-monophosphate-N-acetylneuraminic acid hydroxylase: Catalytic function and regulation, *Eur. J. Biochem.* **206**:269–277.
Shaw, L., Schneckenburger, P., Schlenzka, W., Carlsen, J., Christiansen, K., Jürgensen, D., and Schauer, R., 1994, Cytidine-5′-monophosphate-N-acetylneuraminic acid hydroxylase from mouse liver and pig submandibular glands: Interaction with membrane-bound and soluble cytochrome b_5-dependent electron transport chains, *Eur. J. Biochem.* **219**:1001–1011.

Sherblom, A. P., Bharathan, S., Hall, P. J., Smagula, R. M., Moody, C. E., and Anderson, G. W., 1988, Bovine serum sialic acid: Age-related changes in type and content, *Int. J. Biochem.* **20:**1177–1183.

Shukla, A. K., and Schauer, R., 1986, Analysis of sialidase and N-acetylneuraminate pyruvate lyase substrate specificity by high-performance liquid chromatography, *Anal. Biochem.* **158:**158–164.

Shukla, A. K., Schröder, C., Nöhle, U., and Schauer, R., 1987, Natural occurrence and preparation of O-acetylated 2,3-unsaturated sialic acids, *Carbohydr. Res.* **168:**199–209.

Siegelman, M., Van de Rijn, M., and Weissman, I. L., 1989, Mouse lymph node homing receptor cDNA clone encodes a glycoprotein revealing tandem interaction domains, *Science* **243:** 1165–1172.

Sjoberg, E. R., Manzi, A. E., Khoo, K. H., Dell, A., and Varki, A., 1992, Structural and immunological characterization of O-acetylated GD2—Evidence that GD2 is an acceptor for ganglioside O-acetyltransferase in human melanoma cell, *J. Biol. Chem.* **267:**16200–16211.

Smirnova, G. P., Kochetkov, N. K., and Sadovskaya, V. L., 1987, Gangliosides of the starfish *Aphelasterias japonica,* evidence for a new linkage between two N-glycolylneuraminic acid residues through the hydroxy group of the glycolic acid residue, *Biochim. Biophys. Acta* **920:**47–55.

Song, Y., Kitajima, K., Inoue, S., and Inoue, Y., 1991, Isolation and structural elucidation of a novel type of ganglioside, deaminated neuraminic acid (KDN)-containing glycosphingolipid, from rainbow trout sperm—The first example of the natural occurrence of KDN-ganglioside, (KDN) GM3, *J. Biol. Chem.* **266:**21929–21935.

Stamenkovic, I., and Seed, B., 1990, The B-cell antigen CD22 mediates monocyte and erythrocyte adhesion, *Nature* **345:**74–77.

Stamenkovic, I., Sgroi, D., Aruffo, A., Sy, M. S., and Anderson, T., 1991, The lymphocyte-B adhesion molecule CD22 interacts with leukocyte common antigen CD45RO on T-cells and α-2,6 sialyltransferase, CD75, on B-cells, *Cell* **66:**1133–1144.

Stamenkovic, I., Sgroi, D., and Aruffo, A., 1992, CD22 binds to α-2,6-sialyltransferase-dependent epitopes on COS cells, *Cell* **68:**1003–1004.

Stickl, H., Huber, W., Faillard, H., Becker, A., Holzhauser, R., and Graeff, H., 1991, Veränderung der Acylneuraminsäuregehalte auf T-Lymphozyten und im Plasma bei Erkrankung an Mamma-Karzinom, *Klin. Wochenschr.* **69:**5–9.

Strecker, G., Wieruszeski, J. M., Michalski, J.-C., Alonso, C., Boilly, B., and Montreuil, J., 1992a, Characterization of Le^x, Le^y and a Le^y antigen determinants in KDN-containing O-linked glycan chains from *Pleurodeles waltlii* jelly coat eggs, *FEBS Lett.* **298:**39–43.

Strecker, G., Wieruszeski, J.-M., Michalski, J.-C., Alonso, C., Leroi, Y., Boilly, B., and Montreuil, J., 1992b, Primary structure of neutral and acidic oligosaccharide-alditols derived from the jelly coat of the Mexican *Axolotl.* Occurrence of oligosaccharides with fucosyl(α1-3)-fucosyl(α1-4)-3-deoxy-D-*glycero*-D-*galacto*-nonulosonic acid and galactosyl(α1-4)fucosyl-(α1-2)-galactosyl(β1-4)-N-acetylglucosamine sequences, *Eur. J. Biochem.* **207:**995–1002.

Strecker, G., Wieruszeski, J.-M., Plancke, Y., and Michalski, J.-C., 1993, Comparative study of the O-linked carbohydrate chains released from the jelly coat of amphibian eggs, *Glycoconj. J.* **10:**345.

Strittmatter, P., Spatz, L., Corcoran, D., Rogers, M. J., Setlow, B., and Redline, R., 1974, Purification and properties of rat liver microsomal stearoyl-CoA desaturase, *Proc. Natl. Acad. Sci. USA* **71:**4565–4569.

Suguri, T., Kelm, S., Schauer, R., and Reuter, G., 1993, Detection of N-acetyl-9-O-acetylneuraminic acid on human lymphocytes, *Glycoconj. J.* **10:**331.

Suzuki, M., Suzuki, A., Yamakawa, T., and Matsunaga, E., 1985, Characterization of 2,7-anhydro-N-acetylneuraminic acid in human wet cerumen, *J. Biochem.* **97:**509–515.

Svensson, L., 1992, Group-C rotavirus requires sialic acid for erythrocyte and cell receptor binding, *J. Virol.* **66:**5582–5585.

Tanaka, H., Ito, F., and Iawasaki, T., 1992, Purification and characterization of a sialidase from *Bacteroides fragilis* SBT3182, *Biochem. Biophys. Res. Commun.* **189:**524–529.

Terada, T., Kitazume, S., Kitajima, K., Inoue, S., and Ito, F., 1993, Synthesis of CMP-deaminoneuraminic acid (CMP-KDN) using the CMP-3-deoxynonulosonat cytidylyltransferase from rainbow trout testis—Identification and characterization of a CMP-KDN synthetase, *J. Biol. Chem.* **268:**2640–2648.

Teufel, M., Roggentin, P., and Schauer, R., 1989, Properties of sialidase isolated from *Actinomyces viscosus* DSM 43798, *Biol. Chem. Hoppe-Seyler* **370:**435–443.

Thurin, J., Herlyn, M., Hindsgaul, O., Strömberg, N., Karlsson, K.-A., Elder, D., Steplewski, Z., and Koprowski, H., 1985, Proton NMR and fast-atom bombardment mass spectrometry analysis of the melanoma-associated ganglioside 9-O-acetyl GD3, *J. Biol. Chem.* **260:**14556–14563.

Tiemeyer, M., Yasuda, Y., and Schnaar, R. L., 1989, Ganglioside-specific binding protein on rat brain membranes, *J. Biol. Chem.* **264:**1671–1681.

Tiemeyer, M., Swank-Hill, P., and Schnaar, R. L., 1990, A membrane receptor for gangliosides is associated with central nervous system myelin, *J. Biol. Chem.* **265:**11990–11999.

Tiemeyer, M., Swiedler, S. J., Ishihara, M., Moreland, M., Schweingruber, H., Hirtzer, P., and Brandley, B. K., 1991, Carbohydrate ligands for endothelial leukocyte adhesion molecule-1, *Proc. Natl. Acad. Sci. USA* **88:**1138–1142.

Titani, K., Takio, K., Kuwada, M., Nitta, K., Sakakibara, F., Kawauchi, H., Takayanagi, G., and Hakomori, S.-i., 1987, Amino acid sequence of sialic acid binding lectin from frog (*Rana catesbeiana*) eggs, *Biochemistry* **26:**2189–2194.

Torres, R. M., Law, C. L., Santos-Argumedo, L., Kirkham, P. A., Grabstein, K., Parkhouse, R.M.E., and Clark, E. A., 1992, Identification and characterization of the murine homologue of CD22, a lymphocyte-B-restricted adhesion molecule, *J. Immunol.* **149:**2641–2649.

Townsend, R. R., Hardy, M. R., Hindsgaul, O., and Lee, Y. C., 1988, High performance anion-exchange chromatography of oligosaccharides using pellicular resins and pulsed amperometric detection, *Anal. Biochem.* **174:**459–470.

Traving, C., Schauer, R., and Roggentin, P., 1993, The primary structure of the "large" sialidase isoenzyme of *Clostridium perfringens* A99 and its comparison with further sialidases, *Glycoconj. J.* **10:**238–239.

Troy, F. A., 1992, Polysialylation from bacteria to brains, *Glycobiology* **2:**5–23.

Tyrrell, D., James, P., Rao, N., Foxall, C., Abbas, S., Dasgupta, F., Nashed, M., Hasegawa, A., Kiso, M., Asa, D., Kidd, J., and Brandley, B. K., 1991, Structural requirements for the carbohydrate ligand of E-selectin, *Proc. Natl. Acad. Sci. USA* **88:**10372–10376.

Uchida, Y., Taikada, Y., and Sugimori, T., 1979, Enzymic properties of a neuraminidase from *Arthrobacter ureafaciens, J. Biochem.* **86:**1573–1585.

Unland, F., and Müthing, J., 1993, Separation of isomeric gangliosides by anion exchange HPLC, *Biol. Chem. Hoppe-Seyler* **374:**961.

Vamecq, J., Mestdagh, N., Henichart, J.-P., and Poupaert, J., 1992, Subcellular distribution of glycosyltransferases in rodent liver and their significance in special reference to the synthesis of N-glycolylneuraminic acid, *J. Biochem.* **111:**579–583.

Van den Berg, T. K., Breve, J.J.P., Damoiseaux, J.G.M.C., Dopp, E. A., Kelm, S., Crocker, P. R., Dijkstra, C. D., and Kraal, G., 1992, Sialoadhesin on macrophages—Its identification as a lymphocyte adhesion molecule, *J. Exp. Med.* **176:**647–655.

van den Eijnden, D. H., and Joziasse, D. H., 1993, Enzymes associated with glycosylation, *Curr. Opin. Struct. Biol.* **3:**711–721.

Vann, W. F., Zapata, G., Roberts, I., Boulnois, G., and Silver, R. P., 1993, Structure and function

of enzymes in sialic acid metabolism in polysialic producing bacteria, in: *Polysialic Acid* (J. Roth, U. Rutishauser, and F. A. Troy II, eds.), Birkhäuser Verlag, Basel, pp. 125–136.
Varki, A., 1992a, Diversity in the sialic acids, *Glycobiology* **2**:24–40.
Varki, A., 1992b, Selectins and other mammalian sialic acid-binding lectins, *Curr. Opin. Cell Biol.* **4**:257–266.
Varki, A., 1993, Biological roles of oligosaccharides—All of the theories are correct, *Glycobiology* **3**:97–130.
Varki, A., and Diaz, S., 1984, The release and purification of sialic acids from glycoconjugates: Methods to minimize the loss and migration of O-acetyl groups, *Anal. Biochem.* **137**:236–247.
Varki, A., and Higa, H. H., 1993, Studies of the O-acetylation and (in)stability of polysialic acid, in: *Polysialic Acid* (J. Roth, U. Rutishauser, and F. A. Troy II, eds.), Birkhäuser Verlag, Basel, pp. 165–170.
Varki, A., and Kornfeld, S., 1980, An autosomal dominant gene regulates the extent of 9-O-acetylation of murine erythrocyte sialic acids: A probable explanation for the variation in capacity to activate the alternate complement pathway, *J. Exp. Med.* **152**:532–544.
Varki, A., Hooshmand, F., Diaz, S., Varki, N. M., and Nedrick, S. M., 1991, Developmental abnormalities in transgenic mice expressing a sialic acid-specific 9-O-acetylesterase, *Cell* **65**:65–74.
Vlasak, R., Krystal, M., Nacht, M., and Palese, P., 1987, The influenza C virus glycoprotein (HE) exhibits receptor-binding (hemagglutinin) and receptor-destroying (esterase) activities, *Virology* **160**:419–425.
Vlasak, R., Luytjes, W., Spaan, W., and Palese, P., 1988, Human and bovine coronaviruses recognize sialic acid-containing receptors similar to those of influenza C viruses, *Proc. Natl. Acad. Sci. USA* **85**:4526–4529.
Vliegenthart, J.F.G., Dorland, L., van Halbeek, H., and Haverkamp, J., 1982, NMR spectroscopy of sialic acids, in: *Sialic Acids—Chemistry, Metabolism and Function* (R. Schauer, ed.), Springer, New York, pp. 127–172.
Vliegenthart, J.F.G., Dorland, L., and van Halbeek, H., 1983, High-resolution, ^{1}H-nuclear magnetic resonance spectroscopy as a tool in the structural analysis of carbohydrates related to glycoproteins, *Adv. Carbohydr. Chem. Biochem.* **41**:209–374.
von Itzstein, M., Wu, W.-Y., Kok, G. B., Pegg, M. S., Dyason, J. C., Jin, B., Van Phan, T., Smythe, M. L., White, H. F., Oliver, S. W., Colman, P. M., Varghese, J. N., Tyan, D. M., Woods, J. M., Bethell, R. C., Hotham, V. J., Cameron, J. M., and Penn, C. R., 1993, Rational design of potent sialidase-based inhibitors of influenza virus replication, *Nature* 363:418–423.
Walz, G., Aruffo, A., Kolanus, W., Bevilacqua, M., and Seed, B., 1990, Recognition by ELAM-1 of the sialyl-Le^x determinant on myeloid and tumor cells, *Science* **250**:1132–1135.
Wang, X. C., Vertino, A., Eddy, R. L., Byers, M. G., Janisait, S. N., Shows, T. B., and Lau, J.T.Y., 1993, Chromosome mapping and organization of the human beta-galactoside alpha 2,6-sialyltransferase gene—Differential and cell-type specific usage of upstream exon sequences in B-lymphoblastoid cells, *J. Biol. Chem.* **268**:4355–4361.
Warren, L., 1963, Distribution of sialic acids in nature, *Comp. Biochem. Physiol.* **10**:153–171.
Warren, L., 1964, N-Glycolyl-8-O-methylneuraminic acid: A new form of sialic acid in the starfish *Asterias forbesi*, *Biochim. Biophys. Acta* **83**:129–132.
Watson, S. R., Fennie, C., and Lasky, L. A., 1991, Neutrophil influx into an inflammatory site inhibited by a soluble homing receptor–IgG chimaera, *Nature* **349**:164–166.
Weinstein, J., Lee, E. U., McEntee, K., Lai, P.-H., and Paulson, J. C., 1987, Primary structure of β-galactoside α2,6-sialyltransferase, *J. Biol. Chem.* **262**:17735–17743.
Wen, D. X., Livingston, B. D., Medzihradszky, K. F., Kelm, S., Burlingname, A. L., and Paulson, J. C., 1992, Primary structure of Galβ1,3(4)GlcNAc α2,3-sialyltransferase determined by mass spectrometry sequence analysis and molecular cloning, *J. Biol. Chem.* **267**:21011–21019.

Wilson, G. L., Fox, C. H., Fauci, A. S., and Kehrl, J. H., 1991, cDNA cloning of the B cell membrane protein CD22: A mediator of B–B cell interactions, *J. Exp. Med.* **173:**137–146.

Wilson, G. L., Najfeld, V., Kozlow, E., Menniger, J., Ward, D., and Kehrl, J. H., 1993, Genomic structure and chromosomal mapping of the human CD22 gene, *J. Immunol.* **150:**5013–5024.

Yeung, M. K., 1993, Complete nucleotide sequence of the *Actinomyces viscosus* T14V sialidase gene: Presence of a conserved repeating sequence among strains of *Actinomyces* ssp., *Infect. Immun.* **61:**109–116.

Yuen, C. T., Lawson, A. M., Chai, W. G., Larkin, M., Stoll, M. S., Stuart, A. C., Sullivan, F. X., Ahern, T. J., and Feizi, T., 1992, Novel sulfated ligands for the cell adhesion molecule E-selectin revealed by the neoglycolipid technology among O-linked oligosaccharides on an ovarian cystadenoma glycoprotein, *Biochemistry* **31:**9126–9131.

Zeng, F. Y., and Gabius, H. J., 1991, Carbohydrate-binding specificity of calcyclin and its expression in human tissues and leukemic cells, *Arch. Biochem. Biophys.* **289:**137–144.

Zeng, F. Y., and Gabius, H. J., 1992a, Mammalian fetuin-binding proteins sarcolectin, aprotinin and calcyclin display differences in their apparent carbohydrate specificity, *Biochem. Int.* **26:**17–24.

Zeng, F. Y., and Gabius, H. J., 1992b, Sialic acid-binding proteins—Characterization, biological function and application, *Z. Naturforsch.* **47c:**641–653.

Zenz, K. I., Roggentin, P., and Schauer, R., 1993, Isolation and comparison of the natural and the recombinant sialidase from *Clostridium septicum* NC0054714, *Glycoconj. J.* **10:**50–56.

Zhou, Q., Moore, K. L., Smith, D. F., Varki, A., McEver, R. P., and Cummings, R. D., 1991, The selectin GMP-140 binds to sialylated, fucosylated lactosaminoglycans on both myeloid and nonmyeloid cells, *J. Cell Biol.* **115:**557–564.

Zimmer, G., Reuter, G., and Schauer, R., 1992, Use of influenza C virus for detection of 9-O-acetylated sialic acids on immobilized glycoconjugates by esterase activity, *Eur. J. Biochem.* **204:**209–215.

Zimmer, G., Suguri, T., Reuter, G., Yu, R. K., and Schauer, R., 1994, Modification of sialic acids by 9-0-acetylation is detected in human leucocytes using the lectin property of influenza C virus, *Glycobiology* **4:**343–349.

Zingales, B., Carniol, C., de Lederkremer, R., and Colli, W., 1987, Direct sialic acid transfer from a protein donor to glycolipids of trypomastigote forms of *Trypanosoma cruzi*, *Mol. Biochem. Parasitol.* **26:**135–144.

Chapter 3

Biological Specificity of Sialyltransferases

Subhash Basu, Manju Basu, and Shib Sankar Basu

1. INTRODUCTION

Sialyltransferases (abbreviated as SATs or STs) are a family of glycosyltransferases involved in the biosynthesis of complex sialoglycoproteins (SGPs) and sialoglycosphingolipids (SGSLs). A separate sialyltransferase (SAT or ST) catalyzes the transfer of sialic acid to a specific oligosaccharide with concomitant formation of a defined anomeric as well as positional linkage (Schachter and Roseman, 1980; Sadler *et al.*, 1982; S. Basu and Basu, 1982). If one accepts the "one linkage, one enzyme" hypothesis (Schachter and Roseman, 1980), then based on the various different sialyl linkages present in cell surface oligosaccharides, genes for many different SATs (M. Basu *et al.*, 1987; S. C. Basu, 1991; Paulson and Colley, 1989) occur in animal cells. The substrate specificity profiles serve as the basis for a general classification of Golgi-bound animal SATs into three major classes: glycoprotein:sialyltransferases, glycolipid:sialyltransferases, and mucin:sialyltransferases. Comparison of their primary structures (as defined by amino acid sequence derived from the cDNAs), as well as three-dimensional structures, will be an important approach for further studies.

In eukaryotic cells, sialic acids (Neu5Ac or Neu5Gc, see Chapter 2) occur essentially as terminal sugars in α2,3 and α2,6 linkages onto galactose residues of *N*-linked oligosaccharides which are attached to proteins (Montreuil, 1975) or to sphingolipids (Wiegandt, 1985; Hakomori, 1984). The α2,6 linkage onto N-acetylgalactosamine as in *O*-linked glycoproteins is ubiquitous (Schachter and

Subhash Basu, Manju Basu, and Shib Sankar Basu Department of Chemistry and Biochemistry, University of Notre Dame, Notre Dame, Indiana 46556.

Biology of the Sialic Acids, edited by Abraham Rosenberg. Plenum Press, New York, 1995.

Roseman, 1980; Olden *et al.*, 1982). Recently, the α2,6 linkage of sialic acid to internal *N*-acetylgalactosamine has also been detected in sialoglycolipids, e.g., GDlα ganglioside (Taki *et al.*, 1986), GT1α ganglioside (K. Nakamura, *et al.*, 1988), GQ1α ganglioside (Ohashi, 1981), and globo-series ganglioside (Levery *et al.*, 1994; see Chapter 6). The α2,8-linked sialic acids occur in polysialyl structures such as N-CAM (Finne *et al.*, 1983; see Chapter 4) and gangliosides (Hakomori, 1981; Wiegandt, 1985). A disialosyl structure in Neu5Acα2, 9Neu5Ac has also been characterized in a glycoprotein isolated from embryonic carcinoma cells (Fukuda *et al.*, 1985).

The recent renewal of interest in SATs is due in part to the critical, and pharmacologically applicable, relevance of sialylated glycoproteins and glycolipids on normal and cancer cell surfaces. Moreover, some SATs govern sialylation of cell surface SGPs and SGSLs of lymphocytes and macrophages and therefore SATs constitute important target enzymes to be regulated during inflammation or tumor metastasis (see Chapter 2). Major advances have occurred in our understanding of the structures and function of some of the mammalian sialoglycoconjugates of the neolacto family in recent years. A family of sialylated carbohydrate-binding protein lectins, selectins, is operative is lymphocyte homing (Bradley *et al.*, 1990; Moore *et al.*, 1991; Varki, 1993) and inflammatory responses (McEver, 1991; Lasky, 1991). The three well-characterized selectins are L-selectin (expressed constitutively on leukocytes), E-selection (expressed by cytokine-activated endothelial cells) and P-selectin (expressed by thrombin-activated platelets and endothelial cells). Both E- and P-selectins bind to a common epitope known as the sialyl-Lewis × (SA-Le^x) antigen [NeuAcα2-3Galβ1-4(Fucα1--3)GlcNAcβ1-R] (Phillips *et al.*, 1990: Walz *et al.*, 1990; Bradley *et al.*, 1990; Lasky, 1991). The structure of a SA-Le^x glycolipid was established by Hakomori and his co-workers (Fukushima *et al.*, 1984) (see Chapter 7).

SA-Le^x and SA-diLe^x type oligosaccharides attached to proteins or lipids (ceramides) may play a major role in cell–cell and cell–molecule recognition events (Bradley *et al.*, 1990). Recently, it has been reported that SA-Le^a which is expressed on colon carcinoma cell surfaces (Magnani, 1991; Rice and Belivacqua, 1989) also facilitates attachment to E-selectin and may contribute to early adhesion events in tumor metastasis. Although this excellent example elucidates the pivotal importance of carbohydrate epitopes in physiological functions, we do not as yet have a clear-cut understanding of these epitopes in the etiology of cancers. Based on the observations in inflammatory and lymphocyte-homing responses, we wonder whether the regulated expression of a particular SAT (SAT-3; CMP-NeuAc:nLcOse4Cer α2-3 sialyltransferase; M. Basu *et al.*, 1982; S. Basu *et al.*, 1993a) is required for sialylation of the neolacto- and polylactosamine backbone-containing GSLs on lymphoreticular cell surfaces. The activity of this enzyme might specifically determine or facilitate their metastasizing potential. The reader may refer profitably to several review articles (Schauer, 1982; M. Basu *et al.*, 1987; Broquet *et al.*, 1991; Varki, 1993). This chapter presents some recent pivotal information on SAT activities that catalyze biosynthesis of sialooligosaccharides, sialoglycoproteins, and sialoglycolipids in eukaryotic cells.

2. SIALYLTRANSFERASES CATALYZING SIALOOLIGOSACCHARIDES AND SIALOGLYCOPROTEIN BIOSYNTHESIS

At least six SATs (Table I) have been characterized as having importance in the biosynthesis of the major sialic acid linkages found in mammalian oligosaccharides and glycoproteins.

Table I
Sialyltransferases Utilize Oligosaccharides and Glycoproteins as Substrates

Substrate	Linkage of sialic acid to terminal sugar	Nomenclature	References
Galβ1-4Glc (lactose)	Galβ1-Glc $\alpha|^{3}_{2}$ NeuAc	?	Jourdian *et al.* (1963), Kaufman and Basu (1966), Carlson *et al.* (1973a), S. Basu (1966)
Galβ1-4GlcNAc { — or NGP	Galβ1-4GlcNAc { — / NGP $\alpha|^{6}_{2}$ NeuAc	ST6N	Roseman *et al.* (1966), Bartholomew *et al.* (1973), Paulson *et al.* (1977a,b), Weinstein *et al.* (1982), Datta and Paulson (1995)
Galβ1-3GlcNAc { — or NGP	Galβ1-3GlcNAc- $\alpha|^{3}_{2}$ NeuAc	ST3N	Weinstein *et al.* (1982), Wen *et al.* (1992), Sasaki *et al.* (1993)
Galβ1-4GlcNAc { — or NGP	Galβ1-4GlcNAc- $\alpha|^{3}_{2}$ NeuAc	ST3N	S. Basu (1966), Dijak *et al.* (1979), Kitagawa and Paulson (1993a,b), Sasaki *et al.* (1993)
Galβ1-3GalNAc { — or OGP	Galβ1-3GalNAc- $\alpha|^{3}_{2}$ NeuAc	ST3O	Carlson *et al.* (1973b), Rearick *et al.* (1979), Joziasse *et al.* (1983), Gillespie *et al.* (1992)
Galβ1-4GlcNAc { — / NGP $\alpha|^{3}_{2}$ NeuAc	NeuAcα2-3Gal- $\alpha|^{8}_{2}$ NeuAc	ST-Pol	Roseman *et al.* (1966), Kundig *et al.* (1971), Troy and McCloskey (1979) Weisgerber *et al.* (1991)
GalNAc-*O*- { — or OGP	3GalNAc- $\alpha|^{6}_{2}$ NeuAc	ST6O	Carlson *et al.* (1973b), Sadler *et al.* (1979), Kurosawa *et al.* (1994a)

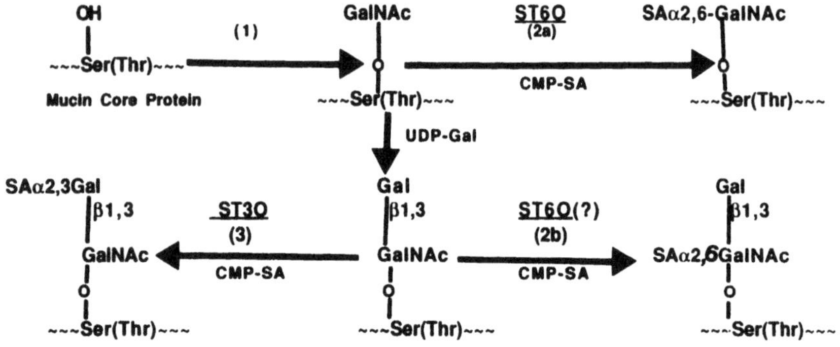

FIGURE 1. Mucin sialyltransferases.

The SAT that catalyzes the synthesis of sialyllactose was first detected in rat mammary tissue by Roseman and his co-workers (Jourdian *et al.*, 1963; Carlson *et al.*, 1973a). In fact, this was the first SAT to be characterized in an animal system, being detected later in embryonic chicken brain (Kaufman and Basu, 1966). However, the SAT activities that catalyze the transfer of sialic acid to lactose and to lactosylceramide appear to be different from the latter by all kinetic parameters (S. Basu, 1966; Kaufman and Basu, 1966; M. Basu *et al.*, 1987; S. C. Basu, 1991). In addition to CMP-NeuAc:lactose α2-3 sialyltransferase, CMP-NeuAc:lactosamine α2-6 sialyltransferase (ST6N) was first characterized by Roseman and his co-workers (Bartholomew *et al.*, 1973) from colostrum. The ST6N enzyme activity is ubiquitous and was purified by Hill and his associates

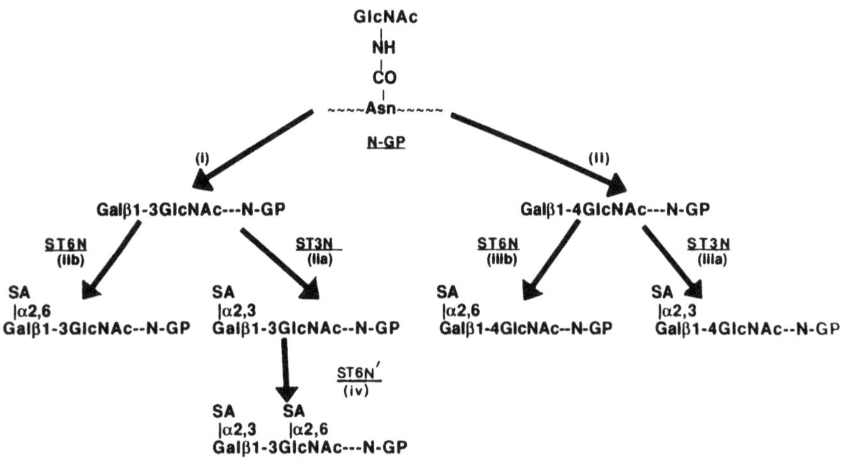

FIGURE 2. *N*-Linked oligosaccharide-bound glycoprotein sialyltransferases. N-GP, glycoproteins with *N*-linked oligosaccharide.

(Paulson *et al.*, 1977a,b) from bovine colostrum, using a novel affinity column matrix, CDP-hexanolamine Sepharose 4B. This affinity column matrix enabled researchers in this field to purify more than one oligosaccharide or glycoprotein SAT to homogeneity (Weinstein *et al.*, 1982; Dijak *et al.*, 1979; Wen *et al.*, 1992a,b; Gillespie *et al.*, 1992). As we will discuss later, the putative amino acid sequence derived from cDNA sequences has also been determined for various SATs, which have proved to be independent gene products.

Using membrane-bound SAT activity from rat liver, Paulson and his co-workers provided evidence for the presence of three different SATs (ST3O, Figure 1; ST3N and ST6N, Figure 2). Using rat liver Golgi-containing SAT (ST6N′, Figure 2, step iv), biosynthesis of a disialylated sequence in *N*-linked oligosaccharides from sialyllactosamine was established (Paulson *et al.*, 1984). Recent efforts of Paulson and his co-workers have resulted in the elucidation of cDNA sequences for these three SATs which will be described in detail in a following section.

3. SIALYLTRANSFERASES CATALYZING SIALOGLYCOLIPID BIOSYNTHESIS

In addition to SGPs (see Chapter 5) the eukaryotic cell surfaces contain SGSLs of different structures (see Chapter 6). All glycosphingolipids known thus far have a common lipophilic structural feature. The lipophilic ceramide moiety containing a sphingosine base [(2*S*:3*R*)-2-amino-*trans*-4-octadecene-1,3-diol] with a fatty acid linked as an amide to the 2-amino group of sphingosine, is widely distributed in the animal world. Sugar units are attached to the C-1 OH group of ceramide to give rise to different families of glycosphingolipids of various biological functions. Based on the first three sugars attached to the ceramide moiety, glycosphingolipids are classified into four distinct families:

1. Ganglio (Gg; GalNAcβ1-4Galβ1-4Glc-ceramide)
2. Lacto (Lc; GlcNAcβ1-3Galβ1-4Glc-ceramide)
3. Globo (Gb; Galα1-4Galβ1-4Glc-ceramide)
4. Muco (Mc; Galβ1-4Galβ1-4Glc-ceramide)

3.1. Ganglio- and Globo-Family Glycolipids

Gangliosides are especially characterized by the presence of one or more sialic acid residues, i.e,. an *N*-acetylated or *N*-,*O*-acetylated or else an *N*-glycoloylated neuraminic acid bound to a sugar residue (galactose) by α-ketosidic linkages (see Chapter 6). Gangliosides from human sources contain mainly *N*-acetylneuraminic acid (Neu5Ac), while *N*-glycoloylneuraminic acid (Neu5Gc) is the predominant sialic acid in gangliosides of other primates as well as horses

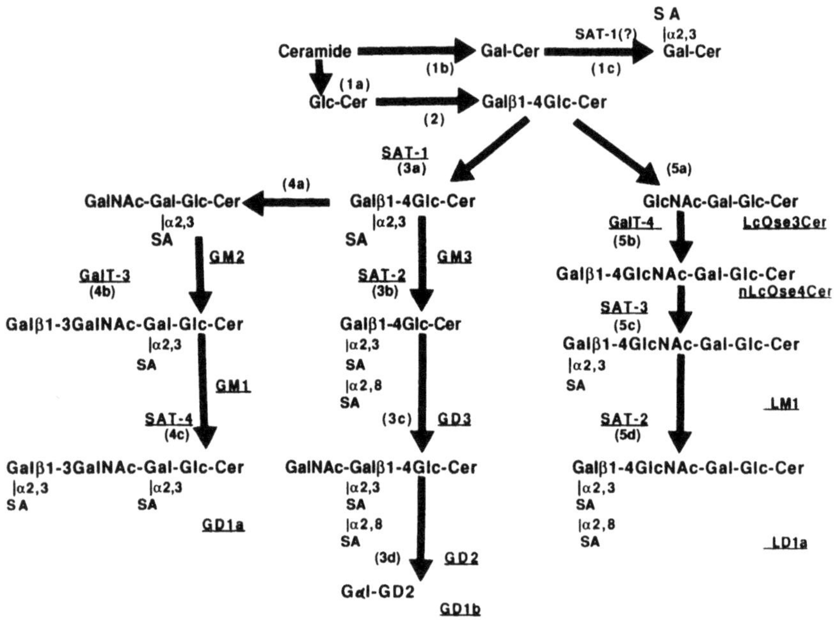

FIGURE 3. Glycosphingolipid sialyltransferases.

and cattle (Wiegandt, 1985). The biosynthesis of SGSLs (Figure 3) (S. C. Basu, 1991) depends on the coordinated addition of monosaccharide units to mono- or oligo-glycosylceramide moieties. Most of the SATs involved in ganglioside biosynthesis were detected originally in a Golgi-rich membrane fraction isolated from embryonic chicken brain (ECB) (S. Basu, 1966; Keenan *et al.*, 1974). The reaction specificity of each glycolipid SAT (Table II) was characterized by the

Table II
Sialytransferases Utilize Glycolipids as Substrates

Substrate	Linkage of sialic acid to terminal sugar	Nomenclature	References
Gal-Cer	NeuAcα2-3GalCer (G7)	SAT-1 (?)	Yu and Lee (1976)
Galβ1-4Glc-Cer (Lc2)	Galβ1-4Glc-Cer α|$^3_2$ NeuAc	SAT-1	S. Basu (1966), Kaufman and Basu (1966), M. Basu *et al.* (1987), Melkerson-Watson and Sweeley (1991)

(*continued*)

Table II
(*Continued*)

Substrate	Linkage of sialic acid to terminal sugar	Nomenclature	References
Galβ1-4GlcNAc-Lc2 (nLc4)	Galβ1-4GlcNAc-Lc2 $\alpha\|^{3}_{2}$ NeuAc (LM1)	SAT-3	M. Basu *et al.* (1982, 1987), S. Basu *et al.* (1993a), Kitagawa and Paulson (1994a)
Galβ1-3GalNAc-Lc2 (Gg4)	Galβ1-3GalNAc-Lc2 $\alpha\|^{3}_{2}$ NeuAc (GM1b)	SAT-4	S. Basu (1966), Kaufman and Basu (1966), Stoffyn and Stoffyn (1980)
Galβ1-3GalNAc-Lc2 (Lc4)	Galβ1-3GlcNAc-Lc2 $\alpha\|^{3}_{2}$ NeuAc (isoLM1)	SAT-3 (?)	Liepkans *et al.* (1988)
Galβ1-4GlcNAc-Lc2 | NeuGc (LM1)	Galβ1-3GlcNAc-Lc2 $\alpha\|^{3}_{2}$ NeuAc $\alpha\|^{8}_{2}$ NeuAc (LD1a)	SAT-2	Higashi *et al.* (1985)
Galβ1-4Glc-CerLc2 $\alpha\|^{3}_{2}$ NeuAc (GM3)	NeuGc$\frac{\alpha}{2\ 3}$Galβ1-4Glc-Cer $\alpha\|^{8}_{2}$ NeuAc (GD3)	SAT-2	Kaufman *et al.* (1968), M. Basu *et al.* (1987), Higashi *et al.* (1985), Sasaki *et al.* (1994), Nara *et al.* (1994), Haraguchi *et al.* (1994)
Galβ1-3GalNAc-GM3 (GM1)	Gal-GalNAc-Gal-Glc-Cer $\alpha\|^{3}_{2}$ $\alpha\|^{3}_{2}$ NeuAc NeuAc	SAT-4	S. Basu (1966), S. Basu and Basu (1982), M. Basu *et al.* (1987), Gu *et al.* (1990a)
Galβ1-4Glc-Cer α|2,3 NeuAc α|2,8 NeuAc (GD3)	Galβ1-4Glc-Cer $\alpha\|^{3}_{2}$ NeuAc $\alpha\|^{8}_{2}$ NeuAc$\frac{\alpha}{8\ 2}$NeuAc (GT3)	STV or SAT-POL	Iber *et al.* (1992), Van Echten and Sandhoff (1993), Cho and Troy (1989)

enzyme's stringency with respect to both the donor and the acceptor structure. The nucleotide sugar donor specificity of a glycosyltransferase appears to be absolute, while specificities concerning the sugar acceptors vary for the same gene product. Membrane-bound or detergent-solubilized SATs from ECB or tumor cells exhibited a broad acceptor specificity, transferring sialic acid at comparable rates to different glycolipids (Table II), glycoproteins, or the free oligosaccharides.

Almost three decades ago, the activity of SAT-1 [CMP-NeuAc: LcOse2Cer α2-3 sialyltransferase catalyzing GM3 biosynthesis (Figure 3, step 3a)] was first characterized in embryonic chicken brain (S. Basu and Kaufman, 1965; S. Basu, 1966; Kaufman and Basu, 1966). The same activity was suggested to catalyze the synthesis of NeuAcα2-3Gal-Cer (G-4 ganglioside) in rat brain (Yu and Lee, 1976). In calf brain cortex SAT-1 was localized in synaptic membranes (Preti *et al.*, 1980). Recently, SAT-1 activity has been purified (40,000-fold) from rat liver, with an apparent molecular weight of about 60,000 (Melkerson-Watson and Sweeley, 1991). Based on *N*-glycanase treatment followed by Western blot analysis, SAT-1 appears to be a glycoprotein. Ceramide may be an essential moiety of the lactosylceramide substrate for the SAT-1 activity (S. Basu, 1966; Kaufman and Basu, 1966; M. Basu *et al.*, 1987). However, shorter fatty acid (C_2–C_8)-containing ceramides appear to have higher K_m values than lactosylceramide containing C_{16} to C_{18}, natural, fatty acids (Li *et al.*, 1994).

A second SAT, SAT-2 (CMP-NeuAc: GM3 or LM1 α2-8 sialyltransferase), was also characterized in ECB [Kaufman *et al.*, 1968 (Figure 3, step 3b); Higashi *et al.*, 1985 (Figure 3, step 5d)] and in rat liver Golgi-rich membrane fraction (Keenan *et al.*, 1974). Purification of SAT-2 (GD3 synthase) to homogeneity has also been reported from rat brain (T. Gu *et al.*, 1990; X.-B. Gu *et al.*, 1990). Unlike SAT-1, SAT-2 does not recognize the ceramide moiety (S. C. Basu, 1991) and it catalyzes transfer of sialic acid to the terminal sialic acid of an oligosaccharide containing lactose or lactosamine (Higashi *et al.*, 1985). A similar activity is perhaps responsible for the synthesis of polysialoglycoproteins in developing brain. These polysialic acid-containing glycoproteins (see Chapter 4) may be involved in neuronal cell adhesion and the pathogenesis of meningitis. In developing brain tissue, a significant portion of all protein-bound sialic acid occurs in the form of polysialic acid (Finne *et al.*, 1983; Finne and Makela, 1985; Margolis and Margolis, 1983). There are as many as 12, and possibly even more sialic acid residues in the longest chains. The sialic acid is bound to a core glycan containing *N*-linked tri- and tetra-antennary-type glycose units. Polysialic acid chains of glycoproteins are similar to gangliosides which contain only di- and trisialosyl units. Neither the functional significance nor the exact biosynthetic mechanism of these polysialo-molecules (GPs and GSLs) is known. A few review articles on polysialic acid-bound glycoconjugates have been published (Rougon, 1993; Rutishauser, 1989). Fish eggs also contain polymers of sialic acid but of another type and bound to a different proteoglycan (Inoue and Matsumura, 1979). Their relationship to the brain polysialic acid is unknown and nothing is known about polysialylation during the early phases of embryonic development. A polysialoglycoprotein antigen (N-CAM) which is implicated in brain cell adhesion has recently been characterized in several animal species (Edelman, 1983; Lemmon *et al.*, 1982). Colominic acid, a mixture of α2-8-linked sialyloligosaccharides found in the culture filtrate of *E. coli* K1, GD3 ganglioside, and an endogenous 20-kDa protein function as acceptors for a polysialyltransferase (SAT-Pol; 20 kDa) isolated from *E. coli* K1 (Cho and Troy, 1989; Troy and McCloskey, 1979). The reader is referred to Chapter 4 for an authoritative treatment of this subject.

3.2. Lacto-Family Glycolipids

Both *N*-acetyl- (Neu5Ac) and *N*-glycoloyl- (Neu5Gc) neuraminic acid attached to the terminal galactose moiety of neolactotetraosylceramide (nLcOse4Cer; Galβ1-4GlcNAcβ1-3Galβ1-4Glc-Cer) and lactotetraosylceramide (LcOse4Cer; Galβ1-3GlcNAcβ1-3Galβ1-4Glc-Cer) are widely distributed on the surface of lymphoid (Weigandt, 1985; Chien *et al.*, 1978; Hakomori, 1981), embryonic (Uemura *et al.*, 1989; Chien and Hogan, 1983), and tumor cells (Hakomori, 1984; Magnani *et al.*, 1982; Nilsson *et al.*, 1985). Further terminal fucosylation of these sialosyl-neolactotetraosylceramide and sialosyl-lactotetraosylceramide

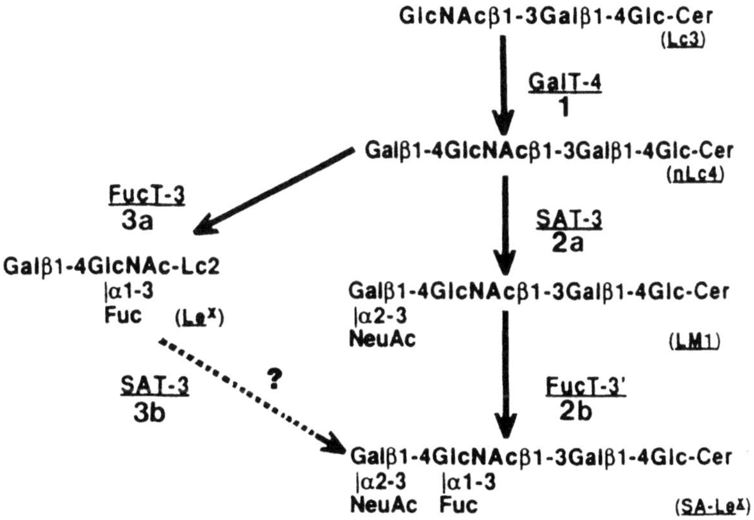

FIGURE 4. Proposed biosynthetic steps for sialyl-Le^x glycolipid.

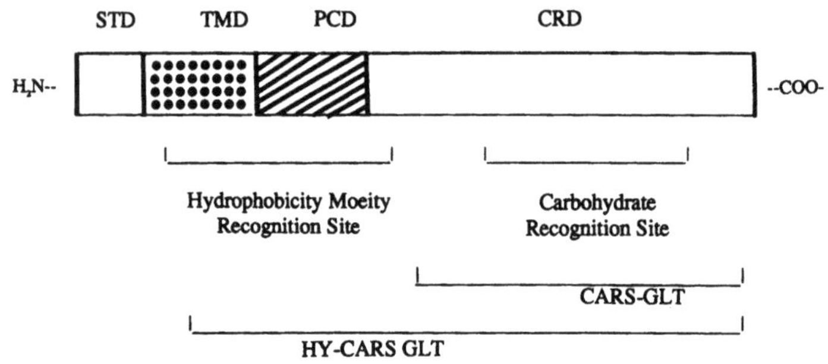

FIGURE 5. A proposed model for CARS and HY-CARS glycosyltransferases (GLTs). STD, short terminal domain; TMD, transmembrane domain; PCD, protease cleavable domain, CRD, carbohydrate moiety recognition domain.

gives rise to SA-Le^x (Figure 4, step 2b) (M. Basu *et al.*, 1991, 1993) and SA-Le^a (Liepkans *et al.*, 1988), respectively.

A novel SAT, SAT-3 (CMP-NeuAc: nLcOse4Cer α2-3 sialyltransferase), was characterized in bovine spleen (Chien *et al.*, 1974), embryonic chicken brain (M. Basu *et al.*, 1982), and human colon carcinoma Colo-205 cells (M. Basu *et al.*, 1988). Recently this SAT-3 activity has been cloned from ECB and Colo-205 (S. Basu *et al.*, 1993a). The purified SAT-3 from embryonic chicken brain as well as expressed SAT-3 activity from human placenta (Kitagawa and Paulson, 1994b) catalyze the transfer of sialic acid to the terminal galactose of the lactosamine moiety attached either to lactosylceramide (Fig. 3, step 5b) or to the mannosyl residues of *N*-linked glycoproteins (Fig. 2, steps iia and iiia). Glycosyltransferases are classified as CARS (containing only carbohydrate recognition sites) and HY-CARS (hydrophobic and carbohydrate recognition sites) (S. Basu *et al.*, 1990; S. C. Basu, 1991). It appears, perhaps, that the expressed SAT-3 (STZ-2) belongs to the CARS family. Success in cloning of SAT-3 activity from human Colo-205 (S. Basu *et al.*, 1993a) and from human placenta (Kitagawa and Paulson, 1994a,b) will provide an opportunity to study the structure–function relationship of this enzyme at the molecular level. Perhaps the CARS and HY-CARS nature of SAT-3 activity will be established conclusively from site-specific mutagenesis studies of its different domains (Figure 5). Biosynthesis of LD1a (NeuAcα2-8NeuGcα2-3Galβ1-4GlcNAc-Gal-Glc-Cer) was achieved by a membrane-bound and detergent-solubilized SAT-2 isolated from embryonic chicken brain (Higashi *et al.*, 1985) (Figure 3, step 5d). From a substrate competition study (Higashi *et al.*, 1985) it appears that both GM3-to-GD3 (Figure 3, step 3c) and LM1-to-LD1a (Figure 3, step 5d) reactions are, perhaps, catalyzed by the same SAT-2 gene product. The cloning of SAT-2 and further studies with the expressed protein would unambiguously establish the accuracy of this conclusion.

4. SIALYLTRANSFERASES CATALYZING MUCIN BIOSYNTHESIS

Animal mucins contain *O*-linked oligosaccharides with sialic acid attached to terminal galactose and internal *N*-acetylgalactosamine by α2-3 and α2-6 linkages, respectively (see Chapter 5). Biosynthesis of bovine and porcine submaxillary mucin oligosaccharides (Figure 1) has been studied extensively (Carlson *et al.*, 1973b; Sadler *et al.*, 1979). The mucin α2-6 SAT (CMP-NeuAc:GalNAc-mucin α2-6 sialyltransferase) has been highly purified from detergent extracts of porcine submaxillary gland by affinity chromatography on CDP-hexanolamine-agarose columns (Sadler *et al.*, 1979, 1982). The best substrate for this partially purified enzyme was found to be GalNAc-mucins prepared from ovine submaxillary mucin or other salivary mucins (Schachter and Roseman, 1980). The SAT (ST6O) probably acts on GalNAc-mucin and/or Galβ1-3GalNAc-mucin (Figure 1). The disaccharide Galβ1-3GalNAc had low acceptor activity whereas antifreeze glycoprotein containing Galβ1-3GalNAc units was an excellent acceptor. The SAT (ST3O) that catalyzes transfer of sialic acid to the terminal galactose moiety of the Galβ1-3GalNAc-disaccharide unit of mucin or antifreeze glycoprotein has been purified from porcine submaxillary tissues (Carlson *et al.*, 1973b; Rearick *et al.*, 1979; Gillespie *et al.*, 1992) and human placenta (Joziasse *et al.*, 1985).

5. POSTTRANSLATIONAL DISTRIBUTION OF SIALYLTRANSFERASES IN GOLGI MEMBRANES

Glycosyltransferases that catalyze the glycosylation of glycoproteins (Schachter *et al.*, 1970) or gangliosides (Keenan *et al.*, 1974) were detected in membrane fractions rich in Golgi bodies. Using immunofluorescent probe-labeled (Berger and Hesford, 1985) and gold-labeled antibodies against α2-6 SAT, the sites of ST6N localization in the Golgi bodies were shown to be in the trans-cisternae and the trans-tubular network in a luminal orientation (Carey and Hirschberg, 1981; Roth *et al.*, 1986). ST6N (47 kDa), primarily localized in the trans-region of the Golgi apparatus, was converted into catalytically active secreted protein (41 kDa) by inserting a cleavable signal peptide from human γ-interferon in place of the NH_2-terminal signal-anchor domain (Colley *et al.*, 1989, 1992). This demonstrates that a Golgi apparatus retention-signal for ST6N perhaps resides in the NH_2-terminal portion of the enzyme, which includes the cytoplasmic tail to stem regions (Alurno, 1991). Sequences within and adjacent to the transmembrane sequence of ST6N are reported (Munro, 1991) to be specific for Golgi retention. Careful studies (Dahdal and Colley, 1993) have revealed that the SAT (ST6N) signal anchor is not necessary or sufficient for Golgi retention. However, appropriately special cytoplasmic and luminal flanking sequences are the important elements of the SAT Golgi retention region. Human ST6N expressed in *Saccharomyces cerevisiae* is retained as an active enzyme in the endoplasmic reticulum (Kruzdorn *et al.*, 1994). Similar studies with other SATs have not yet been reported. Sucrose density gradient fractionation of a highly purified Golgi apparatus from rat liver (Trinchera *et al.*, 1990; Iber *et al.*, 1991, 1992) resulted in the recovery of SAT-1 in several density fractions, whereas SAT-2 was mainly found in less dense sub-Golgi membranes; this indicates that perhaps these two glycolipid:SAT activities are physically separated within the Golgi membranes.

Studies on retention of SATs on the cell surface are of importance because of their possible role in intercellular adhesion (Roseman, 1970). The B-cell antigen CD-75 is reported to be a cell surface SAT (Stamenkovic *et al.*, 1990) (see Chapter 2). The amino acid sequence of ST domains required for plasma membrane retention has not been established. Previous studies showed that a catalytically active ST6N (41–43 kDa) is secreted in serum, perhaps as a result of cleavage of Golgi membrane proteins by the action of a cathepsin D-like protease (Lammers and Jamieson, 1988).

A soluble serum glycoprotein:SAT from inflamed rats has been reported by Mookerjea and his co-workers (Fraser et al., 1980, 1984). Although intestinal lymph SAT was increased by colchicine administration in rats, enterectomy did not prevent the rise of serum SAT, suggesting that the intestine is not a major source of the serum enzyme (Ratnam *et al.*, 1981). However, the exact mechanism of the production of soluble serum glycoprotein: SAT(s) from the membrane-bound Golgi or plasma membrane forms is unknown.

Secreted and plasma membrane-bound proteins expressed by eukaryotic cells are often modified posttranslationally. This generally involves the covalent attachment of a carbohydrate moiety to threonine and/or asparagine residues. While species-, tissue-, and non-specific glycosylation have been clearly recognized, the importance of specific sialylation by defined SATs has been studied for the last three decades exhaustively. Perhaps the intramembrane (or Golgi) localization of any specific SAT is a determinant factor for the biosynthesis of a glycoprotein or glycolipid with a specific sialyloligosaccharidyl sequence (Figure 2 and 3).

6. CLONING OF α2-3 SIALYLTRANSFERASES

Several SAT-catalyzed reactions have been characterized for mucins (two activities, Figure 1: ST3O and ST6O), glycoproteins (two activities, Figure 2: ST3N and ST6N), and glycolipids (four activities, Figure 3: SAT-1, SAT-2, SAT-3, and SAT-4). Recently, tests in several laboratories have resulted in the sequence elucidation of at least five SATs: ST6N (Weinstein *et al.*, 1987; Wang *et al.*, 1990a,b; Wen *et al.*, 1992a; Kurosawa *et al.*, 1994b), ST3O (Gillespie *et al.*, 1992) and ST3N (Wen *et al.*, 1992b), SAT-3 (Kitagawa and Paulson, 1993a,b, 1994a; S. Basu *et al.*, 1993a), and ST6O (Kurosawa *et al.*, 1994a). ST3N (Weinstein *et al.*, 1982; Wen *et al.*, 1992b) from rat liver efficiently utilizes oligosaccharides containing Galβ1-3GlcNAc as well as Galβ1-4GlcNAc structures, whereas a similar ST3N cloned from a human melanoma cell line (Sasaki *et al.*, 1993) showed preferential expression of the SA-Le^x determinant only. The newly cloned SAT-3 is also named STZ (Kitagawa and Paulson, 1993a,b, 1994a). Human placenta cDNA was used (Kitagawa and Paulson, 1994a) as a template for the PCR-based cloning procedure. A new sialyl motif fragment emerged, SMZ, which contained 18 of 21 invariant amino acids that were found in three other cloned SATs (human ST3N, porcine ST30, and human ST6N). The SMZ 150-bp fragment was used to screen a human placenta cDNA library. Six positively hybridized clones (STZ-1 to STZ-6) were isolated. Expression of a plasmid pSTZ-prot A containing the STZ clone in COS-1 cells produced (Kitagawa and Paulson, 1994a) an active SAT, which used oligosaccharide, glycoproteins, and glycolipid acceptor substrates with terminal galactose in the Galβ1-3GalNAc and Galβ1-4GlcNAc sequences but not in the Galβ1-3GlcNAc. The relative level of activity of the cloned human placenta SAT (STZ) showed highest activity with an oligosaccharide [Galβ1-3GalNAc (100%) > asialo-GM1 (60%) > antifreeze GP (50%) > nLcOse4Cer (11%)]. The cloning results confirm that this new member of the SAT gene family perhaps is the enzyme previously described as SAT-3 (CMP-NeuAc: nLcOse4Cer α2-3 sialyltransferase) from embryonic chicken brain (M. Basu *et al.*, 1982). However, two different glycolipid:SATs have been isolated from human colon carcinoma Colo-205 cells: SAT-3 and SAT-4, which catalyze the transfer of sialic acid to nLcOse4Cer (Galβ1-4GlcNAcβ1-3Galβ1-4Glc-Cer) and GgOse4Cer (Galβ1-3GalNAc β1-4Galβ1-4Glc-Cer), respectively (S. Basu *et al.*, 1993a; M. Basu *et al.*, 1993). Nevertheless, two different SAT genes have not been cloned from human colon carcinoma cells as yet. It would be interesting to find that these two activities are expressed in separate species in different ways. Recently, in our laboratory, strong and specific hybridization of the RT-PCR amplified partial cDNA products from respective RNAs of Colo-205(800 bp), ECB(800 bp), and human placenta (800 bp) were obtained with human placenta SAT-3 probe. This result demonstrates their partial homologous nature at nucleotide sequence level. A cDNA library of the human melanoma cell line WM266-4 was constructed in an Epstein–Barr virus-based cloning vector (Sasaki *et al.*, 1993). Expression of a novel α2,3 SAT in the B-Namalwa cell line with the cDNA-containing vector was screened for *Ricinus* agglutinin sensitivity. Expression of the carboxy-terminal catalytic domain of this α2,3 SAT showed substrate specificity different from that of SAT-3 [Galβ1-3(4)GlcNAc α2-3 sialyltransferase]. Expression of this cDNA in Namalwa cells increased the level of SA-Le^x antigens. The cloning approach for a specific SAT based on lectin resistance is novel and useful for isolation of cDNAs encoding other mammalian glycosyltransferases (Sasaki *et al.*, 1993). cDNA encoding a new type of α2,3 SAT (ST3GalA.2; CMP-NeuAc:Galβ1-3GalNAc-R) has been isolated (Lee *et al.*, 1994) from both mouse and rat brain cDNA libraries. The cDNA sequences included an open reading frame coding for 350 amino acids, and the primary structure of this enzyme indicated four domains like those in α2,6 SAT (ST6N). It showed 76% identity in the active domain with that of the previously cloned mouse ST3GalA.1 (CMP-NeuAc:Galβ1-3GalNAc-α2,3 sialyltransferase; Lee *et al.*, 1993). The expression of ST3GalA.2 mRNA is tissue-specific, and it is prominent in brain and liver. The SAT (ST3GalA.2) expressed in COS-7 cells exhibited better activity with glycolipid substrates containing Galβ1-3GalNAc disaccharides at the terminal of the oligosaccharides.

7. CLONING OF α2-6 SIALYLTRANSFERASES

In recent years cDNAs encoding ST6N (CMP-NeuAc:Galβ1-4GlcNAc-R α2,6 sialyltransferase) have been reported from rat liver by Paulson and his co-workers (Weinstein *et al.*, 1987), human placenta (Grundmann *et al.*, 1990), mouse liver (Hamamoto *et al.*, 1993), and a B-cell line (Bast *et al.*, 1992). The human ST6N or Siat-1 (EC 2.4.99.1) gene is located on chromosome 3 (q21–q28) (Wang *et al.*, 1993). Comparative analysis of the human and the previously separated rat SiaT-1 genomic sequences demonstrates major conservation of the intron/exon boundaries throughout the coding domains. Analysis of genomic sequences from human cell lines of the B lineage—Rih, Nalm-6, Jok-1, Ball-1, Dandi, and Louches—showed three upstream exons, Exon (X) and Exons (Y + Z), leading to the synthesis of at least two mRNA isozymic variants of ST6N (a form of SiaT-1). These genomic exons are absent in cDNA coding for ST6N mRNA expressed in Hep G2 human hepatoma cells (Shah *et al.*, 1992). In some B cells, ST6N (SiaT-1) containing only Exon X is expressed. Little is known about ST6N (SiaT-1) regulation in human lymphocytes (Wang *et al.*, 1990a,b, 1993). However, the liver mRNA for ST6N was found (Svensson *et al.*, 1990) to be 4.3 kb versus 4.7 kb in other tissues. The 4.3-kb liver mRNA spans 40kb of genomic DNA and contains six exons. The promoter responsible for the transcription of ST6N is 50-fold more active in a hepatoma cell line (Hep G2) than in CHO cells and contains consensus-binding sites for the liver-restricted transcription factors NHF-1 and DBP as well as the transcription factors AP-1 and AP-2.

The cDNA clones encoding ST6N have also been isolated from chick embryonic cDNA libraries using the conserved amino acid sequence information previously published in other systems (Kurosawa *et al.*, 1994a). The embryonic chicken cDNA sequence revealed an open reading frame coding for 413 amino acids and contained 60% sequence identity with the rat liver enzyme. The expression of a recombinant ST6N in COS-7 cells, containing the cytoplasmic tail, signal anchor domains, and stem regions (replaced with an immunoglobulin signal peptide sequence), resulted in secretion of a catalytically active soluble form of the enzyme into the medium. The expressed chick embryo ST6N gene product exhibited activity only toward the lactosamine (Galβ1-4GlcNAc-R) bound to glycoproteins (Kurosawa *et al.*, 1994b). cDNA clones encoding ST60 (CMP-NeuAc:GalNAc-O-SerGP; EC 2.4.99.3) have also been isolated from chick embryo cDNA libraries using sequence information obtained from the conserved amino acid sequence of the previously cloned SATs (Kurosawa *et al.*, 1994a). The cDNA sequence included an open reading frame coding for 566 amino acids, and the putative amino acid sequence showed only 12% identity with that of ST6N (CMP-NeuAc: Galβ1-4GlcNAc-R α2-6 sialyltransferase) from chick embryo. Secretions of the recombinant ST60 into the medium from COS-7 cells also occurred when the NH_2-terminal part (232 amino acid residues) was replaced with the immunoglobulin signal sequence. The expressed enzyme (ST60) exhibited activity only toward asialo mucin and asialo fetuin. Synthetic GalNAc-Ser-NAc also served as an acceptor for ST60 (Kurosawa *et al.*, 1994b).

Several sialyltransferases cloned to date contain a conserved region, the "Sialylmotif," consisting of 48–49 amino acids in the center of the coding sequence. Mutant sequences of this region of ST6N have been constructed and expressed in COS-1 cells (Datta and Paulson, 1995). Kinetic analysis showed six of eight mutants had 3–12-fold higher Km for CMP-NeuAc relative to the wild type enzyme. These results suggest that the sialylmotif in the sialyltransferase gene family may participate in the binding of donor substrate CMP-NeuAc.

8. CLONING OF α2-8 SIALYLTRANSFERASES

Biosynthesis of the polysialic acid-containing virulence factor of *E. coli* K1 is catalyzed by a membrane-bound (Cho and Troy, 1989; Weisgerber and Troy, 1990) polysialyltransferase (ST-Pol or SAT-Pol; Tables I and II). The enzyme catalyzes transfer of α2-8 sialyl groups to the terminal sialyl group attached to *E. coli* endogenous membrane protein, soluble α2-8 sialyl-containing oligomers (colominic acid), or α2-8 sialyl-linked GD3 ganglioside (Cho and Troy, 1989) (see Chapter 4). The gene encoding the *E. coli* polysialyltransferase (ST-Pol or SAT-Pol) has been identified by subcloning and DNA sequence analysis (Weisgerber *et al.*, 1991). The subcloned DNA fragment codes for a polypeptide with a molecular mass of 47 kDa catalyzing the *in vitro* synthesis of poly-α2-8 *N*-acetylneuraminic acid (polysialyl) groups of exogenous acceptors (Weisgerber *et al.*, 1991).

Recently, a cDNA encoding a GM3-specific α2-8 SAT (SAT-2 or GD3 synthase) has been isolated from an expression cDNA library of human melanoma cell line WM266-4 by enrichment of Namalwa KJM-1 cells highly expressing GD3 (Sasaki *et al.*, 1994). Anti-GD3 antibody is used for fluorescence-activated cell sorting. Transfection of the plasmid pAMo-GD3 expressing SAT-2 (or GD3 synthase) lacking the first putative initiation codon showed expression of this gene and GD3 in several lines (Sasaki *et al.*, 1994). Using expression cloning techniques, Nagai and his coworkers (Nara *et al.*, 1994) and Furukawa and his coworkers (Haraguchi *et al.*, 1994) have isolated cDNAs encoding SAT-2 from human SK-Mel-28 and KF3027-Hyg5 cells, respectively. The sequence of α2,8-sialyltransferase showed a high level of similarity with other SATs at two conserved regions. This SAT-2 catalyzed the formation of the α2–8 linkages in GD3 and GQ1b gangliosides with relative rates, 100 and 16, respectively. As predicted, mRNA of this SAT-2 gene (2.6 kb) is strongly expressed in human melanoma lines.

9. MODULATION OF SIALYLTRANSFERASES IN DEVELOPING TISSUES

Sialylation of glycoproteins and glycolipids changes during development and differentiation in normal tissue. In intestinal tissue, weaning appears to be a key point, acting differently on soluble and membrane-bound SATs (Biol *et al.*, 1987). Expression of SATs responsible for glycolipid biosynthesis in developing tissues has been reviewed (S. Basu, 1966; S. Basu and Basu, 1982; M. Basu *et al.*, 1982). All four SATs of the ganglioside pathway (Figure 3; SAT-1, SAT-2, SAT-3, and SAT-4) appear in embryonic chicken brain at early stages of development (7 to 12 days). Analysis of developmental patterns of ganglioside sialosylation in embryonic chicken brain neurons suggests (Rosenberg *et al.*, 1992) that stable neurite growth is directly related to the biosynthesis of complex (or multi-sialyl-) gangliosides and their expression on the plasma membrane (see Chapter 11). However, mRNA levels of these enzymes have not been established in the developing tissues. A postnatal lowering of α2-8 sialic acid-linked glycoconjugate of N-CAM was related to changes in the α2-8 SAT activity involved in biosynthesis of polysialosyl units of N-CAM in embryonic cells (McCoy *et al.*, 1983).

Expression of α2-6 SAT (ST6N) in *Xenopus* embryos by injection of mRNA prevents the polysialylation of N-CAM (Livingston *et al.*, 1990), perhaps catalyzed by α2-8 SAT (SAT-2 or a similar protein). Expression of the α2-6 SAT (ST6N) was shown to regulate the generation of multiple cell-surface antigens (HB-6, CDW75, and CD76) of human lymphocytes (Bast *et al.*, 1992).

10. REGULATION OF SIALYLTRANSFERASES IN CANCER CELLS

Tumor or cancer cells express a variety of sialofucosylated glycoconjugates (SA-Le^x, SA-diLe^x, etc.). Our recent work (M. Basu *et al.*, 1988, 1991, 1993) suggested expression of a specific α1-3 fucosyltransferase in human colon carcinoma Colo-205 cells which utilized LM1 glycolipid (NeuAcα2-3Galβ1-4GlcNAcβ1-3Galβ1-4Glc-Cer; Figure 4, step 2b), at a much higher rate than nLcOse4Cer (Galβ1-4GlcNAcβ1-3Galβ1-4Glc-Cer; Figure 4, step 3a). Recently a partial cDNA for SAT-3 has also been cloned from Colo-205 cells using the RT-PCR route (S. Basu *et al.*, 1993a,b; S. S. Basu *et al.*, 1995). Colorectal primary carcinomas from 20 Duke's stage C and D patients were examined for the histochemical localization (Matsushita *et al.*, 1991) and contents of various fucosylated *N*-acetyllactosamine oligomers by specific monoclonal antibodies (CSLEX-1, SH-1, and FH-6). The results supported the hypothesis that high-molecular-weight sialylated polylactosamine glycoconjugates produced by colo rectal carcinoma tissues are heterogeneous with regard to their carbohydrate chains and their antigenic structures may change during tumor progression (Saitoh *et al.*, 1992). However, the regulation of SAT-3 or a similar protein related to SAT-3 have not been reported and will be of importance for future investigation. Sialylation of cell surface glycoconjugates is a signal for metastatic potential of some specific cell lines as well as of T hybridomas (M. Basu *et al.*, 1980; Taki *et al.*, 1988; Yogeeswaran and Salk, 1981; Collard *et al.*, 1986). It is believed that a higher level of sialic acid in cells may precede the development of malignancy. It has been shown that inhibitors of oligosaccharide processing can influence the ability of tumor cells to metastasize (Bolscher *et al.*, 1986; Poste and Nicolson, 1980). K1-8110, a specific inhibitor for sialic acid incorporation, has been shown to reduce the incidence of lung metastasis from mouse colon tumor (Kijima-Suda *et al.*, 1986). Recently, similar reduction of metastatic potential of human colorectal carcinoma cell lines has been observed after treatment of the cell line with K1-8110 (analogue of CMP-NeuAc) (Harvey *et al.*, 1992). It has been proposed (Nakamura *et al.*, 1992) that the metabolic flow of glycosphingolipid biosynthesis in HL-60 cells is regulated by GlcNAcT-1 (UDP-GlcNAc:LcOse3Cer β1-3 (GlcNAc-transferase) and SAT-1 (CMP-NeuAc:lactosylceramide α2-3 sialyltransferase). The expression of soluble and cell-bound ST6N (α2-6 SAT) in colonic carcinoma Caco-2 cells correlates with the degree of differentiation (Dall'Olio *et al.*, 1992). Regulation of ST6N gene expression by dexamethasone has been studied by Lau and his co-workers (Wang *et al.*, 1989). SAT levels and ganglioside expression in melanoma and other cultured human cells have also been correlated (Thampoe *et al.*, 1989).

11. PERSPECTIVES

Within the last three decades tremendous advances have been made in our understanding of the structure and biosynthesis of sialylated glycoconjugates by mammalian cells and *E. coli* K12. To understand strict specificity of a glycosyltransferase and the structure of its catalytic products (mucin, glycoprotein, or glycolipid), the availability of nonheterogeneous oligosaccharides biosynthesized by a cloned glycosyltransferase is necessary. This information is valuable in the studies of the therapeutic application of oligosaccharides against the lymphoreticular inflammation process and the metastatic behavior of cancer cells. During the last decade, at least seven or eight different SATs have been cloned and putative amino acid sequences from cDNA sequences are known. As in other glycosyltransferases they all consist of a short NH_2-terminal cytoplasmic domain, a signal-membrane anchor domain, a proteolytically sensitive stem region, and a large COOH-terminal catalytic domain. Based on synthetic substrate specificity studies with purified GalT-3 (UDP-Gal:GM2 β1-3 galactosyltransferase) (Ghosh *et al.*, 1990; S. Basu *et al.*, 1990; S. C. Basu, 1991), GalNAcT-1 (UDP-GalNAc:GM3 β1-4 GalNAc-transferase) (Schaeper *et al.*, 1992), SAT-3 (CMP-NeuAc: nLcOse4Cer α2-3 sialyltransferase) (S. Basu *et al.*, 1993b), and SAT-1 (CMP-NeuAc:lactosylceramide α2-3 sialyltransferase) (Li *et al.*, 1995), we propose to classify glycosyltransferases (GLTs) in two broad families (Figure 5):

1. The GLTs that contain only a carbohydrate recognition site (CARS)
2. The GLTs that contain recognition sites for a substrate containing both hydrophobic and carbohydrate moieties (HY-CARS)

Many careful studies with recombinant SATs as well as other GLTs will be needed to understand structure–function relationships of these enzymes and their roles in the biosynthetic regulation of cell surface glycoconjugates.

Acknowledgments. We thank Mrs. Dorisanne Nielsen and Mrs. Rosemary Patti for their help during preparation of this manuscript. Cloning of SAT-3 from ECB, the work reviewed in this chapter, was supported by United States Public Health Services Grant NS-18005 (The Jacob Javits Award) to S.B.

REFERENCES

Alurno, S., 1991, Sequences within and adjacent to the transmembrane sequence of α2-6SAT specify Golgi retention, *EMBO J.* **10:**3577–3588.

Bartholomew, B. A., Jourdian, G. W., and Roseman, S., 1973, The sialic acids. XV. Transfer of sialic acid to glycoproteins by a sialyltransferase from colostrum, *J. Biol. Chem.* **248:**5751–5762.

Bast, B. J., Zhou, L. J., Freeman, G. J., Colley, K. J., Ernst, T. J., Munro, J. M., and Tedder, T. F., 1992, The HB-6, CDW75, and CD 76 differentiation antigens are unique cell-surface carbohydrate determinants generated by the β-galactoside α2,6-sialyltransferase, *J. Cell Biol.* **116:**423–435.

Basu, M., Basu, S., and Potter, M., 1980, Biosynthesis of blood group related glycosphingolipids in

T- and B-lymphomas and neuroblastoma cells, in: *Cell Surface Glycolipids* (C. C. Sweeley, ed.), Chem. Soc. Symp. Vol. 128, pp. 187–212.

Basu, M., Basu, S., Stoffyn, A., and Stoffyn, P., 1982, Biosynthesis *in vitro* of sialyl α(2-3) neolactotetrosyl ceramide by a sialyltransferase from embryonic chicken brain, *J. Biol. Chem.* **257:**12765–12769.

Basu, M., De, T., Das, K. K., Kyle, J. W., Chon, H. C., Schaeper, R. J., and Basu, S., 1987, Glycolipid glycosyltransferases, *Methods Enzymol.* **138:**575–607.

Basu, M., Das, K. K., Zhang, B., Khan, F. A., and Basu, S., 1988, Biosynthesis of tumor related glycosphingolipids, *Indian J. Biochem. Biophys.* **25:**112–118.

Basu, M., Hawes, J. W., Li, Z., Ghosh, S., Khan, F. A., Zhang, B. J., and Basu, S., 1991, Biosynthesis *in vitro* of SA-Le^x and SA-diLe^x by α1-3fucosyltransferases from colon carcinoma cells and embryonic brain tissues, *Glycobiology* **1:**527–535.

Basu, M., Basu, S. S., Li, Z., Tang, H., and Basu, S., 1993, Biosynthesis and regulation of Le^x and SA-Le^x glycolipids in metastatic human colon carcinoma cells, *Indian J. Biochem. Biophys.* **30:**324–332.

Basu, S., 1966, Ph.D. thesis, University of Michigan, Ann Arbor, Studies on the biosynthesis of gangliosides.

Basu, S., and Basu, M., 1982, Expression of glycosphingolipid glycosyltransferases in development and transformation, in: *The Glycoconjugates,* Vol. III, Academic Press, New York, pp. 265–284.

Basu, S., and Kaufman, B., 1965, Ganglioside biosynthesis in embryonic chicken brain, *Fed. Proc.* **24:**479.

Basu, S., Ghosh, S., Basu, M., Hawes, J. W., Das, K. K., Zhang, B. J., Li, Z., Weng, S. A., and Westervelt, C., 1990, Carbohydrates and hydrophobic recognition sites (CARS and HY-CARS) in solubilized glycosyltransferases, *Indian J. Biochem. Biophys.* **27:**386–395.

Basu, S., Basu, S. S., Basu, M., and Li, Z., 1993a, Cloning of SAT-3 from human colon carcinoma, Colo 205 cells and its role in biosynthesis of SA-Le^x glycoconjugate, *Glycobiology* **3:**531.

Basu, S., Ghosh, S., Basu, S. S., Kyle, J. W., Li, Z., and Basu, M., 1993b, Regulation of expression of cell surface neolactoglycolipids and cloning of embryonic chicken brain GalT-4, *Indian J. Biochem. Biophys.* **30:**315–323.

Basu, S. C., 1991, The serendipity of ganglioside biosynthesis: Pathway to CARS and HY-CARS glycosyltransferases, *Glycobiology* **1:**469–475.

Basu, S. S., Basu, M., and Basu, S., 1995, Isolation of putative cDNA clones for SAT-3 from embryonic chicken brain (ECB) and human colon carcinoma (Colo-205) cell, *Glycoconjugate J.*, in press.

Berger, E. G., and Hesford, F. J., 1985, Localization of galactosyl- and sialyltransferase by immunofluorescence evidence for different sites, *Proc. Natl. Acad. Sci. USA,* **82:**4736–4739.

Biol, M. C., Martin, A., Richard, M., and Louisot, P., 1987, Developmental changes in intestinal glycosyl-transferase activities, *Pediatr. Res.* **22:**250–255.

Bolscher, J.G.M., Schaller, D.C.C., Smets, L. A., van Rooy, H., Collard, J. G., Bruynell, E. A., and Mareel, M.M.K., 1986, Effect of cancer-related and drug-induced alteration in surface carbohydrates on the invasive capacity of mouse and rat cells, *Cancer Res.* **46:**4080–4086.

Bradley, B. K., Swedler, S. J., and Robbins, P. W., 1990, Carbohydrate ligands of the LEC cell adhesion molecules, *Cell* **63:**861–863.

Broquet, P., Baubichon-Cortey, H., George, P., and Louisot, P., 1991, Glycoproteins sialyltransferases in eucaryotic cells, *Int. J. Biochem.* **23:**385–389.

Carey, D. J., and Hirschberg, C. B., 1981, Topography of sialoglycoproteins and sialyltransferases in mouse and rat liver Golgi, *J. Biol. Chem.* **256:**989–993.

Carlson, D. M., Jourdian, G. W., and Roseman, S., 1973a, The sialic acids. XIV. Synthesis of sialyl-lactose by a sialyltransferase from rat mammary gland, *J. Biol. Chem.* **248:**5742–5750.

Carlson, D. M., McGuire, E. J., Jourdian, G. W., and Roseman, S., 1973b, The sialic acids. XVI. Isolation of a mucin sialyltransferase from sheep submamillary gland, *J. Biol. Chem.* **248:**5763–5773.

Chien, J.-L., and Hogan, E. L., 1983, Novel pentahexaosyl ganglioside of the globo series purified from chicken muscle, *J. Biol. Chem.* **258:**10727–10730.

Chien, J.-L., Basu, M., and Basu, S., 1974, Biosynthesis in vitro of an N-acetylglucosamine-containing ganglioside, *Fed. Proc.* **33:**1225.

Chien, J.-L., Li, S.-C., Laine, R. A., and Li, Y.-T., 1978, Characterization of gangliosides from bovine erythrocyte membranes, *J. Biol. Chem.* **253:**4031–4035.

Cho, J. W., and Troy, F. A., 1989, Gangliosides as exogenous acceptors to map the acceptor sugar requirements of the poly-α2,8-sialyltransferase in Escherichia coli k1, Proc. Xth Int. Symp. Glycoconjugates, p. 143.

Collard, J. G., Schigven, T. F., Bikker, A., LaRiviere, G., Bolscher, T.G.M., and Roos, E., 1986, Cell surface sialic acid and invasive and metastatic potential of T-cell hybridomas, *Cancer Res.* **46:**3521–3527.

Colley, K. J., Lee, E. U., Adler, B., Browne, J., and Paulson, J. C., 1989, Conversion of a Golgi apparatus sialyltransferase to a secretory protein by replacement of the NH_2 terminal signal anchor with a signal peptide, *J. Biol. Chem.* **264:**17619–17622.

Colley, K. J., Lee, E. U., and Paulson, J. C., 1992, The signal anchor and stem regions of the beta-galactose alpha-2,6-sialyltransferase may each act in localizing the enzyme to the Golgi apparatus, *J. Biol. Chem.* **267:**7784–7793.

Dahdal, R. Y., and Colley, K. J., 1993, Specific sequences in the signal anchor of the β-galactoside α-2,6-sialyltransferase are not essential for Golgi localization, *J. Biol. Chem.* **268:**26310–26319.

Dall'Olio, F., Malagolini, N., and Berafini-Cessi, F., 1992, The expression of soluble and cell-bound α-2-6SAT in human colonic carcinoma Caco-2 cell correlates with the degree of enterocytic differentiation, *Biochem. Biophys. Res. Commun.* **184:**1405–1410.

Datta, A. K. and Paulson, J. C., 1994, The sialyltransferase "Sialomotif" participates in binding the donor substrate CMP-NeuAc, *J. Biol. Chem.* **270:**1497–1500.

Dijak, W. V., Lasthius, A., and van den Eijnden, D. H., 1979, Glycoprotein biosynthesis in calf kidney. Glycoprotein sialyltransferase activities towards serum glycoproteins and calf Tamm–Horsfall glycoprotein, *Biochim. Biophys. Acta* **584:**129–142.

Edelman, G. M., 1983, Cell adhesion molecules, *Science* **219:**450–457.

Finne, J., and Makela, P. H., 1985, Cleavage of the polysialosyl units of brain glycoproteins by a bacteriophage endosialidase. Involvement of a long oligosaccharide segment in molecular interaction of polysialic acid, *J. Biol. Chem.* **268:**1265–1270.

Finne, J., Finne, U., Deagostini-Bazin, H., and Goridis, C., 1983, Occurrence of α2-8 linked polysialosyl units in a neural cell adhesion surface, *Biochem. Biophys. Res. Commun.* **112:**482–486.

Fraser, I. H., Ratnam, S., Collins, J. M., and Mookerjea, S., 1980, Sialyltransferase activity in intestinal mucosa after colchicine treatment, *J. Biol. Chem.* **255:**6617–6625.

Fraser, I. H., Coolbear, T., Sarkar, M., and Mookerjea, S., 1984, Increase of sialyltransferase activity in the serum and liver of inflamed rats, *Biochim. Biophys. Acta* **799:**102–105.

Fukuda, M. N., Dell, A., Oates, J. E., and Fukuda, M., 1985, Embryonal lactosaminoglycan. The structure of lactosaminoglycans with novel disialosyl (sialylα2-9sialyl) terminals isolated from PA1 embryonal carcinoma cells, *J. Biol. Chem.* **260:**6623–6631.

Fukushima, K., Hirota, M., Terasaki, P. I., Wakisaka, A., Togashi, H., Chia, D., Sugijama, N., Fukushi, Y., Nudelman, E., and Hakomori, S., 1984, Characterization of sialylated Lewis x as a new tumor antigen, *Cancer Res.* **44:**5279–5285.

Ghosh, S., Das, K. K., Daussin, F., and Basu, S., 1990, Effect of fatty acid moiety of phospholipid and ceramide on purified GalT-3 (UDP-Gal:GM2β1-3 galactosyltransferase) activity from embryonic chicken brain, *Indian J. Biochem. Biophys.* **27**:379–385.

Gillespie, W., Kelm, S., and Paulson, J. C., 1992, Cloning expression of the Galβ1-3GalNAc α2-3 sialyltransferase, *J. Biol. Chem.* **267**:21004–21010.

Grundmann, U., Nerlich, C., Rein, T., and Zeltlmeissl, G., 1990, Complete cDNA sequence encoding human β-galactoside α2,6-sialyltransferase, *Nucleic Acids Res.* **18**:667–672.

Gu, T.-J., Gu, X.-B., Arigan, T., and Yu, R. K., 1990, Purification and characterization of CMP-NeuAc:GM1 (Galβ1-4GalNAc-)α2-3 sialyltransferase from rat brain, *FEBS* **275**:83–86.

Gu, X.-B., Gu, T.-J., and Yu, R. K., 1990, Purification to homogeneity of GD3 synthase and partial purification of GM3 synthase from rat brain, *Biochem. Biophys. Res. Commun.* **166**:387–393.

Hakomori, S., 1981, Glycosphingolipids in cellular interaction, differentiation and oncogenesis, *Annu. Rev. Biochem.* **50**:733–764.

Hakomori, S., 1984, Tumor-associated carbohydrate antigens, *Annu. Rev. Immunol.* **2**:103–216.

Hamamoto, T., Kawasaki, M., Kurosawa, N., Nakaoka, T., Lee, Y. C., and Tsuzi, S., 1993, Two step single primer mediated polymerase chain reaction. Application to cloning of putative mouse β-galactosidase α2,6-sialyltransfase cDNA, *Bioorg. Med. Chem.* **1**:141–145.

Haraguchi, M., Yamashiro, S., Yamamoto, A., Furukawa, K., Takamiya, K., Lloyd, K. O., Shiku, H., and Furukawa, K., 1994, Isolation of GD3 synthase gene by expression cloning of GM3 a2,8-sialyltransferase cDNA using anti-GD2 monoclonal antibody, *Proc. Natl. Acad. Sci. USA* **91**:10455–10459.

Harvey, B. E., Toth, C. A., Wagner, H. E., Steele, G. D., and Thomas, P., 1992, Sialyltransferase activity and hepatic tumor growth in a nude mouse model of colorectal cancer metastasis, *Cancer Res.* **52**:1775–1779.

Higashi, H., Basu, M., and Basu, S., 1985, Biosynthesis *in vitro* of disialosylneolactotetraosylceramide by a solubilized sialyltransferase from embryonic brain, *J. Biol. Chem.* **260**:824–828.

Iber, H., Vanechten, G., and Sandhoff, K., 1991, Substrate specificity of α2-3-sialyltransferase in ganglioside biosynthesis of rat liver Golgi, *Eur. J. Biochem.* **195**:115–120.

Iber, H., Zacharias, C., and Sandhoff, K., 1992, The e-series gangliosides GT3, GT2 and GPic are formed in rat liver Golgi by the same set of glycosyltransferase that catalyzes the biosynthesis of asialo a- and b-series gangliosides, *Glycobiology* **2**:137–142.

Inoue, S., and Matsumura, G., 1979, Identification of N-glycolylneuraminyl-a2,8-N-glycolylneuraminyl group in a trout egg glycoprotein by methylation analysis and GC-mass spectrometry, *Carbohyd. Res.* **74**:361–368.

Jourdian, G. W., Carlson, D. M., and Roseman, S., 1963, The enzymatic synthesis of sialyllactose, *Biochem. Biophys. Res. Commun.* **10**:352–358.

Joziasse, H. D., Bergh, M.L.E., ter Hart, H.G.J., Koppen, P. L., Hooghwinkel, G.J.M., and van den Eijnden, D. H., 1985, Purification and enzymatic characterization of CMP-sialic acid: β-galactosyl 1-3N-acetylgalactosaminide α2-3 sialyltransferase from human placenta, *J. Biol. Chem.* **260**:4941–4951.

Kaufman, B., and Basu, S., 1966, Embryonic chicken brain sialyltransferase, *Methods Enzymol.* **8**:365–368.

Kaufman, B., Basu, S., and Roseman, S., 1968, Enzymatic synthesis of disialogangliosides from monosialogangliosides by sialyltransferases from embryonic chicken brain, *J. Biol. Chem.* **243**:5804–5806.

Keenan, T. W., Moore, D. J., and Basu, S., 1974, Ganglioside biosynthesis. Concentration of glycosphingolipid glycosyltransferases in Golgi apparatus from rat liver, *J. Biol. Chem.* **249**:310–315.

Kijima-Suda, I., Miyamoto, Y., Toyoshima, S., Itoh, M., and Osawa, T., 1986, Inhibition of

experimental pulmonary metastasis of mouse colon adenocarcinoma 26 sublines by a sialic acid:nucleoside conjugate having sialyltransferase inhibiting activity, *Cancer Res.* **46:**858–862.
Kitagawa, H., and Paulson, J. C., 1993a, Cloning and expression of human Galβ1,3(4) GlcNAcα2,3-sialyltransferase, *Biochem. Biophys. Res. Commun.* **194:**375–382.
Kitagawa, H., and Paulson, J., 1993b, The sialyltransferase gene family, *Glycobiology* **3:**532.
Kitagawa, H., and Paulson, J. C., 1994a, Cloning a novel α-2,3-sialyltransferase that sialylates glycoprotein and glycolipid carbohydrate groups, *J. Biol. Chem.* **269:**1394–1401.
Kitagawa, H., and Paulson, J. C., 1994b, Differential expression of five sialyltransferase genes in human tissues, *J. Biol. Chem.* **269:**17872–17878.
Kruzdorn, C., Kleene, R. B., Watzele, M., Ivanov, S. X., Hokke, C. H., Kamerling, J. P., and Berger, E. G., 1994, Human β1,4-galactosyltransferase and α2,6 sialyltransferase expressed in *Saccharomyces cerevisiae* are retained as active enzyme in the endoplasmic reticulum, *Eur. J. Biochem.* **220:**809–817.
Kundig, F. D., Aminoff, D., and Roseman, S., 1971, The sialic acids. XII. Synthesis of colominic acid by a sialyltransferase from *E. coli* K-235, *J. Biol. Chem.* **246:**2543–2550.
Kurosawa, N., Hamamoto, T., Lee, Y.-C., Nakaoka, T., Kojima, N., and Tsuji, S., 1994a, Molecular cloning and expression of GalNAcα2,6-sialyltransferase, *J. Biol. Chem.* **269:**1402–1409.
Kurosawa, N., Kawasaki, M., Hamamoto, T., Nakaoka, T., Lee, Y.-C., Arita, M., and Tsuji, S., 1994b, Molecular cloning and expression of chick embryo Galβ1,4GlcNAcα2,6-sialyltransferase. Comparison with the mammalian enzyme, *Eur. J. Biochem.* **219:**375–381.
Lammers, G., and Jamieson, J. C., 1988, The role of a cathepsin D-like activity in the release of Galβ1-4GlcNAcα2-6 sialyltransferase from rat liver Golgi membrane during acute phase response, *Biochem. J.* **256:**623–631.
Lasky, L. A., 1991, Lectin cell adhesion molecules (LEC-CAMs): A new family of cell adhesion proteins involved with inflammation, *J. Cell Biochem.* **45:**139–146.
Lee, Y.-C., Kurosawa, N., Hamamoto, T., Nakaoka, T., and Tsuji, S., 1993, Molecular cloning and expression of Galβ1,3GalNAcα2,3-sialyltransferase. Comparison with the mammalian enzyme, *Eur. J. Biochem.* **216:**377–385.
Lee, Y.-C., Kojima, N., Wada, E., Kurosawa, N., Nakaoka, T., Hamamoto, T., and Tsuji, S., 1994, Cloning and expression of cDNA for a new type of Galβ1-3-GalNAcα2,3sialyltransferase, *J. Biol. Chem.* **269:**10028–10033.
Lemmon, U., Staros, E. B., Perry, H. E., and Gottlieb, D. I., 1982, A monoclonal antibody which binds to the surface of chick brain cell and myotubes: Cell selectivity and properties of the antigen, *Dev. Brain Res.* **3:**349–360.
Levery, S. B., Salyan, M.E.K., Steele, S. J., Kannagi, R., Dasgupta, S., Chien, J. L., Hogan, E. L., Halbeek, H. V., and Hakomori, S., 1994, A revised structure for the disialosyl globoseries gangliosides of human erythrocytes and chicken skeletal muscle, *Arch. Biochem. Biophys.* **3/2:**125–134.
Li, Z., Basu, M., Woodward, J. F., and Basu, S., 1994, Studies on the effect of fatty-acyl chains on SAT-1, a glycolipid glycosyltransferase, *Glycobiology* in press.
Liepkans, V., Alain, J., and Garan, L., 1988, Purification and characterization of a CMP-sialic:LeOse4Cer sialyltransferase from human colorectal carcinoma cell membranes, *Biochemistry* **27:**8683–8688.
Livingston, B. D., De Robertis, E. M., and Paulson, J. C., 1990, Expression of β-galactoside α2,6-sialyltransferase blocks synthesis of polysialic acid in Xenopus embryos, *Glycobiology* **1:**39–44.
McCoy, R. D., Vimr, E. R., and Troy, F. A., 1983, Poly-α-2,8-sialosyl sialyltransferase and the biosynthesis of polysialosyl units in neural cell adhesion molecules, *J. Biol. Chem.* **265:**12695–12699.
McEver, R. P., 1991, Selectins: Novel receptor that mediates l-eukocyte adhesion during inflammation, *Thromb. Haemostas.* **66:**80–87.

Magnani, J. L., 1991, SLe[a] and SLe[x] binds ELAM-1, *Glycobiology* **1:**318–320.

Magnani, J. L., Nilsson, B., Brockhaus, M., Zopf, D., Steplewski, A., Koprowski, A., and Ginsburg, V., 1982, A monoclonal antibody-defined antigen associated with gastrointestinal cancer is a ganglioside containing lacto-N-fucopentaose II, *J. Biol. Chem.* **257:**14365–14369.

Margolis, R. K., and Margolis, R. U., 1983, Distribution and characteristic of polysialosyl oligosaccharides in nervous tissue glycoprotein, *Biochem. Biophys. Res. Commun.* **116:**889–894.

Matsushita, Y., Nakamori, S., Seftor, E. A., Hendrix, M.J.C., and Irimura, T., 1991, Human colon carcinoma cells with increased invasive capacity obtained by selection for sialyl-dimeric Lex antigen, *Exp. Cell Res.* **196:**20–25.

Melkerson-Watson, L. J., and Sweeley, C. C., 1991, Purification to apparent homogeneity by immunoaffinity chromatography and partial characterization of the GM3 ganglioside-forming enzyme, CMP-sialic acid: lactosylceramide α2,3-sialyltransferase (SAT-1) from rat liver Golgi, *J. Biol. Chem.* **266:**4448–4457.

Montreuil, J. B., 1975, Recent data on the structure of the carbohydrate moiety of glycoproteins. Metabolic and biological implications, *Pure Appl. Chem.* **42:**431–477.

Moore, K. L., Varki, A., and McEver, R. P., 1991, GMP-140 binds to a glycoprotein receptor on human neutrophils: Evidence for a lectin like interaction, *J. Cell Biol.* **112:**491–499.

Murno, S., 1991, Sequences within and adjacent to the transmembrane sequence of α2-6SAT specify Golgi situation, *EMBO J.* **10:**3577–3588.

Nakamura, K., Inagaki, F., and Tamai, Y., 1988, A novel ganglioside in dogfish brain. Occurrence of a trisialoganglioside with a sialic acid linked to N-acetylgalactosamine, *J. Biol. Chem.* **263:**9896–9900.

Nakamura, M., Tsunoda, A., Sakoe, K., Gu, J., Nishikawa, A., Taniguchi, N., and Saito, M., 1992, Total metabolic flow of glycosphingolipid biosynthesis is regulated by UDP-GlcNAc:lactosylceramide β1-3N-acetylglucosaminyltransferase and CMP-NeuAc:lactosylceramideα2-3 sialyltransferase in human hematopoietic cell line HL-60 during differentiation, *J. Biol. Chem.* **267:**1–8.

Nara, K., Watanabe, Y., Maruyama, K., Kashahara, K., Nagai, Y., and Sanai, Y. 1994, Expression cloning of a CMP-NeuAc:NeuAca2-3Galb1-4Glcb1-1Cer a2,8-sialyltransferase (GD3 synthase) from human melanoma cells, *Proc. Natl. Acad. Sci. USA* **91:**7952–7956.

Nilsson, O., Månsson, J. E., Lindholm, L., and Holmgren, J., 1985, Sialosyllactotetraosylceramide, a novel ganglioside antigen detected in human carcinomas by a monoclonal antibody, *FEBS Lett.* **182:**398–402.

Ohashi, M., 1981, A new type of ganglioside: Carbohydrate structures of the major gangliosides from brain, in: *Proc. 6th Int. Symp. Glycoconjuages* (T. Yamakawa, T. Osawa, and S. Handa, eds.), Japanese Scientific Press, Tokyo, pp. 33–34.

Olden, K., Parent, J. B., and White, S. L., 1982, Carbohydrate moieties of glycoproteins: A reevaluation of their function, *Biochim. Biophys. Acta* **650:** 209–232.

Paulson, J. C., and Colley, K. J., 1989, Glycosyltransferases. Structure, localization and control of cell type-specific glycosylation, *J. Biol. Chem.* **264:**17615–17618.

Paulson, J. C., Beranek, W. E., and Hill, R. L., 1977a, Purification of a sialyltransferase from bovine colostrum by affinity chromatography on CDP-agarose, *J. Biol. Chem.* **252:**2356–2362.

Paulson, J. C., Rearick, J. I., and Hill, R. L., 1977b, Enzymatic properties of D-galactose (α2-6) sialyltransferase from bovine colostrum, *J. Biol. Chem.* **252:**2363–2371.

Paulson, J. C., Weinstein, J., and de Souza-E-Silva, U., 1984, Biosynthesis of a disialated sequence in N-linked oligosaccharides: Identification of an N-acetylglucosaminide α(2-6) sialyl transferase in Golgi apparatus from rat liver, *Eur. J. Biochem.* **140:**523–530.

Phillips, M. L., Nudelman, E., Graeta, F.C.A., Perez, M., Singhal, A. K., Hakomori, S., and Paulson, J. C., 1990, ELAM-1 mediates cell adhesion by recognition of a carbohydrate ligand, sialyl-Le[x], *Science* **250:**1130–1132.

Poste, G., and Nicolson, G., 1980, Arrest and metastasis of blood-borne tumor cells are modified by fusion of plasma membrane vesicles from highly metastatic cells, *Proc. Natl. Acad. Sci. USA* **77**:399–403.

Preti, A., Fiorilli, A., Lombardo, A., Caimi, L., and Tettamanti, G., 1980, Occurrence of sialyltransferase activity in the synaptosomal membranes prepared from calf brain cortex, *J. Biol. Chem.*, **255**:281–296.

Ratnam, S., Fraser, I. H., Collins, J. M., Lawrence, J. A., Barrowman, J. A., and Mookerjea, S., 1981, Elevated sialyltransferase activity in the intestinal lymph of colchicine-treated rats, *Biochim. Biophys. Acta* **673**:435–442.

Rearick, J. I., Sadler, J. E., Paulson, J. C., and Hill, R., 1979, Enzymatic characterization of β-D-galactoside α2-3 sialyltransferase from porcine submaxillary gland, *J. Biol. Chem.* **254**:4444–4451.

Rice, G. E., and Bevilacqua, M. P., 1989, An inducible endothelial cell surface glycoprotein mediates melanoma adhesion, *Science* **246**:1303–1306.

Roseman, S., 1970, The synthesis of complex carbohydrates by multiglycosyltransferase systems and their potential function in intercellular adhesion, *Chem. Phys. Lipids* **5**:270–297.

Roseman, S., Carlson, D. M., Jourdian, G. W., McGuire, E. J., Kaufman, B., Basu, S., and Bartholomew, B., 1966, Animal sialic acid transferases (sialyltransferases), *Methods Enzymol.* **8**:354–372.

Rosenberg, A., Sauer, A., Noble, E. P., Gross, H.-J., Chang, R., and Brossmer, R., 1992, Developmental patterns of ganglioside sialosylation coincident with neuritogenesis in cultured embryonic chick brain neurons, *J. Biol. Chem.* **267**:10607–10612.

Roth, J., Taatjes, D. J., Weinstein, J., Paulson, J. C., Greenwell, P., and Watkins, W., 1986, Differential subcompartmentation of terminal glycosylation in the Golgi apparatus of the intestinal absorptive and goblet cells, *J. Biol. Chem.* **261**:14307–14312.

Rougon, G., 1993, Structure, metabolism and cell biology of polysialic acid, *Eur. J. Cell Biol.* **61**:197–207.

Rutishauser, U., 1989, Neurobiology of glycoconjugates, in: *Polysialic Acid Regulation of Cell Interactions* (R. U. Margolis and R. K. Margolis, eds.), Plenum Press, New York, pp. 367–382.

Sadler, J. E., Rearick, J. I., Paulson, J. C., and Hild, R. L., 1979, Purification and characterization of two sialyltransferase activities from porcine submamillary glands, in: *Glycoconjugate Research: Proceedings of the Fourth International Symposium on Glycoconjugates*, Vol. II (J. D. Gregory and R. W. Jeanloz, eds.), Academic Press, New York, pp. 763–766.

Sadler, J. E., Beyer, T. A., Oppenheimer, C. L., Paulson, J. C., Prieels, J. P., Rearick, J. I., and Hill, R. L., 1982, Purification of mammalian glycosyltransferases, *Methods Enzymol.* **83**:458–515.

Saitoh, O., Wang, W. C., Lotan, R., and Fukuda, M., 1992, Differential glycosylation and cell surface expression of lysosomal membrane glycoprotein in sublines of a human colon cancer exhibiting different metastatic potentials, *J. Biol. Chem.* **267**:5700–5711.

Sasaki, K., Watanabe, E., Kawsashima, K., Sekine, S., Dohi, T., Oshima, M., Hannai, N., Nishi, T., and Hasegawa, M., 1993, Expression cloning of a novel Galβ(1-3/1-4) GlcNAcα2,3-sialyltransferase using lectin resistance selection, *J. Biol. Chem.* **268**:22782–22787.

Sasaki, K., Kurata, K., Kojima, N., Kurosawa, N., Ohita, S., Hanai, N., Tsuji, S., and Nishi, T., 1994, Expression cloning of a GM3 specific α2-8-sialyltransferase (GD3 synthase), *J. Biol. Chem.* **269**:15950–15956.

Schachter, H., Jabbal, I., Hudgin, R. L., Pinteric, L., McGuire, E. J., and Roseman, S., 1970, Intracellular localization of sugar nucleotide glycoprotein glycosyltransferases in a Golgi-rich fraction, *J. Biol. Chem.* **245**:1090–1100.

Schachter, H., and Roseman, S., 1980, Mammalian glycosyltransferases; Their role in the synthesis and function of complex carbohydrates and glycolipids, in: *The Biochemistry of Glycoproteins and Proteoglycans* (W. J. Lennarz, ed.), Plenum Press, New York, pp. 85–160.

Schaeper, R. J., Das, K. K., Li, Z., and Basu, S., 1992, In vitro biosynthesis of GbOse4Cer-(globoside) and GM2 ganglioside by the (1→3) and (1→4)-N-acetyl β-D-galactosaminyltransferases from embryo chicken brain. Solubilization, purification, and characterization of the transferases, *Carbohydr. Res.* **236**:227–244.

Schauer, R., 1982, Sialic acids, chemistry, metabolism and function, in: *Cell Biology Monographs*, Vol. 10, Springer-Verlag, Berlin.

Shah, S., Lance, P., Smith, T. J., Berenson, C. S., Cohen, S. A., Horvath, P. J., Lau, J.T.Y., and Baumann, H., 1992, n-Butyrate reduces the expression of β-galactoside α2,6-sialyltransferase in Hep G2 cells, *J. Biol. Chem.* **267**:5700–5711.

Stamenkovic, I., Asheim, H. C., Deggerdal, A., Blomhoff, H. K., Smeland, E. B., and Funderud, S., 1990, The B-cell antigen CD 75 is a cell surface sialyltransferase, *J. Exp. Med.* **172**:641–643.

Stoffyn, P., and Stoffyn, A., 1980, Biosynthesis *in vitro* of mono- and di-sialogangliosides from gangliotetraosylceramide by cultural cell lines and young rat brain. Structure of the products and activity and specificity of sialyltransferase, *Carbohydr. Res.* **78**:327–340.

Svensson, E. C., Soreghan, B., and Paulson, J. C., 1990, Organization of the β-galactoside α2,6-sialyltransferase gene, *J. Biol. Chem.* **265**:20863–20868.

Taki, T., Hirabayashi, Y., Ishikawa, H., Ando, S., Kon, K., Tanaka, Y., and Matsumoto, M., 1986, A ganglioside of rat ascites hepatoma AH7974F cells. Occurrence of a novel disialoganglioside (GD1aα) with an unique NeuAcα2-6GalNAc- structure, *J. Biol. Chem.* **261**:3075–3078.

Taki, T., Takamatsu, M., Myoga, A., Tanaka, T., Ando, S., and Matsumoto, M., 1988, Glycolipids of metastatic tissues in liver from colon cancer: Appearance of sialyl-Le^x and Le^x lipids, *Biochem. J.* **103**:998–1003.

Thampoe, I.J.T., Furukawa, K., Vellve, E., and Lloyd, K. O., 1989, Sialyltransferase levels and ganglioside expression in melanoma and other cultured human cancer cells, *Cancer Res.* **49**:6258–6266.

Trinchera, M., Pirovano, B., and Ghidoni, R., 1990, Sub-Golgi distribution in rat liver of CMP-NeuAc GM3 and CMP-NeuAc-GT1b-α-2-8 sialyltransferase and comparison with the distribution of the other glycosyltransferase activity involved in ganglioside biosynthesis, *J. Biol. Chem.* **265**:18242–18247.

Troy, F. A., and McCloskey, M. A., 1979, Role of membranous sialyltransferase complex in the synthesis of surface polymers containing polysialic acid in E. coli, *J. Biol. Chem.* **254**:7377–7387.

Uemura, K., Haltori, H., Ono, K., Ogata, H., and Taketomi, T., 1989, Expression of Forssman glycolipid and blood group-related antigens a, Le^x and Le^y in human gastric cancer and in fetal tissues, *Jpn. J. Exp. Med.* **59**:239–249.

Van Echten, G., and Sandhoff, K., 1993, Ganglioside metabolism: Enzymology, topology and regulation, *J. Biol. Chem.* **268**:5341–5344.

Varki, A., 1993, Biological roles of oligosaccharides: All of the theories are correct, *Glycobiology* **3**:97–130.

Walz, G., Aruffo, A., Kolanus, W., Bevilacqua, M., and Seed, B., 1990, Recognition by ELAM-1 of the sialyl-Le^x determinant on myeloid and tumor cells, *Science* **250**:1132–1135.

Wang, X., O'Hanlon, T. P., Young, R. F., and Lau, J.T.Y., 1989, Regulation of β-galactoside α2-6 sialyltransferase gene expression by dexamethasone, *J. Biol. Chem.* **264**:1854–1859.

Wang, X., O'Hanlon, T. P., Young, R. F., and Lau, J.T.Y., 1990a, Rat β-galactoside α2-6-sialyltransferase genomic organization: Alternate promoters direct the synthesis of liver and kidney transcripts, *Glycobiology* **1**:25–31.

Wang, X., Smith, T. J., and Lau, J.T.Y., 1990b, Transcriptional regulation of liver β-galactoside α2-6-sialyltransferase by glucocorticoids, *J. Biol. Chem.* **265**:17849–17853.

Wang, X. C., Vertino, A., Eddy, R. L., Beyers, M. G., Jani-Gait, S. N., Shows, T. B., and Lau, J.T.Y., 1993, Chromosome mapping and organization of the human β-galactoside α2,6 sialyltransferase gene, *J. Biol. Chem.* **268**:4355–4361.

Weinstein, J., de Souza-E-Silva, U., and Paulson, J. C., 1982, Purification of Galβ1-4GlcNAcα2-6 sialyltransferase and a Galβ1-3(4)GlcNAcα2-3 sialyltransferase to homogeneity from rat liver, *J. Biol. Chem.* **257**:13835–13844.

Weinstein, J., Lee, U. E., McEntee, K., Lai Por-Hsiung, and Paulson, J. C., 1987, Primary structure of β-galactosides α2,6-sialyltransferase. Conversion of membrane-bound enzyme to soluble form by cleavage of the NH_2-terminal signal anchor, *J. Biol. Chem.* **262**:17735–17743.

Weisgerber, C., and Troy, F. A., 1990, Biosynthesis of the polysialic acid capsule in E. coli K1. The endogenous acceptor polysialic acid is a membrane protein of 20 kDa, *J. Biol. Chem.* **265**:1578–1587.

Weisgerber, C., Hamsen, A., and Frosch, M., 1991, Complete nucleotide and deduced protein sequence of CMP-NeuAc: poly alpha-2,8-sialyltransferase of Escherichia coli K1, *Glycobiology* **1**:357–366.

Wen, D. X., Svensson, E. C., and Paulson, J. C., 1992a, Tissue specific alternative splicing of α2-6SAT gene, *J. Biol. Chem.* **267**:2512–2518.

Wen, D. X., Livingstone, B. D., Medzihraszky, K. F., Kelm, S., Burlingame, A. L., and Paulson, J. C., 1992b, Primary structure of Gal-β1,3(4)GlcNAcα2-3 sialyltransferase, *J. Biol. Chem.* **267**:21011–21019.

Wiegandt, H., 1985, Gangliosides, in: *Glycolipids* (H. Wiegandt, eds.), Elsevier, Amsterdam, pp. 199–260.

Yogeeswaran, G., and Salk, P. L., 1981, Metastatic potential is positively correlated with cell surface sialylation of cultured murine tumor cell lines, *Science* **212**:1514–1516.

Yu, R. K., and Lee, S. H., 1976, Biosynthesis of sialogalactosylceramide (G7) by mouse brain microsomes, *J. Biol. Chem.* **251**:198–203.

Chapter 4

Sialobiology and the Polysialic Acid Glycotope

Occurrence, Structure, Function, Synthesis, and Glycopathology

Frederic A. Troy II

1. INTRODUCTION AND PERSPECTIVE

Polysialic acids (polySia)* are a structurally diverse family of linear carbohydrate chains that consist of *N*-acetylneuraminic acid (Neu5Ac) or *N*-glycolylneuraminic acid (Neu5Gc) residues, usually joined internally by α2,8-, α2,9-, or alternating α2,8-/α2,9-ketosidic linkages. 3-Deoxy-D-*glycero*-D-*galacto*-2-nonulosonic acid (KDN) is a unique deaminated form of Sia, and polyKDN chains share many properties in common with polySia (Table I). The finding of poly(Neu5Ac), poly(Neu5Gc), poly(Neu5Ac,Neu5Gc), poly(KDN) chains and their partially *O*-acetylated and *O*-lactylated forms in salmonid fish egg glycoproteins demonstrates the natural occurrence of multiple forms of these unique sugar chains.

The polySia glycotope covalently modifies cell surface glycoconjugates on

*Abbreviations used in this chapter: Sia, sialic acid; Neu5Ac, *N*-acetylneuraminic acid; Neu5Gc, *N*-glycolylneuraminic acid; polySia, α2,8-linked polysialic acid; KDN, 3-deoxy-D-*glycero*-D-*galacto*-2-nonulosonic acid; LPS, lipopolysaccharide; GSL, glycosphingolipid; polyST, CMP-sialic acid:polyα2,8-sialosyl sialytransferase; Endo-N, endo-*N*-acylneuraminidase; DP, degree of polymerization; 4-MU-Neu5Ac, 4-methylumbelliferyl Neu5Ac; 4-MU-KDN, 4-methylumbelliferyl KDN.

Frederic A. Troy II Department of Biological Chemistry, University of California, School of Medicine, Davis, California 95616.

Biology of the Sialic Acids, edited by Abraham Rosenberg. Plenum Press, New York, 1995.

Table I
Natural Occurrence and Structure of Oligo-Polysialic Acid Chains

Source	Reported structure
Bacterial capsular polysaccharides	
Escherichia coli K1	α-(2→8Neu5Acα2-)$_n$; 7-*O*-Ac- and 9-*O*-Ac-; $n >$ 200 Sia
Escherichia coli K92	Alternating α-(2→8Neu5Ac) and α-(2→9Neu5Ac) linkages
Neisseria meningitidis group B	α-(2→8Neu5Ac)$_n$
Neisseria meningitidis group C	α-(2→9Neu5Ac)$_n$
Pasteurella haemolytica A2	α-(2→8Neu5Ac)$_n$
Moraxella nonliquefaciens	α-(2→8Neu5Ac)$_n$
Bacterial lipopolysaccharide (O-antigen) *Legionella pneumophila*	α-(2→4)-5-acetamidino-7-acetamido-8-*O*-acetyl-3,5,7,9-tetradeoxy-D-*glycero*-L-*galacto*-nonulosonic acid (Sia)
Fish egg polysialoglycoproteins	
Lake trout eggs (*Salvelinus namaycush*)	α-(2→8Neu5Acα)$_n$; α-(2→8Neu5Gc)$_n$
Rainbow trout eggs (*Salmo gairdneri*)	α-(2→8Neu5Gc) KDN α-(2→8Neu5Gc)$_n$ α-(2→8KDN)$_n$
Vitelline envelope	KDNα2→3Gal-; KDNα2→8KDN-; KDNα2→6GalNAc-; 9-OAcKDNα2→8KDN
Cortical vesicles	KDNα2→8Neu5Ac-; KDNα2→3GalNAc-; KDNα2→8Neu5Gc-; KDNα2→6GalNAc-9-OAcKDNα2→8Neu5Gc-
Ovarian fluid	poly(KDN)-gp (same as VE)
Kokanee salmon eggs (*Oncorhynchus nerka adonis*)	α-(2→8Neu5Gcα)$_n$; 4-O-Ac, 7-O-Ac-, and 9-O-Ac; KDNα2-(8Neu5Gcα2-)$_n$; 9-O-Ac-KDN-
Jelly coat of sea urchin eggs (*Hemicentrotus pulcherrimus, Strongylocentrotus purpuratus*)	(5→O_{glycolyl}-Neu5Gcα2→)$_n$; n = 4 to >40
Embryonic neural membranes	
Sialogangliosides	α(2→8Neu5Ac)$_n$; n = 3–4
Neural cell adhesion molecules	α(2→8Neu5Ac)$_n$; $n \sim$ 150–200
Extraneural cells and tissues	
Human kidney, heart, muscle, pancreas, lung, thyroid, T lymphocytes	α(2→8Neu5Ac)$_n$; expressed on [E]N-CAM
Electrophorus electricus (electroplax sodium channel)	α(2→8Neu5Ac)$_n$
Adult rat brain (α-subunit sodium channel)	α(2→8Neu5Ac)$_n$
Human cancers (see Table III)	

cells as evolutionarily diverse as those of microbes and humans. Its expression can influence neuronal development and the neuroinvasive potential of neurotropic bacteria and some human brain tumors (Tables II and III). The α2,8-linked polySia capsule on neuropathogenic *Escherichia coli* K1 and *Neisseria meningitidis* serogroup B, for example, is a neurovirulent determinant associated with neonatal meningitis in humans. Structurally identical α2,8-polySia chains cap *N*-linked oligosaccharides on neural cell adhesion molecules (N-CAMs) and *O*-linked oligosaccharides on fish egg polysialoglycoproteins (Table I). The polysialylated N-CAMs are particularly fascinating molecules to study because of the myriad of cellular functions that they appear to regulate, including cell adhesion and cell movement. Polysialylated N-CAMs are also oncodevelopmental antigens that are reexpressed on the surface of several human tumors. Thus, polySia chains function in a remarkably diverse range of important biological contexts. As a consequence, polysialylation has emerged as an exciting new area of sialobiology that continues to impact on contemporary studies in molecular microbiology, neurobiology, oncology, and cell and developmental biology.

The major objective of this review will be to summarize the key findings that relate to the occurrence, structure, function, synthesis, and glycopathology of the polySia "sialotope." Because of the critical importance in understanding the unresolved fundamental problem of how synthesis and surface expression of polySia is regulated, greater emphasis will be placed on biosynthesis of the polySia glycotopes, and the prerequisite α2,8-polysialyltransferase activities in prokaryotic and eukaryotic cells.

2. NATURAL OCCURRENCE AND STRUCTURE OF POLYSIALIC ACID-CONTAINING GLYCOCONJUGATES

The pace of discovery of new polySia structures has been rapid in the past several years. The remarkable diversity in structures of the polySia chains on glycoconjugates (Table I) has enriched our appreciation of the possible function of these novel carbohydrates. Since recent summaries of the occurrence and structure of polySia-containing glycoconjugates have been published (Troy, 1990, 1992; Inoue, 1993; Cho and Troy, 1994a,b), greater emphasis shall be placed on new structures that have only been recently discovered.

2.1. Prokaryotes

2.1.1. Polysialic Acid Capsules on Neuroinvasive *Escherichia coli* K1 and *Neisseria meningitidis*

Homooligomers of Sia residues were first reported in the culture filtrate of *E. coli* K-235 by Barry and Goebel (1957), and were designated "colominic acid." Subsequent studies showed that the Sia residues were joined internally through α2,8-ketosidic linkages (McGuire and Binkley, 1964). Colominic acid was later shown to be a "culture artifact" because the Sia oligomers were derived from polymeric membrane-associated polySia chains by mild acid hydrolysis, as the pH of the culture filtrate dropped to below 5.0 as the cells reached stationary phase (Troy and McCloskey, 1979). Structural analysis confirmed that the capsular polySia chains on cells grown at pH 7.0 contained a degree of polymerization (D.P.) of at least 200 Sia residues (Rohr and Troy, 1980; Table I).

The capsule of *N. meningitidis* serogroup B is structurally and immunologically identical to the α2,8-linked polySia capsule of neuroinvasive *E. coli* K1 strains. These bacterial capsular polysaccharides are neuroinvasive determinants that are associated with neonatal meningitis in humans (Robbins *et al.*, 1974; Troy, 1979). The polySia capsule of *N. meningitidis* serogroup C is also a homopolymer of Sia joined through α2,9 linkages (Bhattacharjee *et al.*, 1976). The Sia residues in the capsule of *E. coli* K92 and Bos 12 consist of alternately linked α2,8- and α2,9-ketosidic linkages (Egan *et al.*, 1977). Some substrains of *E. coli* K1 contain *O*-acetylated Neu5Ac residues, which are reported to increase immunogenicity and decrease the neuroinvasive potential (Higa and Varki, 1988). α2,8-linked polySia capsules have been reported to be expressed on pathogenic strains of *Pasteurella haemolytica* A2 and *Moraxella nonlinquefacies* (Adlam *et al.*, 1987). No polySia chains containing Neu5Gc have been described in bacteria. KDN is a component sugar in the heteropolysaccharide capsule of *Klebsiella ozaenae* K4 (Knirel *et al.*, 1989). We have recently observed KDN immunoreactivity in *E. coli* K1 strains that is sensitive to KDNase SM.*

2.1.2. *O*-Specific Lipopolysaccharide of *Legionella ozaenae*

The newest polySia structure to be described in bacteria is the *O*-specific polysaccharide of *Legionella pneumophili* serogroup 1 lipopolysaccharide (LPS) (Knirel *et al.*, 1994). This unique structure was shown by chemical analysis and ^{1}H- and ^{13}C-NMR spectroscopy to be an α2,4-linked homopolymer of 5-acetamidino-7-acetamido-8-*O*-acetyl-3,5,7,9-tetradeoxy-D-*glycero*-L-*galacto*-nonulosonic acid, a structure that has not been previously described. The LPS O-chains contained 40–75 nonulosonic acid units, which structurally is an unusual sugar for three reasons. First, the sugar is hydrophobic because it lacks any free hydroxyl groups. The OH group on C-8 is quantitatively acetylated, while the OH group usually present on C-7 is substituted by an acetyl group. Second, the *N*-acyl group normally present on C-5 in Neu5Ac or Neu5Gc is substituted by an acetimidoyl group:

$$\begin{array}{ccccc}(-C_5- & N & - & C & -CH_3) \\ & | & & \| & \\ & H & & NH & \end{array}$$

Third, C-9, like C-3, C-5, and C-7 is also deoxy, while the hydroxyl group

*Terada, T., Kitijima, K., Ye, J., Inoue, S., Inoue, Y., and Troy II, F. A. (unpublished results).

chains were attached to core oligosaccharides that were *O*-glycosidically linked to threonine residues on a core polypeptide. Each polypeptide contained about 17 *O*-linked polysialylglycan chains. The apparent molecular weight of polySia-gp(H) was 180,000. The expression of this new polySia structure in place of α2,8-linked polySia is the main structural feature that distinguishes polySia-gp from other known polysialylated glycoproteins. The $(\rightarrow 5\text{-}O_{\text{glycolyl}}\text{-Neu5Gc}\alpha 2\rightarrow)_n$ chains were resistant to exo- and endosialidases (Endo-N) from *Arthrobacter ureafaciens* and bacteriophage K1F, respectively. Discovery of these $(\rightarrow 5\text{-}O_{\text{glycolyl}}\text{-Neu5Gc}\alpha 2\rightarrow)_n$ chains adds a new class of naturally occurring polySia to the structurally diverse family of polysialylated glycoproteins.

The unusual structure of these polySia chains may relate to the functional diversity ascribed to the jelly coat glycoproteins of sea urchin eggs. The structure of a polySia-gp from a different sea urchin species, *Strongylocentrotus purpuratus* (designated polySia-gp(S)), was also determined to ascertain if there were any species-specific differences. The 500-MHz ^{1}H-NMR spectra of the two polySia-gp's were identical, indicating that at this level of molecular detail, the structures were the same. The molecular weight of polySia-gp(S) was larger, however (250,000), and it contained about 25 polySia chains *O*-glycosidically linked to both threonine (2/3) and serine (1/3) residues (Kitazume *et al.*, 1994b).

2.2.3. Neural membranes

2.2.3.a. Polysialylated Neural Cell Adhesion Molecules. The most thoroughly characterized polysialylated glycoproteins in mammalian cells are the N-CAMs. N-CAMs are morphoregulatory glycoproteins that affect a number of cell migration and cell–cell adhesive interactions during embryogenesis, including neurite fasciculation and neuromuscular interactions (Bock *et al.*, 1980; Edelman, 1985; Cunningham *et al.*, 1987; Rutishauser *et al.*, 1988; Troy, 1990, 1992; Rougon, 1993).

N-CAM is a member of the immunoglobulin gene superfamily. A unique structural feature of the embryonic (E) form of N-CAM is the presence of α2,8-linked poly(Neu5Ac) chains that cap *N*-linked oligosaccharides on the N-CAM polypeptide (Hoffman *et al.*, 1982; Finne, 1982; Vimr *et al.*, 1984). Maximal expression of the polySia glycotope on N-CAM is usually found in embryonic tissue, when cells are migratory, but is greatly reduced in adult tissue, where little cell migration occurs (Hoffman *et al.*, 1982; Rutishauser *et al.*, 1988).

The gene for N-CAM in humans is located on chromosome 11q23 (Gold *et al.*, 1987). At least four different N-CAM isoforms have been described, with polypeptide molecular masses of 180, 140, 120, and 110 kDa. These arise by alternative mRNA splicing (Cunningham *et al.*, 1987). Each isoform contains five immunoglobulin-like domains and two fibronectin-like repeats. The 180- and 140-kDa isoforms are linked to the plasma membrane via a transmembrane spanning domain, and each has a cytoplasmic domain. The 120-kDa isoform, in contrast, is attached to the membrane via a glycosylphosphatidylinositol (GPI) anchor and therefore lacks a cytoplasmic domain. The 110-kDa N-CAM is a secreted isoform. The 180-kDa isoform is found primarily on neurons but not on muscle cells, whereas the 140-kDa isoform is found on both (Rutishauser and Goridis, 1986; reviewed in Kern *et al.*, 1993).

In spite of the extensive information regarding the protein moiety of N-CAM, there is surprisingly little detailed structural information about the carbohydrate moieties. Three of the seven potential asparagine-linked attachment sites on all four isoforms of [E]N-CAM appear to be polysialylated (Crossin *et al.*, 1984; Hemperly *et al.*, 1986). The polySia glycosylation sites are located in the central region of the polypeptide chains, in the fifth immunoglobulin-like domain. Synthesis of N-CAM in the presence of tunicamycin prevents *N*-glycosylation and, as a consequence, polysialylation (Hemperly *et al.*, 1986). It has been estimated that, on average, each N-CAM polypeptide contains about 130–150 Sia residues (Livingston *et al.*, 1988). Livingston *et al.* (1988) showed that extended polySia chains containing more than 55 Sia residues (D.P. 55) were expressed on human neuroblastoma cells. This suggests that the earlier estimates of 8–16 Sia residues on (E)N-CAM (Rothbard *et al.*, 1982; Finne, 1982) may have been low because of the lability of these polymers to mildly acidic conditions (Troy and McCloskey, 1979; Varki and Higa, 1993), or contaminating sialidases. The first direct proof that internally linked Sia residues were present in N-CAM was provided by Finne (1982) who showed by mass fragmentography internally linked α2,8-Sia residues. These residues were reportedly attached as outer branches to presumed tri- and tetraantennary chains of *N*-linked oligosaccharides (Hoffman *et al.*, 1982; Finne, 1982; Margolis and Margolis, 1983), although no detailed structural information was presented.

We have recently reexamined the structure of the polysialylglycans isolated from embryonic chick brain N-CAM (Kitazume *et al.*, 1993). Two glycopeptide fractions (GP-1 and GP-2), both containing polySia chains, were obtained by pronase digestion, reductive carboxymethylation, redigestion with pronase, gel filtration, ion exchange, and gel chromatography. Based on structural analyses including chemical, enzymatic, and instrumental methods, two α2,8-linked polySia-containing glycan structures, different from that previously reported for the embryonic mouse brain N-CAM, were found. The smaller glycopeptide (GP-2) contained an α2,8-poly(Neu5Ac) complex-type biantennary glycan chain. The larger glycopeptide (GP-1) was a polysialylated triantennary glycan chain. To date, no α2,9-ketosidic linkages, KDN or Neu5Gc residues have been reported as constituents of N-CAM.

In contrast to the developmentally regulated reduction of polysialylated N-CAM in neural, kidney, heart, and some hematopoietic cells, this glycotope is expressed continuously in the retina, optic nerve, and tectum of the adult salamander and frog (Becker *et al.*, 1993a,b). There is, however, some selective attenuation of polySia expression in the optic nerve and optic fiber pathway in the adult brain of salamander.

A recent mapping of the distribution of polysialylated N-CAM in the central nervous system of the adult rat by immunocytochemistry revealed that this epitope continued to be expressed in discrete areas of the brain and spinal cord (Bonfanti *et al.*, 1992). Immunolabeling of the hypothalamic and thalamic nuclei, regions of the spinal cord, hippocampus, mesencephalic central gray matter, the supraoptic nucleus, and the olfactory bulb, was observed. Kiss *et al.* (1993) have also shown that expression of the embryonic form of N-CAM is retained in adult rats on neurohypophysial astrocytes, and have suggested that this expression may be regulated by neurosecretory axons. Further confirmation that polySia chain expression persists in the CNS of adult rats was provided by the histochemical studies of Seki and Arai (1993a,b). These investigators showed that polySia chains appeared in the olfactory bulb, the piriform cortex, the hippocampal dentate gyrus, the hypothalamus, the medulla, and the dorsal horn in the adult nervous system. Mesencephalic dopaminergic cells in rat brain also contained polySia chains which persisted throughout development (Shults and Kimber, 1992). Accordingly, it now seems clear that the paradigm in which the highly polysialylated embryonal N-CAM undergoes a reduction in polySia during development to the adult form is an oversimplification for many regions of the nervous system. In fact, polysialylated N-CAM is more abundant on adult than embryonic epithelial chicken lens cells (Watanabe *et al.*, 1992). Consequently, it should now be expected rather than unexpected to observe the polySia glycotope on certain adult tissues (see below).

The newest citing of α2,8-polysialylated N-CAM is on several human hematopoietic tumors, including malignant lymphoma (Grogan *et al.*, 1994), acute myeloid leukemia (Scott *et al.*, 1994), and multiple myeloma,* as described below.

2.2.3.b. Sodium Channel Proteins. The α subunit of the sodium channel in the adult rat brain was the first mammalian glycoprotein other than N-CAM to be shown to contain α2,8-linked polySia chains (Zuber *et al.*, 1992). This protein was identified by immunocytochemistry using an anti-polySia antibody (mAb 735) that was specific for α2,8-linked polySia chains. Specificity of the immunoreactivity was confirmed by showing that pretreatment of the channel with an Endo-N abolished immunoreactivity. Thus, N-CAM is not the only polysialylated glycoprotein in brain. Caution must therefore be exercised to not overinterpret the identity of polysialylated proteins based on immunohistochemical or Western blotting methods using only the anti-polySia antibodies and Endo-N.

*Hanneman, L., Guptil, J., Hemperly, J. J., Hersh, E., Tompson, F., Moore, J. J., Richter-Barbuto, J.A.M., Ye, J., McDonald, J. J., Troy, F. A., and Grogan, T. M. (unpublished results).

The first report of α2,8-linked polySia on the α subunit of sodium channel proteins was by James and Agnew (1987), who identified such chains on the voltage-sensitive channel from the electric organ of the eel, *Electrophorus electricus*. The sodium channel protein appears to be the only protein in electroplax membranes that is polysialylated, suggesting the presence of a specific α2,8-polysialyltransferase for the polysialylation of this protein (James and Agnew, 1989).

2.2.4. Extraneural Cells and Tissues

The first report of an α2,8-polysialylated N-CAM in extraneural tissue was by Roth and colleagues (Roth *et al.*, 1987, 1988a; Bitter-Suermann and Roth, 1987; Finne *et al.*, 1987). In a series of exquisite immunocytochemical and immunoblotting experiments, Roth *et al.* (1987) showed that expression of the embryonic form of N-CAM was developmentally regulated in rat and human kidney, a mesodermally derived tissue. The polySia chains in kidney were maximally expressed early in development, suggesting that, as in brain, they may regulate cell–cell adhesive interactions during kidney differentiation and development. As described below, polySia chains were reexpressed in Wilms' tumor, a highly malignant tumor of the kidney (Roth *et al.*, 1988c). Finne *et al.* (1987) also used anti-polySia antibodies to show that polysialylated N-CAM was expressed on newborn heart and muscle tissue. More recently, Lackie *et al.* (1990) showed that in human, rat, and chicken embryos, polysialylated N-CAMs were expressed on cells derived from mesoderm (mesenchymal cells) and endoderm (pancreas, lung epithelium), in addition to neuroectoderm cells, e.g., neural or lung endocrine cells. In the thyroid gland, calcitonin-producing cells, which are endocrine derivatives of the neural crest, also express the embryonic form of polysialylated N-CAM (Nishiyama *et al.*, 1993). Expression of the highly polysialylated N-CAM during development in the rat heart was investigated by Lackie *et al.* (1991). These studies revealed that during the early stages of development, embryonic N-CAM was expressed on myocardial, endocardial, and some atrioventricular cells in the epicardium. Later in development, its appearance in the epicardium decreased. In the adult heart, the only significant expression of polysialylated N-CAM was during innervation.

Polysialylated N-CAM has been reported to be transiently expressed in developing chicken osteoblasts during osteogenesis as lower-molecular-weight (156 and 100,000) species (Lee and Chuong, 1992), and on dermal and smooth muscle cells in the chick during feather development (Marsh and Gallin, 1992). In *Drosophila melanogaster,* expression of α2,8-polySia chains is developmentally regulated, and occurs only during the early stages of development (Roth *et al.*, 1992).

The first identification of naturally occurring oligo-polyα2,8-KDN sequences in mammalian tissue was reported by Kanamori *et al.* (1994). A monoclonal antibody specific for α2,8-linked oligoKDN sequences was first prepared by immunizing mice with a KDN-rich glycoprotein (KDN-gp) containing $(\rightarrow 8\text{KDN}\alpha 2\rightarrow)_n$ →6(KDNα2→3Galβ1→3GalNAcα1→3)GalNAcα1→ residues. The antibody, designated mAb.kdn8kdn, could specifically distinguish $(\rightarrow 8\text{KDN}\alpha 2\rightarrow)_n$ chains from $(\rightarrow \text{Neu5Ac}\alpha 2\rightarrow)_n$ and $(\rightarrow 8\text{Neu5Gc}\alpha 2\rightarrow)_n$ chains. mAb.kdn8kdn was used for the immunohistochemical detection of the KDN glycotope in rat pancreas (Kanamori *et al.*, 1994). We have also used this mAb to detect reactive KDN glycotopes in some human cancer cells and neurotropic bacteria.*

2.2.5. Human Cancers

The first human tumors shown to express α2,8-polysialylated N-CAM were neuroblastomas (Livingston *et al.*, 1987, 1988; Lipinski *et al.*, 1987) and nephroblastomas (Wilms' tumor; Roth *et al.*, 1988c). Anti-polySia and anti-N-CAM antibodies were used in combination with Endo-N, to verify the presence of polySia expression on N-CAM. In one case, structural studies were used to prove that extended polySia chains (D.P. $>$ 55 Sia residues) were expressed on human neuroblastoma cells (Livingston *et al.*, 1988). As summarized in Table II, subsequent immunocytochemical and, in some cases, Western blotting methods were used to identify α2,8-polysialylated N-CAM on a variety of nonneuroendocrine and neuroendocrine human tumors. Included were high-grade tumors such as medulloblastomas and neuroblastomas, and pheochromocytomas, medullary thyroid carcinomas, small-cell lung carcinomas, and pituitary adenomas (Roth *et al.*, 1988c; Heitz *et al.*, 1990; Moolenaar *et al.*, 1990). More recently, Nadasdy *et al.* (1993) reported that congenital mesoblastic nephromas express the highly polysialylated N-CAM.

Hanneman and colleagues† have recently demonstrated that human multiple myeloma cells express the embryonic form of N-CAM. The presence of polysialylated N-CAM on neoplastic plasma cells and not on normal plasma cells suggests that this epitope may play a critical role in myeloma tumorigenicity. Recent studies by Grogan *et al.* (1994) and Scott *et al.* (1994) have demonstrated polysialylated N-CAM in malignant lymphoma and acute nonlymphocytic leukemia. Importantly, the polysialylated N-CAM expressive tumors were found in patients with CNS involvement.

As our methods to detect the polySia glycotope in glycoconjugates become more sensitive (Ye *et al.*, 1994), we can expect that an even greater number of polysialylated tumors will be identified. Also, some tumors judged initially to

*Terada, T., Kitijima, K., Ye, J., Inoue, S., Inoue, Y., and Troy, F. A. (unpublished results).

†Hanneman, L., Guptil, J., Hemperly, J. J., Hersh, E., Tompson, F., Moore, J. J., Richter-Barbuto, J.A.M., Ye, J., McDonald, J. J., Troy, F. A., and Grogan, T. M. (unpublished results).

Table II
Possible Functions of the Polysialic Acid Glycotope

Bacterial polysialic acid capsules
- In neuroinvasive *E. coli* K1 and *N. meningitidis,* functions as a neurotropic determinant; facilitates invasion of the blood–brain barrier and colonization of the meninges of neonatal brains
- Polyanionic shield that masks the somatic O-antigen chains of lipopolysaccharide and renders cell resistant to immune detection and phagocytosis
- Receptor for binding K1-specific bacteriophages

Fish egg polysialoglycoproteins
- Implicated in regulating cell–cell interactions and cell migration during oogenesis
- May function as recognition markers for mediating egg–sperm interaction
- May protect the embryo from osmotic lysis, artificial activation, mechanical destruction, and bacterial infection by retaining Ca^{2+} around the embryo (KDN-gp has affinity for Ca^{2+} ions)
- *O*-Acetylation and KDN cosylation render polySia chains resistant to depolymerization by sialidases; consequence unknown (protective?)
- Expression of an $\alpha 2 \rightarrow 8$-polysialyltransferase is developmentally regulated

Polysialic acid-containing glycoproteins in the jelly coat of sea urchin eggs
- May mediate cell–cell interactions during gastrulation when endodermal cells interact with ectodermal cells
- May be involved in inducing the acrosome reaction in sperm
- Expression of an $\alpha 2,8$-polysialyltransferase is developmentally regulated

Polysialylated neural cell adhesion molecules (N-CAMs)
- Implicated in embryonic neural development and neuronal plasticity; mediates cell adhesive interactions including neurite fasciculation, neuromuscular interactions, and cell migration. Expression of polySia usually decreases N-CAM-mediated cell adhesion
- The amount of polySia on N-CAM is critical for normal morphogenesis and neural development
- Influences cell–cell apposition and regulates contact-dependent cell interactions. PolySia "shield" may simply increase the intercellular space between cells
- Participates in the establishment of neuronal connections and in modulating neurite outgrowth. May activate a second messenger pathway in primary neurons
- May influence the formation of new neural circuits in the dentate gyrus, and in reorganization of the piriform cortex in the adult rat
- Proposed to regulate intramuscular nerve branching during embryogenesis
- Implicated in the normal separation of secondary myotobes from primary myotobes during muscle development
- Influences the interaction of cells of the preimplantation mammalian embryo
- May couple the morphogenic effects of adhesion and synaptic activity-dependent processes
- In bone formation, may mediate the interaction of osteoblasts and regulate skeletogenesis
- May mediate the formation of fiber cell gap junctions and adherence junction during lens cell differentiation
- May control the migration and maturation of dopaminergic cells of the developing mesencephalon
- Participates in the internalization of the Antennapedia homeobox peptide involved in late expression of some homeogenes in the CNS

Expression on tumors
- In developing human kidney, brain (neuroblastomas), and plasma cells, polysialylated N-CAM is an oncodevelopmental antigen

(*continued*)

Table II
(Continued)

- On some human cancers, e.g., natural killer-like and T-cell malignant lymphomas, acute myeloid leukemia, multiple myeloma, and some head and neck tumors, polySia expression may enhance neuroinvasive potential and metastases
- May protect malignant cells from immune surveillance

Sodium channel glycoproteins
- Function unknown. By analogy with bacterial polySia capsules, the polyanionic surface charge over the channel may maintain a solute reservoir and shield channel from toxins

lack polysialylated N-CAM have recently been shown to express this epitope. Bronchial endocrine carcinoid tumors, for example, were initially reported to lack N-CAM (Komminoth *et al.*, 1991). Recently, however, Tome *et al.* (1993) have shown that lung carcinoid tumors express polysialylated N-CAM. Based on this new finding, it does not seem likely that polySia expression can be used unambiguously to differentiate small-cell lung cancers from carcinoid tumors, as originally thought (Komminoth *et al.*, 1991).

Subclonal heterogenity is another important factor that likely accounts for reported differences among a given tumor cell population as to whether the cells express polysialylated N-CAM or not. For example, not all human neuroblastomas nor small-cell lung carcinomas express polySia (Figarella-Branger *et al.*, 1990; Moolenaar *et al.*, 1990; Scheidegger *et al.*, 1994), and a highly metastatic melanoma subclone expressed less N-CAM than the nonmetastatic parental melanoma (Lehmann *et al.*, 1989). N-CAM is also known to be expressed by a subset of T-cell lymphomas (Kern *et al.*, 1993), pheochromocytomas (Jin *et al.*, 1991), and pituitary adenomas (Aletsee-Ufrecht *et al.*, 1990). Only about 10% of patients with malignant lymphoma and acute myeloid leukemia express N-CAM-positive tumors (Grogan *et al.*, 1994; Scott *et al.*, 1994). The fact that expression of polysialylated N-CAM on the surface of human tumor cells is a clonable trait may also contribute to variations in the aggressiveness and metastatic potential of some tumors, as discussed below.

3. FUNCTION OF POLYSIALYLATION

In most cases, the biological role of the polySia glycotopes is unknown. As one might glean from the bewildering structural diversity that continues to emerge, however, there is likely to be a myriad of functional roles. To seek a better understanding of these functions, it is instructive to consider the presumed activities or roles in which these molecules have been implicated, as summarized in Table II.

3.1. Bacterial Pathogenicity

The polySia capsule on neurotropic *E. coli* K1 and *N. meningitidis* serogroups B and C is a neurovirulent determinant associated with meningitis (reviewed in Troy, 1979). While the molecular mechanism is not clear, expression of the capsule appears to help these neuropathogens penetrate the blood–brain barrier and colonize the meninges of neonates (Robbins *et al.*, 1980; Silver and Vimr, 1990). Kim *et al.* (1992) have provided evidence that the *E. coli* K1 capsule is a critical determinant in the development of meningitis in the rat, leading to the suggestion that there may be specific polySia receptors in the brain that may be important in allowing polysialylated cells into the meninges.

The polySia chains may also function to enhance the virulence of encapsulated cells by masking the somatic O-antigen chains of LPS, thereby escaping immune detection. Accordingly, surface expression of the polySia glycotope on neuroinvasive bacteria and tumor cells may be an elaborate survival mechanism that evolved to trick the human immune system. The polySia capsule on *E. coli* K1 provides a barrier against phagocytosis and conceals cell surface components that activate complement (Bortolussi *et al.*, 1979; Horowitz and Silverstein, 1980). α2,8- but not α2,9-polySia chains are poor immunogens because of "antigenic mimicry." The central idea of antigenic mimicry relates to the α2,8-polySia glycotope being poorly immunogenic because of immune tolerance, since these chains are structurally and immunologically identical to the α2,8-polySia expressed on embryonic N-CAM (Finne *et al.*, 1983; Whitfield *et al*, 1984a). In retrospect, the concept of antigenic mimicry helps explain why it has been so difficult to develop an effective vaccine to protect against *E. coli* K1 and *N. meningitidis* serogroup B, but not against *N. meningitidis* serogroup C, which produces the α2,9-linked polySia capsule.

Perhaps the clearest function for the α2,8-polySia capsule on *E. coli* K1 is that it serves as a receptor for polySia-specific bacteriophages. These K1F bacteriophages require surface expression of the capsule for infectivity (Vimr *et al.*, 1984). The phages also serve as the source of the different endo-N-acylneuraminidases that have been described (reviewed by Hallenbeck *et al.*, 1987a).

3.2. Teleost Fish Oogenesis

While the exact molecular involvement of polySia chains on teleost fish PSGPs during oogenesis continues to be studied (Inoue, 1993), these molecules appear to function in a diverse number of cellular processes, including species-specific cell–cell recognition events during fertilization and early embryogenesis (Table II; Inoue and Inoue, 1986; Inoue, 1993), and in providing the egg with a protective barrier (Inoue *et al.*, 1991). As a class, the Salmonidae fish egg PSGPs are localized in the extracellular carbohydrate-enriched capsule, designated the vitelline envelope (VE), and as soluble components in the cortical vesicles that localize in the cortex of unfertilized eggs. After fertilization or egg activation, cortical vesicle exocytosis occurs, and the 200-kDa high-molecular-mass PSGP (H-PSGP) undergoes a rapid proteolysis at the perivitelline space to produce a 9-kDa low-molecular-mass repeating unit (L-PSGP; Kitajima and Inoue, 1988; Song *et al.*, 1990). There is no change in the size of the polySia chains. The presence of PSGPs in the VE may serve initially as a recognition marker for sperm binding. The polyanionic nature of the PSGPs may also influence the three-dimension conformation of the protein. Results supporting the sperm recognition hypothesis were reported by Kanamori *et al.* (1991) who showed that the KDN-containing glycoprotein of the VE of rainbow trout eggs was localized in the second layer of the envelope, and had strong sperm-agglutinating activity. The *O*-linked KDN-oligosaccharide structure described above inhibited sperm agglutination, suggesting that KDN-glycoprotein may specifically recognize sperm and thus may mediate egg–sperm interaction (Inoue *et al.*, 1991).

The polySia gp's in the vitelline envelope have also been postulated to protect eggs from bacterial invasion (Table II). Because of the high viscosity and polyanionic nature of ovarian fluid KDN(Sia)-gp's, these molecules may protect eggs from artificial activation, mechanical destruction, osmotic lysis, and bacterial infection (Kanamori *et al.*, 1989). These molecules have an affinity and high capacity for binding Ca^{2+} ions, which could also be important for fertilization, since Ca^{2+} is required for the hardening or conversion of the VE into the fertilization envelope (Inoue, 1993). Thus, the physiological function of KDN-gp may be to retain Ca^{2+} around the embryo, thereby protecting the embryo from osmosis and bacterial infection (Kitazume *et al.*, 1994a).

Capping of the nonreducing termini of oligo-polySia chains with KDN protects these chains from degradation by exosialidases, and this may be important in egg activation (Nadano *et al.*, 1986). *O*-Acetylation of the poly(Neu5Gc) chains also renders them resistant to depolymerization by bacterial exosialidases and Endo-N (Nadano *et al.*, 19986; Iwasaki *et al.*, 1990). While the significance of the resistance to neuraminidases is not known, maintenance of the net negative surface charge again suggests that polySia may function as a conformational determinant to modulate cell-to-cell recognition processes and interactions, and cell migration.

3.3. Sea Urchin Embryogenesis

Identification of polySia-containing glycoproteins in the jelly coat of sea urchin eggs is a very recent finding (Kitazume *et al.*, 1994b). Consequently, the function of these molecules is mostly speculative, and centers on the key finding that embryos contain a developmentally regulated α2,8-polysialyltransferase. The major polysialylated product is a 38-kDa glycoprotein. Peak activity for polysialylation was found to occur at the mesenchyme blastula/early gastrula stage. These findings led to the hypothesis that the polySia chains may mediate cell–cell interactions, perhaps during gastrulation, when endodermal cells interact with ectodermal cells. They also suggest that some type of polySia structure may be involved in inducing the acrosome reaction in sperm (Kitazume *et al.*, 1994b), since a Sia-containing glycoprotein has been implicated in this process (Keller and Vacquier, 1994). Expression of the polySia glycotope in sea urchin may have many of the same physiological functions as proposed for the fish egg PSGPs.

3.4. Neurobiology

As noted above (Section 2.2.3a) the originally hypothesized function of the polySia glycotope on N-CAM was to modulate cell–cell adhesive interactions and cell migration, principally during embryogenesis. The functional importance of sialyl residues on N-CAM was first shown by Cunningham *et al.* (1983), and by Hoffman and Edelman (1983), who used sialidase to establish that the embryonal form of N-CAM (Sia-rich) was less adhesive than the adult form (Sia-poor). These studies were carried out before it was known that the Sia residues were joined together internally, forming α2,8-polySia chains on the *N*-linked oligosaccharides of N-CAM. With the discovery that Endo-N could selectively remove the polySia chains on N-CAM at physiological pH (Vimr *et al.*, 1984), the functional role of polySia in neural morphogenesis could be studied directly. The initial studies in this area showed that the microinjection of Endo-N into the eyes of 3½-day-old chick embryos produced marked abnormalities in those regions of the retina that contained the highest amounts of N-CAM (Rutishauser *et al.*, 1985). Thickening of the neural epithelium in the posterior eye, a failure of cells in the eye to elongate radially, formation of an ectopic optic fiber layer, and an incomplete association of the pigmented epithelium with the neural retina were the major perturbations associated with the absence of polySia at this stage of development. These results showed that the polySia residues on N-CAM exerted regulatory effects on adhesion between living cells *in situ* by decreasing N-CAM-mediated cell adhesion. These studies also showed that the amount of polySia was a critical determinant for normal morphogenesis and neural development. Thus, polySia represents one of the few defined carbohydrate structures with an identifiable role in a morphogenic pathway.

The functional significance of polysialylated N-CAM in development was further confirmed, again taking advantage of Endo-N to alter polysialylation (Rutishauser *et al.*, 1988; Rutishauser, 1989). PolySia expression was determined to be associated with more plasticity during cell migration and axon outgrowth (Sunshine *et al.*, 1987), and to decrease choline acetyltransferase activity in chick sympathetic ganglion cells in culture (Acheson and Rutishauser, 1988). Again, the Endo-N-catalyzed removal of polySia from the surface of a neuroblastoma/sensory neuron cell hybrid increased cell–substrate adhesion (Rutishauser, 1989; Acheson *et al.*, 1991), while its removal from the axon surface during innervation of chick muscle increased axon fasciculation and decreased nerve branching (Landmesser *et al.*, 1990). These findings led Landmesser *et al.* (1990) to propose that the polySia glycotope on N-CAM may regulate intramuscular nerve branching during embryogenesis. The polySia chains of embryonal N-CAM may prevent cell adhesion by physically preventing cell-to-cell apposition, simply because the polyanionic shield may increase the intercellular space between cells (Yang *et al.*, 1992).

The importance of polySia expression in neuronal sprouting was first revealed by Rosenberg *et al.* (1986) who showed that the 5B4 antigen, a surface antigen expressed on sprouting neurons from fetal but not adult rat brain, contained α2,8-linked polySia chains. Expression of the 5B4 antigen was developmentally regulated, and its appearance on neurons was coincident with neuronal sprouting (Ellis *et al.*, 1985; Wallis *et al.*, 1985). Subsequent studies by Doherty *et al.* (1990) showed that polySia may modulate N-CAM-dependent axonal growth during development, since neurite outgrowth from embryonic chick retinal ganglion cells was inhibited by removal of polySia, or by anti-polySia antibodies. In cultured rat hippocampal neurons, N-CAM-dependent neurite outgrowth was also dependent on polySia, and was inhibited by pertussis toxin or Ca^{2+} channel antagonists (Doherty *et al.*, 1992). These results suggested that polysialylated N-CAM may activate a second messenger pathway in primary neurons.

During postnatal neurogenesis in the adult rat, polysialylated N-CAMs are expressed on newly generated granule cells in the deepest part of the hippocampus in the dentate gyrus, suggesting that this glycotope may influence the formation of new neural circuits (Sedi and Arai, 1991a,b, 1993a,b). In the piriform cortex, polySia expression also persists in the adult, indicating that the embryonal N-CAM may function in reorganization of the adult piriform cortex (Seki and Arai, 1991a,b). During chick lens cell differentiation, epithelial cell N-CAM is more highly sialylated in the adult than the embryonic lens, indicating that polySia may mediate the formation of fiber cell gap and adherence junctions (Watanabe *et al.*, 1992). Polysialylated N-CAMs may also regulate the migration and maturation of mesencephalic dopaminergic cells during development of the mesencephalon in the rat (Shults and Kimber, 1992).

In muscle development, polysialylation of the GPI-anchored isoform of N-CAM appears to be essential for the developmentally regulated separation of secondary myotubes from primary myotubes (Fredette *et al.*, 1993). In mouse embryos, the polysialylated N-CAM isoform is first expressed at the eight-cell stage and becomes reduced on the trophoblast of blastocysts at implantation. These findings suggest that the regulated expression of the polySia glycotope may be essential for cell interactions of the preimplantation mammalian embryo (Kimber *et al.*, 1994). During bone morphogenesis, polysialylated N-CAMs are transiently expressed on the cell surface of osteoblasts, where their function may be to modulate the interaction of osteoblasts and regulate skeletogenesis (Lee and Chuong, 1992).

The possible involvement of polysialylated N-CAM in inducing neuronal morphological differentiation was suggested by Joliot *et al.* (1991, 1992, 1993a,b, 1994) who showed that polySia is the neuronal surface receptor for the neurotropic activity of a homeobox peptide of the Antennapedia (pAntp) gene. Homeoproteins define the morphology of organs during early development, where expression of homeogenes later in development may influence neurite growth (Joliot *et al.*, 1994). A 60-amino-acid peptide corresponding to the homeobox of pAntp was transported through the membrane of cultured neurons and was translocated to the nucleus where it induced neuronal differentiation. Internatlization of the homeobox peptide was dependent on surface expression of α2,8-polySia chains, based on the experimental finding that pretreatment of neurons with Endo-N abolished the neurotropic activity of pAntp (Joliot *et al.*, 1991, 1994). While the molecular details of this provocative finding are obscure, it is possible that the homeopeptide recognizes and binds the extended helical conformer of α2,8-polySia chains on the surface of neurons because its three-dimensional solution conformation is similar to the helical epitope of DNA (Brisson *et al.*, 1992).

3.5. Tumor Metastasis

The hypothesis that polysialylation enhances the metastatic potential of some human tumors was noted above. While much of the evidence in support of this hypothesis is correlative, the available data strongly suggest that polysialylated N-CAM may be of clinical importance in certain malignancies, particularly as a neurovirulent determinant on neurotropic tumors. By analogy with the function of the polySia capsule on neuroinvasive *E. coli* K1 and *N. meningitidis,* surface expression of the polySia glycotope may help some malignant tumors to detach and/or to penetrate the blood–brain barrier and colonize the CNS. Evidence in support of this concept is summarized.

The original finding that polySia was temporally expressed in developing human kidney, not in adult kidney, but was reexpressed in Wilms' tumor led to the important conclusion that polySia was an oncodevelopmental antigen in kidney (Roth *et al.*, 1988c). The observation that long chains of polySia were expressed on some human neuroblastomas, some of which were metastatic to the bone marrow, also suggested that polySia was an oncodevelopmental antigen in brain, since the embryonal form of N-CAM was less frequently expressed on adult brain tissue (Livingston *et al.*, 1988; Glick *et al.*, 1991). Possibly the polyanionic shield on the surface of these and other malignant tumors alters normal cell-to-cell interactions and processes, and may raise the invasive potential (Livingston *et al.*, 1988; Roth *et al.*, 1988b,c; Molenaar *et al.*, 1990). The α2,8-linked polysialotope may also protect malignant cells from immune surveillance.

Michalides *et al.* (1994) noted a significant correlation between polySia expression and aggressiveness of small-cell lung carcinomas, and suggested that the polySia glycotope may function to impair cell–cell and cell–matrix interactions which leads to tumor cell detachment. Nilsson (1992) also reported a positive correlation between the presence of embryonal N-CAM in small-cell lung carcinomas and malignant behavior. PolySia is not expressed on all small-cell lung carcinomas, but rather is a clonable trait (Scheidegger *et al.*, 1994). Its expression on small-cell lung carcinomas is correlated with reduced cell–cell adhesions which leads to a higher incidence of metastases in nude mice.

The functional role of α2,8-polysialylated N-CAM as a neurodeterminant for neurometastases is best illustrated by three recent findings. First, Grogan *et al.* (1994) have reported that this glycotope is a likely neurodeterminant relevant to CNS localization in natural killer-like T-cell malignant lymphoma. Second, there is an association between N-CAM positivity and polySia expression in acute myeloid leukemia (AML) (Scott *et al.*, 1994). Expression of polySia N-CAM may mediate homotypic and heterotypic events in leukemic cells, thereby promoting extramedullary and CNS involvement in AML. Thus, the presence of polysialylated N-CAM expression may be a useful diagnostic marker, along with cytogenetic characteristics, in identifying subtype AML with the potential for extramedullary and CNS involvement (Scott *et al.*, 1994). Third, several solid tumors of the head and neck (adenoid cystic carcinoma, olfactory neuroblastoma, basal and squamous cell carcinoma, olfactory neuroblastoma, basal and squamous cell carcinoma) spread along cranial nerve tracts which permit direct invasion of tumor into the base of the skull and brain. We have initiated studies to determine if polysialylated N-CAM can be used as a molecular marker for assessing the metastatic potential of adult head and neck tumors. Approximately one-third of the tumors examined, mostly squamous cell carcinomas, expressed the embryonic polySia glycotope on N-CAM. Histological evidence of perineural invasion was demonstrated in about two-thirds of the polysialylated N-CAM tumors which showed perineural invasion (Gandour-Edwards *et al.*, 1995). These preliminary findings also suggest that polySia expression may be a more sensitive molecular marker for assessing perineural invasion than the currently used histological methods.

All of the α2,8-polySia chains expressed on human tumors (Table III) are believed to be associated with N-CAM, and in many of the studies cited, this has been confirmed directly. A novel α2,9-linked disialosyl structure $(\rightarrow 9\text{Neu5Ac}\alpha 2\rightarrow)_2$ on human embryonal carcinoma cells was described by Fukuda *et al.* (1985), which has interesting implications for future studies. α2,9-linked Sia chains have not been previously reported in eukaryotes, their only previous citing being restricted to neurotropic *N. meningitidis* serogroup C and *E. coli* Bos12 and K92 strains. Thus, the capping of polylactosamine chains on surface glycoproteins at their nonreducing termini with α2,9-disialosyl residues raises two important possibilities. The first is that the α2,9-disialosyl structure will not be restricted to a single ovarian tetracarcinoma. Rather, it may be a decisive glycotope on other

Table III
Human Tumors Expressing the α2→8-Linked Polysialic Acid Glycotope

Neuroblastomas
Nephroblastomas (Wilms' tumor)
Medulloblastomas
Pheochromocytomas
Medullary thyroid carcinomas
Small-cell lung cancers
Lung carcinoid tumors
Pituitary adenomas
Congenital mesoblastic nephroma
Multiple myelomas
T-cell malignant lymphomas
Leukemias (AML)
Head and neck tumors (principally squamous cell carcinomas)
Malignant melanomas
Human tumor expressing $(\rightarrow 9\text{Neu5Ac}\alpha 2\rightarrow)_2$
Ovarian teratocarcinoma (PA1 embryonal carcinoma cells)

glycoconjugates implicated in normal development, and it may be an oncodevelopmental antigen that is expressed on tumors. The second possibility is that other α2,9-linked polySia chains will be discovered, and that these chains may play an important role in the metastatic phenotype, analogous to their α2,8-polySia counterpart. What is needed to follow up Fukuda's intriguing findings are facile, sensitive, and specific reagents to detect and analyze polyα2,9-sialyl residues.

4. BIOSYNTHESIS OF THE POLYSIALIC ACID GLYCOTOPES

4.1. The CMP-Sia:α2,8-polysialyltransferase Complex in Neuroinvasive *Escherichia coli* K1

4.1.1. Biosynthetic Pathway

In neuroinvasive *E. coli* K1, synthesis of the α2,8-linked polySia capsule is catalyzed by a CMP-Sia:poly-α2,8-sialosyl sialyltransferase (polyST) complex which is postulated to carry out the following reactions:

$$1.\ \text{CMP-Sia} + \text{P-C}_{55} \overset{\text{polyST}}{\rightleftharpoons} \text{Sia-P-C}_{55} + \text{CMP}$$

$$2.\ n(\text{Sia-P-C}_{55}) \xrightarrow{\text{polyST}} (\rightarrow 8\text{Sia}\alpha 2\rightarrow)_n\text{-P-C}_{55} + (n\text{-}1)\text{P-C}_{55}$$

$$3.\ (\rightarrow 8\text{Sia}\alpha 2\rightarrow)_n\text{-P-C}_{55} \xrightarrow[\text{+ endogenous acceptor}]{\text{polyST}} (\rightarrow 8\text{Sia}\alpha 2\rightarrow)_n\text{-acceptor} + \text{P-C}_{55}$$

where P-C_{55} is the glycosyl carrier lipid, undecaprenylphosphate.

Chain synthesis is initiated by the transfer of Sia from CMP-Sia to P-C_{55} (Reaction 1), which functions as an intermediate carrier of Sia residues (Troy *et al.*, 1975). Polysialylation appears to occur on undecaprenylphosphate (Troy *et al.*, 1975; Masson and Holbein, 1983), and the D.P., *n* in Reaction 2, extends up to 100–200 Sia residues (Troy *et al.*, 1993). Such preassembled "activated" polySia chains may be translocated across the inner membrane while linked to P-C_{55}, possibly catalyzed by a translocase. Alternatively, a polysialylated reactive phosphoryl intermediate could be transferred *en bloc* to a translocator, designated "endogenous acceptor" in Reaction 3, which then shuttles the polySia chains across the inner membrane. A 20-kDa membrane protein that is implicated as an endogenous acceptor for this reaction has been identified (Weisgerber and Troy, 1990). Similar sialylated P-C_{55} intermediates have also been described for polySia capsule synthesis in *N. meningitidis* serogroup B (Masson and Holbein, 1983).

PolySia chain growth occurs by the addition of Sia residues to the nonreducing terminus of the growing chain (tail growth) in which the activated linkage in CMP-Sia is used for its own addition (Kundig *et al.*, 1971; Rohr and Troy, 1980). The enzymatic activities responsible for initiating polySia synthesis (an initiase) and for catalyzing chain polymerization (a polymerase) have not been isolated from *E. coli* K1. Such enzymes have recently been identified, however, in the synthesis of trout egg polysialoglycoproteins during oogenesis by Kitazume *et al.* (1994a) (see below). In *E. coli* K1, polySia chain initiation but not chain polymerization requires protein synthesis (Whitfield and Troy, 1984; Whitfield *et al.*, 1984b). A dozen membrane proteins were identified by temperature up-shift experiments that were temporally correlated with polysialylation (Whitfield and Troy, 1984). Subsequent gene cloning experiments confirmed that at least 12–14 proteins were required for surface expression of the polySia capsule (Boulnois *et al.*, 1987; Silver and Vimr, 1990). It also seems likely that at least two gene products are required for polySia chain synthesis in *E. coli* K1. While one transferase could possess two catalytic sites, identification of a presumed undecaprenol/dolichol recognition sequence in the deduced protein sequence of NeuE but not NeuS (Troy, 1992) suggest that NeuE may be an initiase responsible for starting polySia chain synthesis or is somehow involved in polySia chain translocation, while NeuS catalyzes chain polymerization (Troy, 1992; Troy *et al.*, 1993). As described below, NeuE and NeuS are the products of two genes (*neu*E and *neu*S) that are encoded within the *kps* gene cluster of *E. coli* K1. By gene cloning and sequencing experiments, *neu*E and *neu*S have been implicated in polySia chain synthesis (Weisgerber *et al.*, 1991; Steenbergen and Vimr, 1991).

4.1.2. Genetic Organization of the *kps* Gene Cluster

At both the biochemical and genetic levels, the pathway controlling surface expression of the polySia glycotope in neuroinvasive *E. coli* K1 has turned out to be surprisingly complex. The genetic organization of the *kps* gene cluster encoding the genes for K1 capsule expression is shown in Figure 1. *kps* is a designation for "capsular polysaccharide." The cluster maps at 64 units on the *E. coli* K1 chromosome and is absent from common isolates of *E. coli* (Ørskov *et al.*, 1976; Vimr, 1991). The gene complex has been cloned and is contained in ca. 16–19 kb of DNA that consists of three coordinately regulated regions that code for at least a dozen proteins (Silver *et al.*, 1981; Boulnois *et al.*, 1987). A central region of ca. 5–6 kb of DNA (region 2) codes for at least six enzymes required for Sia synthesis (*neu*B, *neu*C), activation (*neu*A), and chain initiation and polymerization (*neu*E/*neu*S). A flanking region of ca. 9 kb (region 1) codes for 5–6 proteins implicated in the export of the capsule from the inner to the outer membrane. On the opposite side of the biosynthetic genes is region 3 (ca. 1.6 kb) that is postulated to code for proteins involved in the energetics of chain synthesis (*kps*T) and in the translocation (*kps*M) of polymers across the inner membrane (Pavelka *et al.*, 1994). The molecular organization of region 1 and 3 genes is architecturally similar in all *E. coli* cells synthesizing group II-type capsules, and they appear to be functionally interchangeable (Roberts *et al.*, 1988; Frosch *et al.*, 1991). In contrast, the biosynthetic genes in region 2 are specific for each type of capsule (Frosch *et al.*, 1991).

In addition to the *kps* gene products that are required for polySia chain synthesis, an inducible catabolic system for Sia, consisting of two new genes,

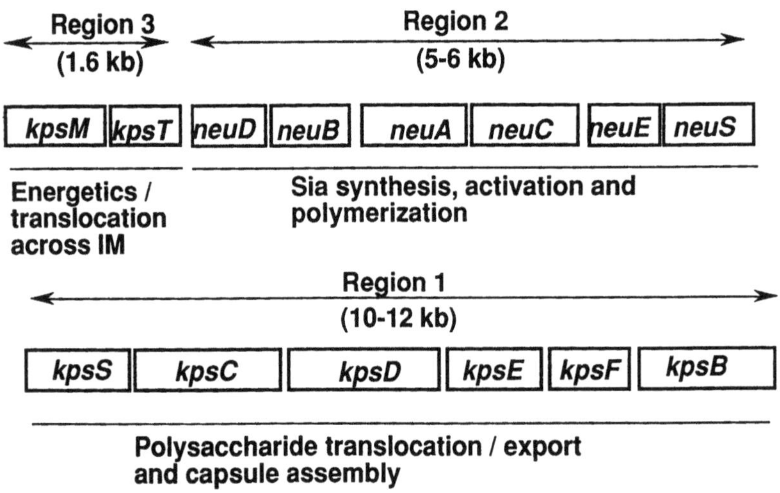

FIGURE 1. Genetic organization of the *kps* gene complex in neuroinvasive *E. coli* K1.

has been described (Vimr and Troy, 1985a,b). *nan*T codes for a Sia-specific permease, and *neu*A for Sia-specific aldolase. Regulation of Sia aldolase, induced by Sia, is necessary for dissimilating Sia, and for modulating the level of metabolic intermediates in the Sia pathway. These findings have been exploited to construct strains that do not degrade [^{14}C]-Sia (*nan*A4 mutation) so that its metabolic fate can be followed, and to synthesize intracellular [^{14}C]polySia chains with high specific activity (Merker and Troy, 1990; Troy *et al.*, 1990a,b; Vimr, 1992). Importantly, this has allowed studies on the membrane topology of the polyST complex and the vectorial translocation of polySia chains across the inner membrane to be carried out (Janas and Troy, 1989; Troy *et al.*, 1990a,c).

Region 2 Gene Cluster: Possible Function of *neu*E and *neu*S Gene Products. *neu*S codes for a 46-kDa protein, designated NeuS, which has α2,8-polyST activity (Weisgerber *et al.*, 1991). A 34-kDa protein implicated earlier as the *neu*S gene product (Steenbergen and Vimr, 1991) had a truncated carboxy-terminus, yet still retained some sialyltransferase activity (Weisgerber *et al.*, 1991). Sequence comparisons revealed that NeuS had no significant transmembrane spanning domains. Secondary structure plots revealed, however, that NeuS contained at least eight potential membrane-spanning amphiphilic β-sheets (Weisgerber *et al.*, 1991), which likely accounts for how the polyST complex is associated with the inner membrane (Troy and McCloskey, 1979).

NeuS has a relatively high similarity score when aligned with DNA polymerases and some other DNA binding proteins (Troy *et al.*, 1993). This implies that these proteins may be evolutionarily related, perhaps because they share the common property of binding to polyanions.

When cloned *neu*S was overexpressed in *E. coli* lacking *kps*, it could not initiate *de novo* polySia chain synthesis. Rather, NeuS could only transfer activated Sia residues to exogenous sialyl oligomer acceptors (Steenbergen and Vimr, 1991). This observation emphasized the possible involvement of other region 2 gene products, e.g., NeuE in polysialylation. The *neu*E gene is located 5′ to *neu*S, and codes for a 22-kDa protein (Weisgerber *et al.*, 1991; Steenbergen and Vimr, 1991). NeuE has two potential membrane-spanning regions (Weisgerber *et al.*, 1991; Troy, 1992). A 20-amino-acid domain in the carboxy-terminal region contains a 13-amino-acid conserved sequence that has homology with polyisoprenol (dolichol) recognition sequences (Troy, 1992; Troy *et al.*, 1993). Presumed dolichol recognition sequences were initially reported by Albright *et al.* (1989) to be present in three glycosyltransferases (ALG 1, ALG 7, and DPM 1) associated with *N*-linked glycosylation in yeast. A comparison of the NeuE sequence with ALG 7 showed 62% similarity (Z value 11.0) with 16.1% identity in 188 amino acid overlap (Troy *et al.*, 1993). ALG 7 is the *N*-acetylglucosaminyltransferase that initiates *N*-linked oligosaccharide synthesis by catalyzing the transfer of GlcNAc-1-P from UDP-GlcNAc to dolichylphosphate (Albright *et al.*, 1989). Thus, while the function of NeuE remains to be determined, the unexpected new finding that NeuE, but not NeuS, contained then a presumed polyisoprenol recognition sequence, led us to hypothesize that if two different enzyme activities were responsible for polysia chain synthesis, then NeuE was likely the sialyltransferase that initiated polysialylation by catalyzing the formation of sialylmonophosphorylundecaprenol from CMP-Sia and undecaprenylphosphate (Reaction 1; Troy, 1992; Troy *et al.*, 1993). NeuE, like NeuS, is also evolutionarily related to DNA polymerases and some other DNA binding proteins (Troy *et al.*, 1993).

It is not known if the 22-kDa *neu*E gene product is the same 20-kDa membrane protein that was identified earlier to be an endogenous acceptor in polySia chain synthesis (Weisgerber and Troy, 1990). It was suggested that this acceptor protein may function in polySia chain initiation and/or as a translocator to transport polySia chains across the inner membrane (Weisgerber and Troy, 1990). These two functions may not be mutually exclusive if NeuE and NeuS function together in a multienzyme complex with undecaprenylphosphate to catalyze polySia chain initiation, polymerization, and translocation across the inner membrane.

Neither NeuE nor NeuS appears to have any relevant homology to the liver α2,3 or α2,6 sialyltransferases (Steenbergen and Vimr, 1991). Whether any homology exists between NeuE or NeuS and the eukaryotic CMP-Sia:poly-α2,8-sialosyl sialyltransferase originally described in a Golgi-enriched fraction from 20-day-old fetal rat brains (McCoy *et al.*, 1985), awaits cloning of this important mammalian enzyme.

4.1.3. Membrane Topology of the α2,8-Polysialyltransferase Complex

The membrane topology of the polyST complex in *E. coli* K1 was determined by using sealed membrane vesicles of defined orientation (Janas and Troy, 1989; Troy *et al.*, 1990b,c; reviewed in Troy, 1992). Enzyme activity in right-side-out vesicles (ROV) that have the same topology as intact cells was compared to that in inside-out vesicles (IOV), which have the opposite orientation. IOV have a unique orientation since enzymes located on the inner surface of the cytoplasmic (inner) membrane appear on the exterior side of the vesicle (Kaback, 1971; Owen and Kaback, 1978). Such enzymes can thus be studied using impermeable reagents like CMP-Sia, and proteolytic enzymes, in the absence of detergents. The sidedness and impermeability of ROV and IOV were verified using voltage-sensitive fluorescent probes. Our results showed no polyST activity in ROV above background. In contrast, IOV showed a fivefold increase in polyST activity. Enzyme activity could also be measured in ROV after sonication or by permeabilization with Triton X-100 or toluene. These findings revealed that the functional domain of the polyST was located on the cytoplasmic surface of the inner membrane. This conclusion was confirmed by partial trypsinolysis, which showed only a slight decrease in polyST activity when ROV were treated with trypsin before inversion. In contrast, more than 90% of the activity was lost when ROV were inverted to IOV before trypsinolysis.

4.1.4. Energetics of Polysialylation in *E. coli* K1

The importance of the proton electrochemical potential gradient, $\Delta\mu H^+$, in activation of polySia chain synthesis was first demonstrated by showing that energy uncouplers which collapsed the gradient, e.g., carbonyl cyanide *m*-chlorophenylhydrazone (CCCP), inhibited activation (Whitfield and Troy, 1984). While uncertainty still exists as to the molecular events in polySia chain expression which are dependent on the membrane potential, it was proposed that the proton motive force may be required for maintenance of a membrane conformation necessary for coupling chain synthesis and translocation (Whitfield and Troy, 1984). More recent studies to further delineate the energetics of polysialylation in sealed inside-out vesicles have shown that both ATP hydrolysis and NADH and lactate oxidation increased polyST activity. Transferase activity was reduced by the nonhydrolyzable ATP analogue, AMP-PNP, and the photoaffinity probe, 8-azido ATP. The addition of NADH partially restored activity. These changes in polyST activity were correlated with changes in $\Delta\mu H^+$, as measured by quenching 9-amino-6-chloro-2-methoxyacridine fluorescence. These results thus confirmed that polyST activity was modulated by $\Delta\mu H^+$, and that polySia chain synthesis required the high-energy phosphoryl potential of ATP. ATP hydrolysis may be used in chain elongation to drive the polymerization reaction forward (Whitfield and Troy, 1984).

Because polySia chain synthesis in inside-out vesicles does not involve translocation of these polymers across the inner membrane, the requirement of energized membranes for full enzymatic activity of the polyST complex was established (Troy, 1992). The proton motive force (both $\Delta\mu H^+$ and $\Delta\Psi$) was also obligatorily coupled to the *in vivo* translocation of the polySia chains across the inner membrane (Troy *et al.*, 1990a, 1991), as summarized below.

4.1.5. Polysialic Acid Chain Translocation across the Inner Membrane

The new finding that the polyST complex was located on the cytoplasmic surface of the inner membrane required that either oligo- or polySia chains must traverse this membrane before being exported to the outer leaflet of the outer membrane. To study directly the translocation of polySia chains across the inner membrane, an *in vivo* radiolabeling procedure was developed (Troy *et al.*, 1990a,c, 1991). In this method, spheroplasts were prepared from *E. coli* K1 cells unable to degrade Sia because of a mutation in Sia aldolase (*nan*A4), and from a translocation-defective mutant, also with the *nan*A4 mutation. The Sia permease was first induced by growing cells in the presence of Sia (Vimr and Troy, 1985b). After pulse labeling with [^{14}C]-Sia, synthesis and translocation were followed kinetically, and the theory of compartmental analysis was applied to determine polySia chain distribution among the different compartments, defined as: total membrane bound ("in total" and "out total"). The "in total" was further defined as "in membrane bound" and "in free." Similarly, these polySia chains translocated across the inner membrane, and which appeared as "out total," were further resolved into the "out membrane bound" and "out free" fractions. The capability of distinguishing between chains that were bound to the cytoplasmic or periplasmic surface of the inner membrane was a key feature of this experimental procedure. This was made possible because translocated chains, in contrast to [^{14}C]polySia chains remaining within the spheroplast, were sensitive to depolymerization by the impermeable enzyme Endo-N. Endo-N is a diagnostic enzyme for identifying α2,8-linked polySia chains (Vimr *et al.*, 1984; Troy *et al.*, 1987). The involvement of the proton motive force in translocation could also be directly tested by examining the effects of CCCP, a modulator of $\Delta\mu H^+$, and valinomycin, a modulator of $\Delta\Psi$, on translocation (Troy *et al.*, 1990a; 1991). Three important conclusions emerged from these studies. First, full-length [^{14}C]polySia chains can be fully polymerized on the cytoplasmic surface of the inner membrane. No [^{14}C]oligoSia chains were detected in any compartment. This unanticipated finding means that polySia chain polymerization *in vivo* must precede translocation, and that full-length chains must be translocated across the inner membrane before being exported to the outer membrane. Second, polySia chain synthesis is not obligatorily coupled to chain translocation, since full-length [^{14}C]polySia chains were also identified intracellularly in a translocation-defective mutant. Third, both $\Delta\mu H^+$ and $\Delta\Psi$ are required for polySia chain translocation since CCCP and valinomycin inhibited transport 85 and 50%, respectively. The inhibitory effect of CCCP and valinomycin was shown to be related to inhibition of polySia chain translocation and not chain synthesis (Troy *et al.*, 1991). While the mechanism linking the proton motive force to polySia chain translocation remains unknown, energized membranes may be necessary for conformational change which allow the polyST activities (NeuE and NeuS) to be integrated with the "superlipid," undecaprenylphosphate, and the "energetics/translocator" complex (NeuE, KpsM and KpsT). KpsM and KpsT are the two gene products encoded by region 3 of the *kps* gene cluster (Figure 1), and belong to the *A*TP-*B*inding *C*assette (ABC)-transporter Family. These two proteins have been implicated in the transport of polySia chains across the inner membrane (Pavelka *et al.*, 1994). A central challenge now is to elucidate the components and mechanism of the membrane machinery that moves these polyanions across the membrane, and to define more fully the energetics.

4.1.6. Polysialic Acid Engineering: Use of the *E. coli* K1 Polysialyltransferase to Create Polysialylated Neoglycosphinogolipids

The polyST exposed on the cytoplasmic surface of inside-out vesicles from *E. coli* K1 can be used as a synthetic reagent to enzymatically engineer the glycosyl moiety of glycosphinogolipids (GSLs), thus creating a new class of oligo- or polysialylated neoGSLs (Cho and Troy, 1994b). While bacteria do not synthesize gangliosides, the *E. coli* K1 enzyme can selectively polysialylate several structurally related GSLs, when added as exogenous sialyl-acceptors. A structural feature common to the preferred sialyl-acceptors (GD3 >GT1a > GQ1b = GT1b > GD2 = GD1b = GD1a > GM1) was the disialyl glycotope, Siaα2,8-Sia, linked α2,3 to Gal. A linear tetrasaccharide with a terminal Sia residue (e.g., GD3) was the minimum-length oligosaccharide recognized by the polyST. Endo-N was used to confirm the α2,8-specific polysialylation of GSLs. Ceramide glycanase was used to release the polysialyllactose chains from the ceramide moiety. Size analysis of these chains showed that 60–80 Sia residues were transferred to the disialyllactose moiety of GD3. Accordingly, the structure of the newly synthesized polysialoganglioside from GD3 is:

$$\text{Sia}\alpha 2\text{-}(8\text{Sia}\alpha 2)_{60-80}\text{-}8\text{Sia}\alpha 2\text{-}3\text{Gal}\beta 1\text{-}4\text{Glc}\beta 1\text{-}1\text{ceramide}$$

Such polysialylated GSLs have not been previously described, although "polysialogangliosides" have been reported in embryonic chicken brain (Greis and Rösner, 1993; Freischutz *et al.*, 1994). In these cases, however, the six or seven Sia residues are distributed between the two nonreducing terminal branches of the sugar chain, indicating that three or four α2,8-linked Sia residues is the maximum chain length thus reported. Given the remarkable structural diversity in α2,8-linked polySia chains that has been described recently in other glycoconjugates (Sato *et al.*, 1993), it is probable that GSLs with oligo-polySia chains may occur naturally. The difficulty in isolating such hydrophilic lipids has likely precluded their identification. Thus, while the significance of these findings with respect to naturally occurring polysialylated gangliosides awaits further study, it seems likely that such novel polySia-GSLs may have important biological and pharmacological functions. This relates particularly to the number of GSLs which are developmentally regulated heterophile antigens that function as surface receptors for bacteria, viruses, and toxins (Stromberg *et al.*, 1988; Karlsson, 1989; Hakomori, 1990). Some GSLs are also tumor-associated antigens which have been implicated in transmembrane signaling, and as mediators for cell–cell interactions (Ladisch, 1987; Karlsson, 1989; Hakomori, 1990; van Echten and Sandhoff, 1993). Some cell adhesion processes also appear to involve a complementary GSL:GSL interaction that is mediated by specific carbohydrate residues in each GSL (Phillips *et al.*, 1990; Walz *et al.*, 1990).

A second practical advantage of the finding that the *E. coli* K1 polyST can polysialylate the glycosyl chains of GSLs is that the exogenous addition of GSL to some eukaryotic membranes can be used to help identify and study α2,8-polyST activities that are expressed at low levels such as to escape detection. Using this strategy, for example, we have identified and characterized an α2,8-polyST activity in the sea urchin *Lytichinus pictus*. The discovery, developmental expression, and characterization of this enzyme were readily facilitated because it also polysialylated GD3, when added as an oxogenous sialyl-acceptor.* These results thus predict that polyST activities in other species may also be discovered by using this experiment protocol (Cho and Troy, 1994b).

4.2. Rainbow Trout Egg Polysialoglycoproteins

4.2.1. Developmental Expression of an α2,8-Polysialyltransferase Activity during Oogenesis

Following the discovery and structural characterization of poly(Neu5Ac), poly(Neu5Gc), poly(Neu5Ac, Neu5Gc), and poly(KDN) chains, and their partially acetylated forms in salmonid fish egg PSGPs (Sato *et al.*, 1993), collaborative studies were initiated to determine how these α2,8-oligo/polySia chains were synthesized during oogenesis in rainbow trout oocytes (Kitazume *et al.*, 1994a). The rationale for looking at both the developmental expression of the polySia glycotope and the prerequisite α2,6-,α2,8-sialyl, and α2,8-polysialyltransferase activities involved in their synthesis was that oogenesis occurs over a 6-month period in trout. Since this developmental period is relatively slow compared to developing chick or rat brains, we reasoned that it might be possible to correlate the temporal expression of the different sialyltransferases predicted to be expressed, and their cognate sialyloligosaccharide structures. This strategy proved successful, and five key findings were made (Kitazume *et al.*, 1994a). First, glycoproteins were identified in ovaries 4 to 6 months prior to ovulation. The amount of PSGP increased with an increase in ovary weight. Second, 3 months prior to ovulation, a second more highly sialylated glycoprotein appeared. Structural studies confirmed that the two glycoproteins were discrete molecular species, designated PSGP(low Sia) and PSGP(high Sia), which differed only in their Sia content. PSGP(low Sia) contained mostly disialyl (Siaα2,8-Siaα2,6-) side chains whereas PSGP(high Sia) contained α2,8-linked oligo/polySia side chains ranging in length from 2 to over 20 Sia residues. The average degree of polymerization ($\langle DP \rangle_{av}$) was six. Compositional analysis verified that only the Sia content was increased in PSGP(high Sia), while the content of the other sugars remained unchanged. Third, biosynthetic studies using CMP-[^{14}C]-Neu5Ac indicated that three sialyltransferase activities were likely responsible

*Cho, J.-W. Troy, F. A., and Lennarz, W. J. (1994; unpublished).

for synthesis of the polysialyl residues of PSGPs, as summarized below, where R = H; Galβ1; GalNAcβ1→4Galβ1; Fucα1→3GalNAcβ1→4Galβ1; GalNAcβ1→4(Neu5Gcα2→3)GalNAcβ1→4Galβ1:

1.

```
                                                         α2,6-
            3GalNAcα1→O-Ser/Thr + CMP-Sia*                 →
           ↗                                              ST
R→3Galβ1
     Sia*α2
           ↘
            6
            3GalNAcα1→O-Ser/Thr + CMP
           ↗
R→3Galβ1
```

2.

```
       Siaα2
           ↘
            6                                            α2,8-
            3GalNAcα1→O-Ser/Thr + CMP-Sia*                 →
           ↗                                              ST
R→3Galβ1
Sia*α2→8Siaα2
           ↘
            6
            3GalNAcα1→O-Ser/Thr + CMP
           ↗
R→3Galβ1
```

3.

```
      Siaα2→8Siaα2
                ↘
                 6                                        α2,8-
                 3GalNAcα1→O-Ser/Thr + CMP-Sia*             →
                ↗                                        PolyST
      R→3Galβ1
(Sia*α2→8)nSiaα2→8Siaα2
                ↘
                 6
                 3GalNAcα1→O-Ser/Thr + CMP
                ↗
      R→3Galβ1
```

The first ST, an α-N-acetylgalactosaminide (R→Galβ1→3GalNAcα1→*O*-Ser/Thr) α2,6-sialyltransferase (α2,6-ST), catalyzed incorporation of the first

Sia residue onto the C-6 position of the proximal GalNAc residues of asialo-PSGP. The second reaction is catalyzed by the α2,6-sialoside[R→Galβ1→3 (Siaα2→6)GalNAcα1→*O*-Ser/Thr) α2,8-sialyltransferase (α2,8-ST). This enzyme is responsible for initiating α2,8-sialylation by catalyzing synthesis of the disialyl moiety of PSGP(low Sia), according to the reaction:

$$\text{Sia}\alpha2\rightarrow6(\text{R}\rightarrow\text{Gal}\beta1\rightarrow3)\text{GalNAc}\alpha1\rightarrow O\text{-Ser/Thr}$$
$$+\text{CMP-Sia}\rightarrow\text{Sia}\alpha2\rightarrow8\text{Sia}\alpha2\rightarrow6(\text{R}\rightarrow\text{Gal}\beta1\rightarrow3)\text{GalNAc}\alpha1\rightarrow O\text{-Ser/Thr.}$$

The α2,8-polysialosyl sialyltransferase (α2,8-polyST) catalyzed the third reaction,

$$(\text{Sia}\alpha2\rightarrow8)_n\text{Sia}\alpha2\rightarrow6(\text{R}\rightarrow\text{Gal}\beta1\rightarrow3)\text{GalNAc}\alpha\rightarrow O\text{-Ser/Thr}$$
$$+\text{CMP-Sia}\rightarrow(\text{Sia}\alpha2\rightarrow8)_{n+1}\text{Sia}\alpha2\rightarrow6(\text{R}\rightarrow\text{Gal}\beta1\rightarrow3)\text{GalNAc}\alpha1\rightarrow O\text{-Ser/Thr}$$

to give rise to PSGP(high Sia) via PSGP(low Sia). Expression of these sialyltransferase activities increased in accordance with the developmental appearance of each PSGP.

The fourth important finding was that structural characterization of the [^{14}C]-Sia-labeled side chains of each PSGP at different stages of development confirmed that synthesis of the disialyl unit containing a single α2,8-Sia residue occurred before α2,8-polysialylation. This finding suggested that α2,8-polysialylation was initiated by an α2,8-ST (Reaction 2) that catalyzed transfer of the first α2,8-Sia residue (the "initiase"). Appearance of the α2,8-ST activity proceeded by 2 or 3 months expression of the α2,8-polyST that catalyzed chain polymerization (Reaction 3; "polymerase").

The α2,6- and α2,8-sialyltransferase activities appeared to be associated with distinct polypeptide chains because they could be differentiated by their stability toward freezing and thawing. Greater than 90% of the α2,6-ST activity, for example, remained after freezing and thawing. In contrast, the α2,8-ST and α2,8-polyST activities were lost after freezing and thawing.

The fifth finding of significance was that the α2,6-ST and α2,8-polyST activities were shown to be localized in the Golgi-derived immature cortical vesicles in oocytes, and subsequently in the cortical vesicles in the mature eggs. This finding indicates that the core region of the glycan units of PSGP is assembled in the RER/Golgi apparatus in oocytes, giving rise to asialo-PSGPs, which are secreted from the Golgi apparatus into the cortical vesicles where sialylation occurs. Thus, asialo-PSGP molecules are packaged in the immature cortical vesicles along with the sialyltransferases and other macromolecules that constitute the "spherical core" (Inoue *et al.*, 1987). Upon fertilization, the cortical vesicles undergo fusion with the plasma membrane, discharging their contents into the perivitelline space where the peptide moiety of the PSGPs are depolymerized into the smallest repeating glycopeptide units, L-PSGP (Inoue and Inoue, 1986; Kitajima *et al.*, 1986; Inoue *et al.*, 1987). Elucidation of the detailed mechanism of polysialylation of PSGP molecules within the cortical vesicles and the molecular mechanisms of sorting and packaging of these enzymes with their physiological substrates, e.g., asialo-PSGP in the cortical vesicles, remains an important unresolved problem currently under study.

4.2.2. Synthesis of CMP-Deaminoneuraminic Acid (CMP-KDN) Using the CTP:CMP-KDN Cytidylyltransferase from Rainbow Trout Testis

Identification and Characterization of a CMP-KDN Synthetase. A logical step in the investigation of KDN-glycoconjugates is to determine how the KDN residues are incorporated into KDN-glycan chains. Presumably, the penultimate step would be synthesis of the activated KDN nucleotide, CMP-KDN, catalyzed by CTP:CMP-KDN cytidylyltransferase (CMP-KDN synthetase). A related enzyme, CTP:CMP-Sia cytidylyltransferase (CMP-Sia synthetase), has been identified in cells and tissues of animals (Kean and Roseman, 1966; Kean, 1970; van Dijk *et al.*, 1973) and bacteria (Warren and Blacklow, 1962; Kean and Roseman, 1966; Vann *et al.*, 1987). Since CMP-KDN synthetase had not been previously reported, we could not rule out the possibility that CMP-Neu5Ac or CMP-Neu5Gc residues were deaminated at the level of the sugar nucleotide or, alternatively, after incorporation into polymer. To resolve these alternate possibilities, studies were initiated to determine if CMP-KDN synthetase activity existed in rainbow trout testis, a tissue where KDN-gangliosides were first discovered (Song *et al.*, 1991) and to use such an enzyme, if found, to synthesize CMP-[^{14}C]-KDN. CMP-[^{14}C]-KDN would thus provide a key substrate for biosynthetic studies to determine how surface expression of KDN-glycoconjugates are regulated.

The results of these studies (Terada *et al.*, 1993) showed that the testis of rainbow trout (*Oncorhynchus mykiss*) contained a CMP-KDN synthetase that catalyzed the formation of CMP-KDN from CTP and KDN, as shown below:

$$\text{CTP} + \text{KDN} \xrightarrow{Mg^{2+}} \text{CMP-KDN} + \text{PPi}$$

V_{max}/K_m studies showed that KDN was a preferred nonulosoic acid substrate compared to Neu5Ac or Neu5Gc (4.4×10^{-3} min^{-1} for KDN versus 2.3 and 1.8×10^{-3} min^{-1} for Neu5Ac and Neu5Gc, respectively). CMP-KDN synthetase activity was maximal at pH 9–10 and at 25°C. The presence of either Mg^{2+} or Mn^{2+} was essential for CMP-KDN synthetase activity. CMP-KDN synthesis was stimulated more than tenfold by 25 nM Mg^{2+}, yet Mg^{2+} stimulated formation of CMP-Neu5Ac and CMP-Neu5Gc only fourfold. In contrast,

Neu5Ac and Neu5Gc were the preferred nonulosonic acid substrates for the calf brain CMP-Sia synthetase. Thus, mammalian CMP-Sia synthetases recognized similar, yet distinctively different, substrate specificity determinants. The expression of CMP-KDN synthetase was shown to be temporally correlated with development, and to parallel the developmental expression of (KDN)GM3 in sperm. The enzyme was used to synthesize CMP-[^{14}C]-KDN, which was characterized by ^{1}H NMR (Terada *et al.*, 1993), and to identify and characterize a CMP-KDN:α2,8-KDN transferase, as described below.

4.2.3. Identification, Characterization, and Developmental Expression of a Novel CMP-KDN:α2,8-KDN Transferase

In rainbow trout PSGPs, KDN residues cap the nonreducing termini of the α2,8-linked oligo-polySia chains, as shown below:

$$\begin{array}{l} \text{KDN}\alpha2\rightarrow8\text{Neu5Gc}\alpha2\longrightarrow8\text{Neu5Gc}\alpha2\rightarrow8\text{Neu5Gc}\alpha2 \\ \qquad\qquad\qquad\qquad\qquad\qquad\qquad\qquad\qquad\searrow \\ \qquad\qquad\qquad\qquad\qquad\qquad\qquad\qquad\qquad\quad {}^{6}_{3}\text{GalNAc}\alpha1\rightarrow\text{Ser/Thr} \\ \qquad\qquad\qquad\qquad\qquad\qquad\qquad\qquad\qquad\nearrow \\ \qquad\qquad\qquad\qquad\qquad\qquad\qquad\qquad\quad \text{R} \end{array}$$

where R→3GalNAcα1→ represents five different oligosaccharide chains (Inoue, 1993, and references therein). Terminal capping of the polySia chains by KDN appears to protect them from the action of exosialidases (Nadano *et al.*, 1986). Such modification would also provide a mechanism to prevent further chain elongation.

Studies were initiated to determine if a CMP-KDN transferase activity, which may be involved in the capping reaction, was present in trout ovaries. These studies revealed that a new glycosyltransferase, which catalyzed the transfer of [^{14}C]-KDN from CMP-[^{14}C]-KDN to the nonreducing termini of the polyNeu5Gc chains forming KDNα2→8(Neu5Gcα2→8)$_n$, was present in the Golgi-derived immature cortical vesicles obtained from trout ovaries 2 months prior to ovulation (Angata *et al.*, 1994). Appearance of the KDN-transferase at this stage of development coincided with the developmental expression of the α2,8-polysialyltransferase required for polySia chain polymerization (Kitazume *et al.*, 1994a). Expression of the KDN-transferase was also temporally correlated with expression of the KDN residues.

The KDN-transferase activity had a pH optimum about 7, and required either Mg^{2+} or Mn^{2+} for maximal activity. Incorporation of KDN into the oligo-polyNeu5Gc chains at their nonreducing termini was shown to prevent the subsequent elongation of the chains. Therefore, synthesis of the oligo-polyNeu5Gc chains preceded incorporation of KDN residues during maturation, and it is the KDN-capping reaction that appears to terminate polySia chain elongation in PSGPs.

Identification of the KDN-transferase is the first step in elucidating the molecular mechanism regulating expression of the different types of KDN-glycan chains in the various KDN-containing glycoconjugates. While this is the first report of a glycosyltransferase that catalyzes termination of α2,8-polysialylation in trout egg PSGPs, we can anticipate that a similar capping mechanism may also terminate α2,8-linked polySia chain elongation in other systems (Angata *et al.*, 1994).

4.3. Sea Urchin Embryo

Characterization and Developmental Expression of a CMP-Sia:α2,8-polysialyltransferase

Concomitant with the demonstration that the jelly coat of two different species of sea urchins contained a new class of naturally occurring polySia chains consisting of $(\rightarrow 5\text{-}O_{\text{glycoly}}\text{-Neu5Gc}\alpha 2\rightarrow)_n$ chains (Kitazume *et al.*, 1994b), a study of the biosynthesis of polySia-containing macromolecules in the developing sea urchin and embryo was undertaken. These studies led to the discovery that embryos of *S. purpuratus* contained a CMP-Sia:α2,8-polysialyltransferase (polyST) which catalyzed synthesis of α2,8-linked polySia chains on an endogenous 38-kDa glycoprotein (Cho *et al.*, 1995).* Expression of both the α2,8-polySia chains and the polyST were developmentally regulated. No significant polysialylation occurred prior to the mesenchyme blastula stage. Activity was maximal at the early gastrula stage of development, and then declined. These results were confirmed using the GSL, GD3, in saturating amounts as an exogenous acceptor. An excess of GD3 ensured that the polyST rather than the endogenous acceptor was limiting. The polyST was maximally active at pH 7.0 and at 20°C. No divalent cation requirement could be demonstrated. Similar to the *E. coli* K1 and polyST, the transferase from *L. pictus* embryos could polysialylate several exogenous ganglioside acceptors in addition to GD3, including GT1b, GQ1b, GD1a, and GM1a (Cho and Troy, 1994b). The enzyme, however, did not sialylate GM3, lactosyl caramide, sea urchin neutral glycolipids and gangliosides, α2,8-linked oligo-polySia (colominic acid), or trout egg PSGPs containing α2,8-linked oligo-polySia residues.

The developmental expression of polySia chain synthesis in sea urchin PSGPs during embryogenesis seems likely to be an excellent experimental system to elucidate what regulates polysialylation. Because the sea urchin embryo is a developmental system composed of three cell types in which the cell fate and tissue rearrangements are extensively understood (Foltz and Lennarz, 1993), this model system may also be ideal for studying the functional role of polySia chains.

*Cho, J.-W., Troy, F. A., and Lennarz, W. J. (1995; manuscript in preparation).

4.4. Fetal Rat and Embryonic Chick Brains

Polysialylation of Neural Cell Adhesion Molecules by a Developmentally Regulated CMP:Sia:α2,8-polysialyltransferase

The initial discovery of a eukaryotic α2,8-polysialyltransferase was made by McCoy *et al.* (1985), who described a Golgi-enriched enzyme from 20-day-old fetal rat brain that was responsible for the polysialylation of N-CAM. The polyST catalyzed the transfer of [^{14}C]-Neu5Ac from CMP-[^{14}C]-Neu5Ac to endogenous acceptor molecules, which migrated in SDS–polyacrylamide gels as embryonic N-CAM. Approximately one-third of the incorporated [^{14}C]-Neu5Ac was sensitive to depolymerization by Endo-N, thus demonstrating that the [^{14}C]-Neu5Ac was incorporated into α2,8-linked polySia chains (Vimr *et al.*, 1984; Troy *et al.*, 1987). This conclusion was confirmed by structural studies that proved that the products of Endo-N digestion were ^{14}C-labeled sialyloligomers (McCoy *et al.*, 1985; Hallenbeck *et al.*, 1987a). PolySia chain synthesis was stimulated threefold when embryonic N-CAM was added as an exogenous acceptor. Nearly two-thirds of the product of this reaction was Endo-N-sensitive, thus showing that embryonic N-CAM can be polysialylated *in vitro* by the rat brain α2,8-polyST. The fact that adult rat brains contain markedly reduced levels of polysialylated N-CAM (Rothbard *et al.*, 1982), and lacked measurable levels of the α2,8-polyST activity, led us to conclude that expression of this enzyme was developmentally regulated (McCoy *et al.*, 1985). The relationship between what regulates synthesis of the polySia chains and what controls the developmental reduction in polySia during the embryonic-to-adult conversion of N-CAM remains unknown. During rat brain development, however, a linkage-specific transition has been suggested for astrocytic progenitor cells (Blass *et al.*, 1994).

While the identification and general properties of the rat brain polyST as described above, have been confirmed by others (Regan, 1991; Halberstadt *et al.*, 1993), and are similar in developing chick brains (Sevigny and Troy, 1994),* there has been a surprising dearth of new information related to the *in vitro* biosynthesis of the polySia glycotope during the past decade. This likely results from the fact that, in spite of considerable effort, nobody has been successful in purifying or cloning a mammalian α2,8-polyST. It now seems likely that cloning by gene transfer has not been successful because polySia chain synthesis requires at least two α2,8-sialyltransferase activities. The first enzyme must begin chain synthesis (an "initiase"), and the second must catalyze α2,8-polySia chain polymerization (a "polymerase"), as shown to be the case for synthesis of the polysialylated glycoproteins in trout eggs (Kitazume *et al.*, 1994a). Alcaraz and Goridis (1991) have reported that polysialylation of N-CAM in AtT-20 cells, presumed to be of neuroendocrine origin, occurs rapidly and in either a late Golgi or post-Golgi compartment. Until the mammalian polysialyltransferase(s) is (are)

*Sevigny, M., and Troy, F. A. (1994; unpublished results).

cloned, it is unlikely that significant progress will be made in understanding the genetic and biochemical mechanisms that regulate surface expression of these multifunctional polysialylated glycoconjugates.

5. IDENTIFICATION OF POLYSIALIC ACID AND OLIGO-POLY KDN STRUCTURES

A *Methods in Enzymology* chapter has recently been published that describes in detail the strategies, reagents, and methods to detect, analyze, and modify the polySia and polyKDN glycotopes (Ye *et al.*, 1994). Accordingly, only a brief summary of these strategies and methods will be reviewed, and the interested reader should consult the above reference for specific details.

5.1. General Strategy: Immunochemical Detection Using Anti-Polysialic Acid and Anti-KDN Antibodies

Reagents to identify the polySia glycotopes have played a key role in advancing our knowledge of the natural occurrence, synthesis, and function of these sugar chains in both prokaryotic and eukaryotic organisms (reviewed in Troy, 1992). The development of highly specific and sensitive prokaryotic-derived reagents for both identifying and depolymerizing α2,8-linked polySia chains on neurovirulent *E. coli* K1 and neural cells was first reported in 1984 (Vimr *et al.*, 1984). These reagents were (1) a polyclonal IgM anti-polySia antibody, designated H.46, prepared by immunizing a horse with *N. meningitidis* serogroup B (Sarff *et al.*, 1975); (2) an α2,8-endo-*N*-acylneuraminidase (Endo-N), a soluble form of a bacteriophage-derived endosialidase produced by infection of *E. coli* K1 with a lytic bacteriophage (Vimr *et al.*, 1984; Hallenbeck *et al.*, 1987a); and (3) an α2,8-polysialyltransferase complex from *E. coli* K1 which could add Sia residues from CMP-Sia (or sialylmonophosphorylundecaprenol) to endogenous or exogenous acceptor substrates containing preexisting α2,8-linked oligo- or polySia chains (Troy and McCloskey, 1979). Thus, the most direct assay to identify polySia is to determine if anti-polySia immunoreactivity is abolished by pretreating the sample with Endo-N. This can be done by dot-blot or Western blot analyses. Confirmation of this conclusion requires structural studies to show that concomitant with the loss of anti-polySia immunoreactivity is the formation of sialyl oligomers. Further confirmation should be provided by showing that the presumed polysialylated glycoconjugate can function as an exogenous acceptor in the *E. coli* K1 polysialyltransferase reaction.

A number of anti-polySia antibodies have now been made that specifically recognize the α2,8-polySia glycotope. A monoclonal IgG2a, designated mAb735, was produced by Bitter-Suermann and colleagues (Frosch *et al.*, 1985). mAb735 has been widely used as a primary antibody to detect polySia, principally by

Roth and colleagues (Roth *et al.*, 1987, 1988a,b; Lackie *et al.*, 1990; Zuber *et al.*, 1992; Komminoth *et al.*, 1994). Another monoclonal antibody, designated mAb12E3, was produced by Seki and is also very selective in recognizing α2,8-polySia residues (Seki and Arai, 1991a,b; 1993a,b). Unfortunately, none of these antibodies are commercially available at present, but hopefully this will change shortly. Nevertheless, the developers of these antibodies have been gracious in providing the scientific community with these valuable reagents. Endo-N is also not commercially available, but interested investigators should contact this author regarding availability.

Several highly specific anti-KDN antibodies have been developed by Inoue and colleagues to detect (KDN)GM3, and oligoKDN expression in both prokaryotic and eukaryotic cells (reviewed in Table I, Ye *et al.*, 1994). mAb.kdn3G is an IgG that reacts most strongly with (KDN)GM3 and less strongly with glycoproteins containing a number of KDNα2,3Galβ1,3GalNAcα1,3[(→8-KDNα2)$_n$→6]GalNAcα1→ chains. Thus, kdn3G specifically recognizes the disaccharide structure, KDNα2,3Galβ1-, and can discriminate between KDN-α2,3Galβ1- and Neu5Acα2,3Galβ1- (Song *et al.*, 1993).

Antibodies that recognize (→8KDNα2)$_n$ oligosaccharides on KDN-containing glycoproteins have been prepared using KDN-gp as the immunogen (Kanamori *et al.*, 1991). These antibodies recognize α2,8-linked KDN chains but not α2,8-linked Neu5Ac or Neu5Gc residues, and have been used to identify new KDN-containing glycoconjugates in bacteria, fish, and, most recently, in rat pancreas (Kanamori *et al.*, 1994).

The continued development of antibodies and specific glycosidases (e.g., Endo-N and KDNases) will surely broaden our repertoire of reagents to identify new classes of polySia chains. It is important to recognize, however, that positive immunoreactivity must be confirmed by pretreatment with a specific glycosidase (Endo-N; KDNase) and rigorous chemical, biochemical, and physical methods. Some of these have been described (Ye *et al.*, 1994).

5.2. Endo-*N*-acylneuraminidase (Endo-N): Specific Molecular Probe to Detect and Structurally Modify α2,8-Linked Polysialic Acid Chains

Endo-N can serve as a specific molecular probe to detect and selectively modify α2,8-polySia chains. The enzyme catalyzes the depolymerization of polySia chains according to the following reaction:

$$(\text{-8Neu5Ac}\alpha\text{2-})_n\text{-X} \rightarrow (\text{-8Neu5Ac}\alpha\text{2-})_{2-4} + (\text{-8Neu5Ac}\alpha\text{2-})_2\text{-X}$$

The soluble form of the enzyme was purified to homogeneity and characterized (Hallenbeck *et al.*, 1987a,b; Troy *et al.*, 1987). The enzyme requires at least five Neu5Ac or Neu5Gc residues for activity. The products of limit digestion from the polySia capsule of *E. coli* K1 of the embryonic form of N-CAM were mainly DP4, with some DP3 and DP1,2 (Happenbeck *et al.*, 1987a). The purified Endo-N derived from bacteriophage K1F infection of *E. coli* K12/K1 hybrids also effectively cleaves the alternating (-8Neu5Acα2-) and (-9Neu5Acα2-) polymers, but not poly(-9Neu5Acα2-) chains (Hallenbeck *et al.*, 1987a). Similar phage-induced enzymes have been described by Kwiatkowski *et al.* (1982), Finne and Makela (1985), and Pelkonen *et al.* (1989) Endo-N has been used to detect polySia residues in bacteria (Vimr *et al.*, 1984), mammalian tissues (Vimr *et al.*, 1984; McCoy *et al.*, 1985; Roth *et al.*, 1987; Finne and Makela, 1985; Lipinski *et al.*, 1987; Acheson and Rutishauser, 1988; Livingston *et al.*, 1988; Moolenaar *et al.*, 1990), fish egg PSGP (Kitajima *et al.*, 1988), and the eel electric organ (James and Agnew, 1987, 1989).

Partial Endo-N digestion of fully polysialylated glycoconjugates results in modified glycoconjugates that contain shorter polySia chains (Weisgerber and Troy, 1990; Cho and Troy, 1994b), a fact that could be of potential usefulness to create new oligosialylated glycoconjugates.

5.3. Discovery and Use of a New Type of Sialidase, "KDNase," to Confirm the Identification of KDN-Containing Glycoconjugates

The isolation and characterization of a novel type of sialidase (KDNase) that specifically hydrolyzes KDN ketosidic but not *N*-acylneuraminyl linkages has been recently described (Kitajima *et al.*, 1994). A KDNase-producing bacterium, identified as *Sphingobacterium multivorum,* was isolated from the sewage pond at a trout hatchery that could grow on KDN-oligosaccharide alditols, KDN-α2,3Galβ1,3GalNAcα1,3(KDNα2-(-8KDNα2)$_n$-6)GalNAcol, as the sole carbon source. The enzyme, designated KDNase SM, exhibited a broad linkage specificity and was able to hydrolyze the KDN residues ketosidically linked α2,3-, α2,6-, and α2,8-. The enzyme does not release Neu5Ac or Neu5Gc from 4-MU-Neu5Ac, *N*-acetylneuraminyllactose, α2,8-oligo-polySia (colominic acid), or other Sia (Neu5Ac or Neu5Gc)-containing glycoconjugates (Kitajima *et al.*, 1994). KDNase SM will thus allow the discovery of new KDN-containing glycoconjugates, identified initially based on immunoreactivity with the anti-KDN antibodies, to be confirmed. Confirmation of the validity of this immunoreactivity has awaited the isolation of a specific KDNase that will selectively cleave KDN ketosidic but not *N*-acylneuraminyl linkages. Importantly, KDNase SM will now allow the function of the KDN glycotope to be studied in a way that heretofore has not been possible.

5.4. Use of the *E. coli* K1 Poly-α2,8-sialyltransferase to Identify Preexisting α2,8-Linked Oligo-Polysialic Acid Chains

The *E. coli* K1 polyST complex catalyzes the transfer of Sia from CMP-Sia to both endogenous and exogenous sialyl acceptors which contain preexisting α2,8-linked Sia residues (Vijay and Troy, 1975; Troy *et al.*, 1975; Troy and McCloskey, 1979; Whitfield *et al.*, 1984a; Whitfield and Troy, 1984; Weisgerber and Troy, 1990; Cho and Troy, 1994a,b). *E. coli* EV11, a mutant derived from a hybrid of *E. coli* K12 and a K1 polySia-expressing strain, is defective *in vitro* in catalyzing the endogenous synthesis of polySia. Inside-out vesicles from EV11 can, however, transfer Neu5Ac from CMP-Neu5Ac to exogenous acceptors (Vimr *et al.*, 1984). Thus, restoration of [^{14}C]polySia synthesis in *E. coli* EV11 membranes by the addition of exogenous acceptors, presumed to contain α2,8-linkages, is a sensitive indicator for determining the presence of preexisting oligo/poly(-8Neu5Acα2-) residues (Vimr *et al.*, 1984; Livingston *et al.*, 1988; Cho and Troy, 1994a,b).

6. CONCLUDING REMARKS AND FUTURE DIRECTIONS

Studies to understand polysialylation of surface glycoconjugates in cells as evolutionarily diverse as neuroinvasive bacteria and human cancer cells have emerged during the past decade as an exciting new area of glycobiology. Look for major advances in the future in our understanding of the function of polysialylation and the biological role that the polySia glycotope plays in nearly every facet of molecular, cell, and developmental biology, and molecular medicine. One of the most crucial challenge will be to determine how synthesis and surface expression of the polySia chains are regulated, and the importance of the three-dimensional conformation of these molecules to their function.

As new reagents and methods are discovered and developed to study polysialylation, we can also expect that new structures, new enzymes (initiating and polymerizing activities), and new biological functions for polysialylated glycoconjugates will continue to appear. We can further anticipate that future studies will reveal an even greater biological role of polySia in nearly every frontier of biomedical science, as the polySia "sialotope" already appears to be a key molecular player in most organ systems. The association of polySia with an increasing number of glycopathologies, notably neuroinvasive human tumors and neurotropic bacteria, predicts an exciting future in this contemporary area of sialobiology, as new discoveries of fundamental importance in understanding the biology of polySia emerge.

Acknowledgments. The seminal contributions of Professors Yasuo and Sadako Inoue and their colleagues to the biology and chemistry of the polysialic acids and polysialylation deserve special recognition and acknowledgment. Much of the author's recent work that is cited in this review was carried out in collaboration with them, and with their outstanding colleagues, especially K. Kitajima,

S. Kitazume, T. Terada, C. Sato, A. Kanamori, I. Tazawa, M. Iwasaki, and Y. Song. Special appreciation is also extended to Dr. J. Ye for her many helpful contributions. The patience and expert secretarial assistance of Ms. Ann Douglass are also acknowledged. Work in the author's laboratory was supported by the National Institutes of Health Grant AI-09352.

REFERENCES

Acheson, A., and Rutishauser, U., 1988, Neural cell adhesion molecule regulates cell contact-mediated changes in choline acetyltransferase activity of embryonic chick sympathetic neurons, *J. Cell Biol.* **106**:479–486.

Acheson, A., Sunshine, J. L., and Rutishauser, U., 1991, NCAM polysialic acid can regulate both cell–cell and cell–substrate interactions, *J. Cell Biol.* **114**:143–153.

Adlam, C., Knight, J. M., Mugridge, A., Williams, J. M., and Lindon, J. C., 1987, Production of colominic acid by *Pasteurella haemolytica* serotype A2 organisms, *FEMS Microbiol. Lett.* **42**:23–25.

Albright, C. F., Orlean, P., and Robbins, P. W., 1989, A 13-amino acid peptide in three yeast glycosyltransferases may be involved in dolichol recognition, *Proc. Natl. Acad. Sci. USA* **86**:7366–7369.

Alcaraz, G., and Goridis, C., 1991, Biosynthesis and processing of polysialylated NCAM by AtT-20 cells, *Eur. J. Cell Biol.* **55**:165–173.

Aletsee-Ufrecht, M. C., Langley, K., Gratzl, O., and Gratzl, M., 1990, Differential expression of the neural cell adhesion molecule NCAM 140 in human pituitary tumors, *FEBS Lett.* **272**:45–49.

Angata, T., Kitajume, S., Terada, T., Kitajima, K., Inoue, S., Troy, F. A., and Inoue, Y., 1994, Identification, characterization and developmental expression of a novel α2,8-KDN-transferase which terminate elongation of α2,8-linked oligo-polysialic acid chain synthesis in trout egg polysialoglycoproteins, *Glycoconj. J.* **11**:493–499.

Barry, G. T., and Goebel, W. F., 1957, Colominic acid, a substrate of bacterial origin related to sialic acid, *Nature* **179**:206.

Becker, C. G., Becker, T., and Roth, G., 1993a, Distribution of NCAM-180 and polysialic acid in the developing tectum mesencephali of the frog Discoglossus pictus and the salamander Pleurodeles waltl, *Cell Tissue Res.* **272**:289–301.

Becker, T., Becker, C. G., Niemann, U., Naujoks, M. C., Gerardy, S. R., and Roth, G., 1993b, Amphibian-specific regulation of polysialic acid and the neural cell adhesion molecule in development and regeneration of the retinotectal system of the salamander Pleurodeles waltl, *J. Comp. Neurol.* **336**:532–544.

Bhattacherjee, A. K., Jennings, H. J., Kenny, C. P., Martin, A., and Smith, I.C.P., 1976, Structural determination of the polysaccharide antigens of Neisseria meningitidis serogroups Y, W-135, and BO, *Can. J. Biochem.* **54**;1–8.

Bitter-Suermann, D., and Roth, J., 1987, Monoclonal antibodies to polysialic acid reveal epitope sharing between invasive pathogenic bacteria, differentiating cells and tumor cells, *Immunol. Res.* **6**:225–237.

Blass, K. S., Reinhardt, M. S., Kindler, R. A., Cleeves, V., and Rajewsky, M. F., 1994, In vitro differentiation of E-N-CAM expressing rat neural precursor cells isolated by FACS during prenatal development, *J. Neurosci. Res.* **37**:359–373.

Bock, E., Yavin, Z., Jorgensen, O. S., and Yavin, E., 1980, Nervous system-specific proteins in developing rat cerebral cells in culture, *J. Neurochem.* **35**:1297–1306.

Bonfanti, L., Olive, S., Poulain, D. A., and Theodosis, D. T., 1992, Mapping of the distribution of polysialylated neural cell adhesion molecule throughout the central nervous system of the adult rat: An immunohistochemical study, *Neuroscience* **49:**419–436.

Bortolussi, R., Ferrieri, P., Bjorksten, B., and Quie, P. G., 1979, Capsular K1 polysaccharide of *Escherichia coli:* Relationship to virulence in newborn rats and resistance to phagocytosis, *Infect. Immun.* **25:**293–298.

Boulnois, G. J., Roberts, I. S., Hodge, R., Hardy, K. R., Jann, K. B., and Timmis, K. N., 1987, Analysis of the K1 capsule biosynthesis genes of Escherichia coli: Definition of three functional regions for capsule production, *Mol. Gen. Genet.* **208:**242–246.

Brisson, J. R., Baumann, H., Imberty, A., Perez, S., and Jennings, H. J., 1992, Helical epitope of the group B meningococcal α(2-8)-linked sialic acid polysaccharide, *Biochemistry* **31:**4996–5004.

Cho, J.-W., and Troy, F. A., 1994a, Polysialic acid expression on neuroinvasive *Escherichia coli* K1 and human cancers: Use of the *E. coli* K1 polysialyltransferase to synthesize polysialylated glycosphingolipid, in: *Complex Carbohydrates in Drug Research, Alfred Benzon Symposium 36* (K. Bock and H. Clausen, eds.), Munksgaard, Copenhagen, pp. 260–275.

Cho, J.-W., and Troy, F. A., 1994b, Polysialic acid engineering: Synthesis of polysialylated neoglycosphingolipid by using the polysialyltransferase from neuroinvasive *Escherichia coli* K1, *Proc. Natl. Acad. Sci. USA* **91:**11427–11431.

Crossin, K. L., Edelman, G. M., and Cunningham, B. A., 1984, Mapping of three carbohydrate attachment sites in embryonic and adult forms of the neural cell adhesion molecule, *J. Cell Biol.* **99:**1848–1855.

Cunningham, B. A., Hoffman, S., Rutishauser, U., Hemperly, J. J., and Edelman, G. M., 1983, Molecular topography of the neural cell adhesion molecule N-CAM: Surface orientation and location of sialic acid-rich and binding regions, *Proc. Natl. Acad. Sci. USA* **80:**3116–3120.

Cunningham, B. A., Hemperly, J. J., Murray, B. A., Prediger, E. A., Brackenbury, R., and Edelman, G. M., 1987, Neural cell adhesion molecule: Structure, immunoglobulin-like domains, cell surface modulation, and alternative RNA splicing, *Science* **236:**799–806.

Doherty, P., Cohen, J., and Walsh, F. S., 1990, Neurite outgrowth in response to transfected N-CAM changes during development and is modulated by polysialic acid, *Neuron* **5:**209–219.

Doherty, P., Skaper, S. D., Moore, S. E., Leon, A., and Walsh, F. S., 1992, A developmentally regulated switch in neuronal responsiveness to NCAM and N-cadherin in the rat hippocampus, *Development* **115:**885–892.

Edelman, G. M., 1985, Cell adhesion molecule expression and the regulation of morphogenesis, *Cold Spring Harbor Symp. Quant. Biol.* **50:**877–889.

Egan, W., Liu, T.-Y., Dorow, D., Cohen, J. S., Robbins, J. D., Gotschlich, E. C., and Robbins, J. B., 1977, Structural studies on the sialic acid polysaccharide antigen of Escherichia coli strain Bos-12, *Biochemistry* **16:**3687–3692.

Ellis, L., Wallis, I., Abreau, E., and Pfenninger, K. H., 1985, Nerve growth cones isolated from fetal rat brain. IV. preparation of a membrane subfraction and identification of a membrane glycoprotein expressed on sprouting neurons, *J. Cell Biol.* **101:**1977–1989.

Figarella-Branger, D. F., Durbec, P. L., and Rougon, G. N., 1990, Differential spectrum of expression of neural cell adhesion molecule isoforms and L1 adhesion molecules on human neuroectodermal tumors, *Cancer Res.* **50:**6364–6370.

Finne, J., 1982, Occurrence of unique polysialosyl carbohydrate units in glycoproteins of developing brain, *J. Biol. Chem.* **257:**11966–11970.

Finne, J., and Makela, P. H., 1985, Cleavage of the polysialosyl units of brain glycoproteins by a bacteriophage endosialidase. Involvement of a long oligosaccharide segment in molecular interactions of polysialic acid, *J. Biol. Chem.* **260:**1265–1270.

Finne, J., Leinonen, M., and Makela, P. H., 1983, Antigenic similarities between brain components and bacteria causing meningitis. Implications for vaccine development and pathogenesis, *Lancet* **2**:355–357.

Finne, J., Bitter-Suermann, D., Goridis, C., and Finne, U., 1987, An IgG monoclonal antibody to group B meningococci cross-reacts with developmentally regulated polysialic acid units of glycoproteins in neural and extraneural tissues, *J. Immunol.* **138**:4402–4407.

Flesher, A. R., Jennings, H. J., Lugowski, C., and Kasper, D. L., 1982, Isolation of a serogroup 1-specific antigen from legionella pneumophila, *J. Infect. Dis.* **145**:224–233.

Foltz, K. R., and Lennarz, W. J., 1993, The molecular basis of sea urchin gamete interactions at the egg plasma membrane, *Dev. Biol.* **158**:46–61.

Fredette, B., Rutishauser, U., and Landmesser, L., 1993, Regulation and activity-dependence of N-cadherin, NCAM isoforms, and polysialic acid on chick myotubes during development, *J. Cell Biol.* **123**:1867–1888.

Freischutz, B., Saito, M., Rahmann, H., and Yu, R. K., 1994, Activities of five different sialyltransferases in fish and rat brains, *J. Neurochem.* **62**:1965–1973.

Frosch, M., Görgen, I., Boulnois, G. J., Timmis, K. N., and Bitter-Suermann, D., 1985, NZB mouse system for production of monoclonal antibodies to weak bacterial antigens: Isolation of an IgG antibody to the polysaccharide capsules of Escherichia coli K1 and group B meningococci, *Proc. Natl. Acad. Sci. USA* **82**:1194–1198.

Frosch, M., Edwards, U., Bousset, K., Kraube, B., and Weisgerber, C., 1991, Evidence for a common molecular origin of the capsule gene loci in gram-negative bacteria expressing group II capsular polysaccharides, *Mol. Microbiol.* **5**:1251–1263.

Fukuda, M. N., Dell, A., Oates, J. E., and Fukuda, M., 1985, Embryonal lactosaminoglycan. The structure of branched lactosaminoglycans with novel disialosyl (sialyl α2-9sialyl) terminals from PAI human embryonal carcinoma cells, *J. Biol. Chem.* **260**:6623–6631.

Gandour-Edwards, R., Deckard-Janatpour, K., Ye, J., Donald, P. J., and Troy, F. A., 1995, Neural cell adhesion molecule(N-CAM) in head and neck malignancies, *Modern Pathology* **8**:101A.

Glick, M. C., Livingston, B. D., Shaw, G. W. Jacobs, J. L., and Troy, F. A., 1991, Expression of polysialic acid on human neuroblastoma, In: Evans, A. E., Knudson, A. G., Seeger, R. C., and D'Angio, G. J. (eds.), *Advances in Neuroblastoma Research,* Vol. 3, Wiley-Liss, New York, pp. 267–274.

Gold, D. P., van Dongen, J. J., Morton, C. C., Bruns, G. A., van den Elsen, P., Geurts van Kessel, A. H., and Terhorst, C., 1987, The gene encoding the epsilon subunit of the T3/T-cell receptor complex maps to chromosome 11 in humans and to chromosome 9 in mice, *Proc. Natl. Acad. Sci. USA* **84**:1664–1668.

Greis, C., and Rösner, H., 1990, Migration and aggregation of embryonic chicken neurons in vitro: Possible functional implication of polysialogangliosides, *Brain Res. Dev. Brain Res.* **57**:223–234.

Grogan, T., Guptil, J., Mullen, J., Ye, J., Hanneman, E., Vela, L., Frutiger, Y., Miller, T., and Troy, F., 1994, Polysialylated NCAM as a neurodeterminant in malignant lymphoma (ML), *Lab. Invest.* **70**:110A (Abstr. 637).

Hakomori, S., 1990, Bifunctional role of glycosphinogolipids. Modulators for transmembrane signaling and mediators for cellular interactions, *J. Biol. Chem.* **265**:18713–18716.

Halberstadt, J. B., Flowers, H., and Glick, M. C., 1993, A method to detect polysialic acid in polymers of 10 or more sialyl residues synthesized in vivo and in vitro, *Anal. Biochem.* **209**:136–142.

Hallenbeck, P. C., Vimr, E. R., Yu, F., Bassler, B., and Troy, F. A., 1987a, Purification and properties of a bacteriophage-induced endo-N-acetylneuraminidase specific for poly-α-2,8-sialosyl carbohydrate units, *J. Biol. Chem.* **262**:3553–3561.

Hallenbeck, P. C., Yu, F., and Troy, F. A., 1987b, Rapid separation of oligomers of polysialic acid by high-performance liquid chromatography, *Anal. Biochem.* **161**:181–186.

Heitz, P. U., Komminoth, P., Lackie, P. M., Zuber, C., and Roth, J., 1990, Demonstration of polysialic acid and N-CAM in neuroendocrine tumors, *Verh. Dtsch. Ges. Pathol.* **74:**376–377.

Hemperly, J. J., Murray, B. A., Edelman, G. M., and Cunningham, B. A., 1986, Sequence of a cDNA clone encoding the polysialic acid-rich and cytoplasmic domains of the neural cell adhesion molecule N-CAM, *Proc. Natl. Acad. Sci. USA* **83:**3037–3041.

Higa, H. H., and Varki, A., 1988, Acetyl-coenzyme A:polysialic acid O-acetyltransferase from K1-positive Escherichia coli. The enzyme responsible for the O-acetyl plus phenotype and for O-acetyl form variation, *J. Biol. Chem.* **263:**8872–8878.

Hoffman, S., and Edelman, G. M., 1983, Kinetics of homophilic binding by embryonic and adult forms of the neural cell adhesion molecule, *Proc. Natl. Acad. Sci. USA* **80:**5762–5766.

Hoffman, S., Sorkin, B. C., White, P. C., Brackenbury, R., Mailhammer, R., Rutishauser, U., Cunningham, B. A., and Edelman, G. M., 1982, Chemical characterization of a neural cell adhesion molecule purified from embryonic brain membranes, *J. Biol. Chem.* **257:**7720–7729.

Horowitz, M. A., and Silverstein, S. C., 1980, Influence of the *Escherichia coli* capsule on complement fixation and on phagocytosis and complement killing by human phagocytes, *J. Clin. Invest.* **65:**82–94.

Inoue, S., and Inoue, Y., 1986, Fertilization (activation)-induced 200- to 9-kDa depolymerization of polysialoglycoprotein, a distinct component of cortical alveoli of rainbow trout eggs, *J. Biol. Chem.* **261:**5256–5261.

Inoue, S., and Iwasaki, M., 1978, Isolation of a novel glycoprotein from the eggs of rainbow trout: Occurrence of disialosyl groups on all carbohydrate chains, *Biochem. Biophys. Res. Commun.* **83:**1018–1023.

Inoue, S., and Iwasaki, M., 1980, Characterization of a new type of glycoprotein saccharides containing polysialosyl sequence, *Biochem. Biophys. Res. Commun.* **93:**162–165.

Inoue, S., Kitajima, K., Inoue, Y., and Kudo, S., 1987, Localization of polysialoglycoprotein as a major glycoprotein component in cortical alveoli of the unfertilized eggs of Salmo gairdneri, *Dev. Biol.* **123:**442–454.

Inoue, S., Kanamori, A., Kitajima, K., and Inoue, Y., 1988, KDN-glycoprotein: A novel deaminated neuraminic acid-rich glycoprotein isolated from vitelline envelope of rainbow trout eggs, *Biochem. Biophys. Res. Commun.* **153:**172–176.

Inoue, S., Iwasaki, M., Kanamori, A., Jerada, T., Kitajima, K., and Inoue, Y., 1991, Structure and function of poly(KDN)-GP and poly(Sia)-GP: Two distinct types of glycoproteins isolated from the vitelline envelope and ovarian fluid of salmonid fish, *Glycoconj. J.* **8:**233.

Inoue, Y., 1993, Glycobiology of fish egg polysialoglycoproteins (PSGP) and deaminated neuraminic acid-rich glycoproteins (KDN-gp), in: *Polysialic Acid: From Microbes to Man* (J. Roth, U. Rutishauser, and F. A. Troy, eds.), Birkhauser Verlag, Basel, pp. 171–181.

Iwasaki, M., and Inoue, S., 1985, Structures of the carbohydrate units of polysialoglycoproteins isolated from the eggs of four species of salmonid fishes, *Glycoconj. J.* **2:**209–228.

Iwasaki, M., Nomoto, H., Kitajima, K., Inoue, S., and Inoue, Y., 1984, Isolation and structures of the third major type of carbohydrate units in polysialoglycoproteins from rainbow trout eggs, *Biochem. Int.* **8:**573–579.

Iwasaki, M., Inoue, S., and Inoue, Y., 1987, Identification and determination of absolute and anomeric configurations of the 6-deoxyaltrose residue found in polysialoglycoprotein of Salvelinus leucomaenis pluvius eggs. The first demonstration of the presence of a 6-deoxyhexose other than fucose in glycoprotein, *Eur. J. Biochem.* **168:**185–192.

Iwasaki, M., Inoue, S., and Troy, F. A., 1990, A new sialic acid analogue, 9-O-acetyl-deaminated neuraminic acid, and α-2,8-linked O-acetylated poly(N-glycolylneuraminyl) chains in a novel polysialoglycoprotein from salmon eggs, *J. Biol. Chem.* **265:**2596–2602.

James, W. M., and Agnew, W. S., 1987, Multiple oligosaccharide chains in the voltage-sensitive

Na^+ channel from *Electrophorus electricus:* Evidence for α-2,8-linked polysialic acid, *Biochem. Biophys. Res. Commun.* **148:**817–826.

James, W. M., and Agnew, W. S., 1989, α-(2,8)-polysialic acid immunoreactivity in voltage-sensitive sodium channel of eel electric organ, *Proc. R. Soc. London Ser. B* **237:**233–245.

Janas, T., and Troy, F. A., 1989, The Escherichia coli K1 poly-α-2,8-sialosyl sialytransferase is topologically oriented toward the cytoplasmic face of the inner membrane, in: *Proc. Xth International Symposium on Glycoconjugates,* Vol. 142, (N. Sharon, H. Lis, D. Duksin, and I. Kahane, eds.), Organizing Committee Press, Jerusalem, pp. 207–208.

Jin, L., Hemperly, J. J., and Lloyd, R. V., 1991, Expression of neural cell adhesion molecule in normal and neoplastic human neuroendocrine tissues, *Am. J. Pathol.* **138:**961–969.

Joliot, A. H., Triller, A., Volovitch, M., Pernelle, C., and Prochiantz, A., 1991, α-2,8-Polysialic acid is the neuronal surface receptor of antennapedia homeobox peptide, *New Biol.* **3:**1121–1134.

Joliot, A., Triller, A., Volovitch, M., and Prochiantz, A., 1992, Are embryonic forms of NCAM homeobox receptors? *C. R. Acad. Sci. Iii,* 59–63.

Joliot, A., Le Roux, I., Volovitch, M., Bloch, G. E., and Prochiantz, A., 1993a, Neurotrophic activity of homeopeptide, *C. R. Seances Soc. Biol. Fil.* **187:**24–27.

Joliot, A., Le Roux, I., Volovitch, M., Bloch, G. E., and Prochiantz, A., 1993b, Neurotrophic activity of an homeobox peptide, *Ann. Genet.* **36:**70–72.

Joliot, A., Le Roux, I., Volovitch, M., Bloch, G. E., and Prochiantz, A., 1994, Neurotrophic activity of a homeobox peptide, *Prog. Neurobiol.* **42:**309–311.

Kaback, H. R., 1971, Bacterial membranes, *Methods Enzymol.* **22:**99–120.

Kanamori, A., Kitajima, K., Inoue, S., and Inoue, Y., 1989, Isolation and characterization of deaminated neuraminic acid-rich glycoprotein (KDN-gp-OF) in the ovarian fluid of rainbow trout (Salmo gairdneri), *Biochem. Biophys. Res. Commun.* **164:**744–749.

Kanamori, A., Inoue, S., Iwasaki, M., Kitajima, K., Kawai, G., Yokoyama, S., and Inoue, Y., 1990, Deaminated neuraminic acid-rich glycoprotein of rainbow trout egg vitelline envelope. Occurrence of a novel α-2,8-linked oligo(deaminated neuraminic acid) structure in O-linked glycan chains, *J. Biol. Chem.* **265:**21811–21819.

Kanamori, A., Kitajima, K., Inoue, Y., and Inoue, S., 1991, Immunochemical probes for KDN and oligo/poly(KDN) structures in glycoconjugates, *Glycoconj. J.* **8:**222–223.

Kanamori, A., Inoue, S., Xu lei, Z., Zuber, C., Roth, J., Kitajima, K., Ye, J., Troy, F. A., and Inoue, Y., 1994, Monoclonal antibody specific for α2,8-linked oligo deaminated neuraminic acid (KDN) sequences in glycoproteins. Preparation and characterization of a monoclonal antibody and its application in immunohistochemistry, *Histochemistry* **101:**333–340.

Karlsson, K. A., 1989, Animal glycosphingolipids as membrane attachment sites for bacteria, *Annu. Rev. Biochem.* **58:**309–350.

Kean, E. L., 1970, Nuclear cytidine 5′-monophosphosialic acid synthetase, *J. Biol. Chem.* **245:**2301–2308.

Kean, E. L., and Roseman, S., 1966, The sialic acids: X. Purification and properties of cytidine 5′-monophosphosialic acid synthetase, *J. Biol. Chem.* **241:**5643–5650.

Keller, S. H., and Vacquier, V. D., 1994, The isolation of acrosome-reaction-inducing glycoproteins from sea urchin egg jelly, *Dev. Biol.* **162:**304–312.

Kern, W. F., Spier, C. M., Miller, T. P., and Grogan, T. M., 1993, NCAM (CD56)-positive malignant lymphoma, *Leuk. Lymphoma* **12:**1–10.

Kim, J. J., Zhou, D., Mandrell, R. E., and Griffiss, J. M., 1992, Effect of exogenous sialylation of the lipooligosaccharide of Neisseria gonorrhoeae on opsonophagocytosis, *Infect. Immun.* **60:**4439–4442.

Kimer, S. J., Bentley, J., Ciemerych, M., Moeller, C. J., and Bock, E., 1994, Expression of N-CAM in fertilized pre- and periimplantation and parthenogenetically activated mouse embryos, *Eur. J. Cell Biol.* **63:**102–113.

Kiss, J. Z., Wang, C., and Rougon, G., 1993, Nerve-dependent expression of high polysialic acid neural cell adhesion molecule in neurohypophysial astrocytes of adult rats, *Neuroscience* **53:**213–221.

Kitajima, K., and Inoue, S., 1988, A proteinase associated with cortices of rainbow trout eggs and involved in fertilization-induced depolymerization of polysialoglycoproteins, *Dev. Biol.* **129:**270–274.

Kitajima, K., Inoue, Y., and Inoue, S., 1986, Polysialoglycoproteins of *Salmonidae* fish eggs. Complete structure of 200-kDa polysialoglycoprotein from the unfertilized eggs of rainbow trout (*Salmo gairdneri*), *J. Biol. Chem.* **261:**5262–5269.

Kitajima, K., Inoue, S., Inoue, Y., and Troy, F. A., 1988, Use of a bacteriophage-derived endo-N-acetylneuraminidase and an equine antipolysialyl antibody to characterize the polysialyl residues in salmonid fish egg polysialoglycoproteins. Substrate and immunospecificity studies, *J. Biol. Chem.* **263:**18269–18276.

Kitajima, K., Kuroyanagi, H., Inoue, S., Ye, J., Troy, F. A., and Inoue, Y., 1994, Discovery of a new type of sialidase, "KDNase," which specifically hydrolyzes deaminoneuraminyl (3-deoxy-D-glycero-D-galacto-2-nonulosonic acid) but not N-acylneuraminyl linkages, *J. Biol. Chem.* **269:**21415–21419.

Kitazume, S., Kudo, M., Kitajima, K., Inoue, S., Ye, J., Cho, J.-W., Troy, F. A., and Inoue, Y., 1993, Structural elucidation of the α-2,8-polysialylglycan chains in neural cell adhesion molecules (N-CAM) in embryonic chick brains and characterization of the α2,8-polysialyltransferase responsible for α-2,8-polysialylation, *Glycoconj. J.* **10:**332.

Kitazume, S., Kitajima, K., Inoue, S., Inoue, Y., and Troy, F. A., 1994a, Developmental expression of trout egg polysialoglycoproteins and the prerequisite α2,6-, and α2,8-sialyl and α2,8-polysialyltransferase activities required for their synthesis during oogenesis, *J. Biol. Chem.* **269:**10330–10340.

Kitazume, S., Kitajima, K., Inoue, S., Troy, F. A., Cho, J.-W. Lennarz, W. J., and Inoue, Y., 1994b, Identification of polysialic acid-containing glycoprotein in the jelly coat of sea urchin eggs. Occurrence of a novel type of polysialic acid structure, *J. Biol. Chem.* **269:**22712–22718.

Knirel, Y. A., Kocharova, N. A., Shashkov, A. S., Kochetkov, N. K., Mamontova, V. A., and Soloveva, T. F., 1989, Structure of the capsular polysaccharide of Klebsiella ozaenae serotype K4 containing 3-deoxy-D-glycero-D-galacto-nonulosonic acid, *Carbohydr. Res.* **188:**145–155.

Knirel, Y. A., Rietschel, E. T., Marre, R., and Zahringer, U., 1994, The structure of the O-specific chain of Legionella pneumophila serogroup 1 lipopolysaccharide, *Eur. J. Biochem.* **221:**239–245.

Komminoth, P., Roth, J., Lackie, P. M., Bitter-Suermann, D., and Heitz, P. U., 1991, Polysialic acid of the neural cell adhesion molecule distinguishes small cell lung carcinoma from carcinoids, *Am. J. Pathol.* **139:**297–304.

Komminoth, P., Roth, J., Saremaslani, P., Matias, G. X., Wolfe, H. J., and Heitz, P. U., 1994, Polysialic acid of the neural cell adhesion molecule in the human thyroid: A marker for medullary thyroid carcinoma and primary C-cell hyperplasia. An immunohistochemical study on 79 thyroid lesions, *Am. J. Surg. Pathol.* **18:**399–411.

Kundig, F. D., Aminoff, D., and Roseman, S., 1971, The sialic acids. XII. Synthesis of colominic acid by a sialyltransferase from *Escherichia coli* K-235, *J. Biol. Chem.* **246:**2543–2550.

Kwiatkowski, B., Boscheck, B., Thiele, H., and Stirm, S., 1982, Endo-N-acetylneuraminidase associated with bacteriophage particles, *J. Virol.* **43:**697–704.

Lackie, P. M., Zuber, C., and Roth, J., 1990, Polysialic acid and N-CAM localisation in embryonic rat kidney: Mesenchymal and epithelial elements show different patterns of expression, *Development* **110:**933–947.

Lackie, P. M., Zuber, C., and Roth, J., 1991, Expression of polysialylated N-CAM during rat heart development, *Differentiation* **47**:85–98.

Ladisch, S., 1987, Tumor cell gangliosides, *Adv. Pediatr.* **34**:45–58.

Landmesser, L., Dahm, L., Tang, J. C., and Rutishauser, U., 1990, Polysialic acid as a regulator of intramuscular nerve branching during embryonic development, *Neuron.* **4**:655–667.

Lee, Y. S., and Chuong, C. M., 1992, Adhesion molecules in skeletogenesis: I. Transient expression of neural cell adhesion molecules (NCAM) in osteoblasts during endochondral and intramembranous ossification, *J. Bone Miner. Res.* **7**:1435–1446.

Lehmann, J. M., Riethmuller, G., and Johnson, J. P., 1989, MUC18, a marker of tumor progression in human melanoma, shows sequence similarity to the neural cell adhesion molecules of the immunoglobulin superfamily, *Proc. Natl. Acad. Sci. USA* **86**:9891–9895.

Lipinski, M., Hirsch, M. R., Deagostini-Bazin, H., Yamada, O., Tursz, T., and Goridis, C., 1987, Characterization of neural cell adhesion molecules (NCAM) expressed by Ewing and neuroblastoma cell lines, *Int. J. Cancer* **40**:81–86.

Livingston, B. D., Jacobs, J., Shaw, G. W., Glick, M. C., and Troy, F. A., 1987, Polysialic acid in human neuroblastoma cells, *Fed. Proc.* **46**:2151.

Livingston, B. D., Jacobs, J. L., Glick, M. C., and Troy, F. A., 1988, Extended polysialic acid chains (n greater than 55) in glycoproteins from human neuroblastoma cells, *J. Biol. Chem.* **263**:9443–9448.

McCoy, R. D., Vimr, E. R., and Troy, F. A., 1985, CMP-NeuNAc:poly-α-2,8-sialosyl sialyltransferase and the biosynthesis of polysialosyl units in neural cell adhesion molecules, *J. Biol. Chem.* **260**:12695–12699.

McGuire, E. J., and Binkley, S. B., 1964, The structure and chemistry of colominic acid, *Biochemistry* **3**:247–251.

Margolis, R. K., and Margolis, R. U., 1983, Distribution and characteristics of polysialosyl oligosaccharides in nervous tissue glycoproteins, *Biochem. Biophys. Res. Commun.* **116**:889–894.

Marsh, R. G., and Gallin, W. J., 1992, Structural variants of the neural cell adhesion molecule (N-CAM) in developing feathers, *Dev. Biol.* **150**:171–184.

Masson, L., and Holbein, B. E., 1983, Physiology of sialic acid capsular polysaccharide synthesis in serogroup B *Neisseria meningitidis*, *J. Bacteriol.* **154**:728–736.

Merker, R. I., and Troy, F. A., 1990, Biosynthesis of the polysialic acid capsule in Escherichia coli K1. Cold inactivation of sialic acid synthase regulates capsule expression below 20°C, *Glycobiology* **1**:93–100.

Michalides, R., Kwa, B., Springall, D., van Zandwijk, N., Koopman, J., Hilkens, J., and Mooi, W., 1994, NCAM and lung cancer, *Int. J. Cancer Suppl.* **8**:34–37.

Moolenaar, C. E., Muller, E. J., Schol, D. J., Figdor, C. G., Bock, E., Bitter-Suermann, D., and Michalides, R. J., 1990, Expression of neural cell adhesion molecule-related sialoglycoprotein in small cell lung cancer and neuroblastoma cell lines H69 and CHP-212, *Cancer Res.* **50**:1102–1106.

Nadano, D., Iwasaki, M., Endo, S., Kitajima, K., Inoue, S., and Inoue, Y., 1986, A naturally occurring deaminated neuraminic acid, 3-deoxy-D-glycero-D-galacto-nonulosonic acid (KDN). Its unique occurrence at the nonreducing ends of oligosialyl chains in polysialoglycoprotein of rainbow trout eggs, *J. Biol. Chem.* **261**:11550–11557.

Nadasdy, T., Roth, J., Johnson, D. L., Bane, B. L., Weinberg, A., Verani, R., and Silva, F. G., 1993, Congenital mesoblastic nephroma: An immunohistochemical and lectin study, *Hum. Pathol.* **24**:413–419.

Nilsson, O., 1992, Carbohydrate antigens in human lung carcinomas, *Apmis Suppl.* **27**:149–161.

Nishiyama, I., Seki, T., Oota, T., Ohta, M., and Ogiso, M., 1993, Expression of highly polysialylated neural cell adhesion molecule in calcitonin-producing cells, *Neuroscience* **56**:777–786.

Nomoto, H., Iwasaki, M., Endo, T., Inoue, S., Inoue, Y., and Matsumura, G., 1982, Structures of carbohydrate units isolated from trout egg polysialoglycoproteins: Short-cored units with oligosialosyl groups, *Arch. Biochem. Biophys.* **218:**335–341.

Ørskov, I., Sharma, V., and Ørskov, F., 1976, Genetic mapping of the K1 and K4 antigens of K(L) antigens and K antigens of 08:K27(A), 08:K8(L) and 09:K57(B), *Acta Pathol. Microbiol. Scand. Sect. B* **84:**125–131.

Owen, P., and Kaback, H. R., 1978, Molecular structure of membrane vesicles from Escherichia coli, *Proc. Natl. Acad. Sci. USA* **75:**3148–3152.

Pavelka, M. J., Hayes, S. F., and Silver, R. P., 1994, Characterization of KpsT, the ATP-binding component of the ABC-transporter involved with the export of capsular polysialic acid in Escherichia coli K1, *J. Biol. Chem.* **269:**20149–20158.

Pelkonen, S., Pelkonen, J., and Finne, J., 1989, Common cleavage pattern of polysialic acid by bacteriophage endosialidases of different properties and origins, *J. Virol.* **63:**4409–4416.

Phillips, M. L., Nudelman, E., Gaeta, F. C., Perez, M., Singhal, A. K., Hakomori, S., and Paulson, J. C., 1990, ELAM-1 mediates cell adhesion by recognition of a carbohydrate ligand, sialyl-Lex, *Science* **250:**1130–1132.

Regan, C. M., 1991, Regulation of neural cell adhesion molecule sialylation state, *Int. J. Biochem.* **23:**513–523.

Robbins, J. B., McCracken, G.H.J., Gotschlich, E. C., Ørskov, F., Ørskov, I., and Hanson, L. A., 1974, *Escherichia coli* K1 capsular polysaccharide associated with neonatal meningitis, *N. Engl. J. Med.* **290:**1216–1220.

Robbins, J. B., Schneerson, R., Egan, W. B., Vann, W. F., and Liu, D. T., 1980, Virulence properties of bacterial capsular polysaccharides: Unanswered questions, in: *The Molecular Basis of Microbial Pathogenicity: Report of the Dahlem Workshop on the Molecular Basis of the Infective Process, Berlin, 1979, October 22–26* (H. Smith, J. J. Skehel, and J. J. Turner, eds.), Verlag Chemie, Weinheim, pp. 115–132.

Roberts, I. S., Mountford, R., Hodge, R., Jann, K. B., and Boulnois, G. J., 1988, Common organization of gene clusters for production of different capsular polysaccharides (K antigens) in Escherichia coli, *J. Bacteriol.* **170:**1305–1310.

Rohr, T. E., and Troy, F. A., 1980, Structure and biosynthesis of surface polymers containing polysialic acid in *Escherichia coli, J. Biol. Chem.* **255:**2332–2342.

Rosenberg, J., Ellis, L., Troy, F., and Kayalar, C., 1986, The 5B4 antigen expressed on sprouting neurons contains α-2,8-linked polysialic acid, *Brain Res.* **395:**262–267.

Rösner, H., 1993, Developmental expression of gangliosides in vivo and in vitro, in: *Polysialic Acid: From Microbes to Man* (J. Roth, U. Rutishauser, and F. A. Troy, eds.), Birkhauser Verlag, Basel, pp. 279–297.

Roth, J., Taatjes, D. J., Bitter-Suermann, D., and Finne, J., 1987, Polysialic acid units are spatially and temporally expressed in developing postnatal rat kidney, *Proc. Natl. Acad. Sci. USA* **84:**1969–1973.

Roth, J., Blaha, I., Bitter-Suermann, D., and Heitz, P. U., 1988a, Blastemal cells of nephroblastomatosis complex share an onco-developmental antigen with embryonic kidney and Wilms' tumor. An immunohistochemical study on polysialic acid distribution, *Am. J. Pathol.* **133:**596–608.

Roth, J., Brada, D., Blaha, I., Ghielmini, C., Bitter-Suermann, D., Komminoth, P., and Heitz, P. U., 1988b, Evaluation of polysialic acid in the diagnosis of Wilms' tumor. A comparative study on urinary tract tumors and non-neuroendocrine tumors, *Virchows Arch B* **56:**95–102.

Roth, J., Zuber, C., Wagner, P., Taatjes, D. J., Weisgerber, C., Heitz, P. U., Goridis, C., and Bitter-Suermann, D., 1988c, Reexpression of poly(sialic acid) units of the neural cell adhesion molecule in Wilms tumor, *Proc. Natl. Acad. Sci. USA* **85:**2999–3003.

Roth, J., Kempf, A., Reuter, G., Schauer, R., and Gehring, W. J., 1992, Occurrence of sialic acids in Drosophila melanogaster, *Science* **256:**673–675.

Rothbard, J. B., Brackenbury, R., Cunningham, B. A., and Edelman, G. M., 1982, Differences in the carbohydrate structures of neural cell-adhesion molecules from adult and embryonic chicken brains, *J. Biol. Chem.* **257:**11064–11069.

Rougon, G., 1993, Structure, metabolism and cell biology of polysialic acids, *Eur. J. Cell Biol.* **61:**197–207.

Rutishauser, U., 1989, Polysialic acid as a regulator of cell interactions, in: *Neurobiology of Glycoconjugates* (R. U. Margolis and R. K. Margolis, eds.), Plenum Press, New York, pp. 367–382.

Rutishauser, U., and Goridis, C., 1986, NCAM: The molecule and its genetics, *Trends Genet.* **2:**72–76.

Rutishauser, U., Watanabe, M., Silver, J., Troy, F. A., and Vimr, E. R., 1985, Specific alteration of NCAM-mediated cell adhesion by an endoneuraminidase, *J. Cell Biol.* **101:**1842–1849.

Rutishauser, U., Acheson, A., Hall, A. K., Mann, D. M., and Sunshine, J., 1988, The neural cell adhesion molecule (NCAM) as a regulator of cell–cell interactions, *Science* **240:**53–57.

Sarff, L. D., McCracken, G. H., Schiffer, M. S., Glode, M. P., Robbins, J. B., Ørskov, I., and Ørskov, F., 1975, Epidemiology of Escherichia coli K1 in healthy and diseased newborns, *Lancet* **1:**1099–1104.

Sato, C., Kitajima, K., Tazawa, I., Inoue, Y., Inoue, S., and Troy, F. A., 1993, Structural diversity in the α2→8-linked polysialic acid chains in salmonid fish egg glycoproteins. Occurrence of poly(Neu5Ac), poly(Neu5Gc), poly(Neu5Ac, Neu5Gc), poly(KDN), and their partially acetylated forms, *J. Biol. Chem.* **268:**23675–23684.

Scheidegger, E. P., Lackie, P. M., Papay, J., and Roth, J., 1994, In vitro and in vivo growth of clonal sublines of human small cell lung carcinoma is modulated by polysialic acid of the neural cell adhesion molecule, *Lab. Invest.* **70:**95–106.

Scott, A. A., Kopecky, K. J., Grogan, T. M., Head, D. R., Troy, F.A.I., Mullen, J., Ye, J., Appelbaum, F. R., Theil, K. S., and Willman, C. L., 1994, CD56: A determinant of extramedullary and central nervous system (CNS) involvement in acute myeloid leukemia (AML), *Lab. Invest.* **70:**120A (Abstr. 695).

SeGall, G. K., and Lennarz, W. J., 1979, chemical characterization of the component of the jelly coat from sea urchin eggs responsible for induction of the acrosome reaction, *Dev. Biol.* **71:**33–48.

Seki, T., and Arai, Y., 1991a, Expression of highly polysialylated NCAM in the neocortex and piriform cortex of the developing and the adult rat, *Anat. Embryol.* **184:**395–401.

Seki, T., and Arai, Y., 1991b, The persistent expression of a highly polysialylated NCAM in the dentate gyrus of the adult rat, *Neurosci. Res.* **12:**503–513.

Seki, T., and Arai, Y., 1993a, Highly polysialylated neural cell adhesion molecule (NCAM-H) is by newly generated granule cells in the dentate gyrus of the adult rat, *J. Neurosci.* **13:**2351–2358.

Seki, T., and Arai, Y., 1993b, Distribution and possible roles of the highly polysialylated neural cell adhesion molecule (NCAM-H) in the developing and adult central system, *Neurosci. Res.* **17:**265–290.

Shults, C. W., and Kimber, T. A., 1992, Mesencephalic dopaminergic cells exhibit increased density of neural cell adhesion molecule and polysialic acid during development, *Brain Res. Dev. Brain Res.* **65:**161–172.

Silver, R. P., and Vimr, E. R., 1990, Polysialic acid capsule of *Escherichia coli* K1, in: *The Bacteria: A Treatise on Structure and Function*, Vol. XI (B. H. Iglewski and V. L. Clark, eds.), Academic Press, New York, pp. 39–60.

Silver, R. P., Finn, C. W., Vann, W. F., Aaronson, W., Schneerson, R., Kretchmer, P. J., and Garon, C. F., 1981, Molecular cloning of the K1 capsular polysaccharide genes of *E. coli*, *Nature* **289:**696–698.

Song, Y., Kitajima, K., and Inoue, Y., 1990, New tandem-repeating peptide structures in polysialoglycoproteins from the unfertilized eggs of kokanee salmon, *Arch. Biochem. Biophys.* **283:**167–172.

Song, Y., Kitajima, K., Muto, H., and Inoue, Y., 1991, Deaminated neuraminic acid (KDN)-containing glycosphingolipids, KDN-gangliosides: Their structure and function, *Glycoconj. J.* **8:**161.

Song, Y., Kitajima, K., and Inoue, Y., 1993, Monoclonal antibody specific to α-2→3-linked deaminated neuraminyl β-galactosyl sequence, *Glycobiology* **3:**31–36.

Steenbergen, S. M., and Vimr, E. R., 1991, Overexpression, membrane localization, and sequencing of the polysialyltransferase from Escherichia coli K1, *Glycoconj. J.* **8:**145.

Stromberg, N., Ryd, M., Lindberg, A. A., and Karlsson, K. A., 1988, Studies on the binding of bacteria to glycolipids. Two species of Propionibacterium apparently recognize separate epitopes on lactose of lactosylceramide, *FEBS Lett.* **232:**193–198.

Sunshine, J., Balak, K., Rutishauser, U., and Jacobson, M., 1987, Changes in neural cell adhesion molecule (NCAM) structure during vertebrate neural development, *Proc. Natl. Acad. Sci. USA* **84:**5896–5990.

Terada, T., Kitazume, S., Kitajima, K., Inoue, S., Ito, F., Troy, F. A., and Inoue, Y., 1993, Synthesis of CMP-deaminoneuraminic acid (CMP-KDN) using the CTP:CMP-3-deoxynonulosonate cytidylyltransferase from rainbow trout testis. Identification and characterization of a CMP-KDN synthetase, *J. Biol. Chem.* **268:**2640–2648.

Tome, Y., Hirohashi, S., Noguchi, M., Matsuno, Y., and Shimosato, Y., 1993, Comparison of immunoreactivity between two different monoclonal antibodies recognizing peptide and polysialic acid chain epitopes on the neural cell adhesion molecule in normal tissues and lung tumors, *Acta Pathol. Jpn.* **43:**168–175.

Troy, F. A., 1979, The chemistry and biosynthesis of selected bacterial capsular polymers, *Annu. Rev. Microbiol.* **33:**519–560.

Troy, F. A., 1990, Polysialylation of neural cell adhesion molecules, *Trends Glycosci. Glycotechnol.* **2:**430–449.

Troy, F. A., 1992, Polysialylation: From bacteria to brains, *Glycobiology* **2:**5–23.

Troy, F. A., and McCloskey, M. A., 1979, Role of a membranous sialyltransferase complex in the synthesis of surface polymers containing sialic acid in *Escherichia coli:* Temperature-induced alteration in the assembly process, *J. Biol. Chem.* **254:**7377–7378.

Troy, F. A., Vijay, I. K., and Tesche, N., 1975, Role of undecaprenyl phosphate in synthesis of polymers containing sialic acid in *Escherichia coli, J. Biol. Chem.* **250:**156–163.

Troy, F. A., Hallenbeck, P. C., McCoy, R. D., and Vimr, E. R., 1987, Detection of polysialosyl-containing glycoproteins in brain using prokaryotic-derived probes, *Methods Enzymol.* **138:**169–185.

Troy, F. A., Janas, T., Janas, T., and Merker, R. I., 1990a, Transmembrane translocation of polysialic acid chains across the inner membrane of neuroinvasive E. coli K1, *Glycoconj. J.* **8:**152.

Troy, F. A., Janas, T., and Merker, R. I., 1990b, Topology of the polysialyltransferase complex in the inner membrane of E. coli K1, *FASEB J.* **4:**3189.

Troy, F. A., Janas, T., and Merker, R. I., 1990c, Topology of the poly-α-2,8-sialyltransferase in *E. coli* K1 and energetics of polysialic acid chain translocation across the inner membrane, *Glycoconj. J.* **7:**383.

Troy, F. A., Janas, T., Janas, T., and Merker, R. I., 1991, Vectorial translocation of polysialic acid chains across the inner membrane of neuroinvasive E. coli K1, *FASEB J.* **5:**A1548.

Troy, F. A., Cho, J.-W., and Ye, J., 1993, Polysialic acid capsule synthesis and chain translocation in neuroinvasive *E. coli:* "Activated" intermediates and a possible functional role for undecaprenol, in: *Polysialic Acid: From Microbes to Man* (J. Roth, U. Rutishauser, and F. A. Troy, eds.), Birkhauser Verlag, Basel, pp. 93–111.

van Dijk, W., Ferwerda, W., and van den Eijnden, D. H., 1973, Subcellular and regional distribution of CMP-N-acetylneuraminic acid synthetase in the calf kidney, *Biochim. Biophys. Acta* **315:**162–175.

van Echten, G., and Sandhoff, K., 1993, Ganglioside metabolism. Enzymology, topology, and regulation, *J. Biol. Chem.* **268:**5341–5344.

Vann, W. F., Silver, R. P., Abeijon, C., Chang, K., Aaronson, W., Sutton, A., Finn, C. W., Lindner, W., and Kotsatos, M., 1987, Purification, properties, and genetic location of Escherichia coli cytidine 5′-monophosphate N-acetylneuraminic acid synthetase, *J. Biol. Chem.* **262:**17556–17562.

Varki, A., and Higa, H., 1993, Studies of the O-acetylation and (in)stability of polysialic acid, in: *Polysialic Acid: From Microbes to Man* (J. Roth, U. Rutishauser, and F. A. Troy, eds.), Birkhauser Verlag, Basel, pp. 165–170.

Vijay, I. K., and Troy, F. A., 1975, Properties of membrane-associated sialyltransferase of Escherichia coli, *J. Biol. Chem.* **250:**164–170.

Vimr, E. R., 1991, Map position and genomic organization of the kps cluster for polysialic acid synthesis in Escherichia coli K1, *J. Bacteriol.* **173:**1335–1338.

Vimr, E. R., 1992, Selective synthesis and labeling of the polysialic acid capsule in Escherichia coli K1 strains with mutations in nanA and neuB, *J. Bacteriol.* **174:**6191–6197.

Vimr, E. R., and Troy, F. A., 1985a, Identification of an inducible catabolic system for sialic acids (nan) in *Escherichia coli, J. Bacteriol.* **164:**845–853.

Vimr, E. R., and Troy, F. A., 1985b, Regulation of sialic acid metabolism in *Escherichia coli:* Role of N-acylneuraminate pyruvate-lyase, *J. Bacteriol.* **164:**854–860.

Vimr, E. R., McCoy, R. D., Vollger, H. F., Wilkison, N. C., and Troy, F. A., 1984, Use of prokaryotic-derived probes to identify poly(sialic acid) in neonatal neuronal membranes, *Proc. Natl. Acad. Sci. USA* **81:**1971–1975.

Wallis, I., Ellis, L., Suh, K., and Pfenninger, K. H., 1985, Immunolocalization of a neuronal growth-dependent membrane glycoprotein, *J. Cell Biol.* **101:**1990–1998.

Walz, G., Aruffo, A., Kolanus, W., Bevilacqua, M., and Seed, B., 1990, Recognition by ELAM-1 of the sialyl-Lex determinant on myeloid and tumor cells, *Science* **250:**1132–1135.

Warren, L., and Blacklow, R. S., 1962, The biosynthesis of cytidine 5′-monophospho-N-acetylneuraminic acid by an enzyme from *Neisseria meningitidis, J. Biol. Chem.* **237:**3527–3534.

Watanabe, M., Kobayashi, H., Yao, R., and Maisel, H., 1992, Adhesion and junction molecules in embryonic and adult lens cell differentiation, *Acta Ophthalmol. Suppl.* **1992:**46–52.

Weisgerber, C., and Troy, F. A., 1990, Biosynthesis of the polysialic acid capsule in Escherichia coli K1. The endogenous acceptor of polysialic acid is a membrane protein of 20 kDa, *J. Biol. Chem.* **265:**1578–1587.

Weisgerber, C., Hansen, A., and Frosch, M., 1991, Complete nucleotide and deduced protein sequence of CMP-NeuAc:poly-α-2,8 sialosyl sialyltransferase of Escherichia coli K1, *Glycobiology* **1:**357–365.

Whitfield, C., and Troy, F. A., 1984, Biosynthesis and assembly of the polysialic acid capsule in Escherichia coli K1: Activation of sialyl polymer synthesis in inactive sialyltransferase complexes requires protein synthesis, *J. Biol. Chem.* **259:**12776–12780.

Whitfield, C., Adams, D. A., and Troy, F. A., 1984a, Biosynthesis and assembly of the polysialic acid capsule in Escherichia coli K1: Role of a low-density vesicle fraction in activation of the endogenous synthesis of sialyl polymers, *J. Biol. Chem.* **259:**12769–12775.

Whitfield, C., Vimr, E. R., Costerton, J. W., and Troy, F. A., 1984b, Protein synthesis is required for in vivo activation of polysialic acid capsule synthesis in *Escherichia coli* K1, *J. Bacteriol.* **159:**321–328.

Yang, P., Yin, X., and Rutishauser, U., 1992, Intercellular space is affected by the polysialic acid content of NCAM, *J. Cell Biol.* **116:**1487–1496.

Ye, J., Kitajima, K., Inoue, Y., Inoue, S., and Troy, F. A., 1994, Identification of polysialic acids in glycoconjugates, *Methods Enzymol.* **230**:460–484.

Zuber, C., Lackie, P. M., Catterall, W. A., and Roth, J., 1992, Polysialic acid is associated with sodium channels and the neural cell adhesion molecule N-CAM in adult rat brain, *J. Biol. Chem.* **267**:9965–9971.

Chapter 5

Biochemistry and Oncology of Sialoglycoproteins

Veer P. Bhavanandan and Kiyoshi Furukawa

1. GENERAL INTRODUCTION

Sialoglycoproteins are defined as proteins containing one or more sialyl oligosaccharides covalently bound via their reducing end to the polypeptide chain. Most, but not all, of the glycosylated proteins in animals contain sialic acid and therefore qualify as sialoglycoproteins. Some examples of glycoproteins lacking sialic acid are ovalbumin, ribonuclease, and antifreeze glycoprotein. Sialoglycoproteins are ubiquitous in animals, either as components of cellular membranes of extracellular fluids such as serum, spinal fluid, saliva, respiratory mucous, gastric juice, sweat, and semen. The intrinsic membrane glycoproteins of the cell have their carbohydrate moieties usually projecting into the cytoplasm, or in the case of the plasma membrane, the exterior of the cell. The cell membrane sialoglycoproteins and sialoglycolipids (gangliosides) enrich the cell surface in sialyl residues which are important determinants in the social behavior of the cell.

Veer P. Bhavanandan Department of Biochemistry and Molecular Biology, The Milton S. Hershey Medical Center, Pennsylvania State University, Hershey, Pennsylvania 17033. **Kiyoshi Furukawa** Department of Biochemistry, Institute of Medical Science, University of Tokyo, Tokyo 108, Japan; *present address:* Department of Biosignal Research, Tokyo Metropolitan Institute of Gerontology, Tokyo 173, Japan.

Biology of the Sialic Acids, edited by Abraham Rosenberg. Plenum Press, New York, 1995.

2. CLASSIFICATION AND NOMENCLATURE

Sialoglycoproteins tend to be classified according to the nature of the predominant oligosaccharide–protein linkages. Those in which the majority of the oligosaccharides are linked, via terminal GlcNAc, to the amide group of asparagine are referred to as Asn- or *N*-linked glycoproteins, *N*-glycosylated proteins, *N*-glycans, or serum-type glycoproteins since the majority of the serum components are sialoglycoproteins of this class. Those in which the oligosaccharides are linked via GalNAc to the hydroxyl group of serine and threonine are referred to as Ser/Thr- or *O*-linked glycoproteins, *O*-glycosylated proteins, *O*-glycans, or mucin-type glycoproteins since this is the primary mode of linkage in mucins. The majority of the known glycoproteins contain one or the other type of linkage as the predominant one and therefore do not present a problem in their classification. For example, glycophorin A having 15 oligosaccharides linked to Ser/Thr residues and only one oligosaccharide linked to Asn is clearly an *O*-glycosylated protein. However, there are several glycoproteins in which both linkages are almost equally represented and therefore a clear distinction cannot be easily made. Examples of such *N, O*-glycosylated proteins include fetuin, immunoglobulin A, and human chorionic gonadotropin. However, based on their serum source the former two are generally considered as serum-type, *N*-glycosylated proteins. The nomenclature is further complicated by the fact that other types of *O*-glycosidic linkages between saccharide, albeit nonsialylated, and amino acids exist. These include xylose–serine linkage in proteoglycans, single GlcNAc linked to Ser/Thr in nuclear/cytoplasmic proteins, galactose linked to the δ hydroxyl group of hydroxylysine in collagens, and mannose linked to Ser (Thr) in yeast mannan and arabinose linked to hydroxyproline in plant glycoproteins (Lis and Sharon, 1993). Thus, strictly the term *O*-linked glycoproteins and *O*-glycosylated proteins should include all of the above and the term *O*-glycan should also include polysaccharides such as glycogen, cellulose, and galactans. While an insider is generally able to understand the sometimes confusing terminology, for investigators and students of this field who are not familiar with glycoconjugates this could be problematic and confusing.

3. OLIGOSACCHARIDE STRUCTURES OF SIALOGLYCOPROTEINS

The identity of the monosaccharide and amino acid involved in linkage is not the only characteristic by which the classes of sialoglycoproteins could be distinguished. The composition and structure of the saccharides linked to Asn or Ser/Thr are also markedly different. Compositionally the most distinctive feature is that the former always contain Man and seldom contain GalNAc, the latter always contain GalNAc and never Man. Therefore, it is possible to obtain tentative information on the nature of the sialoglycoprotein by analysis of the monosaccharide composition.

3.1. Asn-Linked Saccharides

The saccharide chains linked to Asn show enormous structural variations and are usually classified into three groups (Figure 1) referred to as the high mannose type, complextype, and hybrid type, all of which contain the pentasaccharide Manα1→6(Manα1→3)Manβ1→4GlcNAcβ1→4GlcNAc as a common core structure (boxed in Figure 1a). However, only the complex-type and hybrid-type sugar chains contain siliac acids. High-mannose-type sugar chains contain only Man and GlcNAc residues, and the heterogeneity in these sugar chains is the result of variations in the locations and numbers of Manα1→2 residues linked to the three terminal α-mannose residues of the heptasaccharide: Manα1 → 6(Manα1 → 3)Manα1 → 6(Manα1 → 3)Manβ1 → 4GlcNAcβ1 → 4GlcNAc (Figure 1a).

Complex-type sugar chains exhibit a wide spectrum of structures, since both

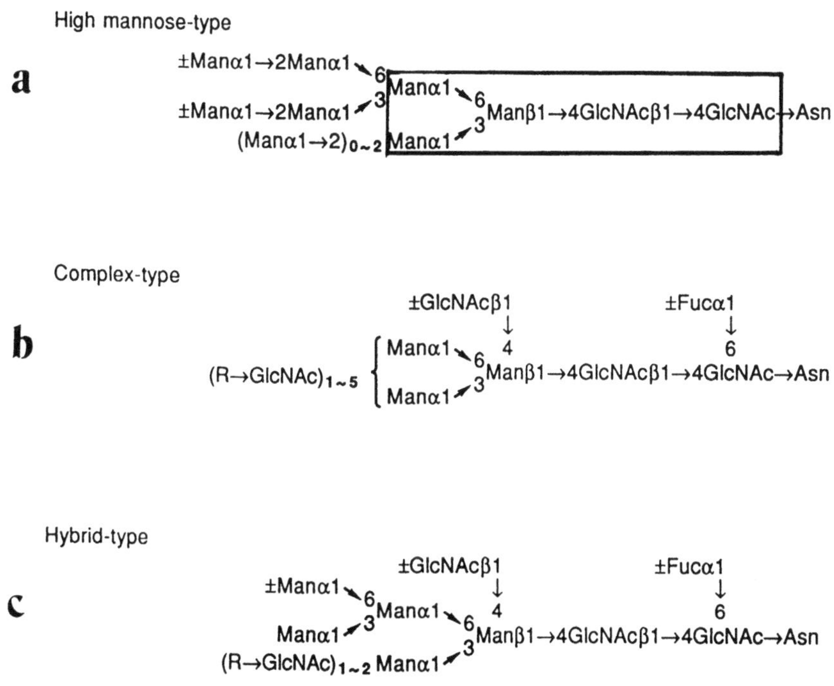

FIGURE 1. Three major types of Asn-linked saccharide structures found in glycoproteins. The core pentasaccharide is shown boxed in panel a.

the core and the outer chain moieties exhibit structural variations. Variations in the core include fucosylation of the C-6 position of the proximal GlcNAc residue, and *N*-acetylglucosaminylation of the C-4 position of the β-mannosyl residue; this latter sugar residue is referred to as the bisecting GlcNAc (Figure 1b). The outer chains, which are devoid of mannose, are formed by the addition of up to five residues to the two α-mannosyl residues of the pentasaccharide core, thus resulting in mono-, bi-, tri-, tetra-, and penta-antennary complex-type sugar chains (Figure 2). Next, β-galactose residues are attached to each of the GlcNAc residues to form type 1 chains (Galβ1→3GlcNAc) or type 2 chains (Galβ1→4GlcNAc). The type 2 chain is more widely observed in glycoprotein sugar chains than the type 1 chain. Recently, β-*N*-acetylgalactosamine has been shown to occur in place of the galactose residues as in the *N*-linked sugar chains of most bovine mammary epithelial glycoproteins (Sato *et al.*, 1993b). Finally, variations of the outer chains occur because of differences in the chain termination by the linkage of α-galactose, α-fucose, α-*N*-acetylgalactosamine, sialic acid, and/or sulfate (Figure 3). In the *N*-linked sugar chains, sialic acid residues are most commonly present in α2→3 and α2→6 linkages such as Neu5Ac-α2→3Gal, Neu5Acα2→6Gal, and Neu5Acα2→6GlcNAc. Glycoproteins obtained from CHO cells and human placenta have sialyl oligosaccharides containing exclusively α2,3-sialyl linkages (Takeuchi *et al.*, 1988; Endo *et al.*, 1988). The Neu5Acα2→6GlcNAc sequence is found in the sugar chains of bovine prothrombin, blood coagulation factor IX, and fetuin expressed solely in the Neu5Acα2→3Galβ1→3GlcNAc branch (Mizuochi *et al.*, 1980). Since the occurrence of the above type 1 chain is limited, th Neu5Acα2→6GlcNAc sequence is rare in glycoproteins. Fucosylation of the Gal and/or the GlcNAc residue of type 1 and type 2 chains by different linkages gives rise to Lewis (Le)-type and ABO-type blood group determinants (Figure 3). Sialylation followed by fucosylation of type 1 and type 2 chains results in formation of the sialyl Le^a and sialyl Le^x structures. In human leukocytes, sialyl Le^a and sialyl Le^x determinants bind selectins expressed on the blood vessel endothelial cells and platelets, causing various cellular interactions and inducing inflammation (Bevilacqua and Nelson, 1993). CD11/CD18 and CD45 are adhesion molecules expressed on leukocytes. Structural studies of the *N*-linked sugar chains of these molecules revealed that CD11/CD18, but not CD45, contain the oligosaccharides with Le^x and sialyl Le^x determinant (Asada *et al.*, 1991; Sato *et al.*, 1993a). The sialyl Le^x determinants of CD11/CD18 were shown to bind to E-selectin *in vitro,* suggesting their involvement in the selectin-mediated adhesion process (Kotovuori *et al.*, 1993). In contrast, CD45 containing exclusively oligosaccharides with the Neu5Ac-α2→6Gal structures (Sato *et al.*, 1993b), serves as a ligand for CD22, a sialic acid binding lectin, found on the surface of B lymphocytes (Sgroi *et al.*, 1993). It is very interesting that even though the CD11/CD18 and CD45 adhesion molecules are produced by the same cells, the sialylation of these glycoproteins is

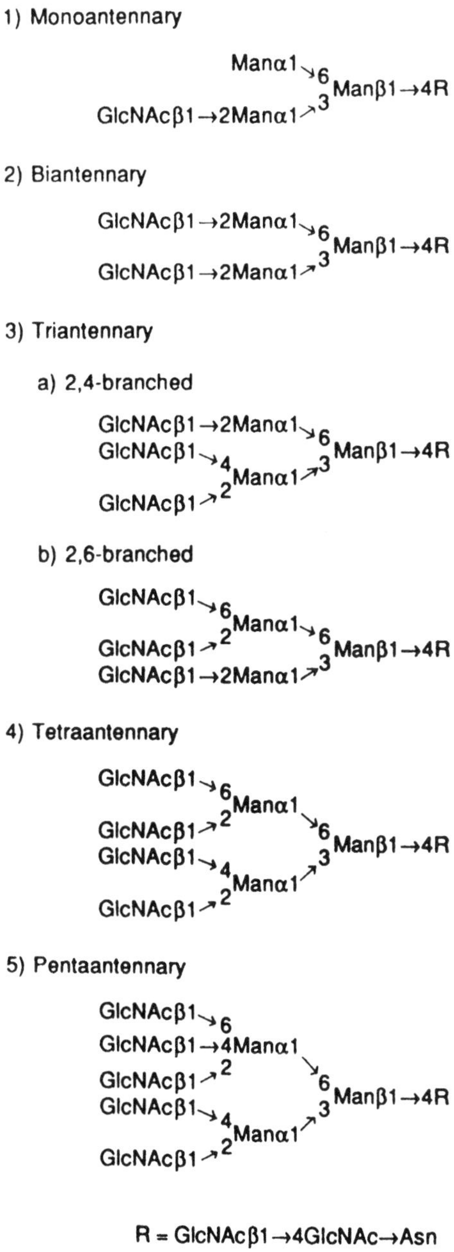

FIGURE 2. Branch variations in the complex-type Asn-linked oligosaccharides which give rise to multiantennary structures.

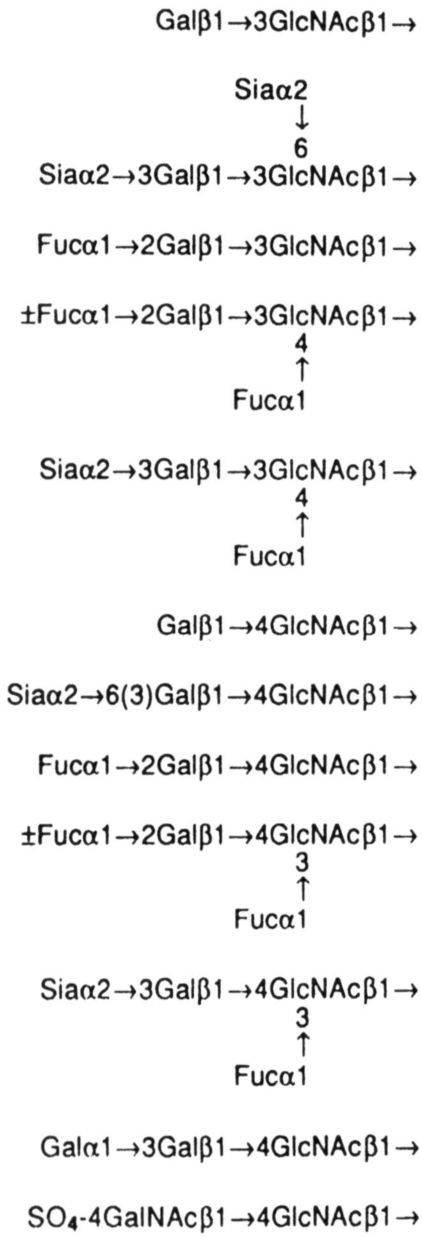

FIGURE 3. List of peripheral saccharide structures that terminate the complex-type Asn-linked oligosaccharides of glycoproteins.

very different. Currently, there is no explanation for the differential sialylation of these glycoproteins but it is possible that this is brought about by sialyltransferases whose specificities are influenced by peptide sequences (see Chapter 3). Glycosyltransferases such as the β-*N*-acetylgalactosaminyltransferase in bovine pituitary gland which, in addition to saccharides, also recognize specific peptide sequences in the acceptor molecule are known (see Section 5.2).

The third group of Asn-linked saccharides consist of sugar chains with structural features of both the high mannose type and complex type, and are therefore referred to as the hybrid type (Figure 1c). One or two α-mannose residues are linked to the Manα1→6Man branch of the core, and the outer chain structures of complex type are expressed on the Manα1→3Man side of the core. Recently, a novel hybrid-type sugar chain having the Manα1→Manα1→3Man-α1→6Man outer chain has been reported to occur in bovine mammary epithelial glycoproteins (Sato *et al.*, 1993b). The formation of this hybrid-type sugar chain cannot be explained by any of the currently known biosynthetic pathways.

3.2. Ser/Thr-Linked Saccharides

A general structure for the saccharides linked to Ser/Thr could be depicted as follows:

Peripheral	Backbone (not always present)	Core
Sialic acid, fucose	Gal	Gal, GalNAc
Sulfate, GalNAc	GlcNAc	GlcNAc

The seven different core structures so far found in this class of saccharides are presented in Figure 4. While core structures 1 to 4 show broad species and organ distribution, core structures 5, 6, and 7 have been found so far only in a few glycoproteins such as human rectal adenocarcinoma glycoprotein (Kurosaka *et al.*, 1983), salivary gland mucin of swiftlet (Wieruszeski *et al.*, 1987), and bovine submaxillary mucin (Chai *et al.*, 1992). In a general way, the numbering approximately reflects the frequency of occurrence, core structures 1 and 2 being the most common and core 7 being rare. The simplest sialyl saccharide found in these glycoproteins is the disaccharide SAα2→6GalNAc (the sialyl Tn antigen structure) since sialic acid by itself has not been found directly linked to protein (cf. GlcNAc-Ser/Thr in nuclear/cytoplasmic glycoproteins). It has been proposed that the addition of sialic acid to the linkage GalNAc blocks synthesis of the above core structures and thereby results in the termination of elongation of the saccharide chain. The major portion (~80%) of the carbohydrate in ovine submaxillary mucin exists as this disaccharide (Murthy and Horowitz, 1966). It is clear from the core structures that Ser/Thr-linked saccharides can be unbranched or branched. The *O*-linked saccharide chains are terminated typically,

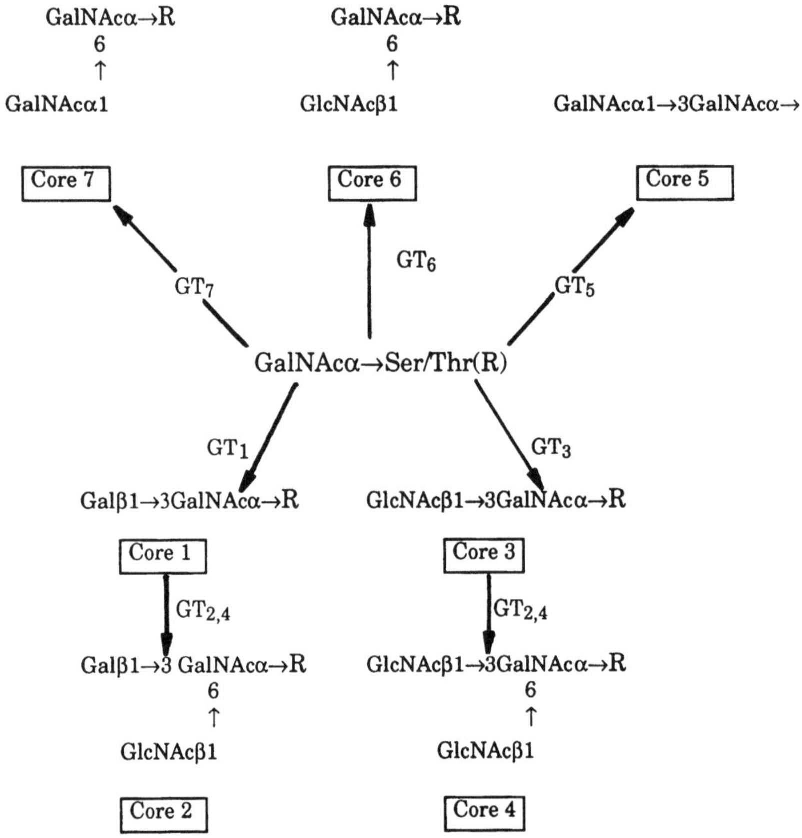

FIGURE 4. Biosynthetic pathway for the formation of core structures of Ser/Thr-linked oligosaccharides in glycoproteins. GT_1 to GT_7 refer to the glycosyltransferases that catalyze the reaction; for example, GT_1 is UDP Gal: GalNAcα → R, β1 → 3 galactosyltransferase.

but not exclusively, by addition of sialic acids, fucose, and sulfate residues. Sialic acids and fucose are always linked by α anomeric linkages. Occasionally the saccharide chains are terminated by α-linked Gal or GalNAc. Sulfation when it occurs is on galactose, GlcNAc, or GalNAc residues, either terminal or internal. The combination of different core structures, chain elongation, formation of branch chains, and variation of terminations give rise to a bewildering variety of saccharides in this class of sialoglycoproteins. The peripheral saccharide structures in this class of sialoglycoproteins also include those that confer blood group status such as ABO or Lewis, as discussed above for the Asn-linked saccharides.

The number of oligosaccharides per molecule of protein in this class of glycoproteins range from a few (three?) to a few hundred occurring as dense clusters along the protein backbone. In fact there is no information on the exact number of oligosaccharide units for any epithelial mucin molecule which usually has a high content (50 to 80% by weight) of carbohydrate compared to other sialoglycoproteins. Further, it appears that unlike most proteins, which are transcribed to yield a homogeneous population, mucin core proteins could consist of variable numbers of repeating sequences (tandem repeats) giving rise to a varying number of glycosylation sites (further discussed below). Thus, the presence of a very heterogeneous population of saccharide units and the variation in the peptide core are probably both equally responsible for the polydisperse nature of mucin preparations noted by both physical and chemical techniques.

There are two clearly distinguishable subclasses of *O*-linked sialoglycoproteins having a high (>50%) carbohydrate content. Group 1 consists of mucins which are the major components of epithelial mucus secretions. These secretory mucins generally have poorly defined, but apparently very high molecular weights (on the order of millions), and in solution yield highly viscous solutions or gels. Group 2 consists primarily of cell surface membrane-associated sialoglycoproteins such as erythrocyte membrane glycophorins (Cartron and Rahuel, 1992), human white blood cell-associated leukosialin (CD43, sialophorin) (Fukuda, 1991a), platelet membrane-associated CD42b (glycocalcin, GP1b) (Okumura *et al.*, 1976), mouse macrosialin (CD68) (Fukuda, 1991a), human milk fat granule membrane glycoproteins (Shimizu and Yamauchi, 1982) and decay-accelerating factor (CD55) of human complement (Medof *et al.*, 1987). In fact, many of the still poorly characterized leukocyte glycoproteins such as the leukocyte common antigen CD45RA, CD96, and PSGL-1 probably belong to this subclass of *O*-linked glycoproteins (Thomas, 1989). Recently, the interest in this subclass of mucin-type glycoprotein has greatly intensified because of the realization that they constitute the primary ligands of selectins. Endothelial cell-associated mucin-type glycoproteins such as GlyCAM-1, CD34, and MAdCAM-1 have all been reported to be ligands for L-selectin (Shimizu and Shaw, 1993, and references therein). The leukocyte membrane mucin-type glycoprotein PSGL-1 has been shown to be a ligand for P-and E-selectin (Sako *et al.*, 1993). The expression of this subclass of membrane-associated cell surface *O*-linked glycoproteins is markedly increased in malignantly transformed cells (see Section 6.2).

Mannose is invariably found as a component of several purified secretory mucins (e.g., see Van Nieuw Amerongen *et al.*, 1987) but because of the difficulties in entirely ruling out the co-purification of small amounts of *N*-glycosylated proteins and in establishing purity of mucin preparations, the significance of this has not been very clear (Strous and Dekker, 1992). In fact, earlier definition of mucins suggested the absence of Asn-linked saccharides and therefore mannose as a distinct feature (Pigman, 1977). The recent detection of potential *N*-glycosylation sites in the deduced amino acid sequence of several mucin core proteins provides stronger evidence for the presence of Asn-linked saccharides as a true component of mucins (Eckhardt *et al.*, 1991; Xu *et al.*, 1992; Verma and Davidson, 1993). However, to conclude that these sites are actually glycosylated in the final mature product, direct proof is necessary. In contrast to the situation in epithelial mucins (members of group 1 discussed above), the presence of Asn-linked saccharides in membrane-associated mucin-type glycoproteins (members of group 2) such as glycophorin (Tomita and Marchesi, 1975) and leukosialin (Fukuda, 1991a) has been established unequivocally. The biosynthetic studies on MUC1 mucin in breast cells (Hilkens and Buijs, 1988) as well as direct analysis of metabolically labeled MUC1 mucin from H.EP.2 cells (DiIulio and Bhavanandan, unpublished results) have established the presence of Asn-linked saccharide in this membrane-associated mucin-type glycoprotein as well.

3.3. Polylactosaminoglycans

The glycosaminoglycan keratan sulfate consists of long chains of repeating disaccharide units made up of galactose linked via β→4 to GlcNAc either of which may be sulfated in position 6 (Bhavanandan and Meyer, 1966). Two types of keratan sulfate chains can be distinguished based primarily on the nature of the linkage of the polysaccharide chains to the protein. In corneal keratan sulfate, the polysaccharide chains are linked to Asn, and in skeletal keratan sulfate, they are linked to Ser/Thr. Similar polysaccharide chains of repeating *N*-acetyllactosamine units commonly referred to as polylactosaminoglycans are also found in various glycoproteins as part of either *N*-or *O*-linked sugar chains and in glycolipids. The polylactosaminoglycan chain can be either linear or branched. In the linear form, the *N*-acetyllactosamine structures are linked to each other by β-1,3 linkages, while in the branched form the C-6 position of some galactose residues in the repeating units are substituted with additional *N*-acetyllactosamine. These linear and branched structures constitute the blood group i and I antigenic determinants, respectively (Watanabe *et al.*, 1979).

In the Asn-linked glycoproteins, the polylactosaminoglycan chains are attached to the trimannosyl core and show variations in the number and location of the repeating units. The backbone of polylactosaminoglycans consists of the type 2 Galβ1→GlcNAc structure, but at the distal portions of the chains Galβ1→3-GlcNAc structures of type 1 are often found. ABH and Lewis blood group antigenic determinants are usually expressed on the type 1 and type 2 chains attached directly to the trimannosyl core. The nonreducing terminal galactose residues are usually sialylated and most of the sialyl polylactosaminoglycans contain the Neu5Acα2→3Gal group, even though Neu5Acα2→Gal groups are also known to occur.

In a variety of cells, polylactosaminoglycans are mainly carried by lysosomal membrane glycoproteins, LAMP-1 and LAMP-2 (Do *et al.*, 1990; Carlsson and Fukuda, 1992), portions of which are located on cell surfaces (Lippincott-

Schwartz and Fambrough, 1987; Carlsson and Fukuda, 1992). Since the terminal structures of polylactosaminoglycans can express the sialyl Le^x and sialyl Le^a determinants, which are believed to be the natural ligands involved in selectin-mediated cell adhesion, the expression levels of LAMPs may be biologically important.

4. PROTEIN CORE OF *O*-LINKED SIALOGLYCOPROTEINS

The information available on the nature of the core proteins of the *O*-glycosylated proteins is still sketchy and the true sizes of the protein backbone of epithelial mucins are unknown (Strous and Dekker, 1992). Prior to the 1980s there was hardly any information concerning the core proteins of this class of glycoconjugates because of difficulties in deglycosylating and in applying the classical techniques of peptide sequence determination. The problem was particularly serious in the case of the epithelial mucins because of the apparent large size of the native molecules and estimated protein core size in excess of 200,000. Thus, usually only partial peptide sequences could be obtained; for example, in the case of ovine submaxillary mucin, the sequence of only 62 amino acids out of an estimated total 650 amino acids was obtained by analyzing tryptic peptides (Hill *et al.*, 1977a,b). The situation was somewhat better for the lower-molecular-weight, membrane-associated, members of this class of glycoconjugates. The entire peptide sequence of glycophorin was reported by Tomita and Marchesi (1975). This sequence, for the first time, established the presence of a stretch of hydrophobic amino acids of sufficient length to traverse the lipid bilayer. The sequence also confirmed a feature that has been predicted for *O*-glycosylated protein, specifically mucins, but not proven, namely, the presence of consecutively glycosylated serine and threonine residues. However, two recent studies found that the previously proposed cluster of six glycosylation sites in glycophorin is an error and that only four of these are occupied (Nakada *et al.*, 1993; Pisano *et al.*, 1993). However, the clustered saccharide domains provide a dense array of clustered *O*-linked saccharides presumably with important functional implications. Structurally the presence of a high density of sialylated and sometimes sulfated, and therefore negatively charged saccharides results in an extended conformation of the molecule (Gottschalk, 1960; Gottschalk and Thomas, 1961). Interestingly, it appears that the removal of sialic acid and hence the charge does not dramatically alter the conformation, as was previously believed, and the molecules with only a high density of single GalNAc residues still retain the extended rod shape (Shogren *et al.*, 1989). On the cell surface these extended *O*-linked glycoproteins would provide multiple saccharide binding sites for specific receptors. The ligands for lectins, antibodies, and selectins in many instances are likely to be multivalent saccharide structures. This was previously demonstrated to be the case for the lectin wheat germ agglutinin (Bhavanandan and Katlic, 1979; Furukawa *et al.*, 1986). This lectin strongly interacts with glycophorin T1 glycopeptide containing three sialylated saccharides on consecutive amino acids, but does not bind to this same saccharide when present individually as in glycopeptides isolated from fetuin, or if released from the T1 glycopeptide by β-elimination. Similarly, it has been demonstrated that three GalNAc residues present as a cluster on the peptide sequence SerThrThr are essential for recognition and binding by the MLS 128 monoclonal antibody (Nakada *et al.*, 1991).

Sequences of several other membrane-associated heavily *O*-glycosylated proteins have since been obtained. A notable one is that of leukosialin (CD43) having 385 amino acids, of which 239 constitute the mucinlike extracellular domain estimated to carry between 70 and 85 Ser/Thr-linked oligosaccharides (Fukuda, 1991a). A new feature not present in glycophorin is the tandem repeat of amino acid sequences. In leukosialin four repeats of 18 amino acid sequences are present. Tandemly repeated sequences are now considered a standard feature of mucins and mucin-type glycoproteins; however, it may be premature to make a definite conclusion in view of the limited number of full-length core peptide sequences available.

Amino acid analysis of mucin proteins reveals high levels of serine, threonine, proline, glycine, alanine, aspartic acid, glutamic acid, and low levels of aromatic and sulfur-containing amino acids. Chemical and enzymatic deglycosylation have yielded low values for molecular masses, ranging from about 60 kDa to >200 kDa for submaxillary, tracheal, and breast/pancreatic tumor MUC1 mucins (Hill *et al.*, 1977a; Burchell *et al.*, 1987; Bhavanandan and Hegarty, 1987; Woodward *et al.*, 1987; Lan *et al.*, 1990). In contrast, some of the estimates based on *in vitro* translation and biosynthetic studies indicate higher values of 160 kDa for human intestinal mucin (Gum *et al.*, 1989) and MUC1 mucin (Hilkens and Buijs, 1988), 400 kDa for human bronchial mucin (Marianne *et al.*, 1987), and 900 kDa for human gastric mucin (Dekker *et al.*, 1991). The discrepancies are not fully explicable at present but the lower values obtained by deglycosylation are believed to be related to breakage of peptide bonds followed probably by amino-terminal modification since usually no new amino-termini are detectable after deglycosylation (Hill *et al.*, 1977a; Bhavanandan and Hegarty, 1987; Woodward *et al.*, 1987).

Recent investigations using molecular biological techniques have provided new insights concerning the core protein of mucins. However, the information is still fragmentary since the majority of the deduced amino acid sequences are from partial cDNA clones. Currently, only a few full-length mucin glycoprotein sequences deduced from cDNA clones have been reported. These include the human and mouse MUC1 mucin (Gendler *et al.*, 1990; Ligtenberg *et al.*, 1990; Spicer *et al.*, 1991), the frog integumentary mucin glycoprotein FIM (Probst *et al.*, 1992), a low-molecular-weight human salivary mucin (Bobek *et al.*, 1993), and canine tracheobronchial mucin (Verma and Davidson, 1993). The cDNA sequence of a bovine submaxillary mucinlike protein previously reported to be full length (Bhargava *et al.*, 1990) was subsequently found to be incomplete (Woitach, Kiel, and Bhavanandan, unpublished results). A number of other partial cDNAs of mucin core proteins have been sequenced revealing several characteristic features. A feature noted in many, but not all, core proteins is a variable number of tandem repeats (Table I). These repeats, located in the central region of the molecule, vary in size from 6 to 169 and are enriched in serine and threonine and small amino acid residues such as proline, glycine, and alanine. It appears that these repeats provide the attachment sites for the bulk of the Ser/Thr-linked saccharides in the molecule, but in the human salivary MG2 mucin almost half of the carbohydrate is estimated to be located outside of the tandem repeat region (Bobek *et al.*, 1993). The deduced (partial or full) peptide sequences which have so far not revealed a tandem repeat feature include bovine submaxillary mucin (Bhargava *et al.*, 1990; Woitach, J. W., Keil, R., and Bhavanandan, V. P., unpublished results), canine tracheobronchial mucin (Verma and Davidson, 1993), rat intestinal mucin (Xu *et al.*, 1992), and certain clones of human tracheobronchial mucin (Dufosse *et al.*, 1993). The regions outside of the tandem repeats, constituting the N- and C-terminal ends of the core protein, have a wider representation of amino acids. Many of the C-terminal (Bhargava *et al.*, 1990; Eckhardt *et al.*, 1991; Xu *et al.*, 1992) and a few N-terminal segments (Gum *et al.*, 1993; Probst *et al.*, 1992) are enriched in cysteine, an unexpected feature since, typically, purified mucins have very little or no cysteine. In one case, that of human small intestinal MUC2 mucin, a cysteine-rich segment is located in the interior of the molecule, where it separates tandem repeats (Toribara *et al.*, 1991; Gum *et al.*, 1993). The regions outside of the tandem repeats also contain several potential sites for *N*-glycosylation; there are none in the tandem repeat segments. Biosynthetic studies indicate that *N*-glycosylation actually occurs, and precedes *O*-glycosylation of rat gastric mucin (Dekker and Strous, 1990). Since the subsequent processing events have not been fully elucidated, whether these saccharides are removed by processing or remain in the final mature mucin molecule is still not clear. Finally, putative transmembrane domains are present in peptide sequences of membrane-associated *O*-glycosylated proteins such as leukosialin and MUC1 glycoprotein, but not in the sequences of secreted mucins which have been deduced to date.

At present, seven different human mucin glycoprotein genes, termed MUC1 to MUC7, have been described (Table I). Of these the full-length sequence for the MUC1 mucin associated with breast, pancreatic, laryngeal, and other carcinomas is of special interest. The sequence reveals 40 to 90 tandem repeats of 20 amino acids each and additional repeats which differ in a few amino acids. Two signal sequence variants were detected in the N-terminus segment indicating

Table I
Tandemly Repeated Amino Acid Sequences Found in *O*-Glycosylated Proteins

		References
Human *O*-glycosylated proteins		
Glycophorin[a]	None	Tomita and Marchesi (1975)
Leukosialin[a]	TSGPPVTMATDSLETSTG	Fukuda (1991a)
MUC1[a]	PDTRPAPGSTAPPAHGVTSA	Gendler *et al.* (1990), Ligtenberg *et al.* (1990), Williams *et al.* (1990)
MUC2[a]	PTTTPITTTTTVTPTPTPTGTQT	Toribara *et al.* (1991)
MUC3	HSTPSFTSSITTTETTS	Gum *et al.* (1990)
MUC4	TSSASTGHATPLPVTD	Porchet *et al.* (1991)
MUC5	TTSTTSAP	Crepin *et al.* (1990)
MUC6	SPFSSTGPMTATSFQTTTTYPTPSHPOTTLPT HVPPFSTSLVTPSTGTVITPTHAQMATSASH STPTGTIPPPTTLKATGSTHTAPPMTPTTSTS QAHSSFSTAKTSTSLHSHTSSTHHPEVTSTTT ITPNPTSTGTSTPVAHTTSATSSRLPTPFTTH SPPTGS	Toribara *et al.* (1993)
MUC7	TTAAPPTPSATTPAPPSSSAPPE	Bobek *et al.* (1993)
Nonhuman *O*-glycosylated proteins		
Mouse MUC1[a]	PATSAPXDSTSSPVHSGTSS	Spicer *et al.* (1991)
Rat intestinal M-2 (RMUC 176)	TTTPDV	Khatri *et al.* (1993)
Rat intestinal MLP (mucinlike protein)	PSTPSTPPPST	Xu *et al.* (1992)
Rat airway mucinlike protein	TTTTIITI	Tsuda *et al.* (1993)
Frog integumentary mucin-A·1 (spasmolysin)	VPTTPETTT	Hoffmann (1988)
Frog integumentary mucin-B·1	GESTPAPSETT	Probst *et al.* (1992)
Porcine submaxillary mucin	GAGPGTTASSVGVTETARPS VAGSTTGTVSGASGSTGSSSG SPGATGASIGPETSRISVAGSS GAPAVSSGASQATS	Timpte *et al.* (1988)
Bovine submaxillary mucin	None	Bhargava *et al.* (1990)
Canine tracheobronchial mucin[a]	None	Verma and Davidson (1993)

[a]Full-length sequences available; all others are partial sequences deduced from cDNA.

alternate splicing. The C-terminus segment contains a 28-amino-acid hydrophobic region and 68 amino acids beyond the putative transmembrane segment which probably constitutes the cytoplasmic tail. Five potential *N*-glycosylation sites are also present in the nonrepetitive C-terminal segment. Five potential *O*-glycosylation sites are present within each tandem repeat of 20 amino acids and an equal proportion of additional sites in the nonrepetitive domains. However, there is no information as to which of these sites are occupied in the mature MUC1 mucin. The other six listed in Table I are tandem repeats from partial cDNA clones coding for secreted epithelial mucins. The human intestinal mucin coded by the MUC2 gene consists of 51 to 115 tandem repeats of 23 amino acids in the middle, with additional incompletely conserved repeats at the *N*-terminal end (Toribara *et al.*, 1991). Multiple cysteine residues, including several Cys-Cys dipeptides, are present in the sequence but not all of these are confined to the C-terminal segment. The N-terminal segment has a number of potential *N*-glycosylation sites and most of the aromatic amino acids. A different human intestinal mucin core protein coded by the MUC3 gene has tandem repeats of 17 amino acids. It is proposed that this core protein predominates in the acidic intestinal mucin while the MUC2 core protein predominates in the neutral intestinal mucin (Gum *et al.*, 1990). The MUC4 and MUC5 genes code for human tracheobronchial mucin core proteins with 16 and 8 amino acid tandem repeats, respectively. In the case of MUC4, 38 repeats are present but only 8 of the 16 amino acids are perfectly conserved in them (Porchet *et al.*, 1991). MUC6 codes for a gastric mucin core protein which has the longest repeat, of 169 amino acids, so far reported (Toribara *et al.*, 1993). MUC7 codes for a small salivary mucin glycoprotein of 120–150 kDa and the central region of the core protein consists of six perfect tandem repeats of 23 amino acids of which 9 are Thr and Ser (Bobek et al, 1993). A putative signal peptide of 20 hydrophobic amino acids in the N-terminal end and five potential *N*-glycosylation sites are also present in the MUC7 sequence. The tandem repeats of these seven human mucin proteins, varying between 8 and 169 amino acid residues, all have high proportions of hydroxy amino acids but there is no sequence homology at either the amino acid or nucleotide levels. In fact the only other common amino acids in all of them are proline and alanine. A point of interest is that while all reported tandem repeats have potential *O*-glycosylation sites, there is a preponderance of threonine in some with the MUC2 tandem repeat having only threonine. The significance of this is unclear at the present, but the possibility that mucins may be coded by more than one gene should be considered. In fact there is already evidence for the presence of more than one cDNA clone in human tracheobronchial (Dufosse *et al.*, 1993), rat intestinal (Khatri *et al.*, 1993), and bovine submaxillary mucin (Woitach, Keil, and Bhavanandan, unpublished results) cDNA libraries. The lack of sequence homology at either the nucleotide or amino acid levels between even closely related molecules such as human and rat gastric mucins or human and mouse MUC1 mucin-type glycoproteins is surprising. So far, the greatest sequence homology in mucin core proteins has been noted in the cysteine-rich C-terminal domains outside of the tandem repeats. There is an 82% sequence similarity at the protein level including exact matches of 30 cysteine residues in the C-terminal domains of porcine and bovine submaxillary mucins (Bhargava *et al.*, 1990; Eckhardt *et al.*, 1991). The amino acid homology in the transmembrane and C-terminal domains of the human and mouse MUC1 mucin is 87% (Spicer *et al.*, 1991). There are also lesser amino acid sequence similarities between the C-terminal domains of the bovine and porcine submaxillary mucins, MUC2 mucin, and frog integumentary mucin B.1 (Probst *et al.*, 1992). One other sequence similarity is the 73% homology between the C-terminal 700 amino acids of human intestinal MUC2 mucin and the rat intestinal mucinlike peptide (MLP) (Xu *et al.*, 1992). The high degree of sequence conservation in the C-terminus of both the secretory and membrane-associated mucins suggests some functional importance. However, as mentioned above, most purified secretory mucins are poor in cysteine, and therefore it is necessary to clearly demonstrate first that the cysteine-rich domains are intact in the fully processed mature mucin. The role of disulfide interactions in the polymeric structures of secretory mucins has been controversial. The rheological properties of most, but not all, mucins appear to be dependent on S-S bridge formation. Eckhardt *et al.* (1991) have suggested that in the case of the porcine submaxillary mucin the cysteine-rich segments are lost during purification as a result of proteolysis. However, the cysteine content of bovine submaxillary mucin prepared under stringent conditions, to avoid proteolysis, does not differ significantly from that of normal preparations (Woitach and Bhavanandan, unpublished results). For the MUC1 mucins the conserved cytoplasmic domain is thought to be important in linking the transmembrane glycoprotein to cytoskeletal elements such as actin (Spicer *et al.*, 1991). A striking difference between the secretory mucins and the transmembrane *O*-glycosylated proteins (glycophorin, leukosialin, MUC1 mucin, CD 34) is the lack of cysteine residues in the C-terminal regions (cytoplasmic tail) of the latter. This suggests that intra- and/or intermolecular disulfide bonds are not essential for the function of the membrane-associated mucin-type glycoproteins.

Not all mucin core protein sequences known at present have tandem repeats. The mucin sequences that lack tandem repeats, however, do have repetitive peptide motifs. Thus, the partial sequence of bovine submaxillary mucin has repeats of the sequence GTTVAPGSSNT (Bhargava *et al.*, 1990) and canine tracheobronchial mucin contains the motifs TPTPTP and TTTTPV repeated 13 and 19 times, respectively (Verma and Davidson, 1993). Dufosse *et al.* (1993) recently reported a family of cDNA clones isolated from a human tracheobronchial library showing no tandem repeats but a new type of peptide organization. These clones contain degenerate 87-base-pair tandem repeats which encode for nonrepetitive peptide sequences. A new feature noted in these deduced sequences, which lack tandem repeats, is the distinct alternating hydrophilic and hydrophobic domains.

The loci of several mucin genes have now been mapped. MUC1 has been mapped in the region of 22q on chromosome 1 (Swallow *et al.*, 1987a), MUC2 to chromosome 11 p15 (Griffiths *et al.*, 1990), MUC3 to chromosome 7 (Gum *et al.*, 1990), and MUC4 to chromosome 3 (Porchet *et al.*, 1991). Thus, the mucin genes are not clustered in the genome.

5. BIOSYNTHESIS OF SIALOGLYCOPROTEINS

5.1. General Concepts

The protein backbones of all sialylated glycoproteins are synthesized on polyribosomes on the cytoplasmic face of rough endoplasmic reticulum and cotranslationally translocated into the lumen where glycosylation can occur. The incorporation of glycosyl residues into the polypeptide chain requires the presence of the appropriate glycosyltransferases and the donor sugar–nucleotide. The general reaction catalyzed by glycosyltransferase is

$$\text{Sugar–Nucleotide} + \text{Acceptor} \rightarrow \text{Sugar–Acceptor} + \text{Nucleotide}$$

The glycosyltransferases involved in glycoconjugate biosynthesis are membrane-bound enzymes and when purified require detergents for optimal activity. The glycosyltransferase has an absolute donor (sugar–nucleotide) specificity and the name, e.g., galactosyltransferase, indicates this specificity, i.e, the requirement for UDP-Gal. The glycosyltransferase also has specificity for the acceptor to which it will transfer the sugar and for the type of linkages (e.g., $\alpha 1 \rightarrow 3$, $\beta 1 \rightarrow 4$) that will be formed. Therefore, it is the specificity of the glycosyltransferase that primarily directs and controls the formation of a particular structure. In general, one glycosyltransferase is required for every type of sugar linkage (e.g., sugar $\beta 1 \rightarrow$ x R, where x is position on the acceptor R) that is present in a glycoconjugate. There is no direct genetic control of the synthesis of saccharide structures but indirect control is exerted via expression of specific glycosyltransferase activities as is well illustrated in the synthesis of the blood group ABO(H) determinants. Occasional exceptions to the above rule are known; thus, an external factor may influence/regulate synthesis, such as the requirement for lactalbumin in the formation of the lactose (Gal$\beta 1 \rightarrow$4Glc linkage). In some instances one enzyme may catalyze the formation of two different linkages, while in others there appears to be more than one enzyme for the synthesis of the same linkage, e.g., SA$\alpha 2 \rightarrow$6GalNAc (Kurosawa *et al.*, 1994). Several glycosyltransferases have been cloned and studied at the genetic level. The amino acid sequences reveal common structural domains in these enzymes but only limited amino acid sequence homology. The glycosyltransferases specifying blood group A or B status differ in a few single-base substitutions which cause changes in four amino acid residues (Yamamoto *et al.*, 1990). The activated sugar nucleotides involved are derivatives of either uridine or guanidine diphosphates, with the exception of sialic acid which is activated as the cytosine monophosphate derivative. All of the sugar nucleotides are biosynthesized on the cytoplasmic side of the membrane and therefore, to be useful, must be transported to the lumen of RER and Golgi, where the glycosyltransferases are located. Recent research has shown that this translocation is facilitated by transporter proteins (antiports) present in the membranes which exchange free nucleotide with sugar–nucleotide on a one-to-one basis (e.g., one molecule of UMP for one UDP-Gal). The details of the regulation of glycoprotein biosynthesis remain to be elucidated. Some specific observations that need explanation include the following:

1. The basis for the existence of extreme diversity of both Asn- and Ser/Thr-saccharides. In the case of the Asn-linked saccharides since all of the structures are derived from the common precursor, $Glc_3Man_9GlcNAc_2$-Asn, there must be fine-tuning at the processing level. The peptide sequence and overall protein structure/conformation are believed to play major roles in this.
2. The factors that determine the glycosylation of the same protein in different animal species or in different organ/tissues of the same animal.
3. The differences noted in the glycosylation patterns in different proteins in the same cell.
4. The variations (heterogeneity) of the saccharides at a single glycosylation site in a protein as found in ovalbumin (the so-called glyco forms).

5.2. Biosynthesis of Glycoproteins Containing Asn-Linked Sugar Chains

Since the discovery of the involvement of lipid-linked oligosaccharide, the biosynthesis of Asn-linked sugar chains has been extensively studied (Kornfeld, 1982; Beyer and Hill, 1982). In the biosynthesis of Asn-linked sugar chains, a precursor oligosaccharide consisting of $Glc_3Man_9GlcNAc_2$ is first assembled on a lipid, dolichol, and then transferred to polypeptides by oligosaccharidyltransferase in the rough endoplasmic reticulum. The asparagine residue in the Asn-Xaa-Ser/Thr, in which Xaa can be any amino acid except proline, are potential sites for glycosylation. Since not all of the sites with this sequence are glycosylated, this tripeptide sequence is not sufficient for protein glycosylation. It is suggested that the accessibility of the sequence, in folded peptides, to the oligosaccharidyltransferase may be important. The three glucose residues are cleaved off from the peptide-bound oligosaccharide by the sequential action of α-glucosidases I and II, respectively, resulting in the high-mannose structure, $Man_9GlcNAc_2$. After translocation of glycoproteins to the Golgi apparatus, the oligosaccharide is further processed to the $Man_5GlcNAc_2$ structure by the action of an α-1,2-mannosidase (α-mannosidase I) which cleaves four α-1,2-linked mannose residues. At this stage a GlcNAc residue may be transferred to the Manα1→3Man arm of the heptasaccharide by *N*-acetylglucosaminyltransferase I (GlcNAcTI), leading to hybrid-type sugar chains. Alternatively, two mannose residues may be removed by the action of a second Golgi apparatus-associated mannosidase (α-mannosidase II), and a GlcNAc residue transferred to the Man-α1→6Man arm by GlcNAcTII, leading to the complex-type sugar chains. To these biantennary sugar chains another GlcNAc residue may be added via β-1,4 or β-1,6 linkages by the action of GlcNAcTVI or GlcNAcTV (Cummings *et al.*, 1982), forming 2,4- or 2,6-branched triantennary sugar chains. Tetraantennary sugar chains are produced by the further addition of GlcNAc to the triantennary sugar chains by GlcNAcTVI or GlcNAcTV. Action of GlcNAcTIII and α-1,6-fucosyltransferase on hybrid-type and complex-type sugar chains results in the formation of bisected sugar chains and the fucosylated core, respectively.

In the next step, galactose is transferred to each of the GlcNAc residues, except for bisecting GlcNAc, of the complex-type sugar chains by β-1,3-galactosyltransferase or β-1,4-galactosyltransferase to form type 1 or type 2 structures discussed above. Although the type 1 structure has been found in several glycoproteins, the β-1,3-galactosyltransferase activity responsible for the synthesis of this structure has not been detected. Recently a new variation was found in the Asn-linked sugar chains in which GalNAc instead of Gal is added to the GlcNAc residues. The sequence SO_4→4GalNAcβ1→4GlcNAc in Asn-liked saccharides was first detected in mammalian glycohormones (Green *et al.*, 1985) and later in many other glycoproteins (Sato *et al.*, 1993b, and references therein). This unusual saccharide structure is apparently important for the clearance of these glycoproteins from the circulation since it is the ligand for a lectin present on hepatic reticuloendothelial cells (Fiete *et al.*, 1991). In bovine pituitary gland, *N*-acetylgalactosaminylation of glycohormone sugar chains is catalyzed by a β-*N*-acetylgalactosaminyltransferase, which in addition to the GlcNAcβ→-Man→ also recognizes the Pro-Xaa-Arg/Lys sequences located six to nine amino acids away from the N-terminus of putative glycosylation sites (Smith and Baenziger, 1992). However, human urinary kallidinogenase, which is synthesized in the kidney and contains sugar chains with the GalNAcβ1→4GlcNAc structures (Tomiya *et al.*, 1993), lacks the proposed tripeptide recognition motif. This suggests the presence of other β-1,4-*N*-acetylgalactosaminyltransferases with different acceptor specificities.

Finally, the galactose residues are sialylated by sialyltransferases, resulting in the formation of Neu5Acα2→6Gal and Neu5Acα2→3Gal structures. A detailed discussion of sialyltransferases can be found in Chapter 3. An α-2,6-sialyltransferase purified from rat liver can transfer sialic acid to the Galβ1→4GlcNAc group but not to the Galβ1→3GlcNAc group, while α-2,3-sialyltransferase from the same source can transfer sialic acid to type 1 and type 2 chains (Weinstein *et al.*, 1982). Kinetic studies revealed that the α-2,6-sialyltransferase transfers sialic acid to the galactose residue of the Galβ1→4GlcNAcβ1→2Manα1→3Man branches, while α-2,3-sialyltransferase transfers sialic acid to the galactose residue of the Galβ1→4GlcNAcβ1→2Manα1→6Man branches (Joziasse *et al.*, 1987). The cDNAs that encode these sialyltransferases have been cloned (Weinstein *et al.*, 1987; Wen *et al.*, 1992). Since the *in vitro* assay of the α-2,3-sialyltransferase activity showed that the Galβ1→3GlcNAcβ→ sequence is the preferred acceptor compared to the Galβ1→4GlcNAcβ→ sequence, it is assumed that there is another α-2,3-sialyltransferase which more efficiently catalyzes the synthesis of the Neu5Acα2→3Galβ1→4GlcNAc structure (Weinstein *et al.*, 1982). In accordance with this assumption, the cDNA that encodes the α-2,3-sialyltransferase with an acceptor specificity different from that described above has been cloned from a human melanoma cell cDNA library and shown to be involved in the synthesis of the sialyl Le^x determinant in Namalwa cells by gene transfection experiments (Sasaki *et al.*, 1993).

Thus, specific carbohydrate structures are synthesized by the combined action of glycosyltransferases, each of which has a strict acceptor specificity and therefore determines the structure of the product. Differences in expression of the glycosyltransferases in cells result in differences in the carbohydrate structures produced by different cells. Systematic structural analysis of the Asn-linked sugar chains of γ-glutamyltranspeptidase (γ-GTPs) purified from rat, bovine, mouse, and human liver and kidney revealed that the major sugar chains of all kidney enzymes contain bisecting GlcNAc (for review, see Furukawa and Kobata, 1992). The liver enzymes from the four species are devoid of these structures. Further, the sugar chains of mouse and human kidney but not mouse and human liver γ-GTPs contained Le^x antigenic determinant. These findings strongly suggest that the tissue- and species-specific glycosylation is mostly related to the differential expression of the transferases. In other studies, the Asn-linked sugar chains of the blood coagulation factor XIII purified from human plasma and recombinant BHK cells were examined in detail (Hironaka *et al.*, 1992). Both factor VIII preparations contained high-mannose-type and bi-, tri-, and tetraantennary complex-type sugar chains in similar ratios but the outer chain structures were different. Some of the complex-type sugar chains in human plasma factor VIII contained blood group A and/or H determinant, while in the recombinant product these were absent. In addition, a small proportion, less than 10%, of the complex-type sugar chains of recombinant factor VIII had the Galα1→3Gal structure which was not detected in the sugar chains of the human plasma factor VIII. The expression of the Galα1→3Gal group is known to be species-specific and occurs in New World monkeys and nonprimates but not in humans or Old World monkeys (Galili *et al.*, 1988). The cDNA that encodes α-1,3-galactosyltransferase has been cloned from a bovine cDNA library (Joziasse *et al.*, 1989) and from a mouse F9 cell cDNA library (Larsen *et al.*, 1989).

Two genomic DNAs, homologous to the bovine cDNA, have been isolated from a human genomic library but these turned out to be nonfunctional pseudogenes (Larsen *et al.*, 1990; Joziasse *et al.*, 1991), confirming the absence of the Gal-α1→3Gal determinants in human glycoproteins. Thus, the expression of car bohydrate structures is directly controlled by the gene expression of individual glycosyltransferases.

5.3. Biosynthesis of Glycoproteins Containing Ser/Thr-Linked Sugar Chains

In contrast to the biosynthesis of Asn-linked saccharides, all of the steps in the synthesis of Ser/Thr-linked saccharides appear to occur by transfer of individual monosaccharides directly from their sugar nucleotide. Thus, all *O*-linked chains in mucin glycoproteins are initiated by the transfer of single GalNAcα from UDP-GalNAc to serine or threonine residues in the polypeptide chain. The reaction is catalyzed by polypeptide GalNAc transferase (UDP-GalNAc:polypeptide α-*N*-acetylgalactosaminyltransferase]. Two important issues concerning this initial step still remain to be fully resolved: (1) what factor(s) determine which serine and threonine residues in a protein are glycosylated?, and (2) what is the subcellular location of the polypeptide:GalNAc transferase and when does chain initiation occur? There is no consensus on either issue. Concerning the first issue, extensive research utilizing apomucins and their proteolytic fragments as well as synthetic peptide has failed to reveal a consensus primary amino acid sequence surrounding the glycosylated hydroxy amino acid (Aubert *et al.*, 1976; Hill *et al.*, 1977b; Briand *et al.*, 1981). The available data indicate that peptides containing clusters of serine/threonine residues and those containing proline residues in the vicinity of serine/threonine such as at "−1" and "+3" positions are preferred substrates (Wilson *et al.*, 1991). However, the absence of proline close to 14 out of the 15 glycosylated serine or threonine residues in glycophorin rules out the need for proline in the neighborhood of Ser/Thr as an absolute requirement. Nonetheless, the presence of proline in the vicinity may confer a favorable peptide conformation and presentation of serine/threonine for *O*-glycosylation since it is known that proline residues confer a rigid rodlike conformation on peptides. A serine or threonine residue next to an already *N*-acetylgalactosaminylated serine/threonine may be similarly preferred as a substrate for new chain initiation as a result of the conformational change that had occurred by the addition of the first GalNAc. The attachment of GalNAc residues to peptides is known to alter certain polypeptides from a globular to an extended, rigid conformation (Shogren *et al.*, 1989; Gerken *et al.*, 1989). Based on the corrected amino acid sequence of glycophorin, an attempt was made to formulate rules to determine which serine/threonine in protein would be glycosylated (Pisano *et al.*, 1993). However, the generality of the findings is doubtful since unlike glycophorin, all but one of the tandem repeat amino acid sequences found in mucin glycoproteins so far are rich in proline, the exception being the TTTTIITI repeat in rat airway mucin (Table I). Recent studies on porcine submaxillary gland UDP-GalNAc:polypeptide *N*-acetylgalactosaminyl transferase suggest that the specificity of the enzyme is influenced by the amino acid sequences adjacent to serine or threonine residues to be glycosylated (Wang *et al.*, 1993). In summary, the indications are that even though there is no consensus primary amino acid sequence for *O*-glycosylation comparable to that for *N*-glycosylation, there may be preferences based on signals governed by peptide conformation.

Concerning the timing and localization of the initiation of *O*-glycosylation, again there is no consensus. However, the majority of available evidence indicates that the bulk of the initiation occurs late; that is, after the release of the nascent peptide from the polysomes. Strong evidence in support of this is the localization of the polypeptide GalNAc transferase in the Golgi apparatus, with no detectable activity in rough endoplasmic reticulum, and the demonstration of a transport protein for UDP-GalNAc in Golgi vesicles (Abeijon and Hirschberg, 1987; Roth, 1984; Schweizer *et al.*, 1994). Additionally, it has been noted that less than 1.3% of the potential *O*-linked chains have been initiated in isolated fetuin peptidyl-RNA (Johnson and Heath, 1986). Initiation of *O*-glycosylation of leukosialin was shown to occur only in an early Golgi compartment; the initiated chains were then very rapidly elongated (Piller *et al.*, 1989). Studies on *O*-glycosylated proteins of viruses also confirm that the addition of linkage GalNAc is a late posttransitional event occurring in the Golgi apparatus (Nieman *et al.*, 1982; Johnson and Spear, 1983). Some studies have reported the transfer of GalNAc to the nascent polypeptide as a cotranslational event (Strous, 1979; Jokinen *et al.*, 1985). In the case of rat asialoglycoprotein ASGP-1, it was found that initiation of *O*-glycosylation occurred throughout the endomembrane system, from rough endoplasmic reticulum to the plasma membrane (Spielman *et al.*, 1988). The possibility remains that there are differences in the temporal aspects of the biosynthesis of *O*-glycosylated proteins in different tissues/cells and particularly in the synthesis of different types of molecules such as secreted mucins, membrane-associated *O*-glycosylated protein in normal cells, and *O*-glycosylated proteins in malignant cells. It is not surprising that the details of the synthesis of a mucin molecule having a very high carbohydrate content (70–80% by weight) as hundreds of separate saccharides are very different from those of the synthesis of glycoproteins having a low carbohydrate content and only a few *N*- and/or *O*-glycosyl units. Specifically the glycosylation of clusters of serine/threonine would present different problems than those of glycosylating lone (individual) Asn or Ser/Thr residues.

There is general agreement that, once the *O*-glycosyl chain is initiated, elongation and termination occurs rapidly as the molecule traverses the Golgi and *trans*-Golgi network to the secretory granules or cell surface. In pulse–chase experiments generally it has not been possible to observe discrete intermediates probably because of the rapidity of the process. The chain elongation occurs by sequential addition of monosaccharides by highly specific glycosyltransferases and follows general patterns (Schachter and Brockhausen, 1992). Addition of Galβ→, GlcNAcβ→, or GalNAcα→ to positions 3 or 6 or both of the Ser/Thr-linked GalNAc will give rise to the core types 1 to 7 as illustrated in Figure 4. Schachter (1986) has postulated that the substitution of position 3 occurs before 6 (the 3 before 6 rule). Thus, core types 1 or 3 are first formed by introduction of Gal or GlcNAc, respectively, to position 3 of linkage GalNAc, and subsequent addition of GlcNAc to position 6 leads to core types 2 and 4. It was suggested that the same β1→6*N*-acetylglucosaminyltransferase catalyzes the synthesis of core 2 and 4 from core 1 and 3, respectively (Brockhausen *et al.*, 1985). In subsequent studies, it was found that the core 2 transferase activity but not core 4 activity was present in human leukocytes (Brockhausen *et al.*, 1991). Thus, the situation is apparently more complex, and recently it has been suggested that there may be three enzyme isoforms involved in the biosynthesis of the core 2 and 4 structures (Bierhuizen and Fukuda, 1993). Since the above activities cannot transfer GlcNAc to unsubstituted GalNAc, the synthesis of core 6 must be catalyzed by a different enzyme such as that reported by Yazawa *et al.* (1986). Cores 5 and 7 are synthesized probably by two separate GalNAc transferases which transfer GalNAcα to positions 3 and 6 of the linkage GalNAc, respectively.

In addition to the above transferases, α-*N*-acetylgalactosaminide α2→6 sialyltransferase can also act directly on GalNAc to yield the disaccharide SA-α2→6GalNAc. This is a final product since it is not an acceptor for the enzymes responsible for synthesis of core 1 and 3 *in vitro* experiments. Even though the 3 before 6 rule would predict that core 6 and 7 structures are not substituted at positions 3 of GalNAc, such structures have been detected, albeit rarely, further illustrating the flexibility of the specificities of glycosyltransferases. Typically, core 2 and 4 are further galactosylated on the β1→6-linked GlcNAc before addition of terminal sugars. Chain termination occurs by the addition of sialic acid, fucose, and occasionally Gal, GalNAc, or GlcNAc all in α-anomeric linkage. Apparently, the α-linkage somehow prevents further addition of sugars and thereby inhibits chain elongation. Sulfation also occurs late in the biosynthesis of *O*-glycosyl chains and sulfate may be added to either the terminal or internal residues. In structures that have a long backbone, such as the repeating (Galβ1→4GlcNAβ1→3) units, the synthesis of the repeating sequences takes place prior to sialylation or fucosylation. These repeating backbone structures are not unique to *O*-linked chains but also occur in Asn-linked chains and in some glycolipids (see Section 3.3).

An important point about *N*- or *O*-glycosylation is that the same protein can be differently glycosylated when expressed in different cell types. This was clearly demonstrated for Thy-1 antigen from brain and thymus where it was established that the structures of the saccharides linked to the three Asn differed, but in a site-specific manner. Thus, in both cases Asn_{23} contained only oligomannose structures and Asn_{74} contained the complex sialylated saccharides (Parekh *et al.*, 1987). In the case of *O*-linked saccharides, striking differences in glycosylation are found in leukosialin saccharides of T cells, activated T cells, and neutrophils (Fukuda, 1991a). Resting T cells have the smaller core 1 structures whereas activated T cells and neutrophils have core 2 and elongated core 2 (polyactosamino) structures, respectively.

5.4. Biosynthesis of Polylactosaminoglycan Chains

Elongation of polylactosaminoglycans is initiated by addition of GlcNAc to the C-3 position of the galactose residue of the LacNAc group. The Galβ1→4-GlcNAc and Galβ1→Glc but not the Galβ1→3GlcNAc were found to be good acceptors for β-*N*-acetylglucosaminyltransferase (Piller and Cartron, 1983; Van den Eijnden *et al.*, 1988). Using desialylated α_1-acid glycoprotein, fetuin and transferrin as acceptors for the elongation enzyme, GlcNAcTVII, have been shown to have branch specificity. Oligosaccharides with the Galβ1→4GlcNAc-β1→6(Galβ1→4GlcNAcβ1→2)Man group are more effective than those without this group (Van den Eijnden *et al.*, 1988). The branch specificity of the transferase is in accordance with the location of the polylactosamine structure on the 2,6-branched sugar chains of many glycoproteins. Recently, this glycosyltransferase has been purified to near homogeneity from calf serum (Kawashima *et al.*, 1993).

6. ONCOLOGY OF SIALOGLYCOPROTEINS

Studies in the 1960s demonstrated that sialidase treatment of cancer cells reduced their electrophoretic mobility (Abercrombie and Ambrose, 1962), caused substantial decreases in oncogenicity (Sanford, 1967), and increased immunogenicity in animals and protected them from subsequent challenges (Currie *et al.*, 1968). These observations sparked extensive investigations of the sialoglycoconjugates of malignant cells. Overall changes on the cell surface glycoconjugates were also evident from early studies on the differences in lectin binding and lectin-induced agglutination of cancer cells compared to normal cells. Soon many investigators documented that both quantitative and qualitative changes occur in cell surface sialoglycoproteins and gangliosides as a result of malignant transformation (Hakomori, 1989). These aberrations were considered to be the basis for the altered social behavior of cancer cells, such as uncontrolled cell growth, altered cell adhesion, immunological resistance, invasiveness, and metastatic spread. Attenuation of the sugar chains, which is most widely observed in glycolipids (Hakomori, 1973), is also induced in the sugar chains of glycoproteins produced by transformed cells (Yamashita *et al.*, 1983). Since gangliosides are outside the scope of this chapter (see Chapters 6 and 7), only the alterations in the malignancy-associated sialoglycoproteins, both serum-type and mucin-type, will be discussed. For both classes of sialoglycoproteins, malignancy-associated changes which are of a general nature (i.e., overall, composite changes) and of a specific nature (changes on individual purified glycoproteins) have been documented (see review of Bhavanandan, 1991).

6.1. Malignancy-Associated Alterations in *N*-Glycosylated Sialoglycoproteins

One of the most prominent transformation-associated alterations in the sugar chains of glycoproteins is the increase of large Asn-linked sugar chains in the cell surface glycoproteins (see review by Kobata, 1989). This so-called Warren–Glick phenomenon was discovered by comparing the gel filtration patterns of the glycopeptides obtained by pronase digestion of metabolically labeled normal and malignant cell glycoproteins. Initially the increased molecular weight of the glycopeptides was considered to be related to a higher content of sialic acid in the sugar chains. However, subsequent studies revealed that the larger glycopeptides are not the result solely of increased sialylation but also of increased *N*-acetylglucosaminylation resulting in additional outer chains in the complex-type sugar chains. Detailed structural studies of the oligosaccharides isolated from BHK cells and polyoma- or RSV-transformed BHK cells proved that the increase of the GlcNAcβ1→6 branch attached to the Manα1→6Man arm of the trimannosyl cores is the structural basis for the Warren–Glick phenomenon (Yamashita *et al.*, 1984; Pierce and Arango, 1986). This was further confirmed by studies on a number of malignant cell lines transformed with different agents (Hiraizumi *et al.*, 1990; Santer *et al.*, 1984; Yousefi *et al.*,1991).

Three transformed cell lines, MT1, MTPY, and MTAg, established from mouse NIH 3T3 cells by transfection with the SV40 or polyoma virus early gene segments (Segawa and Yamaguchi, 1986), showed no difference in anchorage-independent growth in soft agar gel, which is believed to be one of the characteristic features of transformed cells. However, when these cells were transplanted subcutaneously into athymic mice, they showed a marked difference in the tumorogenic potentials. MTAg and MTPY cells produced tumors 1 week and 2 months after inoculation, respectively, but MT1 and parental 3T3 cells did not form tumors even after 1 year (Figure 5). Furthermore, of the three cell lines, only MTAg showed pulmonary colonization when injected intravenously into athymic mouse. Therefore, the *in vitro* properties of transformed cells do not

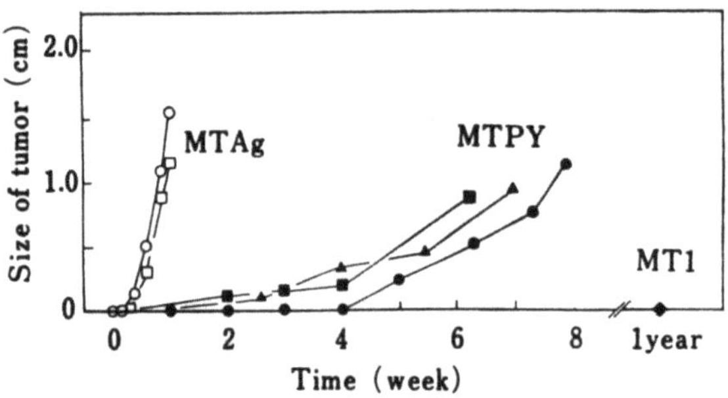

FIGURE 5. Tumor growth in athymic mice of NIH3T3 and its three transformed cell lines, MTAg, MTPY, and MT1. About 10^7 cells were injected subcutaneously and tumor growth monitored weekly by measuring the mean geometric diameter (cm). MTAg (○, □), MTPY (●, ■, ▲), and MT1 (♦) cells were established by transfection with the polyoma middle T gene linked to the polyoma membrane attachment gene, SV40 large T gene linked to the polyoma membrane attachment gene and SV40 large gene, respectively (Segawa and Yamaguchi, 1986). While MTAg and MTPY cells formed tumors, MT1 and the parental NIH3T3 cells did not form tumors even after 1 year.

always reflect the *in vivo* properties. Structural analysis of the *N*-linked glycopeptides from these cells revealed that only about 20% of the glycopeptides from 3T3 and MT1 cells had 2,6-branched sugar chains compared to 31 and 39% of the glycopeptides from MTPY and MTAg cells, respectively (Asada *et al.*, 1992), indicating a proportionality between the increased 2,6-branching and tumorigenicity and/or metastatic potential. The increased expression of the 2,6-branched sugar chains is also correlated with tumor-forming activities of various other transformed cells (Hiraizumi *et al.*, 1990, 1992). The increased cell surface binding of fluorescein-labeled L-PHA, which specifically binds to the 2,6-branched sugar chains, was found to be correlated with the metastatic potentials of mouse mammary carcinoma cells, mouse lymphoma cells, transformed rat fibroblasts, and human breast cancers (Dennis *et al.*, 1987; Laferte and Dennis, 1989). Thus, the results of several independent studies establish that the Warren–Glick phenomenon caused by the increase of 2,6-branched sugar chains is actually related to the *in vivo* tumor-forming and metastatic potentials of transformed cells rather than transformation *per se*.

In accordance with the above structural studies, the specific activity of GlcNAcTV, which synthesizes the β-1,6 branch, has been shown to be elevated two- to three-fold in the transformed cells (Yamashita *et al.*, 1985; Pierce *et al.*, 1987; Yousefi *et al.*, 1991). In contrast, no change in the specific activities of other glycosyltransferases involved in the synthesis of the *N*-linked sugar chains was observed in transformed cells (Yamashita *et al.*, 1985). These findings confirm that the increase in the β-1,6-branched chains results from the increased specific activity of GlcNAcTV in transformed cells, although the molecular mechanism causing the change in the GlcNAcTV activity in transformed cells remains to be elucidated. GlcNAcTV has been purified from rat kidney and human lung carcinoma, and its cDNA clone has been isolated from a rat (Shoreibah *et al.*, 1993) and human cDNA libraries (Saito *et al.*, 1994). The expression of the rat cDNA in COS-7 cells showed a significant increase of this enzyme activity (Shoreibah *et al.*, 1993). Using cell lines transfected with the GlcNAcTV cDNA, it will be interesting to determine whether changes in GlcNAcTV activity can directly influence cell adhesion and metastatic activity.

Several lectin-resistant cell lines have been shown to acquire reduced tumorigenic and/or metastatic potentials as compared to the parental malignant cells. The wheat germ agglutinin (WGA)-resistant L18 cell line established from metastatic B16 F10 mouse melanoma cells (Lin *et al.*, 1982) showed reduced tumor formation and metastasis compared to parental cells when injected subcutaneously into C57BL/6j mice (Table II, Furukawa and Bhavanandan, unpublished data). Structural analysis of the sugar chains of WGA-resistant MDW4 cells established from highly metastasizing mouse MDAY-D2 tumor cells showed that the mutant cells contained only ungalactosylated sugar chains and lacked the tetraantennary complex-type, polylactosamine, sugar chains which are the major sugar chains of the parental cells (Dennis *et al.*, 1986). These findings indicate the importance of the sialylated tetraantennary sugar chains with polylactosamine units in tumor metastasis. In contrast, the *N*-linked sugar chains obtained from the WGA-resistant cell line, Wa4-b1, and its parental B16 mouse melanoma F1 cell line showed that both cell lines contain almost similar amounts of high-mannose-type and bi-, tri-, and tetraantennary complex-type structures. The only difference was that the tri- and tetraantennary branches of the saccharides from the lectin mutant were not sialylated but heavily fucosylated (Finne *et al.*, 1980). Taken together, these results suggest that the expression of the β-1,6 branch itself is not sufficient for transformed cells to acquire a metastatic ability. Since low

Table II
Tumorigenicity and Metastatic Potentials of B16 Mouse Melanoma F10 and Its WGA-Resistant L18 Cells[a]

		B16 F10 cells	L18 cells
Tumorigenicity	20 days[b]	10/10, 10/10[c]	6/10, 5/10
	30 days	All dead	7/10, 6/10
Metastatic potentials (lung, lymph node)		9/10, 9/10	2/10, 2/10

[a]Furukawa and Bhavanandan (unpublished results).
[b]Days after subcutaneous transplantation of 5×10^4 cells into C57BL/6 mouse.
[c]Number of animals that formed tumors out of a total of 10. Data from two experiments are shown.

sialylation in transformed cells has been shown to be associated with loss of metastasis (Yamamura *et al.*, 1991), the reduced sialylation caused by the fucosylation of the outer chain moieties in the mutant is considered to be a major factor in reducing the metastatic ability. Consistent with these observations, the different forms of the *N*-linked sugar chains have been shown to be involved in several recognition steps of metastasis. B16 mouse melanoma cells treated with tunicamycin, which blocks the oligosaccharide transfer to peptides, showed a reduced ability to adhere to vascular endothelial cells (Irimura *et al.*, 1981). Treatment of B16 mouse melanoma cell lines with swainsonine, which inhibits the processing of the *N*-linked sugar chains, resulted in inhibition of pulmonary colonization via reduced attachment to lung cells (Humphries *et al.*, 1986) and in *in vitro* invasion of basement membranes (Yagel *et al.*, 1989). Then, what is the role of the altered carbohydrate structures in transformed cells? Two possible roles are indicated by the following studies.

1. Polylactosaminoglycans may mediate cell adhesion. It is found that sialyl Le^x and sialyl Le^a determinants are expressed preferentially on the nonreducing ends of polylactosaminoglycan chains (Spooncer *et al.*, 1984) which are carried primarily by lysosome-associated membrane proteins, LAMP-1 and LAMP-2, respectively (Do *et al.*, 1990; Carlsson and Fukuda, 1992). Even though LAMPs are major constituents of lysosomal membranes, a small portion of these molecules are present on the cell surface (Lippincott-Schwartz and Fambrough, 1987; Carlsson and Fukuda, 1992). It was found that LAMP-1 is the major L-PHA-reactive glycoprotein in metastatic tumor cells (Chen *et al.*, 1988; Laferte and Dennis, 1989). In recent studies, analysis of the glycoproteins from MTAg cells revealed three major L-PHA-reactive bands one of which was tentatively identified as LAMP-1 (Asada, M., Segawa, K., Kobata, A., Endo, T., and Furukawa, K., unpublished). Tumor cells are believed to adhere to vascular endothelial cells by selectin-mediated interactions (Rice and Bevilacqua, 1989; Hession *et al.*, 1990) and the LAMPs are thought to provide some of the sialyl Le^x and sialyl Le^a ligands for this interaction. In fact, genetic manipulation of the level of LAMP-1 expression on colonic carcinoma cells clearly demonstrated that the increase of cell surface LAMP-1 is proportional to the extent of cell surface sialyl Le^x and the binding to E-selectin (R. Sawada *et al.*, 1993). These studies indicate that there is a strong correlation between the metastatic potential of tumor cells and adhesion of these cells to endothelial cells mediated by the sialyl Le^x determinants expressed on LAMP-1 of the tumor cell surface. Since GlcNAcTVII, which is involved in the synthesis of polylactosaminoglycans as described earlier, transfers GlcNAc effectively to the terminal galactose residue of the β-1,6 branch of the pentasaccharide Galβ1→4GlcNAcβ1→6(Galβ1→4GlcNAc-β1→2)Man (Van den Eijnden *et al.*, 1988), the significance of the increased expression of the 2,6-branched sugar chains in transformed cells may be to provide the necessary polylactosamine chains and hence selectin ligands.

2. The altered glycosylation may impair protein functions. In the case of the fibronectin receptor, it is known that the proper glycosylation of the receptor is important for binding to fibronectin. Akiyama *et al.* (1989) showed that human fibroblasts cultured in the presence of 1-deoxymannojirimycin, an inhibitor of Golgi α-mannosidase-I, does not bind well to fibronectin in spite of the presence of the receptor on the cell surface. The fibronectin receptor isolated from the drug-treated cells is also defective in the binding to immobilized fibronectin. These results indicate that the processed (mature) structures of the Asn-linked sugar chains are required for a functional fibronectin receptor. Analysis of the carbohydrate structures of the fibronectin receptor isolated from B16 mouse melanoma cells and the WGA-resistant cell line, which shows diminished binding to fibronectin and laminin (Tao and Johnson, 1982), revealed that the outer chains of the Asn-linked saccharides of the fibronectin receptor is important for binding to fibronectin (Kawano *et al.*, 1993). The parental cell, whose fibronectin receptor contained the sialylated tetraantennary structures, showed strong binding to fibronectin while the mutant cell expressing the receptor without these structures showed reduced binding. Further, the fibronectin receptor is one of the major L-PHA-reactive bands in MTAg cell glycoproteins, and the malignant MTAg cells but not the control 3T3 cells dissociate easily from the fibronectin substrate on treatment with anti-fibronectin receptor antibody (Asada, M., Segawa, K., Kobata, A., Endo, T., and Furukawa, K., unpublished). Therefore, the structural alterations occurring on the sugar chains of the fibronectin receptor may cause the transformed cells to adhere poorly, which is critical for the cells in the process of escaping from the primary tumor.

6.2. Malignancy-Associated Alterations in *O*-Glycosylated Sialoglycoproteins

The examination of the proteolytic fragments from normal and cancer cells provided evidence for the enrichment of higher-molecular-weight mucin-type sialoglycopeptides in a number of malignant cell lines. This was clearly demonstrated by studies on mouse and human melanoma cells; for example, pronase digestion of [^{3}H]glucosamine-labeled human melanoma cells and melanocytes followed by Bio-Gel P-10 gel filtration showed that five times more melanoma glycopeptides eluted at the void volume compared to the melanocyte glycopeptides (Bhavanandan *et al.*, 1977, 1981). Similarly increased production of mucin-type sialoglycopeptides by Morris hepatoma (Furukawa *et al.*, 1986), human breast cancer cells (Chandrasekaran and Davidson, 1979), and rat hepatomas AH66 and AH 130 cells (Funakoshi *et al.*, 1974; Nakada *et al.*, 1975) compared to their nonmalignant counterparts was demonstrated suggesting that this is also a common phenomenon associated with malignancy like the Warren–Glick phenomenon discussed above. Unfortunately, there were a number of limitations to these (earlier) studies which made it difficult to assess the significance of the findings. For example, the normal cells such as hepatocytes or melanocytes used in some of these studies are not the most appropriate controls. Further, for studies on tissue culture cells it is difficult to determine whether the observed increase is malignancy-related or growth-related. More importantly, since the proteolytic fragments were a composite of many cellular glycoproteins whose identities are unknown, it is not possible to determine the functional implications of the changes. Additional evidence that the increased production of *O*-glycosylated protein was a general characteristic of all cancer cells came from subsequent immunochemical studies. The major oligosaccharides from the mucin-type glycopeptides of mouse and human melanomas were identical to those of glycophorin. Thus, polyclonal and monoclonal antiglycophorin antibodies were used to examine malignant and nonmalignant cells (Barsoum *et al.*, 1985; Barsoum and Bhavanandan, 1989). The presence of glycophorinlike epitopes was demonstrated, in varying concentrations, on a variety of cells including erythroleukemia, melanoma, and carcinomas of breast, cervix, larynx, nasopharynx, and colon (Barsoum *et al.*, 1984). However, these epitopes were not confined to malignant cells since cultured fibroblasts and HBL-100 (normal mammary) cells also gave positive immunofluorescences, albeit weak, suggesting quantitative rather than qualitative differences in the production of these glycoproteins.

The known changes in the *O*-glycosylated proteins associated with malignancy can be categorized into three types. The first are changes that occur in the peripheral sugars and backbone regions as a result of either incomplete synthesis or neosynthesis (Hakomori, 1984). This type of changes occurs typically in the lacto-series structures carrying blood group ABH, Lewis, Ii antigens and are thus common to both *N*- and *O*-linked glycoproteins as well as glycolipids. Examples of this type of change are the expression of sialyl Le^a structures in gastrointestinal and pancreatic cancer-associated mucin glycoproteins (Magnani *et al.*, 1983) and the elongation of the backbone structures of Le^x and Le^Y antigens in malignant and premalignant colon cells (Kim *et al.*, 1986). The second type are changes that occur in the core region such as expression of the usually cryptic core structures as a result of blocked synthesis. The classic example of this is the incomplete synthesis and accumulation of GalNAcα→(Tn antigen) and Galβ1→3GalNAcα→(T antigen) structures on *O*-glycosylated proteins in adenocarcinomas (Springer, 1984). In addition to these, there are also increased levels of the disaccharide Neu5Ac2→6GalNAcα→(sialyl Tn antigen) in tumor-associated mucin glycoproteins (Johnson *et al.*, 1986). This can be explained by the conversion of some of the accumulated Tn antigen to sialyl Tn antigen by a specific α2→6 sialyltransferase. These Neu5Acα 2 →GalNAcα→ structures will accumulate since, as mentioned above, this disaccharide is a poor substrate for the chain-elongating glycosyltransferases. The third type of change is the increased production of *O*-glycosylated proteins containing clusters of saccharides, and therefore resistant to proteolytic attack. This is illustrated by the above-discussed isolation of increased quantities of pronase-resistant mucin-type glycopeptides from cancer cells.

A great number of monoclonal antibodies have now been generated that distinguish between malignant cells and their nonmalignant counterparts based on qualitative (structural) and/or quantitative differences in cell surface *O*-glycosylated proteins. Some of the antibodies, such as MLS102, B72-3, and N19-9, are directed against carbohydrate epitopes on these proteins which have resulted as a consequence of the above-mentioned aberrations in glycosylation (Hakomori, 1989). The majority of these antibodies are, however, directed against peptide epitopes which are apparently cryptic on the *O*-glycosylated proteins of the nonmalignant cells. Even though there is no direct proof, it is believed that more of the peptide epitopes of mucin-type glycoproteins are exposed on malignant cell surfaces as a result of impaired glycosylation. Thus, it is possible that blocked synthesis leading to premature chain elongation would lead to (1) expression of short saccharide chains such as GalNAcα→ and Neu5Acα2→6GalNAcα→ epitopes, (2) exposed peptide epitopes which would otherwise have been covered by the larger saccharide structures, and (3) changed conformation of peptide regions creating new epitopes. Similarly, clustering of saccharides which are normally more evenly distributed on the core protein would also result in exposure of peptide epitopes on the malignancy associated *O*-glycosylated proteins and, in addition, create new multivalent saccharide epitopes. In fact, the monoclonal antibodies MLS102 and MLS128 with specificities toward malignant cells have been shown to be against clusters of GalNAcα→ and Neu5Acα 2→6GalNAc, respectively (Kurosaka *et al.*, 1988; Nakada *et al.*, 1993). However, one cannot at present entirely rule out the possibility that the peptide epitopes on malignant cells could also be generated by abnormal gene expression and/or mutations (Yonezawa *et al.*, 1991). It appears that some of the antibodies require both *O*-linked saccharides and peptide structures either because the epitope consists of both as may be the case with the Ca-1 antibody (Ashall *et al.*, 1982) or because glycosylation influences peptide epitope conformation as illustrated by the expression of blood group MN antigens and FD-6 antigens (Matsuura *et al.*, 1989). Currently, many of the monoclonal antibodies that recognize malignancy-associated *O*-glycosylated proteins are being exploited for the development of clinically useful diagnostic and therapeutic reagents for the management of cancer (Hilkens, 1988; Hilgers *et al.*, 1989; Hakomori, 1989).

Investigations of the allotransplantable ascites subline of TA3 mammary adenocarcinoma provided the first clear evidence of the association of one specific mucin-type glycoprotein with malignancy. A heavily *O*-glycosylated protein, named epiglycanin, was found to be present in large quantities on the surface of the allotransplantable TA3-Ha and TA3-MM sublines, but not on the non-allotransplantable TA3-St subline (Codington and Haavik, 1992). This molecule consists of a single polypeptide chain of about 1300 amino acid residues, to which about 550 short saccharide chains are attached via serine or threonine. Carbohydrate represents 75–80% of the glycoprotein of 500 kDa. About 80% of the saccharides in epiglycanin consist of monosaccharides (GalNAc) and disaccharides (Galβ1→3GalNAc), the balance being short sialylated saccharides (Van den Eijnden *et al.*, 1979). Based on the mannose content of the purified glycoprotein, it is estimated that one or two Asn-linked saccharide chains may be present in the molecule. The amino acid composition is typical of that of *O*-glycosylated proteins, with Ser and Thr constituting about 48%, and Pro, Gly, Ala, Glu, and Asp accounting for a further 42%. This glycoprotein is produced by TA3-Ha cells only when grown in the ascites and not when cultured *in vitro* or grown as solid tumors. The *in vitro* cultured cells, however, regain their ability to produce epiglycanin when passaged in the ascites. The implication of this and the relationship of this phenomenon to the tumorigenicity of the cells are not fully understand. Epiglycanin is estimated to account for about 1% of the cell's dry weight and its high concentration on the cell surface (estimated to be about 4×10^6 molecules/cell) was demonstrated by lectin and antibody absorption experiments. Further, shadow-casting electron microscopy shows epitectin molecules on the surface of TA4-Ha and TA3-MM cells as long filaments (450–500 nm in length; Miller *et al.*, 1977). Epiglycanin has been shown to be shed from the cell surface since it is detected in the ascites fluid as well as in the serum of mice bearing the ascites cells. Intracellular epiglycanin is present in two types of exocytotic vesicles one of which consists of large multivesicular sacs with diameters up to 3 mm. It is suggested that epiglycanin can be secreted into the extracellular milieu via a new mode involving the multivesicular sacs (Watkins *et al.*, 1991). Epiglycanin has been proposed to protect the tumor cells from the host immune system by masking of the H-2 histocompatibility antigens (Codington and Frim, 1983) and by immunosuppression (Fung and Longenecker, 1991).

An *O*-glycosylated protein analogous to epiglyanin, and called ascites sialoglycoprotein (ASPG-1), has been identified on ascites 13762 rat mammary adenocarcinoma cells (Carraway and Spielman, 1986). Based on leucine labeling, ASGP-1 is estimated to constitute at least 0.5% of the total cell protein (Sherblom *et al.*, 1980a). Compared to epiglycanin, ASGP-1 appears to be heavily sialylated since sialidase treatment of 13762 cells substantially reduced ruthenium red and ferritin labeling (Sherblom *et al.*, 1980b). ASGP-1 is easy to isolate since centrifugation of the tumor cell membranes in a mixture of 4 M guanidine HCl and cesium chloride yields in a single step ASPG-1 of greater than 95% purity. The two different (MAT-B1 and MAT-C1) ascites sublines of 13762 cells produce ASGP-1 molecules with carbohydrate contents of 67 and 73%, respectively. Consequently, the molecules from the two sublines have different molecular size and density. The amino acid and carbohydrate compositions of ASGP-1 are very different from those of epiglycanin. In contrast to the small saccharides of epiglycanin, the saccharides of ASGP-1 are larger than tetrasaccharides and mostly sialylated and further, in the case of MAT-B1 species, are also sulfated (Sherblom and Carraway, 1980). It has been postulated that at the cell surface ASGP-1 occurs as a heterodimeric complex with a transmembrane sialoglycoprotein, termed ASGP-2. The ASGP-1 shed into the ascites fluid, apparently by a proteolytic cleavage, is free of ASGP-2. The two glycoproteins, ASGP-1 and ASGP-2, are generated from a single precursor protein. ASGP-2 is a 120-kDa sialoglycoprotein rich in Asn-linked saccharides and has two cysteine-rich domains with sequences related to the epidermal growth factor family of proteins (Sheng *et al.*, 1992). The resistance of the 13762 cells to lysis by natural killer cells and rat spleen lymphocytes was dependent on the expression of the ASGP-1/ASGP-2 complex on the cell surfaces. For example, cells treated with tunicamycin, which inhibits *N*-glycosylation of ASGP-2 and hence cell surface expression of ASGP-1, were significantly more susceptible to natural killer cell-mediated lysis (Bharathan *et al.*, 1990). Studies on a series of 13762 NK cultured cells selected for differing metastatic potential demonstrated a correlation with the presence of ASGP and metastasis (Steck and Nicolson, 1983).

Interest in malignancy-associated *O*-glycosylated proteins was greatly stimulated when a number of investigators produced monoclonal antibodies that were found to react with this family of glycoconjugates. Many of these antibodies were generated against human milk fat globule (HMFG) glycoproteins (Ceriani *et al.*, 1983; Burchell *et al.*, 1987), and others were obtained by immunizing with human tumor cells, cell membranes, or partially purified sialoglycoproteins (Colcher *et al.*, 1981; Ashall *et al.*, 1982; Kufe *et al.*, 1984). The sialoglycoproteins recognized by these antibodies have been called by different names including HMFG-1 antigen, PAS-O mucin, MAM-6, episialin, epithelial membrane antigen, DF-3 antigen, Ca antigen, epitectin, polymorphic urinary mucin (PUM), and polymorphic epithelial mucin (PEM). Cloning and sequencing studies have revealed that the core protein of these sialoglycoproteins is the product coded for by a gene now referred to as the MUC1 gene. This gene codes for a core protein that appears to be expressed also in a variety of normal human glandular epithelia (Peat *et al.*, 1992) and found in normal body fluids such as milk, sweat, and urine (Swallow *et al.*, 1986). Studies involving a large number of monoclonal antibodies specific for MUC1 glycoprotein reveal a strong preferential binding to cancer cells, indicating overexpression. The basis of the overexpression of the MUC1 epitopes by malignant cells is not fully understood, but it is believed that both increased synthesis, e.g., there is a more than tenfold increase of MUC1 mRNA in breast carcinoma (Zaretsky *et al.*, 1990), and increased exposure of peptide epitopes as a result of aberrant glycosylation (discussed elsewhere) are contributory factors. Based on the metabolic labeling experiments and the studies on epiglycanin and ASGP-1, it appears that in general an increased quantity of mucin-type glycoprotein, including the MUC1 product, may be produced by malignant cells compared to the counterpart nonmalignant cells. However, direct quantitative determinations of mucin glycoprotein in cancer cells and their normal counterparts by a colorimetric assay or other nonimmunological methods have not been performed (Bhavanandan, 1991). In normal epithelial cells, MUC1 glycoprotein is mainly present at the apical surface (Zotter *et al.*, 1988). In tumor cells this glycoprotein is spread out over the entire cell surface because of the loss of polarization (Hilkens *et al.*, 1992).

The human MUC1 gene encodes a core protein composed of variable numbers of 20-amino-acid tandem repeats, an *N*-terminal domain containing a signal sequence and a splice site yielding two alternative products, a C-terminal segment containing a potential transmembrane sequence and a cytoplasmic tail (Gendler *et al.*, 1990; Ligtenberg *et al.*, 1990; Williams *et al.*, 1990). It has been suggested that the cytoplasmic tail might be interacting specifically with cytoskeletal elements (Parry *et al.*, 1990; Spicer *et al.*, 1991). The extensive polymorphism of the human MUC1 gene (Swallow *et al.*, 1987b) is explained as resulting from variations in the number of tandem repeats. The tandem repeat consists of 25% Ser and Thr serving as potential *O*-glycosylation sites, and a further 25% is constituted by Pro. The nonrepetitive domains contain about 27% Ser and Thr, providing additional potential *O*-glycosylation sites. The nonrepetitive segment between the tandem repeats and the putative transmembrane sequence also contains five Asn in the consensus sequence for potential glycosylation.

The saccharide structures of the MUC1 family of glycoproteins are very poorly characterized. The currently available limited information suggests both tissue-dependent variations (Bardales *et al.*, 1989; Hull *et al.*, 1989) and malignancy-associated alteration in the saccharides of MUC1 glycoproteins (Hanisch *et al.*, 1990). It has been proposed that a shift from larger to smaller saccharides is responsible for exposing new peptide epitopes in MUC1 glycoprotein of cancer cells that are masked in nonmalignant cells (Burchell *et al.*, 1989); however, the evidence for this is very weak.

The MUC1 glycoprotein produced by the human laryngeal carcinoma (H.Ep.2) cell line and named CA antigen or epitectin has been investigated in some detail (Ashall *et al.*, 1982; Bramwell *et al.*, 1983; Bardales *et al.*, 1989; Qin and Bhavanandan, unpublished results). Epitectin was originally discovered after noting the ability of the Ca1 monoclonal antibody recognizing it to distinguish between malignant and nonmalignant segregants of hybrid cells (Ashall *et al.*, 1982). Thus, in the pair of CGL1 and CGL3 cells derived from the fusion of human cervical carcinoma cells (HeLa cells) and human diploid fibroblasts (Stanbridge *et al.*, 1982), the ability to produce epitectin is linked to the ability of the cells to grow progressively in vivo in nude mice. Additional evidence for the link between epitectin production and tumorigenicity comes from observations on the effect of long-term culture of H.Ep.2 cells. After 80 or more generations in culture the production of epitectin by H.Ep.2 cells was significantly reduced and this was accompanied by a decrease in the ability of these cells to produce tumors in nude mice (DiIulio *et al.*, 1994). Gel filtration of epitectin purified by affinity chromatography shows very heterogeneous elution profiles typical of epithelial mucins (Bardales *et al.*, 1989). Mucin-type characteristics are also evident from the heterogeneous populations of flexible strands of epitectin molecules visualized by electron microscopy (Bramwell *et al.*, 1986). However, on SDS-PAGE the epitectin preparations are surprisingly (compared to secreted epithelial mucins) homogeneous showing two distinct, reasonably sharp bands with estimated molecular masses of 350 and 390 kDa. A buoyant density of about 1.4 g/ml for this glycoprotein is suggestive of a carbohydrate content of about 50% by weight. The bulk of the carbohydrate is present as di-, tri-, and tetrasaccharides of the general formula $(\text{NeuAc}\alpha)_{0,1,2}\rightarrow[\text{Gal}\beta1\rightarrow3\text{GalNAc}]\alpha\rightarrow$. Other minor saccharides detected include GalNAcα→(Tn antigen), SAα2→6-GalNAcα→ (sialyl Tn antigen), and the hexasaccharide NeuAc→Gal(NeuAc→Gal→GlcNAc)→GalNAcα→. A small amount of mannose is present suggesting the presence of at least one Asn-linked oligosaccharide. Proteolytic fragmentation followed by HPLC and lectin affinity chromatography revealed the presence of peptide regions with saccharide clustering and regions free of glycosylation (Bardales, DiIulio, and Bhavanandan, unpublished results). Sequential extractions with ionic buffers and detergents indicate that epitectin is an intrinsic membrane glycoprotein which is not released by treatment with phospholipase C. It is, therefore, most likely held by a transmembrane peptide segment as suggested by the deduced amino acid sequence.

It is now abundantly clear that *O*-glycosylated proteins are intimately associated with cell surfaces and that in malignant cells there are both quantitative and qualitative changes in these molecules. However, information concerning the function of these molecules in normal and malignant cells is sparse (see Carraway *et al.*, 1992). It is generally accepted that the secretory mucins, which form viscoelastic gels on epithelial surfaces that are in direct contact with the environment, provide lubrication and protection. Thus, the respiratory tract mucous secretions serve to trap and clear inhaled pathogens (virus, bacteria), particulate pollutants (such as pollen or dust), and corrosive agents (such as sulfur dioxide) via the mucociliary transport system, thereby maintaining the sterility of the lungs (Kaliner, 1991). Similarly, vaginal/cervical mucins are important for maintaining the sterility of the uterus. The gastrointestinal mucus impedes the diffusion of H^+ ions into the epithelium and prevents digestion of the epithelial cells by acids as well as by enzymes (Allen, 1983; Bhaskar *et al.*, 1992). In lower forms of animals, such as earthworms, slugs, frogs, and fish, secretions that cover the outer body surfaces function as mechanical protectant, hydrodynamic lubricant, and probably also as antiparasitic and antibacterial agents (see Ingram, 1980; Bevins and Zasloff, 1990). In serving these functions, the secretory mucins play two distinct roles: that of a mechanical protectant/lubricant gel and that of specific receptors for pathogenic organisms. The ability of the mucins to form viscous solutions or a gel is clearly dependent on the saccharides, but a saccharide–protein glycoconjugate is not essential since a pure polysaccharide, such as hyaluronic acid, has this property. In mucins, a core protein molecule is heavily substituted with small saccharides and thus resembles polysaccharides in an overall sense. Examples of some such "simple" mucins are armadillo and sheep submaxillary mucins in which the major portion of the carbohydrates are *O*-linked GalNAc and SAα2→6GalNAc, respectively. It has been demonstrated that attachment of GalNAc residues leads to stiffening of the core polypeptide chain resulting in highly hydrophilic, random-coil molecules (Shogren *et al.*, 1989). Addition of an extra sugar, particularly a charged sugar such as sialic acid, would increase hydrophilicity and gel-forming ability (Gottschalk and Thomas, 1961; Maeji *et al.*, 1987). Thus, it seems unlikely that larger oligosaccharides and the enormous structural diversity of the saccharides noted in the majority of mucins are essential for the protective/lubricative role of mucins. More than 60 separate saccharides were identifiable in human tracheobronchial mucins and undoubtedly more will be found by improvements in techniques of fractionation (Woodward, Ringler, Davidson, and Bhavanandan, unpublished results). An attractive explanation for this complexity is that these different saccharide structures are needed as specific receptors for capturing and eliminating pathogens (virus, bacteria, parasites). Varki (1993) in a recent review refers to this as the "decoy" function of oligosaccharides of mucins, as opposed to the "traitorous" function of mostly glycolipid-based saccharides on cell surfaces. There is no clear evidence that gel-forming secretory mucins are produced by malignant cells. All available evidence indicates that malignancy-associated *O*-glycosylated proteins are primarily intrinsic membrane molecules, even though some portions are shed/secreted into the extracellular milieu. The observation that epiglycanin is packaged in multivesicular sacs is important and should be pursued to determine whether it is actively secreted and if this glycoprotein has the ability to yield viscous solutions and gels.

It is speculated that the membrane-associated/anchored mucin-type glycoproteins of malignant cells also play protective roles. Some of these were mentioned above with respect to possible functions of epiglycanin and ASGP-1. It has been suggested that mucin-type glycoproteins associated with the surface of cancer cells may serve to shield these cells from an abnormal extracellular environment such as low pH, altered osmolarity and hydrolytic enzymes that would otherwise be destructive (Bramwell *et al.*, 1983). The high glycolytic activity of cancer cells is likely to result in low pH within the tumor, and cell surface mucin-type glycoprotein could play a role similar to that of gastric mucin (Bhaskar *et al.*, 1992). There may be a common function for the intrinsic mucin-type glycoproteins associated with erythrocytes, leukocytes, platelets, and a wide variety of noncancerous cells. An understanding of any such common or of specific functions of glycophorin, leukosialin, CD42b, GYCAM-1, CD34, and the range of MUC1 glycoprotein-positive normal cells (Zotter *et al.*, 1988) would greatly accelerate the elucidation of the role of this class of glycoconjugate associated with the surface of cancer cells. On the cell surface, these extended rigid molecules are likely to project high above the plasma membrane into the lumen and thereby shield the surface and regulate the interaction with other cells, membranes, and soluble components. For MUC1 glycoprotein a variety of effects have been noted in *in vitro* experiments. Shimizu *et al.* (1990) found that the human milk glycoprotein (HMFG) was able to markedly reduce the proliferation rate of BALB/c 3T3 cells in culture in a concentration-dependent and reversible manner, suggesting a role in the regulation of cell growth. Ligtenberg *et al.* (1992) have proposed that MUC1 glycoproteins are antiadhesion molecules which function by masking adhesion molecules that are present on the cell membrane, as well as by charge repulsion. When normal mammary epithelial cells and melanoma cells which do not express exogenous MUC1 were transfected with cDNA encoding MUC1 glycoprotein, the resulting transfectants (expressing MUC1 glycoprotein at levels similar to carcinoma cells) did not aggregate as efficiently as the control (nontransfected) cells. This could be one of the factors that influence the metastatic potential of tumor cells. MUC1 glycoprotein has been implicated in the protection of tumor cells from the immune system by different mechanisms. It could simply cause steric hindrance and interfere with the recognition of surface antigens as proposed to be the case with epiglycanin. Purified MUC1 glycoprotein has been shown to inhibit the cytotoxic action of eosinophils toward target cells (Hayes *et al.*, 1990). Cytotoxic T lymphocytes derived from patients with breast cancer were found to be able to recognize MUC1 glycoprotein as their target in a major histocompatibility complex (MHC)-unrestricted fashion (Jerome *et al.*, 1991). It is suggested that the highly repetitive nature of the MUC1 epitopes allows cross-linking of T-cell receptors directly, accounting for the lack of MHC restriction. In recent studies, van de Wiel-van Kemenade *et al.* (1993) transfected MUC1-minus melanoma cells with MUC1 cDNA and found that the transfected cells were significantly less susceptible to lysis by cytotoxic lymphocytes compared to the parental cells. Finally, the MUC1 glycoprotein shed from the cells into the circulation may also act as blocking factors and inhibit lysis of MUC1-expressing cells by the cytotoxic T cells as noted in studies with soluble MUC1 glycoprotein isolated from milk (Barnd *et al.*, 1989).

The metastatic process is very complex and involves numerous steps including detachment of cells from the primary tumor, motility, extravasation, escape from immunosurveillance, and adhesion/invasion of target organs. Various cell surface saccharide structures and whole glycoconjugate molecules are thought to be important in these steps. Numerous investigators have examined the correlation between alterations in the sialylation of cell surface glycoconjugates and metastatic potential of tumor cells. While some studies indicated that highly metastatic cells had elevated levels of sialic acid (Yogeswaran and Tao, 1980; Altevogt *et al.*, 1983) others did not find significant differences in the cell surface sialic acid of metastatic variants (Passaniti and Hart, 1988). The biphasic effect of cell surface sialic acids on cancer cell adhesiveness recently reported by T. Sawada *et al.* (1993) could be an explanation for some of the conflicting findings on the correlation between sialic acid levels and invasiveness and metastasis of tumor cells. The increased levels of GlcNAcβ→6Manα→6Manβ→ branches in Asn-linked oligosaccharides and of GlcNAcβ→6 branches in Se/Thr-linked oligosaccharides have been found to be correlated with tumor progression and increased metastatic potential of tumor cells as discussed above (Dennis *et al.*, 1987; Laferte and Dennis, 1989). Studies on T24H-ras-infected fibroblasts and SP1 mammary carcinoma cells have shown increased levels of the glycosyltransferases that are responsible for the synthesis of the above structures (Yousefi *et al.*, 1991). Recently, the involvement of cancer cell surface *O*-glycosylated proteins in metastasis has received special attention because of the belief that these molecules could have multiple roles in the complex process of metastatic spread of cancer. For example, *O*-glycosylated proteins such as the MUC1 glycoprotein when present at some optimum level could promote cell–cell adherence necessary for the growth of the tumor, but at high levels it could cause poor adhesion and escape into the circulation. Once in the circulation, the cells with high levels of *O*-glycosylated proteins would be in an advantageous position to escape destruction by the immune system. The ability of these glycoproteins to suppress NK cell cytotoxicity could be one of the factors in this process of evading the immune system. In fact, a number of investigations have shown that the production of high levels of mucin glycoproteins by tumors is correlated to poor prognosis in cancer patients (Symond and Vickery, 1976; Ater *et al.*, 1984). In colorectal cancer, the presence of a high level of sialyl Tn antigen on *O*-glycosylated proteins of tumor cells was found to be a predictor of poor prognosis suggesting that in addition to quantitative differences structural alterations in the molecule are also important (Itzkowitz *et al.*, 1990). A number of investigations have indicated that high-molecular-weight *O*-glycosylated proteins are differentially expressed on cultured metastatic human cancer cells (Irimura *et al.*, 1988; Hoff *et al.*, 1989; Bresalier *et al.*, 1991). Other studies involving a large number of human colorectal carcinoma specimens have also demonstrated a strong association with increased expression of this class of glycoproteins (in this case, MUC1 glycoprotein) and progression to advanced stages and metastasis of human colorectal carcinoma (Nakamori *et al.*, 1994). In this study, since aberrant transcription of MUC1 core protein did not appear to be the reason for the increased expression, altered processing or increased stability of the mature MUC1 glycoprotein is thought to be responsible for the higher levels. Altered processing of MUC1, such as premature termination of saccharide chains, would explain both the increased exposure of peptide epitopes (Zotter *et al.*, 1988) and the increase in Tn and sialyl Tn antigens detected by various antibodies (Springer, 1984; Itzkowitz *et al.*, 1990). These are not the only changes in mucin glycoprotein-associated carbohydrate antigens. Thus, the levels of sialyl-dimeric Le^x antigen expressed on *O*-glycosylated proteins in colon carcinoma cells that metastasized to the liver were clearly higher than those on the primary tumor cells (Hoff *et al.*, 1989; Matsushita *et al.*, 1991). These findings are very important since the sialyl Le^x structure is implicated to be a major ligand for the lectinlike adhesion molecules, E- and P-selectins (Irimura *et al.*, 1993). Interactions between selectins and ligands such as sialyl Le^x saccharide on the tumor cell surface could mediate adhesion to endothelial cells and to platelets involved in the process of extravasation (Irimura *et al.*, 1993). Sawada *et al.* (1994) found that the efficiency of the E-selectin-mediated binding of colon cancer cells to endothelial cells correlated with the metastatic potential of the cancer cells. The E-selectin-mediated binding of the tumor cells could be inhibited by glycoproteins such as leukosialin and LAMP-1 which contain sialyl Le^x structures, but not by other glycoproteins. The colon carcinoma-associated overexpression of sialyl Le^x was found to be the result of altered glycosylation of MUC1 core protein (Hanski *et al.*, 1993). Thus, evidence is accumulating for an intimate involvement of *O*-glycosylated sialoglycoproteins in general and MUC1 sialoglycoprotein in particular with the process of tumorigenicity and metastasis. However, at present very little is known concerning the factors that control the expression of this class of glycoconjugates at the level of the gene as well as the processing and glycosylation of mucin-type glycoproteins in normal and tumor cells. Irimura and co-workers (Irimura *et al.*, 1990; Dohi *et al.*, 1993; Shirotani *et al.*, 1994) have identified a soluble factor, named mucomodulin, in human colon tissues which stimulates the production of MUC1 glycoprotein by upregulation of the mRNA levels. Purification and characterization of mucomodulin and understanding of its role in the regulation of MUC1 gene expression would greatly enhance our understanding of the role of the high-molecular-weight *O*-glycosylated proteins in malignancy.

ACKNOWLEDGMENTS. The investigations of one of us (V.P.B.) in this field have been supported by grants from the USPHS, most recently by grant CA38797 awarded by the National Cancer Institute. We thank Ms. Esha Bhavanandan and Ms. Filomena Cramer for their secretarial assistance.

REFERENCES

Abeijon, C., and Hirschberg, C. B., 1987, Subcellular site of synthesis of the N-acetylgalactosamine (α1→0) serine (or threonine) linkage in rat liver, *J. Biol. Chem.* **262**:4153–4159.

Abercrombie, M., and Ambrose, E. J., 1962, The surface properties of cancer cells: A review, *Cancer Res.* **22**:525–548.

Akiyama, S. K., Yamada, S. S., and Yamada, K. M., 1989, Analysis of the role of glycosylation of the human fibronectin receptor, *J. Biol. Chem.* **264**:18011–18018.

Allen, A., 1983, Mucus—A protective secretion of complexity, *Trends Biochem. Sci.* **8**:169–173.

Altevogt, P., Fogel, M., Cheingsong-Popov, R., Dennis, J. W., Robinson, P., and Schirrmacher, V., 1983, Different patterns of lectin binding and cell surface sialylation detected on related high- and low-metastatic tumor lines, *Cancer Res.* **43**:5138–5144.

Asada, M., Furukawa, K., Kantor, C., Gahmberg, C. G., and Kobata, A., 1991, Structural study of the sugar chains of human leukocyte cell adhesion molecules CD11/CD18, *Biochemistry* **30**:1561–1571.

Asada, M., Furukawa, K., Segawa, K., and Kobata, A., 1992, Biological significance of altered glycosylation of the cell surface glycoproteins in the transformed cells, *Cell Struct. Funct.* **17**:477 (Abstract).

Ashall, F., Bramwall, M. E., and Harris, H., 1982, A new marker for human cancer cells. 1. The Ca antigen and the Ca1 antibody, *Lancet* **7**:1–6.

Ater, J. L., Gooch, W. M., Bybee, B. L., and O'Brien, R. T., 1984, Poor prognosis for mucin-producing Wilms' tumor, *Cancer* **53**:319–323.

Aubert, J., Biserte, G., and Loucheux-Lefebvre, M. H., 1976, Carbohydrate–peptide linkage in glycoproteins, *Arch. Biochem. Biophys.* **175**:410–418.

Bardales, R., Bhavanandan, V. P., Wiseman, G., and Bramwell, M. E., 1989, Purification and characterization of the epitectin from human laryngeal carcinoma cells, *J. Biol. Chem.* **264**:1980–1987.

Barnd, D. L., Lan, M. S., Metzgar, R. S., and Finn, O. K., 1989, Specific major histocompatibility complex-unrestricted recognition of tumor-associated mucins by human cytotoxic T cells, *Proc. Natl. Acad. Sci. USA* **86**:7159–7163.

Barsoum, A. L., and Bhavanandan, V. P., 1989, Detection of glycophorin A-like glycoproteins on the surface of cultured human cells, *Int. J. Biochem.* **21**:635–654.

Barsoum, A. L., Czuczman, M., Bhavanandan, V. P., and Davidson, E. A., 1984, Epitopes immunologically related to glycophorin A on human malignant and nonmalignant cells in culture, *Int. J. Cancer* **34**:789–795.

Barsoum, A. L., Bhavanandan, V. P., and Davidson, E. A., 1985, Monoclonal antibodies to cyanogen bromide fragments of glycophorin A, *Mol. Immunol.* **22**:361–367.

Bevilacqua, M. P., and Nelson, R. M., 1993, Selectins, *J. Clin. Invest.* **91**:379–287.

Bevins, C. L., and Zasloff, M. A., 1990, Peptides from frog skin, *Annu. Rev. Biochem.* **59**:395–414.

Beyer, T. A., and Hill, R. L., 1982, Glycosylation pathways in the biosynthesis of nonreducing terminal sequences in oligosaccharides of glycoproteins, in: *The Glycoconjugates* III (M. I. Horowitz, ed.), Academic Press, New York, pp. 25–45.

Bharathan, S., Moriarty, J., Moody, C. E., and Sherblom, A. P., 1990, Effect of tunicamycin on sialomucin and natural killer susceptibility of rat mammary tumor ascites cells. *Cancer Res.* **50**:5250–5256.

Bhargava, A. K., Woitach, J. J., Davidson, E. A., and Bhavanandan, V. P., 1990, Cloning and cDNA sequence of a bovine submaxillary gland mucin-like protein containing two distinct domains, *Proc. Natl. Acad. Sci. USA* **87**:6798–6802.

Bhaskar, K. R., Garik, P., Turner, B. S., Bradley, J. D., Bansil, R., Stanley, H. E., and Lamont, J. T., 1992, Viscous fingering of HCl through gastric mucin, *Nature* **360**:458–462.

Bhavanandan, V. P., 1991, Cancer-associated mucins and mucin-type glycoproteins, *Glycobiology* **1**:493–503.

Bhavanandan, V. P., and Hegarty, J. D., 1987, Identification of the mucin core protein by cell-free translation of messenger RNA from bovine submaxillary glands, *J. Biol. Chem.* **262**:5913–5917.

Bhavanandan, V. P., and Katlic, A. W., 1979, The interaction of wheat germ agglutinin with sialoglycoproteins; The role of sialic acid, *J. Biol. Chem.* **254**:4000–4008.

Bhavanandan, V. P., and Meyer, K., 1966, Mucopolysaccharides: N-acetyl glucosamine- and galactose-6 sulfates from keratosulfate, *Science* **151**:1404–1405.

Bhavanandan, V. P., Umemoto, J., Banks, J. R., and Davidson, E. A., 1977, Isolation and partial characterization of sialoglycopeptides produced by a murine melanoma, *Biochemistry* **16**:4426–4437.

Bhavanandan, V. P., Katlic, A. W., Banks, J., Kemper, J. G., and Davidson, E. A., 1981, Partial characterization of sialoglycopeptides produced by cultured human melanoma cells and melanocytes, *Biochemistry* **20**:5586–5594.

Bierhuizen, M.F.A., and Fukuda, M., 1993, β-1,6-N-acetylglucosaminyl transferases: Enzymes critically involved in oligosaccharide branching, *Trends in Glycoscience and Glycotechnology* **6**:17–28.

Bobek, L. A., Tsai, H., Biesbrock, A. R., and Levine, M. J., 1993, Molecular cloning, sequence and specificity of expression of the gene encoding the low molecular weight human salivary mucin (Muc7), *J. Biol. Chem.* **268**:20563–20569.

Bramwell, M. E., Bhavanandan, V. P., Wiseman, G., and Harris, H., 1983, Structure and function of the Ca antigen, *Br. J. Cancer* **48**:177–183.

Bramwell, M. E., Wiseman, G., and Shotton, D. M., 1986, Electron-microscopic studies of the Ca antigen, epitectin, *J. Cell Sci.* **86**:249–261.

Bresalier, R. S., Niv, Y., Byrd, J. C., Duh, Q., Toribara, N. W., Rockwell, R. W., Dahiya, R., and Kim, Y. S., 1991, Mucin production by human colonic carcinoma cells correlates with their metastatic potential in animal models of colon cancer metastasis, *J. Clin. Invest.* **87**:1037–1045.

Briand, J. P., Andrews, S. P., Cahill, E., Conway, N. A., and Young, J. D., 1981, Investigation of the requirements for O-glycosylation by bovine submaxillary gland UDP-N-acetylgalactosamine: polypeptide N-acetylgalactosamine transferase using synthetic peptide substrates, *J. Biol. Chem.* **256**:12205–12207.

Brockhausen, I., Matta, K. L., Orr, J., and Schachter, H., 1985, Mucin synthesis. UDP-GlcNAc: GalNAc-Rβ 3-N-acetylglucosaminyltransferase and UDP-GlcNAc:GlcNAcβ1→3GalNAc-R (GlcNAc to GalNAc) β6-N-acetyl glucosaminyltransferase from pig and rat colon mucosa, *Biochemistry* **24**:1866–1860.

Brockhausen, I., Kuhns, W., Schachter, H., Matta, K. L., Sutherland, D. R., and Baker, M. A., 1991, Biosynthesis of O-glycans in leukocytes from normal donors and from patients with leukemia: Increase in O-glycan core 2 UDP-GlcNAc: Galβ3→GalNAcR (GlcNAc to GalNAc) β(1→6)-N-acetyl glucosaminyltransferase in leukemic cells, *Cancer Res.* **51**:1257–1263.

Burchell, J., Gendler, S., Taylor-Papadimitriou, J., Girling, A., Lewis, A., Millis, R., and Lamport, D., 1987, Development and characterization of breast cancer reactive monoclonal antibodies directed to the core protein of the human milk mucin, *Cancer Res.* **47**:5476–5482.

Burchell, J., Papadimitriou, J. T., Boshell, M., Gendler, S., and Duhig, T., 1989, A short sequence, within the amino acid tandem repeat of a cancer-associated mucin, contains immunodominant epitopes, *Int. J. Cancer* **44**:691–696.

Carlsson, S. R., and Fukuda, M., 1992, The lysosomal membrane glycoprotein lamp-1 is transported to lysosomes by two alternative pathways, *Arch. Biochem. Biophys.* **296**:630–639.

Carraway, K. L., and Spielman, J., 1986, Structural and functional aspects of tumor cell sialomucins, *Mol. Cell. Biochem.* **72**:109–120.

Carraway, K. L., Fregien, N., Carraway, K. L., III, and Carraway, C.A.C., 1992, Tumor sialomucin complexes as tumor antigens and modulators of cellular interactions and proliferation, *J. Cell Sci.* **103**:299–307.

Cartron, J. P., and Rahuel, C., 1992, Human erythrocyte glycophorins: Protein and gene structure analysis, *Transfus. Med. Rev.* **11**:63–92.

Ceriani, R. L., Peterson, J. A., Leo, J. Y., Moncada, R., and Blank, E. W., 1983, Characterization of cell surface antigens of human mammary epithelial cells with monoclonal antibodies prepared against human milk fat globule, *Somat. Cell Genet.* **9**:415–427.

Chai, W., Hounsell, E. F., Cashmore, G. C., Rosaqnkiewicz, J. R., Bauer, C. J., Fenney, J., Feizi, T., and Lawson, A. M., 1992, Neutral oligosaccharides of bovine submaxxilary mucin, *Eur. J. Biochem.* **203**:257–268.

Chandrasekaran, E. V., and Davidson, E. A., 1979, Sialoglycoproteins of human mammary cells: Partial characterization of sialoglycopeptides, *Biochemistry* **18**:5615–5620.

Chen, J. W., Cha, Y., Yuksel, K. U., Gracy, R. W., and August, J. T., 1988, Isolation and sequencing of a cDNA clone encoding lysosomal membrane glycoprotein mouse LAMP-1, *J. Biol. Chem.* **263**:8754–8758.

Codington, J. F., and Frim, D. M., 1983, Cell Surface macromolecular and morphological changes related to allotransplantability in the TA3 tumor, *Biomembranes* **11**:207–258.

Codington, J. F., and Haavik, S., 1992, Epiglycanin—A carcinoma-specific mucin-type glycoprotein of the mouse TA3 tumor, *Glycobiology* **2**:173–180.

Colcher, D., Horan, H. P., Nuti, M., and Schlom, J., 1981, A spectrum of monoclonal antibodies reactive with human mammary tumor cells, *Proc. Natl. Acad. Sci. USA* **78**:3199–3203.

Crepin, M., Porchet, N., Aubert, J. P., and Degand, P., 1990, Diversity of the peptide moiety of human airway mucins, *Biorheology* **27**:471–484.

Cummings, R. D., Trowbridge, I. S., and Kornfeld, S., 1982, A mouse lymphoma cell line resistant to the leukoagglutinating mannoside-β1,6 N-acetylglucosaminyltransferase, *J. Biol. Chem.* **257**:13421–13427.

Currie, G. A., Van Doorninck, W., and Bagshawe, K. D., 1968, Effect of neuraminidase on the immunogenicity of early mouse tropoblast, *Nature* **219**:191–192.

Dekker, J., and Strous, G. J., 1990, Covalent oligomerization of rat gastric mucin occurs in the rough endoplasmic reticulum, is N-glycosylation dependent, and precedes initial O-glycosylation, *J. Biol. Chem.* **265**:18116–18122.

Dekker, J., Aelmans, P. H., and Strous, G. J., 1991, The oligomeric structure of rat and human gastric mucins, *Biochem. J.* **277**:423–427.

Dennis, J. W., Laferte, S., Fukuda, M., Dell, A., and Carver, J. P., 1986, Asn-linked oligosaccharides in lectin-resistant tumor-cell mutants with varying metastatic potential, *Eur. J. Biochem.* **161**:359–373.

Dennis, J. W., Laferte, S., Waghorne, M. L., Breitman, M. L., and Kerbek, R. S., 1987, β1→6 branching of Asn-linked oligosaccharides is directly associated with metastasis, *Science* **236**:582–585.

DiIulio, N. A., Yamakami, K., Washington, S., and Bhavanandan, V. P., 1994, Effect of long-term culture of a human laryngeal carcinoma cell line on epitectin production and tumorigenicity in athymic mice, *Glycosylation Dis.* **1**:21–30.

Do, K.-Y., Smith, D. F., and Cummings, R. D., 1990, LAMP-1 in Cho cells is a primary carrier of poly-N-acetyllactosamine chains and is bound preferentially by a mammalian S-type lectin, *Biochem. Biophys. Res. Commun.* **173**:1123–1128.

Dohi, D. F., Sutton, R. C., Frazier, M. L., Nakamori, S., McIsaac, A. M., and Irimura, T., 1993, Regulation of sialomucin production in colon carcinoma cells, *J. Biol. Chem.* **268**:10133–10138.

Dufosse, J., Porchet, N., Audie, J.-P., Duperat, V. G., Laine, A., Van-Seuningen, I., Marrakchi, S.,

Degand, P., and Aubert, J.-P., 1993, Degenerate 87-base-pair tandem repeats create hydrophilic/hydrophobic alternating domains in human mucin peptides mapped to 11p15, *Biochem. J.* **293**:329–337.

Eckhardt, A. E., Timpte, C. S., Abernathy, J. L., Zhao, Y., and Hill, R. L., 1991, Porcine submaxillary mucin contains a cystine-rich, carboxylterminal domain in addition to a highly repetitive, glycosylated domain, *J. Biol. Chem.* **266**:9678–9687.

Endo, T., Ohbayashi, H., Hayashi, Y., Ikehara, Y., Kochibe, N., and Kobata, A., 1988, Structural study on the carbohydrate moiety of human placental alkaline phosphatase, *J. Biochem.* **103**:182–187.

Fiete, D., Srivastava, V., Hindsgaul, O., and Baenziger, J. U., 1991, A hepatic reticuloendothelial cell receptor specific for SO_4→GalNAcβ1,4 GlcNAcβ1,2 Manα that mediates rapid clearance of lutropin, *Cell* **67**:1103–1110.

Finne, J., Tao, T. W., and Burger, M. M., 1980, Carbohydrate changes in glycoproteins of a poorly metastasizing wheat germ agglutinin-resistant melanoma clone, *Carbohydr. Res.* **40**:2580–2587.

Fukuda, M., 1991a, Leukosialin, a major O-glycan-containing sialoglycoprotein defining leukocyte differentiation and malignancy, *Glycobiology* **1**:347–356.

Funakoshi, I., Nakada, H., and Yamashina, I., 1974, The isolation and characterization of glycopeptides from plasma membranes of an ascites hepatoma, AH 66, *J. Biochem.* **76**:319–333.

Fung, P.Y.S., and Longenecker, B. M., 1991, Specific immunosuppressive activity of epiglycanin, a mucin-like glycoprotein secreted by a murine mammary adenocarcinoma (TA3-Ha), *Cancer Res.* **51**:1170–1176.

Furukawa, K., and Kobata, A., 1992, Protein glycosylation, *Curr. Opin. Biotechnol.* **3**:554–559.

Furukawa, K., Minor, J. E., Hegarty, J. D., and Bhavanandan, V. P., 1986, Interactions of sialoglycoproteins with wheat germ agglutinin-Sepharose of varying ratio of lectin to sepharose: Use for the purification of mucin glycoproteins from membrane extracts, *J. Biol. Chem.* **261**:7755–7761.

Galili, U., Shohet, S. B., Kobrin, E., Stults, C. L., and Macher, B. A., 1988, Man, apes, and old world monkeys differ from other mammals in the expression of alpha-galactosyl epitopes on nucleated cells, *J. Biol. Chem.* **263**:17755–17762.

Gendler, S. J., Lancaster, C. A., Taylor-Papadimitriou, J., Duhig, T., Peat, N., Burchell, J., Pemberton, L., Lalani, E., and Wilson, D., 1990, Molecular cloning and expression of human tumor-associated polymorphic epithelial mucin, *J. Biol. Chem.* **265**:15286–15293.

Gerken, T. A., Butenhof, K. J., and Shogren, R., 1989, Effects of glycosylation on the conformation and dynamics of O-linked glycoproteins: Carbon-13 NMR studies of ovine submaxxilary mucin, *Biochemistry* **28**:5536–5543.

Gottschalk, A., 1960, Correlation between composition, structure, shape and function of a salivary mucoprotein, *Nature* **186**:949–951.

Gottschalk, A., and Thomas, M.A.W., 1961, Studies on mucoproteins. V. The significance of N-acetylneuraminic acid for the viscosity of ovine submaxillary gland glycoprotein, *Biochim. Biophys. Acta* **46**:91–98.

Green, E. D., van Halbeek, H., Boime, I., and Baenziger, J. U., 1985, Structural elucidation of the disulfated oligosaccharide from bovine lutropin, *J. Biol. Chem.* **260**:15623–15630.

Griffiths, B., Matthews, D. J., West, L., Attwood, J., Povey, S., Swallow, D. M., Gum, J. R., and Kim, Y. S., 1990, Assignment of the polymorphic intestinal mucin gene (Muc2) to chromosome-11p15, *Ann. Hum. Genet.* **54**:270–277.

Gum, J. R., Byrd, J. C., Hicks, J. W., Toribara, N. W., Lamport, D.T.A., and Kim, Y. S., 1989, Molecular cloning of human intestinal mucin cDNA's, *J. Biol. Chem.* **264**:6480–6487.

Gum, J. R., Hicks, J. W., Swallow, D. M., Lagac, R. L., Byrd, J. C., Lamport, D.T.A., Siddiki, B., and Kim, Y. S., 1990, Molecular cloning of cDNAs from a novel human intestinal mucin gene, *Biochem. Biophys. Res. Commun.* **171**:407–415.

Gum, J. R., Jr., Hicks, J. W., Toribara, N. W., Rothe, E.-M., Lagace, R. E., and Kim, Y. S., 1993, The human MUC2 intestinal mucin has cysteine-rich subdomains located both upstream and downstream of its central repetitive region, *J. Biol. Chem.* **267**:21375–21383.

Hakomori, S., 1973, Glycolipids of tumor membranes, *Adv. Cancer Res.* **18**:265–315.

Hakamori, S., 1984, Tumor-associated carbohydrate antigens, *Annu. Rev. Immunol.* **2**:103–126.

Hakomori, S., 1989, Aberrant glycosylation in tumors and tumor-associated carbohydrate antigens, *Adv. Cancer Res.* **52**:257–331.

Hanisch, F., Pete-Katalinic, J., Egge, H., Dabrowski, U., and Uhlenbruck, G., 1990, Structures of acidic O-linked polylactosaminoglycans on human skim milk mucins, *Glycoconjugate J.* **7**:525–543.

Hanski, C., Drechsler, K., Hanisch, F.-G., Sheehan, H., Manske, M., Ogorek, D., Klussmann, E., Hanski, M.-L., Blank, M., Xing, P.-X., McKenzie, I.F.C., Devine, P. L., and Riecken, E.-O., 1993, Altered glycosylation of the MUC-1 protein core contributes to the colon carcinoma-associated increase of mucin-bound sialyl-Lewis expression, *Cancer Res.* **53**:4082–4088.

Hayes, D. F., Silberstein, D. S., Rodrique, S. W., and Kufe, D. W., 1990, DF3 antigen, a human epithelial cell mucin, inhibits adhesion of eosinophils to antibody-coated targets, *J. Immunol.* **145**:960–962.

Hession, C., Osborn, L., Goff, D., Chi-Rosso, G., Vassallo, C., Pasek, M., Pittack, C., Tizard, R., Goelz, S., McCarthy, K., Hopple, S., and Lobb, R., 1990, Endothelial leukocyte adhesion molecule 1: Direct expression cloning and functional interactions, *Proc. Natl. Acad. Sci. USA* **87**:1673–1677.

Hilgers, J., Zotter, S., and Kenemans, P., 1989, Polymorphic epithelial mucin and CA 125-bearing glycoprotein in basic and applied carcinoma research, *Cancer Rev.* **11–12**:3–10.

Hilkens, J., 1988, Biochemistry and function of mucins in malignant disease, *Cancer Rev.* **11–12**:25–54.

Hilkens, J., and Buijs, F., 1988, Biosynthesis of Mam-6, an epithelial sialomucin, *J. Biol. Chem.* **263**:4215–4222.

Hilkens, J., Ligtenberg, M.J.L., Vos, H. L., and Litvinov, S. V., 1992, Cell membrane associated mucins and adhesion-modulating property, *Trends Biochem. Sci.* **17**:359–363.

Hill, H. D., Reynolds, J. A., and Hill, R. L., 1977a, Purification, composition, molecular weight, and subunit structure of ovine submaxillary mucin, *J. Biol. Chem.* **252**:3791–3798.

Hill, H. D., Schwyzer, M., Steinman, H. M., and Hill, R. L., 1977b, Ovine submaxillary mucin—Primary structure and peptide substrates of UDP-N-acetylgalactosamine: Mucin transferase, *J. Biol. Chem.* **252**:3799–3804.

Hiraizumi, S., Takasaki, S., Shiroki, K., Kochibe, N., and Kobata, A., 1990, Transfection with fragments of the adenovirus 12 gene induces tumorigenicity-associated alteration of N-linked sugar chains in rat cells, *Arch. Biochem. Biophys.* **280**:9–19.

Hiraizumi, S., Takasaki, S., Ohuchi, N., Harada, Y., Nose, M., Shozo, M., and Kobata, A., 1992, Altered glycosylation of membrane glycoproteins associated with human mammary carcinoma, *Jpn. J. Cancer Res.* **83**:1063–1072.

Hironaka, T., Furukawa, K., Esmon, P. C., Fournel, M. A., Sawada, S., Kata, M., Minaga, T., and Kobata, A., 1992, Comparative study of the sugar chains of factor Viii purified from human plasma and from the culture media of recombinant baby hamster kidney cells, *J. Biol. Chem.* **267**:8012–8020.

Hoff, S. D., Matsushita, Y., Ota, D. M., Cleary, K. R., Yamori, T., Hakomori, S., and Irimura, T., 1989, Increased expression of sialyl-dimeric Le antigen in liver metastases of human colorectal carcinoma, *Cancer Res.* **49**:6883–6888.

Hoffmann, W., 1988, A new repetitive protein from *Xenopus laevis* skin highly homologous to pancreatic spasmolytic polypeptide, *J. Biol. Chem.* **263**:7686–7690.

Hull, S. R., Bright, A., Carraway, K. L., Abe, M., Hayes, D. F., and Kufe, D. W., 1989,

Oligosaccharide difference in the DF3 sialomucin antigen from normal human milk and the BT-20 human breast carcinoma cell line, *Cancer Commun.* **1**:261–267.

Humphries, M. J., Matsumoto, K., White, S. L., and Olden, K., 1986, Oligosaccharide modification by swainsonine treatment inhibits pulmonary colonization by B16-F10 murine melanoma cells, *Proc. Natl. Acad. Sci. USA* **83**:1752–1756.

Ingram, G. A., 1980, Substances involved in the natural resistance of fish to infection: A review, *J. Fish Biol.* **16**:23–30.

Irimura, T., Gonzalez, R., and Nicolson, G. L., 1981, Effects of tunicamycin on B16 metastatic melanoma cell surface glycoproteins and blood-borne arrest and survival properties, *Cancer Res.* **41**:3411–3418.

Irimura, T., Carlson, D. A., Price, J., Yamori, T., Giavazzi, R., Ota, D. M., and Cleary, K. R., 1988, Differential expression of a sialoglycoprotein with an approximate molecular weight of 900,000 on metastatic human colon carcinoma cells growing in culture and in tumor tissues, *Cancer Res.* **48**:2353–2360.

Irimura, T., McIsaac, A. M., Carlson, D. A., Yagita, M., Grimm, E. A., Menter, D. G., Ota, D. M., and Cleary, K. R., 1990, Soluble factor in normal tissues that stimulates high-molecular-weight sialoglycoprotein production by human colon carcinoma cells, *Cancer Res.* **50**:3331–3338.

Irimura, T., Nakamori, S., Matsushita, Y., Taniuchi, Y., Todoroki, N., Tsuji, T., Izumi, Y., Kawamura, Y., Hoff, S. D., Cleary, K. R., and Ota, D. M., 1993, Colorectal cancer metastasis determined by carbohydrate-mediated cell adhesion: Role of sialyl-Lex antigens, *Cancer Biol.* **4**:319–324.

Itzkowitz, S. H., Bloom, E. J., Kokal, W. A., Modin, G., Hakomori, S., and Kim, Y. S., 1990, A novel mucin antigen associated with prognosis in colorectal cancer patients, *Cancer* **66**:1960–1966.

Jerome, K. R., Barnd, D. L., Bendt, K. M., Boyer, C. M., Taylor-Papadimitriou, J., McKenzie, I.F.C., Bast, R. C., and Finn, O. J., 1991, Cytotoxic T-lymphocytes derived from patients with breast adenocarcinoma recognize an epitope present on the protein core of mucin molecule preferentially expressed by malignant cells, *Cancer Res.* **51**:2908–2916.

Johnson, D. C., and Spear, P. G., 1983, O-linked oligosaccharides are acquired by herpes simplex virus glycoproteins in the Golgi apparatus, *Cell* **32**:987–999.

Johnson, V. G., Schlom, J., Paterson, A. J., Bennett, J., Magnani, J. L., and Colcher, D., 1986, Analysis of a human tumor-associated glycoprotein (Tag-72) identified by monoclonal antibody B72.3, *Cancer Res.* **46**:850–857.

Johnson, W. V., and Heath, E. C., 1986, Evidence for posttranslational O-glycosylation of fetuin, *Biochemistry* **25**:5518–5525.

Jokinen, M., Andersson, L. C., and Gahmberg, C. G., 1985, Biosynthesis of the major human red blood cell sialoglycoprotein, glycophorin A, *J. Biol. Chem.* **260**:11314–11321.

Joziasse, D. H., Schiphorst, W.E.C.M., van den Eijnden, D. H., Van Kuik, J. A., Van Halbeek, H., and Vliegenthart, J.F.G., 1987, Branch specificity of bovine colostrum CMP-sialic acid: N-acetyllactosaminide $\alpha2\rightarrow6$ sialyltransferase, *J. Biol. Chem.* **262**:2025–2033.

Joziasse, D. H., Shaper, J. H., van den Eijnden, D. H., Van Tunen, A. J., and Shaper, N. L., 1989, Bovine $\alpha1\rightarrow3$-galactosyltransferase: Isolation and characterization of a cDNA clone, *J. Biol. Chem.* **265**:14290–14297.

Joziasse, D. H., Shaper, J. H., Jabs, E. W., and Shaper, N. L., 1991, Characterization of an $\alpha1\rightarrow3$ galactosyltransferase homologue on human chromosome 12 that is organized as a processed pseudogene, *J. Biol. Chem.* **266**:6991–6998.

Kaliner, M. A., 1991, Human nasal respiratory secretions and host defense, *Am. Rev. Respir. Dis.* **144**:S52–S56.

Kawano, T., Takasaki, S., Tao, T.-W., and Kobata, A., 1993, Altered glycosylation of β1 integrins associated with reduced adhesiveness to fibronectin and laminin, *Int. J. Cancer* **53**:91–96.

Kawashima, H., Yamamoto, K., Osawa, T., and Irimura, T., 1993, Purification and characterization of UDP-GlcNAc:Galβ1→4Glc(NAc) β1,3-N-acetyl glucosaminyltransferase (poly-N-acetyllactosamine extension enzyme) from calf serum, *J. Biol. Chem.* **268:**27118–27126.

Khatri, I. S., Forstner, G. G., and Forstner, J. F., 1993, Suggestive evidence for two different mucin genes in rat intestine, *Biochem. J.* **294:**391–399.

Kim, Y. S., Yuan, M., Itzkowitz, S. H., Sun, Q., Kaizu, T., Palekar, A., Trump, B. F., and Hakomori, S., 1986, Expression of LeY and extended LeY blood group-related antigens in human malignant, premalignant and nonmalignant colonic tissues, *Cancer Res.* **46:**5985–5992.

Kobata, A., 1989, Altered glycosylation of surface glycoproteins in tumor cells and its clinical application, *Pigment Cell Res.* **2:**304–308.

Kornfeld, S., 1982, Oligosaccharide processing during glycoprotein biosynthesis, in: *The Glycoconjugates* (M. I. Horowitz, ed.), Academic Press, New York, Vol. III, pp. 3–23.

Kotovuori, P., Tontti, E., Pigott, R., Shepherd, M., Kisco, M., Hasegawa, A., Renkonen, R., Nortamo, P., Altieri, D. C., and Gahmberg, C. G., 1993, The vascular E-selectin binds to the leukocyte integrins CD11/CD18, *Glycobiology* **3:**131–136.

Kufe, D., Inghirami, G., Abe, M., Hayes, D., Justi-Wheeler, H., and Schlom, J., 1984, Differential reactivity of a novel monoclonal antibody (DF3) with human malignant versus benign breast tumors, *Hybridoma* **3:**223–232.

Kurosaka, A., Nakajima, H., Funakoshi, I., Matsuyama, M., Nagayo, T., and Yamashina, I., 1983, Structures of the major oligosaccharides from a human rectal adenocarcinoma glycoprotein, *J. Biol. Chem.* **258:**11594–11598.

Kurosaka, A., Kitagawa, H., Fukui, S., Numata, Y., Yamashina, I., Nakada, H., Funakoshi, I., Kawasaki, T., Ogawa, T., and Iijima, H., 1988, A monoclonal antibody that recognizes a cluster of a disaccharide, Neu5Ac2→6GalNAc, in mucin-type glycoproteins, *J. Biol. Chem.* **262:**8724–8726.

Kurosawa, N., Kojima, N., Inoue, M., Hamamoto, T., and Tsuji, S., 1994, Cloning and expression of Galβ1,3 GalNAc specific GalNAcα2,6 sialyltransferase, *J. Biol. Chem.* **269:**19048–19053.

Laferte, S., and Dennis, J. W., 1989, Purification of two glycoproteins expressing β1-6 branched Asn-linked oligosaccharides from metastatic tumor cells, *Biochem. J.* **259:**569–576.

Lan, M. S., Hollingsworth, M. A., and Metzgar, R. S., 1990, Polypeptide core of a human pancreatic tumor mucin antigen, *Cancer Res.* **50:**2997–3001.

Larsen, R. D., Rajan, V. P., Ruff, M. M. Kukowsaka-Latallo, J., Cummings, R. D., and Lowe, J. B., 1989, Isolation of a cDNA encoding a murine UDP galactose:beta-D-galactosyl-1,4-N-acetyl-D-glucosaminide alpha-1,3-galactosyltransferase: Expression cloning by gene transfer, *Proc. Natl. Acad. Sci. USA* **86:**8227–8331.

Larsen, R. D., Rivera-Marrero, C. A., Ernst, L. K., Cummings, R. D., and Lowe, J. B., 1990, Frameshift and nonsense mutations in a human genomic sequence homologous to a murine Udp-Gal:Beta-D-Gal(1,4)-D-GlcNAc alpha(1,3)-galactosyltransferase cDNA, *J. Biol. Chem.* **265:** 7055–7061.

Ligtenberg, M.J.L., Vos, H. L., Gennissen, A.M.C., and Hilkens, J., 1990, Episialin, a carcinoma-associated mucin, is generated by a polymorphic gene encoding splice variants with alternative termini, *J. Biol. Chem.* **265:**5573–5578.

Ligtenberg, M.J.L., Buijs, F., Vos, H. L., and Hilkens, J., 1992, Suppression of cellular aggregation by high levels of epsialin, *Cancer Res.* **52:**2318–2324.

Lin, L.-H., Stern, J. L., and Davidson, E. A., 1982, Clones from cultured, B16 mouse-melanoma cells resistant to wheat germ agglutinin and with altered production of mucin-type glycoproteins, *Carbohydr. Res.* **111:**257–271.

Lippincott-Schwartz, J., and Fambrough, D. M., 1987, Cycling of the integral membrane glycoprotein, LEP100, between plasma membrane and lysosomes: Kinetic and morphological analysis, *Cell* **49:**669–677.

Lis, H., and Sharon, N., 1993, Protein glycosylation. Structure and functional aspects—Review, *Eur. J. Biochem.* **218:**1–27.

Maeji, N. J., Inoue, Y., and Chujo, R., 1987, Preliminary communication: The role of the N-acetyl group in determining the conformation of 2-acetamido-2-deoxy-D-galactopyranosyl-threonine-containing peptides, *Carbohydr. Res.* **162:**4–8.

Magnani, J. L., Steplewski, Z., Koprowski, H., and Ginsburg, V., 1983, Identification of the gastrointestinal and pancreatic cancer-associated antigen detected by monoclonal antibody 19-9 in the sera of patients as a mucin, *Cancer Res.* **43:**5489–5492.

Marianne, T., Perini, J. M., Lafitte, J., Houdret, N., Pruvot, F. R., Lamblin, G., Slayter, H. S., and Roussel, P., 1987, Peptides of human bronchial mucus glycoproteins, *Biochem. J.* **248:**189–195.

Matsushita, Y., Hoff, S. D., Nudelman, E. D., Ohtaka, M., Hakomori, S., Ota, D. M., Cleary, K. R., and Irimura, T., 1991, Metastatic behavior and cell surface properties of HT-29 human colon carcinoma variant cells selected for their differential expression of sialyl-dimeric LeX antigen, *Clin. Exp. Metast.* **9:**283–299.

Matsuura, H., Greene, T., and Hakomori, S., 1989, An α-N-acetyl galactosaminylation at the threonine residue of a defined peptide sequence creates the oncofetal peptide epitope in human fibronectin, *J. Biol. Chem.* **264:**10472–10476.

Medof, M. E., Lublin, D. M., Holers, V. M., Ayers, D. J., Getty, R. R., Leykam, J. K., Atkinson, J. P., and Tykocinski, M. L., 1987, Cloning and characterization of cDNAs encoding the complete sequence of decay-accelerating factor of human complement, *Proc. Natl. Acad. Sci. USA* **84:**2007–2011.

Miller, S. C., Hay, E. D., and Codington, J. F., 1977, Ultrastructural and histochemical differences in cell surface properties of strain specific and non-strain specific TA3 adenocarcinoma cells, *J. Cell Biol.* **72:**511–529.

Mizouchi, T., Yamashita, K., Fukikawa, K., Titani, K., and Kobata, A., 1980, The structures of the carbohydrate moieties of bovine blood coagulation factor X, *J. Biol. Chem.* **255:**3526–3531.

Murthy, V.L.N., and Horowitz, M. I., 1966, Alkali-reductive cleavage of ovine submaxillary mucin, *Carbohydr. Res.* **6:**266–275.

Nakada, H., Funakoshi, I., and Yamashina, I., 1975, The isolation and characterization of glycopeptides and mucopolysaccharides from plasma membranes of an ascites hepatoma, AH 130, *J. Biochem.* **78:**863–872.

Nakada, H., Numata, Y., Inoue, M., Tanaka, N., Kitagawa, H., Funakoshi, I., Fukui, S., and Yamashina, I., 1991, Elucidation of an essential structure recognized by an anti-GalNAcα-Ser (Thr) monoclonal antibody (MLS 128), *J. Biol. Chem.* **266:**12402–12405.

Nakada, H., Inoue, M., Numata, Y., Tanaka, N., Funakoshi, I., Fukui, S., Mellors, A., and Yamashina, I., 1993, Epitopic structure of Tn glycophorin A for an anti-Tn antibody (MLS 128), *Proc. Natl. Acad. Sci. USA* **90:**2495–2499.

Nakamori, S., Ota, D. M., Cleary, K. R., Shirotani, K., and Irimura, T., 1994, MUC1 mucin expression as a marker of progression and metastasis of human colorectal carcinoma, *Gastroenterology* **106:**353–361.

Nieman, H., Boschek, B., Evans, D., Rosig, M., Tamura, T., and Klenk, H., 1982, Post-translational glycosylation of coronavirus glycoprotein E1—Inhibition by monensin, *EMBO J.* **1:**1499–1504.

Okumura, T., Lombart, C., and Jamieson, G. A., 1976, Platelet glycocalicin II: Purification and characterization, *J. Biol. Chem.* **251:**5950–5955.

Parekh, R. B., Tse, A.G.D., Dwek, R. A., Williams, A. F., and Rademacher, T. W., 1987, Tissue-specific N-glycosylation, site specific oligosaccharide patterns and lentil lectin recognition of rat Thy-1, *EMBO J.* **6:**1233–1244.

Parry, G., Beck, J. C., Moss, L., Barley, J., and Ojakian, G. K., 1990, Determination of apical membrane polarity in mammary epithelial cell cultures: The role of cell–cell, cell–substratum, and membrane–cytoskeleton interactions, *Exp. Cell Res.* **188:**302–311.

Passaniti, A., and Hart, G. W., 1988, Cell surface sialylation and tumor metastasis, *J. Biol. Chem.* **263:**7591–7603.

Peat, N., Gendler, J. J., Lalani, E., Duhig, T., and Taylor-Papadimitriou, J., 1992, Tissue-specific expression of a human polymorphic epithelial mucin (MUC1) in transgenic mice, *Cancer Res.* **52:**1954–1960.

Pierce, M., and Arango, J., 1986, Rous sarcoma virus-transformed baby hamster kidney cells express higher levels of asparagine-linked tri- and tetra-antennary glycopeptides containing [GlcNAc-β(1,6)Man-α(1,6)Man] and poly-N-acetyllactosamine sequences than baby hamster kidney cells, *J. Biol. Chem.* **261:**10772–10777.

Pierce, M., Arango, J., Tahir, S. H., and Hindsgaul, O., 1987, Activity of UDP-GlcNAc: Alpha-mannoside beta (1,6)N-acetylglucosaminyltransferase (GnT V) in cultured cells using a synthetic trisaccharide acceptor, *Biochem. Biophys. Res. Commun.* **146:**679–684.

Pigman, W., 1977, Submaxillary and sublingual glycoprotein, in: *The Glycoconjugates* (M. I. Horowitz and W. Pigman, eds.), Academic Press, New York, Vol. 1, pp. 129–135.

Piller, F., and Cartron, J.-P., 1983, UDP-GlcNAc:Galβ1→4GlcNAcβ1→3 N-acetyl glucosaminyltransferase, *J. Biol. Chem.* **258:**12293–12299.

Piller, V., Piller, F., Klier, F. G., and Fukuda, M., 1989, O-glycosylations of leukosialin in K562 cells, *Eur. J. Biochem.* **183:**123–135.

Pisano, A., Redmond, J. W., Williams, K. L., and Gooley, A. A., 1993, Glycosylation sites identified by solid phase Edman degradation: O-linked glycosylation motifs on human glycophorin A, *Glycobiology* **3:**429–435.

Porchet, N., Nguyen, V. C., Dufosse, J., Audie, J. P., Guyonnet-Duperat, V., Gross, M. S., Denis, C., Degand, P., Bernheim, A., and Aubert, J. P., 1991, Molecular cloning and chromosomal localization of a novel human tracheobronchial mucin cDNA containing tandemly repeated sequences of 48 base pairs, *Biochem. Biophys. Res. Commun.* **175:**414–422.

Probst, J. C., Hauser, F., Joba, W., and Hoffmann, W., 1992, The polymorphic integumentary mucin B.1 from Xenopus laevis contains the short consensus repeat, *J. Biol. Chem.* **267:**6310–6316.

Rice, G. E., and Bevilacqua, M. P., 1989, An inducible endothelial cell surface glycoprotein mediates melanoma adhesion, *Science* **246:**1303–1306.

Roth, J., 1984, Cytochemical localization of terminal N-acetylgalactosamine residues in cellular compartments of intestinal goblet cells: Implications of topology of O-glycosylation, *J. Cell Biol.* **98:**399–406.

Saito, H., Nishikawa, A., Gu, J., Ihara, Y., Soejima, H., Wada, Y., Sekiya, C., Niikawa, N., and Taniguchi, N., 1994, cDNA cloning and chromosomal mapping of human N-acetylglucosaminyltransferase V, *Biochem. Biophys. Res. Commun.* **198:**318–327.

Sako, D., Chang, X.-J., Barone, K. M., Vachino, G., White, H. M., Shaw, G., Veldman, G. M., Bean, K. M., Ahren, T. J., Furle, B., Cumming, D. A., and Larsen, G. R., 1993, Expression cloning of a functional glycoprotein ligand for P-selectin, *Cell* **75:**1179–1186.

Sanford, B. H., 1967, An alteration in tumor histocompatibility induced by neuraminidase, *Transplantation* **5:**1273–1279.

Santer, U. V., Gilbert, F., and Glick, M. C., 1984, Change in glycosylation of membrane glycoproteins after transfection of NIH 3T3 with human tumor DNA, *Cancer Res.* **44:**3730–3735.

Sasaki, K., Watanabe, E., Kawashima, K., Sekine, S., Dohi, T., Oshima, M., Hanai, N., Nishi, T., and Hasegawa, M., 1993, Expression cloning of a novel Galβ(1→3/1→4)GlcNAc α2,3-sialyltransferase using lectin resistance selection, *J. Biol. Chem.* **268:**22782–22787.

Sato, T., Furukawa, K., Autero, M., Gahmberg, C. G., and Kobata, A., 1993a, Structural study of the sugar chains of human leukocyte common antigen CD45, *Biochemistry* **32:**12694–12703.

Sato, T., Furukawa, K., Greenwalt, D. E., and Kobata, A., 1993b, Most bovine milk fat globule membrane glycoproteins contain asparagine-linked sugar chains with GalNAcβ1→4GlcNAc groups, *J. Biol. Chem.* **114:**890–900.

Sawada, R., Lowe, J. B., and Fukuda, M., 1993, E-selectin-dependent adhesion efficiency of colonic carcinoma cells is increased by genetic manipulation of their cell surface lysosomal membrane glycoprotein-1 expression levels, *J. Biol. Chem.* **268:**12675–12681.

Sawada, R., Tsuboi, S., and Fukuda, M., 1994, Differential E-selectin-dependent adhesion efficiency in sublines of a human colon cancer exhibiting distinct metastatic potentials, *J. Biol. Chem.* **269:**1425–1431.

Sawada, T., Ho, J.J.L., Sagabe, T., Yoon, W.-H., Chung, Y.-S., Sowa, M., and Kim, Y. S., 1993, Biphasic effect of cell surface sialic acids on pancreatic cancer cell adhesiveness, *Biochem. Biophys. Res. Commun.* **195:**1096–1103.

Schachter, H., 1986, Biosynthetic controls that determine the branching and microheterogeneity of protein-bound oligosaccharides, *Biochem. Cell Biol.* **64:**163–181.

Schachter, H., and Brockhausen, I., 1992, The biosynthesis of serine (threonine)-N-acetylgalactosamine-linked carbohydrate moieties, in: *Glycoconjugates—Composition, Structure and Function* (H. J. Allen and E. C. Kisailus, eds.), Dekker, New York, pp. 263–404.

Schweizer, A., Clausen, H., Van Meer, G., and Hauri, H.-P., 1994, Localization of O-glycan initiation, sphingomyelin synthesis, and glucosylceramide synthesis in Vero cells with respect to the endoplasmic reticulum–Golgi intermediate compartment, *J. Biol. Chem.* **269:**4035–4041.

Segawa, K., and Yamaguchi, N., 1986, Characterization of the chimeric SV40 large T antigen which has a membrane attachment sequence of polyoma virus middle T antigen, *Virology* **155:**334–344.

Sgroi, D., Varki, A., Braesch-Andersen, S., and Stamenkovic, I., 1993, CD22, a B cell-specific immunoglobulin superfamily member, is a sialic acid-binding lectin, *J. Biol. Chem.* **268:**7011–7018.

Sheng, Z., Wu, K., Carraway, K. L., and Fregien, N., 1992, Molecular cloning of the transmembrane component of the 13762 mammary adenocarcinoma sialomucin complex, *J. Biol. Chem.* **267:**16341–16446.

Sherblom, A. P., and Carraway, K. L., 1980, Sulfate incorporation into the major sialoglycoprotein of the MAT-B1 subline of the 13762 rat ascites mammary adenocarcinoma, *Biochemistry* **19:**1213–1219.

Sherblom, A. P., Buck, R. L., and Carraway, K. L., 1980a, Purification of the major sialoglycoproteins of 13762 MAT-B1 and MAT-C1 rat ascites mammary adenocarcinoma cells by density gradient centrifugation in cesium chloride and guanidine hydrochloride, *J. Biol. Chem.* **255:** 783–790.

Sherblom, A. P., Huggins, J. W., Chesnut, R. W., Buck, R. L., Ownby, C. L., Dermer, G. B., and Carraway, K. L., 1980b, Cell surface properties of ascites sublines of the 13762 rat mammary adenocarcinoma, *Exp. Cell Res.* **126:**417–426.

Shimizu, M., and Yamauchi, K., 1982, Isolation and characterization of mucin-like glycoprotein in human milk fat globule membrane, *J. Biol. Chem.* **91:**515–524.

Shimizu, M., Tanimoto, H., Azuma, N., and Yamauchi, K., 1990, Growth inhibition of Balb/C 3T3 cells by a high-molecular-weight mucin-like glycoprotein of human milk fat globule membrane, *Biochem. Int.* **20:**147–154.

Shimizu, Y., and Shaw, S., 1993, Mucins in the mainstream, *Nature* **366:**630–631.

Shirotani, K., Taylor-Papadimitriou, J., Gendler, S. J., and Irimura, T., 1994, Transcriptional regulation of MUC1 gene in colon carcinoma cells by a soluble factor: Identification of a regulatory element, *J. Biol. Chem.* **269:**15030–15035.

Shogren, R., Gerken, T. A., and Jentoft, N., 1989, Role of glycosylation on the conformation and chain dimensions of O-linked glycoproteins: Light-scattering studies of ovine submaxillary mucin, *Biochemistry* **28:**5525–5536.

Shoreibah, M., Perng, G.-S., Adler, B., Weinstein, J., Basu, R., Cupples, R., Wen, D., Browne, J. K., Buckhaults, P., Fregien, N., and Pierce, M., 1993, Isolation, characterization, and expression of a cDNA encoding N-acetyl glucosaminyltransferase V, *J. Biol. Chem.* **268:** 15381–15385.

Smith, P. L., and Baenziger, J. U., 1992, Molecular basis of recognition by the glycoprotein hormone-specific N-acetylgalactosamine-transferase, *Proc. Natl. Acad. Sci. USA* **89:**329–333.

Spicer, A. P., Parry, G., Patton, S., and Gendler, S. J., 1991, Molecular cloning and analysis of the mouse homologue of the tumor-associated mucin, MUC1, reveals conservation of potential O-glycosylation sites, transmembrane, and cytoplasmic domains and a loss of minisatellite-like polymorphism, *J. Biol. Chem.* **266:**15099–15109.

Spielman, J., Hull, S. R., Sheng, Z., Kanterman, R., Bright, A., and Carraway, K. L., 1988, Biosynthesis of a tumor cell surface sialomucin, *J. Biol. Chem.* **263:**9621–9629.

Springer, G. F., 1984, T and Tn, general carcinoma autoantigens, *Science* **224:**1198–1206.

Spooncer, E., Fukuda, M., Klock, J. C., Oates, J. E., and Dell, A., 1984, Isolation and characterization of polyfucosylated lactosaminoglycan from human granulocytes, *J. Biol. Chem.* **259:**4792–4801.

Stanbridge, E. J., Der, C. J., Doerson, C., Nishimi, R. Y., Wilkinson, J. E., Peehl, D. M., and Weissman, B. E., 1982, Human cell hybrids: Analysis of transformation and tumorigenicity, *Science* **215:**252–259.

Steck, P. A., and Nicolson, G. L., 1983, Cell surface glycoproteins of 13762Nf mammary adenocarcinoma clones of differing metastatic potentials, *Exp. Cell Res.* **147:**255–267.

Strous, G. J., and Dekker, J., 1992, Mucin-type glycoproteins, *Crit. Rev. Biochem. Mol. Biol.* **27:**57–92.

Strous, G.J.A., 1979, Initial glycosylation of proteins with acetylgalactosaminyl serine linkages, *Proc. Natl. Acad. Sci. USA.***76:**2694–2698.

Swallow, D. M., Griffiths, B., Bramwell, M., Wiseman, G., and Burchell, J., 1986, Detection of the urinary 'pum' polymorphism by the tumor-binding monoclonal antibodies Ca1, Ca2, Ca3, HMFG1, and HMFG2, *Dis. Markers* **4:**247–254.

Swallow, D. M., Gendler, S., Griffiths, B., Kearney, A., Povey, S., Sheer, D., Palmer, R. W., and Taylor-Papadimitriou, J., and Bramwell, M., 1987a, The hypervariable gene locus PUM, which codes for the tumor associated epithelial mucins, is located on chromosome 1, within the region 1q21-24, *Ann. Hum. Genet.* **51:**289–294.

Swallow, D. M., Gendler, S., Griffiths, B., Corney, G., Taylor-Papadimitriou, J., and Bramwell, M. E., 1987b, The human tumor-associated epithelial mucins are coded by an expressed hypervariable gene locus PUM, *Nature* **328:**82–84.

Symond, D. L., and Vickery, A. L., 1976, Mucinous carcinoma of the colon and rectum, *Cancer* **37:**1891–1900.

Takeuchi, M., Takasaki, S., Miyazaki, H., Kato, T., Hoshi, S., Kochibe, N., and Kobata, A., 1988, Comparative study of the asparagine-linked sugar chains of human erythropoietins purified from urine and the culture medium of recombinant Chinese hamster ovary cells, *J. Biol. Chem.* **263:**3657–3663.

Tao, T.-W., and Johnson, L. K., 1982, Altered adhesiveness of tumor cell surface variants with reduced metastasizing capacity. Reduced adhesiveness to vascular wall components in culture, *Int. J. Cancer* **30:**763–766.

Thomas, M. L., 1989, The leukocyte common antigen family, *Annu. Rev. Immunol.* **7:**339–369.

Timpte, C. S., Eckhardt, A. E., Abernethy, J. L., and Hill, R. L., 1988, Porcine submaxillary gland apomucin contains tandemly repeated, identical sequences of 81 residues, *J. Biol. Chem.* **263:**1081–1088.

Tomita, M., and Marchesi, V. T., 1975, Amino-acid sequence and oligosaccharide attachment sites of human erythrocyte glycophorin, *Proc. Natl. Acad. Sci. USA* **72:**2964–2968.

Tomiya, N., Awaya, J., Kurono, M., Hanzawa, H., Shimada, I., Arata, Y., Yoshida, T., and Takahashi, N., 1993, Structural elucidation of a variety of GalNAc-containing N-linked oligosaccharides from human urinary kallidinogenase, *J. Biol. Chem.* **268:**113–126.

Toribara, N. W., Gum, J. R., Culhane, P. J., Lagace, R. E., Hicks, J. W., Petersen, G. M., and Kim, Y. S., 1991, MUC-2 human small intestinal mucin gene structure, *J. Clin. Invest.* **88:**1005–1013.

Toribara, N. W., Roberton, A. M., Ho, S. B., Kuo, W.-L., Gum, E., Hicks, J. W., Gum, J. R., Jr., Byrd, J. C., Siddiki, B., and Kim, Y. S., 1993, Human gastric mucin, *J. Biol. Chem.* **268:**5879–5885.

Tsuda, T., Gallup, M., Jany, B., Gum, J., Kim, Y. S., and Basbaum, C., 1993, Characterization of a rat airway cDNA encoding a mucin-like protein, *Biochem. Biophys. Res. Commun.* **195:**363–373.

van den Eijnden, D. H., Evans, N. A., Codington, J. F., Reinhold, V., Silber, C., and Jeanloz, R. W., 1979, Chemical structure of epiglycanin, the major glycoprotein of the TA3-Ha ascites cell, *J. Biol. Chem.* **254:**12153–12159.

van den Eijnden, D. H., Koenderman, A.H.L., and Schiphorst, W.E.C.M., 1988, Biosynthesis of blood group i-active polylactosaminoglycans, *J. Biol. Chem.* **263:**12461–12471.

van de Wiel-van Kemenade, E., Ligtenberg, M.J.L., de Boer, A. J., Buijs, F., Vos, H. L., Melief, C.J.M., Hilkens, J., and Figdor, C. G., 1993, Episialin (MUC1) inhibits cytotoxic lymphocyte–target cell interaction, *J. Immunol.* **151:**767–776.

Van Nieuw Amerongen, A., Oderkerk, C. H., Roukema, P. A., Wolf, J. H., Lisman, J.J.W., and Vliegnthart, J.F.G., 1987, Primary structure of O- and N-glycosidic carbohydrate chains derived from murine submandibular mucin. *Carbohydr. Res.* **164:**43–51.

Varki, A., 1993, Biological roles of oligosaccharides: All of the theories are correct, *Glycobiology* **3:**97–130.

Verma, M., and Davidson, E. A., 1993, Molecular cloning sequencing of a canine tracheobronchial mucin cDNA containing a cysteine-rich domain, *Proc. Natl. Acad. Sci. USA* **90:**7144–7148.

Wang, Y., Agarwal, N., Eckhardt, A. E., Stevens, R. D., and Hill, R. L., 1993, The acceptor substrate specificity of porcine submaxillary UDP-GalNAc:polypeptide N-acetylgalactosaminyltransferase is dependent on the amino acid sequences adjacent to serine and threonine residues, *J. Biol. Chem.* **268:**22979–22983.

Watanabe, K., Hakomori, S., Childs, R. A., and Feizi, T., 1979, Characterization of a blood group I-active ganglioside, *J. Biol. Chem.* **254:**3221–3228.

Watkins, S. C., Slayter, H. S., and Codington, J. F., 1991, Intracellular pathway of a mucin-type membrane glycoprotein in mouse mammary tumor cells, *Carbohydr. Res.* **213:**185–200.

Weinstein, J., de Souza-e-Silva, U., and Paulson, J. C., 1982, Sialylation of glycoprotein oligosaccharides N-linked to asparagine, *J. Biol. Chem.* **257:**13845–13853.

Weinstein, J., Lee, E. U., McEntee, K., Lai, P.-H., and Paulson, J. C., 1987, Primary structure of β-galactoside 2,6-sialyltransferase, *J. Biol. Chem.* **262:**17735–17743.

Wen, D. X., Livingston, B. D., Medzihradszky, K. F., Kelm, S., Burlingame, A. L., and Paulson, J. C., 1992, Primary structure of Galβ1,2(4)GlcNAc α2,3-sialyltransferase determined by mass spectrometry sequence analysis and molecular cloning, *J. Biol. Chem.* **267:**21011–21019.

Wieruszeski, J. M., Michalski, J. C., Montreuil, J., Strecker, G., Peter-Katalinic, J., Egge, H., Van Halbeek, H., Mutsaers, J.H.G.M., and Vliegenthart, J.F.G., 1987, Structure of the monosialyl oligosaccharides derived from salivary gland mucin glycoproteins of the Chinese switlet, *J. Biol. Chem.* **262:**6650–6658.

Williams, C. J., Wreschner, D. H., Tanaka, A., Tsarfaty, I., Keydar, I., and Dion, A. S., 1990, Multiple protein forms of the human breast tumor-associated epithelial membrane antigen (EMA) are generated by alternative splicing and induced by hormonal stimulation, *Biochem. Biophys. Res. Commun.* **170:**1331–1338.

Wilson, I.B.H., Gravel, Y., and Heijne, G. V., 1991, Amino acid distributions around O-linked glycosylation sites, *Biochemistry* **275:**529–534.

Woodward, H. D., Ringler, N. J., Selvakumar, R., Simet, I. M., Bhavanandan, V. P., and Davidson, E. A., 1987, Deglycosylation studies on tracheal mucin glycoproteins, *Biochemistry* **26:**5315–5322.

Xu, G., Huan, L., Khatri, J. A., Wang, D., Bennick, A., Fahim, R.E.F., Forstner, G. G., and Forstner, J., 1992, cDNA for the carboxyl-terminal region of a rat intestinal mucin-like peptide, *J. Biol. Chem.* **267:**5401–5407.

Yagel, S., Feinmesser, R., Waghorne, C., Lala, P. K., Breitman, M. L., and Dennis, J. W., 1989, Evidence that GlcNAcβ1→6 branched Asn-linked oligosaccharides on metastatic tumor cells facilitate invasion of basement membranes, *Int. J. Cancer* **44:**685–690.

Yamamoto, F., Clausen, H., White, T., Marken, J., and Hakomori, S., 1990, Molecular genetic basis of the histo-blood group ABO system, *Nature* **345:**229–233.

Yamamura, K., Takasaki, S., Ichihashi, M., Mishima, Y., and Kobata, A., 1991, Increase of sialylated tetraantennary sugar chains in parallel to the higher lung-colonizing abilities of mouse melanoma clones, *J. Invest. Dermatol.* **97:**735:741.

Yamashita, K., Hitoi, A., Taniguchi, N., Yokosawa, N., Tsukada, Y., and Kobata, A., 1983, Comparative study of the sugar chains of δ-glutamyl transpeptidases purified from rat liver and rat AH-66 hepatoma cells, *Cancer Res.* **43:**5059–5063.

Yamashita, K., Ohkura, T., Tachibana, Y., Takasaki, S., and Kobata, A., 1984, Comparative study of the oligosaccharides released from baby hamster kidney cells and their polyoma transformant by hydrazinolysis, *J. Biol. Chem.* **259:**10834–10840.

Yamashita, K., Tachibana, Y., Ohkura, T., and Kobata, A., 1985, Enzymatic basis for the structural changes of asparagine-linked sugar chains of membrane glycoproteins of baby hamster kidney cells induced by polyoma transformation, *J. Biol. Chem.* **260:**3963–3969.

Yazawa, S., Abbas, S. A., Madiyalakan, R., Barlow, J. J., and Matta, K. L., 1986, N-acetylglucosaminyltransferases related to the synthesis of mucin-type glycoproteins in human ovarian tissue, *Carbohydr. Res.* **149:**241–242.

Yogeeswaran, G., and Tao, T., 1980, Cell surface sialic acid expression of lectin-resistant variant clones of B16 melanoma with altered metastasizing potential, *Biochem. Biophys. Res. Commun.* **95:**1452–1460.

Yonezawa, S., Byrd, J. C., Dahiya, R., Ho, J.J.L., Gum, J. R., Griffiths, B., Swallow, D. M., and Kim, Y. S., 1991, Differential mucin gene expression in human pancreatic and colon cancer cells, *Biochem. J.* **276:**599–605.

Yousefi, S., Higgins, E., Daoling, Z., Pollex-Kruger, A., Hindsgaul, O., and Dennis, J. W., 1991, Increased UDP-GlcNAc:Galβ1→3GalNAc-R(GlcNAc to GalNAc) β-1,6-N-acetylglucosaminyltransferase activity in metastatic murine tumor cell lines, *J. Biol. Chem.* **266:**1772–1782.

Zaretsky, J. Z., Weiss, M., Tsarfaty, I., Hareuveni, M., Wreschner, D. H., and Keydar, I., 1990, Expression of genes coding for Ps2, C-Erbb2, estrogen receptor and the H23 breast tumor-associated antigen. A comparative analysis in breast cancer, *FEBS Lett.* **265:**46–50.

Zotter, S., Hageman, P. C., Lossnitzer, A., Mooi, W. J., and Hilgers, J., 1988, Tissue and tumor distribution of human polymorphic epithelial mucin, *Cancer Rev.* **11–12:**55–101.

Chapter 6

Cellular Biology of Gangliosides

Yoshitaka Nagai and Masao Iwamori

1. INTRODUCTION

Sialic acid-containing glycosphingolipids, or gangliosides, are hybrid molecules composed of a hydrophilic sialyl oligosaccharide and a hydrophobic ceramide moiety, the latter consisting of a sphingoid base linked to a fatty acid. Though molecular species of the fatty acids and sphingosine in the ceramide component of gangliosides are heterogeneous, more significant heterogeneity is observed in the oligosaccharide moiety, which is primarily oriented exofacially in the cell surface, positionally advantaged to interact with neighboring cells and extracellular materials. In the animal kingdom, the lowest order in which gangliosides as well as sialic acids generally are detected is the Deuterostomia, in particular those more developed phylogenetically than the Echinodermata, but not those of the Prostomia (Hoshi and Nagai, 1975; Sugita *et al.*, 1982; Wiegandt, 1985; Dennis *et al.*, 1985), although recent subtle analysis has revealed that sialic acids and their polymeric form occur also in the insect *Drosophila melanogaster* (Roth *et al.*, 1992). Thus, gangliosides seem to be concerned with evolutionarily acquired functions pertinent to higher animals. In vertebrates, they are ubiquitous constituents of all tissues and cells, and their oligosaccharide structures undergo alteration during cellular development, differentiation, ontogenesis, and aging,

---

Yoshitaka Nagai Mitsubishi Kasei Institute of Life Sciences, Machida City, Tokyo, and the Glycobiology Research Group, Frontier Research Program, The Institute of Physical and Chemical Research (Riken), Wako City, Saitama, Japan. **Masao Iwamori** Department of Biochemistry, Faculty of Medicine, The University of Tokyo, Tokyo, Japan.

Biology of the Sialic Acids, edited by Abraham Rosenberg. Plenum Press, New York, 1995.

in keeping with their functional importance (Hakomori, 1990). In fact, by application of the monoclonal antibody technique, several gangliosides were shown to be the tumor-associated antigens, which are utilized as the target molecules for immunotherapy, as well as for the diagnosis of several tumors (Saleh *et al.*, 1992). Also, from severe mental and physical symptoms caused by ganglioside-degrading enzyme defects in patients with GM2 and GM1 gangliosidosis, gangliosides are recognized as being essential in homeostasis of biological functions, while the abnormal morphology of neuronal cells, probably caused by accumulation of gangliosides which was observed in patients with GM1 gangliosidosis, triggered active research on the bioactivity of exogenous gangliosides (Purpura and Baker, 1977) in the induction and promotion of neurite outgrowth in neuronal cells *in vivo* and *in vitro*. In addition, gangliosides serve as the characteristic receptors for a variety of bacterial toxins (Van Heyningen, 1974; Takamizawa *et al.*, 1986), viruses (Markwell *et al.*, 1981), and adhesive proteins (Polley *et al.*, 1991). However, since the same sialo-oligosaccharide structures serving as receptors for viruses and adhesive proteins may possibly be found in glycoproteins and since there are several experimental difficulties inherent in their highly amphiphilic nature, a real function for gangliosides as receptors in signal transduction is not clearly understood at present. A role for gangliosides in the fibronectin–integrin receptor has been elucidated as acting in support of the function of the $\alpha 5\beta 1$ integrin receptor (Zheng *et al.*, 1993). Sialic acid-derived properties are applicable to gangliosides, that is, the properties of a donor group of negative charge in the cell, interaction and trapping of divalent cations, and masking of galactose thereby preventing interaction with galactose-binding receptors (Morell *et al.*, 1971). The study of gangliosides, in particular their cell biological function, in comparison with those of proteins and nucleic acids, is at an early stage and has become a most fascinating field in biochemistry and molecular and cellular biology.

2. STRUCTURE AND METABOLISM OF GANGLIOSIDES

Ganglioside biosynthesis takes place primarily in the endoplasmic reticulum and the Golgi apparatus, and begins with the condensation of serine and palmitoyl CoA to produce 3-dehydrosphinganine followed by rapid reduction of the carbonyl at position 3 to a hydroxyl, yielding sphinganine (Braun *et al.*, 1970; Merrill *et al.*, 1988). After the *N*-acylation of sphinganine, a 4, 5-*trans* double bond is introduced into the alkyl chain of the base to yield the abundant ceramide, *N*-acyl 4-sphingenine (D-*erythro*-1,3-dihydroxy-2-amino-*trans*-4-octadecene] (Merrill *et al.*, 1986). The conclusive evidence for direct introduction of the double bond at the level of dihydroceramide (*N*-acyl sphinganine), but not at the level of the free sphingoid base, was later provided by the application to cultured cells of fumonisin B1 as an inhibitor of *N*-acylation of sphingoid bases (Roth *et al.*, 1992). The molecular species of the fatty acid component in ceramide are heterogeneous, being composed of saturated, unsaturated, or 2-hydroxylated fatty acids with chain lengths of 14 to 26 carbon atoms (Iwamori and Nagai, 1979), and the sphingoid base also is heterogeneous, with chain lengths of 14 to 20 carbon atoms (Karlsson, 1970). Stepwise transfer of carbohydrate to the ceramides from the nucleotide derivatives of carbohydrates is mediated by glucosyltransferase in the endoplasmic reticulum (Lipsky and Pagano, 1985), galactosyltransferase I and sialyltransferase I in the *cis*-Golgi cisternae (Trinchera and Ghidoni, 1989; Yamato and Yoshida, 1982), *N*-acetylglucosaminyltransferase and sialyltransferase II in the medial Golgi (Pohlentz *et al.*, 1988), galactosyltransferase II and *N*-acetylgalactosaminyltransferase as well as sialyltransferases IV and V in the *trans*-Golgi (Chege and Pfeffer, 1990; Van Echten *et al.*, 1990; Young *et al.*, 1990). In accordance with the topographical localization of glycosyltransferases in the Golgi network, monensin, which interferes with vesicular transport within the Golgi, and brefeldin A, which disrupts vesicular docking with the Golgi, have been shown to reduce the synthesis of gangliosides with longer-chain carbohydrates and to increase GlcCer, LacCer, GM3, and GD3 concomitantly in neural cells, indicating that the glycosyltransferases responsible for the synthesis of gangliosides whose carbohydrate chains extend beyond GM3 and GD3, are located in the distal compartment of the Golgi apparatus (Saito *et al.*, 1984; Van Echten and Sandhoff, 1989; Van Echten *et al.*, 1990; Young *et al.*, 1990). The structure of ceramides is thought to be important for the reaction specificity of glycosyltransferases from the observations that the fatty acyl component of the ceramides of gangliosides in the central nervous system are preferentially comprised of stearic acid, and that the ratio of C_{20}-sphingenine to C_{18}-sphingenine of gangliosides in the central nervous tissues increases depending on the increase of sialic acids (Yohe *et al.*, 1976; Iwamori and Nagai, 1979). Meanwhile the major fatty acids of the ceramide moiety of IV^3NeuAc-nLc_4Cer and IV^6NeuAc-nLc_4Cer in human erythrocytes are lignoceric and palmitic acids, respectively (Kannagi *et al.*, 1982). The above structural characteristics in the hydrophobic ceramides of gangliosides are related to the stability of the plasma membrane and the reactivity of the oligosaccharide moieties of gangliosides.

The oligosaccharide structure synthesized by specific glycosyltransferases defines the composition, sequence, linkage position, and anomeric configuration of the component carbohydrates in gangliosides (see Chapter 3). Among 100 carbohydrates in nature, only 6, that is, glucose (Glc), galactose (Gal), *N*-acetylglucosamine (GlcNAc), *N*-acetylgalactosamine (GalNAc), fucose (Fuc), and sialic acid (NeuAc or NeuGc), contribute to construction of gangliosides. Depending on the substrate specificities of glycosyltransferases, six basic types of carbohydrate chain are made: gala-, lacto-, neolacto-, globo-, isoglobo-, and ganglio-series oligosaccharides have been characterized to date in gangliosides from the cells and tissues of vertebrates (Figure 1). As shown in Figure 1,

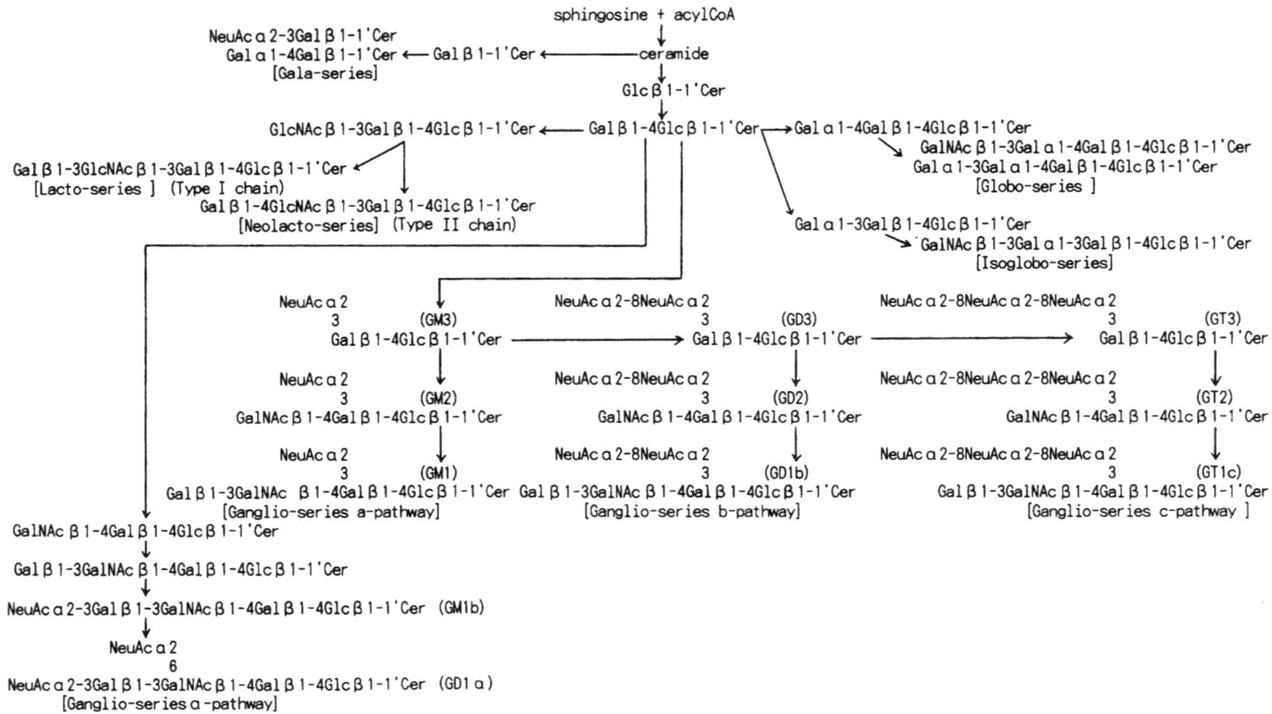

Figure 1. Major metabolic pathways of gangliosides. Although gangliosides with NeuAcα2-6GalNAc are tentatively classified into the ganglio-series α-pathway, because of coexistence of GM1b and GD1α (Iwamori *et al.*, 1988), sialytion to the 6-position of GAlNAc residue in GM1b is not as yet proven.

Table I
Gangliosides in Vertebrates

a. Gangliosides with short carbohydrate chains[a]	
1. NeuAcα2-3Galβ1-4Glcβ1-1′Cer	GM3
2. NeuGcα2-3Galβ1-4Glcβ1-1′Cer	
3. Ac-O-4NeuGcα 2-3Galβ1-4Glcβ1-1′Cer	
4. $NeuNH_2$α2-3Galβ1-4Glcβ1-1′Cer[b]	
5. KDNα (2-keto-3-deoxy-D-glycero-D-galactononoic acid)2-3Galβ1-4Glcβ1-1′Cer	
b. Gala-series gangliosides	
1. NeuAcα2-3Galβ1-1′Cer	GM4
2. NeuGcα2-3Galβ1-1′Cer	
c. Ganglio-series gangliosides	
a-pathway	
1. NeuAcα 2 3 GalNAcβ1-4Galβ1-4Glcβ1-1′Cer	GM2
2. NeuGcα 2 3 GalNacβ1-4Galβ1-4Glcβ1-1′Cer	
3. NeuAcα 2 3 Galβ1-3GalNAcβ1-4Galβ1-4Glcβ1-1′Cer	GM1 (GM1a)
4. NeuGcα 2 3 Galβ1-3GalNAcβ1-4Galβ 1-4Glcβ1-1′Cer	
5. NeuAcα 2 3 Fucα1-2Galβ1-3GalNAcβ1-4Galβ1-4Glcβ1-1′Cer	FucosylGM1

(*continuea*)

Table I (*Continued*)

6.	NeuGcα 2 3 Fucα1-2Galβ1-3GalNAcβ1-4Galβ1-4Glcβ1-1′Cer	
7.	NeuAcα 2 3 Fucα1-3Gal1-3GalNac1-4Gal1-4Glc1-1′Cer	
8.	NeuAcα 2 3 Galα1-3Galβ1-3GalNacβ1-4Galβ1-4Glcβ1-1′Cer	Gal-GM1
9.	NeuAcα 2 3 GalNacβ1-4Galβ1-3GalNAcβ1-4Galβ1-4Glcβ1-1′Cer	GalNAc-GM1
10.	NeuAcα 2 3 Fucα1-3GalNAcβ1-3Galβ1-3GalNAcβ1-4Galβ 1-4Glcβ1-1′Cer	
11.	NeuAcα 2 3 GalNAcα 1-3GalNAcβ1-3Galβ 1-3GalNAcβ1-4Galβ1-4Glcβ1-1′Cer	
12.	Fucα 1 NeuAcα 2 2 3	Galα1-3Galβ 1-3GalNAcβ 1-4Galβ1-4Glcβ 1-1′Cer
13.	NeuAcα 2 3 Galβ1-3Galα1-3Galβ1-3GalNAcβ1-4Galβ1-4Glcβ1-1′Cer	
14.	NeuAcα 2 3 Galα1-3Galβ1-3Galα1-3Galβ1-3GalNAcβ1-4Galβ1-4Glcβ 1-1′Cer	
15.	NeuGcα 2 3 Galβ1-4GlcNAcβ1-3Galβ1-3GalNAcβ1-4Galβ1-4Glcβ 1-1′Cer	

16.	Fucα 1 NeuAcα 2 2 3 Galβ1-4GlcNAcβ1-3Galβ1-3GalNAcβ1-4Galβ1-4Glcβ1-1′Cer	
17.	NeuGcα 2 3 Galα1-3Galβ1-4GlcNAcβ1-3Galβ1-3GalNAcβ1-4Galβ1-4Glcβ1-1′Cer	
18.	NeuGcα 2 3 Galα1-3 $(Gal\beta 1\text{-}4GlcNAc\beta)_2$1-3Galβ1-3GalNAcβ1-4Galβ1-4Glcβ1-1′Cer	
19.	NeuAcα 2 3 NeuAcα2-3Galβ 1-4GlcNAcβ1-3Galβ1-3GalNAcβ1-4Galβ1-4Glcβ1-1′Cer	
20.	NeuAcα 2 NeuAcα 2 3 3 Galβ1-3GalNAcβ1-4Galβ1-4Glcβ1-1′Cer	GD1a
21.	NeuAcα 2 NeuGcα 2 3 3 Galβ1-3GalNAcβ1-4Galβ1-4Glcβ1-1′Cer	
22.	NeuGcα 2 NeuAcα 2 3 3 Galβ1-3GalNAcβ1-4Galβ1-4Glcβ1-1′Cer	
23.	NeuGcα 2 NeuGcα 2 3 3 Galβ1-3GalNAcβ1-4Galβ 1-4Glcβ1-1′Cer	
24.	Ac-O-9NeuAcα 2 NeuAcα 2 3 3 Galβ1-3GalNAcβ 1-4Galβ1-4Glcβ1-1′Cer	
25.	NeuAcα 2 NeuAcα 2 3 3 GalNAcβ1-4Galβ1-3GalNAcβ1-4 Galβ 1-4Glcβ1-1′Cer	GalNAc-GD1a
26.	NeuAcα 2-8NeuAcα 2 NeuAcα 2 3 3 Galβ1-3GalNAcβ1-4Galβ1-4Glcβ1-1′Cer	GT1a

(continued)

Table I *(Continued)*

b-pathway		
27.	NeuAcα2-8NeuAcα2-3Galβ1-4Glcβ1-1′Cer	GD3
28.	NeuGcα2-8NeuAcα2-3Galβ1-4Glcβ1-1′Cer	
29.	NeuAcα2-8NeuGcα2-3Galβ1-4Glcβ1-1′Cer	
30.	NeuGcα2-8NeuGcα2-3Galβ1-4Glcβ1-1′Cer	
31.	Ac-O-9NeuAcα2-8NeuAcα2-3Galβ1-4Glcβ1-1′Cer	
32.	HSO_3-O-8NeuAcα2-8NeuGcα2-3Galβ1-4Glcβ1-1′Cer	
33.	HSO_3-O-8NeuGcα2-8NeuGcα2-3Galβ1-4Glcβ1-1′Cer	
34.	NeuAcα2-8NeuAcα 2	
	3	
	GalNAcβ1-4Galβ1-4Glcβ1-1′Cer	GD2
35.	Ac-O-9NeuAcα2-8NeuAcα 2	
	3	
	GalNAcβ1-4Galβ1-4Glcβ1-1′Cer	
36.	NeuAcα 2-8NeuAcα 2	
	3	
	Galβ1-3GalNAcβ1-4Galβ1-4Glcβ1-1′Cer	GD1b
37.	NeuAcα2-8NeuAcα 2	
	3	
	Fucα1-2Galβ1-3GalNAcβ1-4 Galβ1-4Glcβ1-1′Cer	FucosylGD1b
38.	NeuAcα2-8NeuAcα 2	
	3	
	Galα1-3Galβ1-3GalNAcβ1-4Galβ1-4Glcβ1-1′Cer	
39.	NeuAcα2-8NeuAcα 2	
	3	
	Galα1-3Galα1-3Galβ1-3GalNAcβ1-4Galβ1-4Glcβ1-1′Cer	
40.	Fucα 1 NeuAcα2-8NeuAcα 2	
	2 3	
	Galα1-3Galβ 1-3GalNAcβ1-4Galβ1-4Glcβ1-1′Cer	
41.	NeuAcα2-8NeuAcα 2	
	3	
	NeuAcα2-3Galβ1-3GalNAcβ1-4Galβ1-4Glcβ 1-1′Cer	GT1b

42.	Ac-9-O-NeuAcα2-8NeuAcα 2 3 NeuAcα2-3Galβ1-3GalNAcβ1-4Galβ1-4Glcβ1-1′Cer	
43.	NeuAcα2-8NeuAcα2-8NeuAcα2-8NeuAcα 2 3 3 Galβ1-3GalNAcβ1-4Galβ1-4Glcβ1-1′Cer	GQ1b
c-pathway		
44.	NeuAcα2-8NeuAcα2-8NeuAcα2-3Galβ1-4Glcβ1-1′Cer	GT3
45.	Ac-9-O-NeuAcα2-8NeuAcα2-8NeuAcα2-3Galβ1-4Glcβ1-1′Cer	
46.	NeuAcα2-8NeuAcα2-8NeuAcα 2 3 GalNAcβ1-4Galβ1-4Glcβ1-1′Cer	GT2
47.	NeuAcα2-8NeuAcα2-8NeuAcα 2 3 Galβ1-3GalNAcβ1-4Galβ1-4Glcβ1-1′Cer	GT1c
48.	NeuAcα2-8NeuAcα2-8NeuAcα 2 3 NeuAcα2-3Galβ1-3GalNAcβ1-4Galβ1-4Glcβ1-1′Cer	GQ1c
49.	NeuAcα2-8NeuAcα2-8NeuAcα 2 3 NeuAcα2-8NeuAcα2-3Galβ1-3GalNAcβ1-4Galβ1-4Glcβ1-1′Cer	GP1c
α-pathway		
50.	NeuAcα 2 6 Galβ1-3GalNACβ1-4Galβ1-4Glcβ1-1′Cer	
51.	NeuAcα 2 6 NeuAcα 2-3Galβ1-3GalNAcβ1-4Galβ1-4Glcβ1-1′Cer	GD1α (GD1e)
52.	NeuAcα 2-8NeuAcα 2 6 NeuAcα2-3Galβ1-3GalNAcβ1-4Galβ1-4Glcβ1-1′Cer	
53.	NeuAcα 2 6 NeuAcα2-8NeuAcα2-3Galβ1-3GalNAcβ1-4Galβ1-4Glcβ1-1′Cer	

(continued)

Table I *(Continued)*

54.	NeuAcα 2-8NeuAcα 2	
	6	
	NeuAcα 2-8NeuAcα2-3Galβ1-3GalNAcβ1-4Galβ1-4Glcβ 1-1′Cer	
55.	NeuAcα 2 NeuAcα 2	
	6 3	
	NeuAcα2-3Galβ1-3 GalNAcβ1-4Galβ1-4Glcβ1-1′Cer	Chol-1α-a (GT1aα)
56.	NeuAcα 2-8NeuAcα 2	
	NeuAcα2-$_6$ 3	
	NeuAcα2-3Galβ1-3 GalNACβ1-4Galβ1-4Glcβ1-1′Cer	Chol-1α-b (GQ1bα)
The other ganglio-series gangliosides		
57.	NeuAcα2-3Galβ1-3GalNAcβ1-4Galβ1-4Glcβ1-1′Cer	GM1b
58.	NeuGcα2-3Galβ1-3GalNAcβ1-4Galβ1-4Glcβ 1-1′Cer	
59.	NeuAcα 2-8NeuAcα2-3Galβ1-3GalNAcβ1-4Galβ1-4Glcβ 1-1′Cer	
60.	NeuGcα 2-8NeuAcα2-3Galβ1-3GalNAcβ1-4Glcβ1-1′Cer	
61.	NeuAcα 2-8NeuGcα2-3Galβ1-3GalNAcβ1-4Galβ1-4Glcβ 1-1′Cer	
62.	NeuGcα 2-8NeuGcα2-3Galβ1-3GalNAcβ1-4Galβ1-4Glcβ 1-1′Cer	
63.	NeuAcα 2	
	3	
	GalNAcβ1-4Galβ1-3GalNAcβ1-4Galβ1-4Glcβ1-1′Cer	
64.	NeuGcα 2	
	3	
	GalNAcβ1-4Galβ1-3GalNAcβ1-4Galβ1-4Glcβ1-1′Cer	
65.	NeuGcα 2	
	3	
	Galβ1-3GalNAcβ1-4Galβ1-3GalNAcβ1-4Galβ1-4Glcβ1-1′Cer	
d. Globo-series gangliosides		
1.	NeuAcα2-3GalNAcβ1-3Galα1-4Galβ1-4Glcβ1-1′Cer	
2.	NeuAcα2-3Galβ1-3GalNAcβ1-3Galα1-4Galβ1-4Glcβ1-1′Cer	SSEA-4
3.	NeuGcα2-3Galβ1-3GalNAcβ1-3Galα1-4Galβ1-4Glcβ1-1′Cer	
4.	NeuAcα2-8NeuAcα2-3Galβ1-3GalNAcβ1-3Galα1-4Galβ1-4Glcβ1-1′Cer	

5. NeuGcα2-8NeuGcα2-3Galβ1-3GalNAcβ1-3Galα1-4Galβ1-4Glcβ1-1′Cer
6. NeuAcα 2
6
NeuAcα2-3Galβ1-3GalNAcβ1-3Galα1-4Galβ1-4Glcβ1-1′Cer

e. Isoglobo-series gangliosides
1. NeuACα2-3Galβ1-3GalNAcβ1-3Galα1-3Galβ1-4Glcβ1-1′Cer

f. Lacto-series gangliosides
1. NeuAcα 2-3Galβ1-3GlcNAcβ1-3Galβ1-4Glcβ1-1′Cer
2. Fucα 1
4
NeuAcα2-3Galβ1-3GlcNAcβ1-3Galβ1-4Glcβ1-1′Cer Sialyl Le^a
3. NeuAcα 2
3
GalNAcβ1-4Galβ1-3GlcNAcβ1-3Galβ1-4Glcβ 1-1′Cer
4. NeuAcα 2-3Galβ1-3Galβ1-3GlcNAcβ1-3Galβ1-4Glcβ1-1′Cer
5. NeuAcα 2
6
Galβ1-3GlcNAcβ1-3Galβ1-4Glcβ1-1′Cer
6. NeuACα 2
6
NeuAcα 2-3Galβ1-3GlcNAcβ1-3Galβ1-4Glcβ1-1′Cer
7. NeuAcα 2
6
Fucα 1-2Galβ1-3GlcNAcβ1-3Galβ1-4Glcβ1-1′Cer
8. NeuAcα 2
Fucα 1-_4 6
NeuAca 2-3Galβ1-3GlcNAcβ1-3Galβ1-4Glcβ1-1′Cer

(*continued*)

Table I (*Continued*)

g. Neolactoganglio-series gangliosides

1. GalNAcβ 1
   4
   NeuAcα2-3Galβ1-4GlcNAcβ1-3Galβ1-4Glcβ1-1′Cer
2. NeuAcα 2 GalNAcβ 1
   3 4
   GalNAcβ1-4Ga lβ1-4GlcNAcβ1-3Galβ1-4Glcβ1-1′Cer

h. Gangliosides with the other carbohydrate backbones

1. NeuAcα2-3Galβ1-3GalNAcβ1-4Glcβ 1-1′Cer
2. Fucα 1 GalNAcβ 1
   3 4
   NeuAcα2-3Galβ1-4Gl cNAcβ1-3Galα1-3Galβ1-4Glcβ1-1′Cer
3. Fucα 1
   3
   NeuAcα2-3Galβ1-4GlcNAcβ1-3Galα1-3Galβ1-4Glcβ1-1′Cer
4. NeuAcα 2
   6
   NeuAcα2-3Galβ1-3GalNAcβ1-3Galα1-4Galα1-4Glcβ1-1′Cer
5. NeuAcα2-?GlcNAcβ1-3Galβ1-4Glcβ1-1′Cer

i. Neolacto-series gangliosides

1. NeuAcα2-3Galβ1-4GlcNAcβ1-3Galβ1-4Glcβ1-1′Cer — Sialylparagloboside
2. NeuGcα2-3Galβ1-4GlcNAcβ1-3Galβ1-4Glcβ1-1′Cer
3. NeuAcα2-6Galβ1-4GlcNAcβ1-3Galβ1-4Glcβ1-1′Cer — CD76
4. NeuAcα2-8NeuAcα2-3Galβ1-4GlcNAcβ1-3Galβ1-4Glcβ1-1′Cer

5. NeuGcα2-8NeuGcα2-3Galβ1-4GlcNAcβ1-3Galβ1-4Glcβ1-1′Cer
6. NeuAcα2-8NeuGcα2-3Galβ1-4GlcNAcβ1-3Galβ1-4Glcβ1-1′Cer
7. NeuAcα2-8NeuAcα2-8NeuAcα2-3Galβ1-4GlcNAcβ1-3Galβ1-4Glcβ1-1′Cer
8. NeuAcα2-3GalNAcβ1-3Galβ1-4GlcNAcβ1-3Galβ1-4Glcβ1-1′Cer
9. NeuAcα 2
3
GalNAcβ1-4Galβ1-4GlcNAcβ1-3Galβ1-4Glcβ1-1′Cer
10. Fucα 1
3
NeuAcα2-3Galβ1-4GlcNAcβ1-3Galβ1-4Glcβ1-1′Cer Sialyl Le[x]
11. Galβ1-4GlcNAcβ 1
6
NeuAcα2-6Galβ1-4GlcNAcβ1-3Galβ1-4Glcβ1-1′Cer
12. NeuAcα2-3Galβ1-4GlcNAcβ1-3Galβ1-4GlcNAcβ1-3Galβ1-4Glcβ1-1′Cer
13. NeuGcα2-3Galβ1-4GlcNAcβ1-3Galβ1-4GlcNAcβ1-3Galβ1-4Glcβ1-1′Cer
14. NeuAcα2-6Galβ1-4GlcNAcβ1-3Galβ1-4GlcNAcβ1-3Galβ1-4Glcβ1-1′Cer
15. NeuAcα2-8NeuAcα2-3Galβ 1-4GlcNAcβ 1-3Galβ 1-4GlcNAcβ 1-3Galβ 1-4Glcβ 1-1′Cer
16. NeuAcα2-3Galβ1-4GlcNAcβ1-(3Galβ1-4GlcNAcβ1)$_2$-3Galβ1-4Glcβ1-1′Cer

[a]References for structures other than those listed by Wiegandt (1985) and Stults *et al.* (1989) are as follows: a-5 (Song *et al.*, 1991), c-12 (Bouchours *et al.*, 1987), c-15,17 (Nohara *et al.*, 1990), c-19 (Nohara *et al.*, 1992), c-35 (Sjoberg *et al.*, 1992), c-55 (Ando *et al.*, 1992), c-56 (Hirabayashi *et al.*, 1992), f-3 (Fredman *et al.*, 1989), f-7 (Prieto and Smith, 1985), i-5,6 (Furukawa *et al.*, 1988), i-8 (Thorn *et al.*, 1992), i-24 (Taki *et al.*, 1992a), i-25 (Taki *et al.*, 1992b), and i-30,31 (Nudelman *et al.*, 1987).
[b]$NeuNH_2$: N-deacetylated sialic acid.

transfer of either GlcNac, Gal, GalNAc, or NeuAc to LacCer is a key step for extension of the oligosaccharide chains giving rise to the different series of gangliosides. Depending on the metabolic pathway, the ganglio-series of gangliosides are subdivided into four different subseries, the a-, b-, c-, and α-series. As shown in Table I, the number of gangliosides so far determined in vertebrates is 126, that is, 65 for the ganglio, 32 for the neolacto, 8 for the lacto, 6 for the globo, 2 for the gala, 1 for the isoglobo, 2 for the neolactoganglio, and 10 for the others. Gangliosides with sialic acid analogues, other than *N*-glycolyl, 4-*O*-acetyl-*N*-glycolyl, and 9-*O*-acetyl-*N*-acetylneuraminic acids, and with 2-keto-3-deoxy-D-*glycero-galacto*-nononic acid [deaminoneuraminic acid (KDN), Kitajima *et al.*, 1986; Song *et al.*, 1991] in Table I, are also presumably present in vertebrates. A ganglioside containing de-*N*-acetylated neuraminic acid, de-*N*-acetylated GM1, exists in bovine brain, which can bind to the cholera toxin B-subunit (Hidari *et al.*, 1993).

Quite different structures of gangliosides from those in the vertebrate are found in the invertebrates. Although the sialic acid residue of gangliosides in vertebrates is located at the terminal of the core oligosaccharide chain as the monosialosyl or polysialosyl group, gangliosides with sialic acid substituted with galactose or glucose at the C-4 or C-8 hydroxyl group are contained in the tissues of Echinodermata (Sugita, 1979; Kochetkov and Smirnova, 1983). External sialic acid of the disialosyl group is linked to the *N*-glycoloyl group of an internal sialic acid (structure 14 in Table II) (Smirnova *et al.*, 1987) in GD3-like gangliosides from this species.

3. PHYSICOCHEMICAL PROPERTIES OF GANGLIOSIDES

In general, the polarity of glycolipids increases according to the number of carbohydrates in the molecules and is reflected in their mobility on thin-layer chromatography (TLC). However, in the case of gangliosides, their mobilities on TLC does not always correspond to the number of carbohydrates. For example, gangliosides with a disialyl residue at the internal galactose of the gangliotetraose backbone migrate on TLC with a neutral irrigating solvent system to a lower position than those with a monosialyl residue at the internal galactose, even though they both have the same carbohydrate composition. As shown in Figure 2, GD1b (lane 20) and GT1b (lane 22) migrate to positions lower than those of GD1a (lane 16) and GT1a (lane 21), respectively. In addition, the negative charge of *N*-glycolylneuraminic acid-containing gangliosides seems to be stronger than that of *N*-acetylneuraminic acid-containing gangliosides, since the former developed with an ammonium hydroxide-containing solvent system are always lower than the latter, locating observed between lanes 5 and 6 for sialyl neolactotetraosyl ceramide, lanes 6 and 9 for sialyl neolactohexaosyl ceramides,

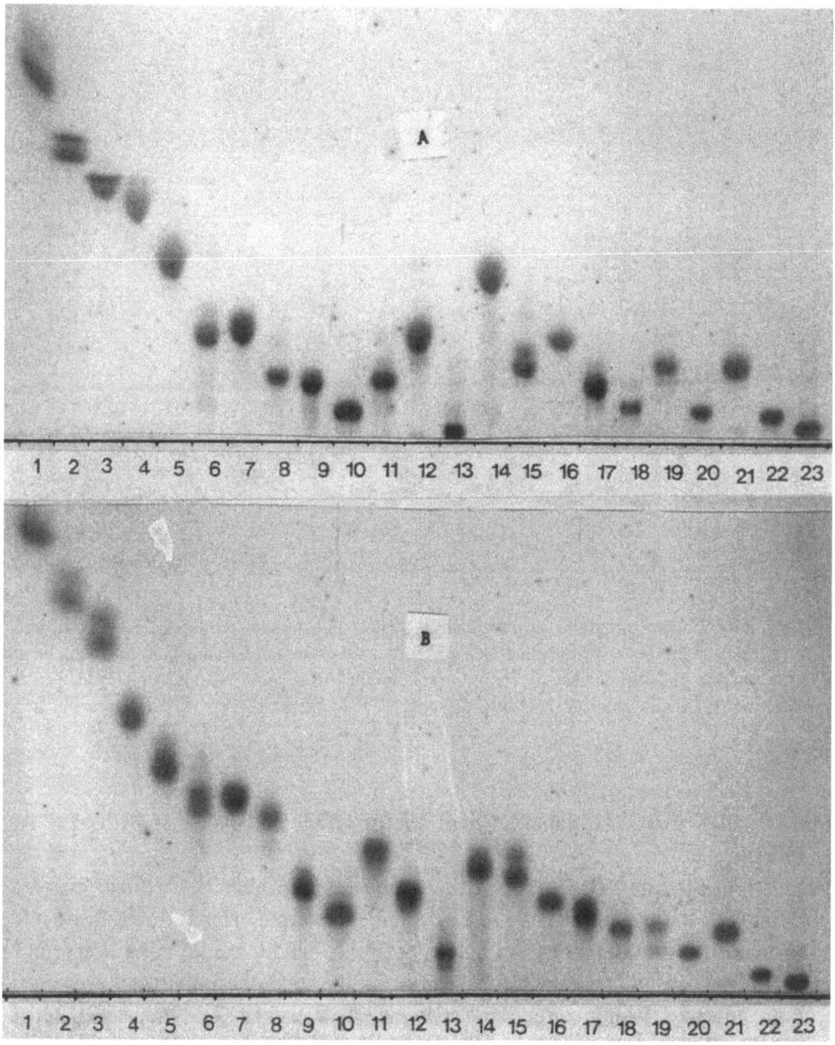

FIGURE 2. TLC chromatograms of several gangliosides. 1, GM4; 2, GM3; 3, GM3(NeuGc); 4, GM2; 5, IV[3]NeuAc-nLc$_4$Cer; 6, IV[3]NeuGc-nLc$_4$Cer; 7, GM1; 8, GM1(NeuGc); 9, VI[3]NeuAc-nLc$_6$Cer; 10, VI[3]NeuGc-nLc$_6$Cer; 11, GalNAc-GM1; 12, Fuc-GM1; 13, VIII[3]NeuAc-nLc$_8$Cer; 14, GD3; 15, GD3(NeuGc,NeuGc); 16, GDla; 17, GDla (NeuAc,NeuGc); 18, GDla(NeuGc,NeuGc); 19, GD2; 20, GD1b; 21, GT1a; 22, GT1b; 23, GQ1b. Gangliosides shown without parentheses contain *N*-acetylneuraminic acid. Solvent systems are (A) chloroform–methanol–ammonia–water (60:40:3:6, by vol.) and (B) chloroform–methanol–0.5% $CaCl_2$ (55:45:10, by vol.).

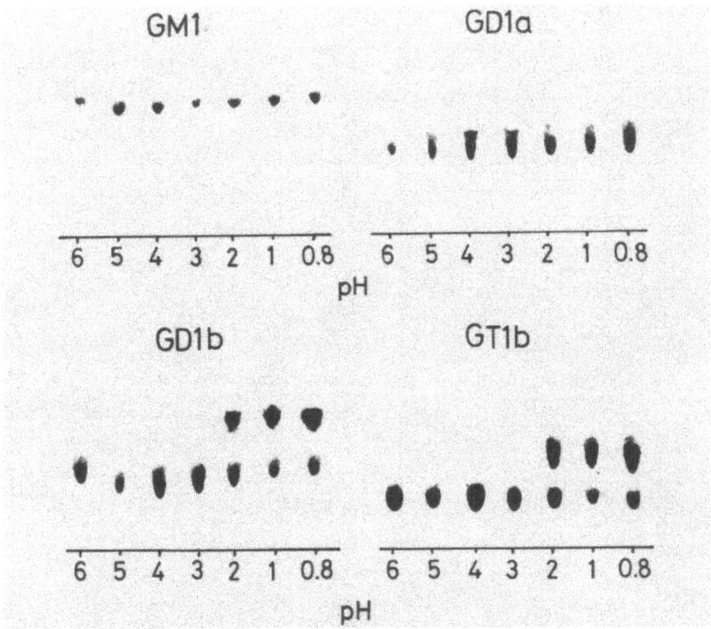

FIGURE 3. TLC chromatograms of gangliosides treated with buffers. GM1, GD1a, GD1b, and GT1b (1 μmol) in 1 ml of 0.1 M glycine-HCl and citrate buffer with different pHs were combined at 25°C for 4 hr and dialyzed against distilled water overnight. After lyophilization, gangliosides were examined on TLC with chloroform–methanol–0.5% $CaCl_2$ (55:45:10, by vol.) (Iwamori *et al.*, 1978).

lanes 7 and 8 for GM1, lanes 14 and 15 for GD3, and lanes 16, 17, and 18 for GD1a.

Gangliosides in aqueous dispersion show unique physicochemical properties related to the presence of sialic acid, that is, internal ester formation, micellar formation, and interaction with divalent ions. Under acidic condition, gangliosides, particularly those with disialyl residues, form the inner ester (lactone) structure, thereby losing or reducing their negative charge. The inner ester linkage is readily formed by exposing gangliosides in aqueous solution to pH lower than 2 (Figure 2), and is cleaved at pH 9, restoring the negative charge. The inner ester linkage is proven to be between the carboxyl group of the external sialic acid residue and the C-9-hydroxyl group of the internal sialic acid residue (Acquotti *et al.*, 1987; Levery *et al.*, 1990). Although the inner ester structure is formed between the carboxyl group of the internal sialic acid and C-2-hydroxyl group of the galactose under treatment with glacial acetic acid or trichloroacetic acid–phosphotungstic acid (Gross *et al.*, 1980; Ando *et al.*, 1989), the formation and breakdown of internal ester between the internal and the external sialic acids of the disialyl group in nature may function as a regulator of negative charge in microenvironments in the cell surface (Riboni *et al.*, 1986), as well as in the presentation of a new antigen. Monoclonal antibodies that specifically recognize ganglioside lactones were established (Bouchon *et al.*, 1992; Kawashima *et al.*, 1994). They reacted with freshly prepared gangliosides from brain (Bouchon *et al.*, 1992), and with human melanoma tissue cells but not with isolated cells from the tissue (Kawashima *et al.*, 1994), strongly suggesting natural occurrence of the epitopes.

Gangliosides in aqueous dispersion form a micellar structure, whose particle mass was shown to be 340 kDa for GM1, 300 kDa for GD1a, and 250 kDa for GT (Yohe *et al.*, 1976) by ultracentrifugation, and 532 kDa for GM1 and 417 kDa for GD1a by laser-light scattering (Corti *et al.*, 1980), respectively. The number of monomers in micelles of GM1, GD1a, and GT was shown to be 225, 165, and 120, respectively, by ultracentrifugation. Though the predicted number of GT molecules in the micelle is less than that by actual measurement, increasing hydrophobicity in the ceramide related to the increased content of C_{20}-sphingenine compensates for the hydrophilic properties of the additional sialyl groups, stabilizing the micelle (Yohe *et al.*, 1976). Thus, an increase in C_{20}-sphingenine also corresponds to an increase in sialic acid content of gangliosides in brain. The critical micelle concentration (cmc) of gangliosides is also different depending on the number of sialic acid residues in the molecules, being reported to be in the range of $2–5 \times 10^{-8}$ M for monosialyl-, 10^{-6} M for disialyl-, and 10^{-5} M for trisialylgangliosides, respectively (Ulrich-Bott and Wiegandt, 1984; Formisano *et al.*, 1979; Mraz *et al.*, 1980). Gangliosides in micelles spontaneously transfer to liposomes, in which gangliosides incorporated are localized only in the outer leaflet of the phospholipid bilayer, mimicking the situation found in plasma membrane (Felgner *et al.*, 1981; Masserini and Freire, 1987). The cmc of gangliosides affects the binding of gangliosides to the membrane. Below the cmc, the binding process proceeds rapidly and the gangliosides are stably and irreversibly associated with the membrane, while above the cmc, binding is slow and gangliosides on the membrane are releasable by washing and exchangeable with newly added gangliosides (Toffano *et al.*, 1980; Kanda *et al.*, 1982). Possible involvements of membrane protein and energy-dependent rearrangement are also considered to be factors in the binding process of gangliosides to cells (Callies *et al.*, 1977). Free sialic acid binds calcium ion (Czarniecki and Thornton, 1977; Jacques *et al.*, 1980), and interaction of calcium ions with gangliosides is characteristic. The binding of calcium and magnesium ions with gangliosides takes place in a competitive manner, and exchange of magnesium ions on gangliosides is observed by addition of calcium ions, indicating the high affinity of calcium ion for gangliosides (Behr and Lehn, 1973; Hayashi and Katagiri, 1974). ^{13}C-NMR of GM1 by titration with paramagnetic europium showed that the carboxyl and glyceryl side chain of sialic acid, *N*-acetylgalactosamine, and terminal galactose are intimately involved in the cation binding (Sillerud *et al.*, 1978). Thus, gangliosides, because of sialic acid, possess several unique physicochemical properties which expectedly are involved in critical biological functions.

4. SPECIES, TISSUE, AND CELLULAR LOCALIZATION OF GANGLIOSIDES

The structural, developmental, and species- and gamete-specific characterization of gangliosides from the sea urchin and starfish have been performed (Table II) (Isono and Nagai, 1966; Hoshi and Nagai, 1970, 1975; Ohsawa and Nagai, 1975; Sugita, 1979; Kochetkov and Smirnova, 1983; Smirnova *et al.*, 1987), and the general absence of gangliosides in insects and Mollusca (most highly developed Protostomia) was reported (Sugita *et al.*, 1982; Dennis *et al.*, 1985). Later, however, occurrence in an insect (*D. melanogaster*) of bound sialic acid was reported, though its origin remains unclear (Roth *et al.*, 1992). Phylogenetic study of the metabolism and distribution of gangliosides in animals other than the vertebrates is insufficient to elucidate evolutionary relationships in ganglioside composition.

4.1. Gangliosides in Nervous Tissues

Comparative study of brain gangliosides in several species of vertebrates has been performed extensively and has shown that ganglio-series gangliosides are the predominant components in mammals, birds, fishes, and frogs, but the metabolic pathways differ among these species. Gangliosides belonging to the a- and b-pathways are the major components in brains of mammals and birds, but are minor in fish, in which those of the c-subseries are predominant (Ando *et al.*, 1978; Hilbig and Rahmann, 1980; Seybold and Rahmann, 1985), while those of the α-subseries are the major species in frog brain (Ohashi, 1981). However, the expression of gangliosides derived via different metabolic pathways is not restricted among species. The c- and α-subseries gangliosides are contained in chick and bovine brains, respectively (Hirabayashi *et al.*, 1988, 1990), and their expression is characteristically altered during development. By immunohistochemical staining of chick embryo with antibodies against the c-series gangliosides, positive cells are detected in the ventral commissural region of the neural tube after 2.5 days of fertilized egg incubation, extending throughout the neural tube after 3 days, and in trace amount after hatching. In addition to the neural tube-derived tissues, those derived from neural crest and acoustic ganglia express c-series gangliosides (Obata and Tanaka, 1988). In the early stage of neural cell differentiation of chick brain, GD3 and GT3 (as well as GD2) appear

Table II
Gangliosides in Invertebrates[a]

1.	NeuAcα2-6Glcβ1-1′Cer
2.	NeuGcα2-6Glcβ1-1′Cer
3.	HSO_3-O-8NeuAcα2-6Glcβ1-1′Cer
4.	HSO_3-O-8NeuGcα2-6Glcβ1-1′Cer
5.	NeuAcα2-8NeuAcα 2-6Glcβ1-1′Cer
6.	NeuAcα2-8NeuAcα 2-6Glc1-6Glc1-1′Cer
7.	NeuAcα2-6Glc1-8NeuAcα2-6Glc1-1′Cer
8.	HSO_3-NeuGcα2-6Glcβ1-8NeuGcα2-6Glcβ1-1′Cer
9.	CH_3-O-8NeuGcα2-3Galβ1-4Glcβ 1-1′Cer
10.	Aral-6/3Gal1-3Gal1-4NeuAc2-3Galβ1-4Glcβ1-1′Cer
11.	Araβ 1-6Galβ 1-4(CH_3-O-8)NeuGcα2-3Galβ1-4Glcβ1-1′Cer
12.	Araβ 1-6Galβ 1-4NeuGcα2-3Galβ1-4Glcβ1-1′Cer
13.	Galβ 1
	8
	Araβ1-6Galβ1-4NeuGcα2-3Galβ1-4Glcβ1-1′Cer
14.	CH_3-O-8NeuGcα 2-5(CH_3-O-8)NeuGcα2-3Galβ1-4Glcβ1-1′Cer
15.	Ara_f 1
	3
	Ara_f1-3Galα1-6Galβ1-4NeuAcα2-3Galβ1-4Glcβ1-1′Cer
16.	Ara_f1-3Galα1-4(CH_3-O-8)NeuAc2-3Gal1-3Gal1-4NeuAc2-3Galβ1-4Glcβ 1-1′Cer
17.	Fucα 1
	3
	NeuAcα 2-3Galβ1-4GlcNAcβ1-3Galβ1-4GlcNAcβ1-3Galβ1-4Glcβ 1-1′Cer
18.	Fucα 1
	3
	NeuAcα 2-6Galβ1-4GlcNAcβ1-3Galβ 1-4GlcNAcβ 1-3Galβ1-4Glcβ1-1′Cer

(continued)

Table II (*Continued*)

19.	Fucα 1 Fucα 1 3 3 NeuAcα2-3Galβ 1-4GlcNAcβ1-3Galβ1-4GlcNAcβ1-3Galβ1-4Glcβ1-1′Cer	
20.	Fucα 1 3 NeuAcα 2-3Galβ1-4GlcNAcβ1-3Galβ1-4GlcNAcβ1-3Galβ1-4GlcNAcβ1-3Galβ1-4Glcβ1-1′Cer	
21.	Fucα 1 3 NeuAcα 2-3Galβ1-4GlcNAcβ1-3Galβ1-4GlcNAcβ1-(3Galβ1-4GlcNAcβ 1)$_2$-3Galβ1-4Glcβ1-1′Cer	CDw65
22.	Fucα 1 2 NeuAcα 2-3Galβ1-3GalNAcα1-3Galβ1-4GlcNAcβ1-3Galβ1-4Glcβ1-1′Cer	
23.	Galβ1-4GlcNAcβ 1 6 NeuAcα2-3Galβ1-4GlcNAcβ1-3Galβ1-4GlcNAcβ1-3Galβ1-4Glcβ1-1′Cer	
24.	Galβ 1-4GlcNAcβ 1 6 NeuAcα 2-6Galβ1-4GlcNAcβ1-3Galβ1-4GlcNAcβ1-3Galβ1-4Glcβ1-1′Cer	
25.	NeuAcα 2-6Galβ1-4GlcNAcβ 1 6 NeuAcα 2-6Galβ 1-4GlcNAcβ 1-3Galβ 1-4GlcNAcβ 1-3Galβ 1-4Glcβ 1-1′Cer	
26.	Fucα1-2Galβ 1-4GlcNAcβ 1 6 NeuAcα 2-3Galβ1-4GlcNAcβ1-3Galβ1-4GlcNAcβ1-3Galβ1-4Glcβ1-1′Cer	
27.	NeuAcα2-3Galβ1-4GlcNAcβ 1 6 NeuAcα 2-3Galβ1-4GlcNAcβ1-3Galβ1-4GlcNAcβ1-3Galβ1-4Glcβ1-1′Cer	
28.	Galα 2-3Galβ1-4GlcNAcβ 1 6 NeuAcα 2-3Galβ1-4GlcNAcβ1-3Galβ1-4GlcNAcβ1-3Galβ1-4Glcβ1-1′Cer	

29. Fucα 1-$_2$
Gal NAcα1-3Galβ1-4GlcNAcβ 1
6
NeuAcα 2-3Galβ1-4GlcNAcβ1-3Galβ1-4GlcNAcβ1-3Galβ1-4Glcβ1-1′Cer

30. Fucα 1-$_2$
Galα1-3Galβ1-4GlcNAcβ 1
6
NeuAcα 2-3Galβ1-4GlcNAcβ1-3Galβ1-4GlcNAcβ1-3Galβ1-4Glcβ1-1′Cer

31. NeuAcα2-3Galβ1-4GlcNAcβ 1
6
NeuAcα2-3Galβ1-4GlcNAcβ1-3Galβ1-4GlcNAcβ1-3Galβ1-4GlcNAcβ1-3Galβ1-4Glcβ1-1′Cer

32. NeuAcα2-3Galβ1-4GlcNAcβ 1
NeuAcα2-3Galβ1-4GlcNAcβ 1-$_6$ 6 (3)
NeuAcα2-3Galβ1-4GlcNAcβ1-3Galβ1-4GlcNAcβ1-3(6)Galβ1-4GlcNAcβ1-3Galβ1-4Glcβ1-1′Cer

[a]References for structures listed by Wiegandt (1985) are as follows: 11–13, Sugita (1979); 14, Smirnova *et al.* (1987); 15, 16, Kochetkov and Smirnova (1983).

as the major components and then synthesis of the c-series gangliosides is initiated to give the maximum concentration of GQ1c at 11–14 days of incubation. With synapse formation, the c-series gangliosides are reduced in concentration and compensated for by the b-series gangliosides, GT1b and GQ1b, and the a-series ganglioside GD1a is greatly increased. Then, ganglioside components in myelin, such as GM1 and GM4, are increased in relative concentration during the period of myelin formation (Rahmann, 1985). In adult chick brain, the majority of gangliosides are comprised of the a- and b-series, and the c-series gangliosides are contained only in a trace amount. In fetal rat brain, the c-series ganglioside, GT3, also may be detected in the growth cones from the forebrain and brain stem in 9-*O*-acetylated form (Igarashi *et al.*, 1990; Hirabayashi *et al.*, 1989) and the developmental alteration in the ganglioside composition except for that of GQ1c is similar to that of chick brain. GD3 is the major component at day 14 of the prenatal period (E14; E1 being the day on which a vaginal plug has been detected), and a dramatic increase in the b-series gangliosides from E16 to E18, and a gradual increase in the a-series gangliosides from E18 to E22 compensate for a relative reduction of GD3 during the development of rat brain (Yu *et al.*, 1988). Also, in human brain, active synthesis of a-series gangliosides follows that of b-series gangliosides and reportedly is claimed to be associated with synapse formation (Svennerholm *et al.*, 1989). The programmed expression of gangliosides in the developing brain suggests that gangliosides play a role in the systematic construction of neural cells. In fact, anti-GD3 antibody can block the interaction of epithelial cells with mesenchymal cells (Sariola *et al.*, 1988). Since the concentration of GD3 in the growth cone is higher than that in the synaptosome (Igarashi *et al.*, 1990) and GD3 is expressed in proliferative and migratory regions of embryonic chick and murine brains (Blum and Barnstable, 1987; Rösner *et al.*, 1985), GD3 seems to be essential for the organizational migration of neuronal cell bodies. In addition, characteristic expression of 9-*O*-acetylated derivatives of GD3 and GT3 in fetal and adult rat brains is suggestive of their possible functional involvement in this process along with GD3. Antibody to 9-*O*-acetyl GD3 reacts with proliferating cells of the external granule layer of the cerebellum and becomes unreactive as the cells have matured and have begun to migrate (Stallcup *et al.*, 1983; Levine *et al.*, 1984; Varki *et al.*, 1991; Hirabayashi *et al.*, 1989). Likewise, the α- and c-series gangliosides, which are the major gangliosides in frog and fish brains, respectively, and which are altered in concentration and distribution during the embryonic development of mammalian brain, may play a role in the evolutionary development of brain, and still may function as recognition molecules in adult mammalian brain. A characteristic distribution of the major gangliosides, as well as of the minor gangliosides, in the different layers of brain (Kotani *et al.*, 1993; Kracun *et al.*, 1984) also indicates functional importance of the sialyloligosaccharide structure in development. The following distribution of gangliosides in rat cerebellar cortex has been observed by immunofluorescence with monoclonal antibodies to gangliosides: GM1 in myelin and in glial cells, GD1a in the molecular layer, GD1b and GQ1b in the granular layer, GD1b in the surface of the granule cell bodies, GT1b in the molecular and granular layers, and GQ1b in the cerebellar glomerulus (Kotani *et al.*, 1993).

To date, numerous monoclonal antibodies have been established against major as well as minor gangliosides (for example, Table III), and are providing a useful cell biological tool not only for analysis of the cellular distribution of gangliosides, but for identification of cell types and lineages as well. Alteration in the ganglioside composition in the brains of mice with genetic mutations indicates the characteristic localization of gangliosides. In the quaking mutant with deficient myelinogenesis, GM4 and GM1 are selectively reduced, indicating their distribution in the myelin sheath (Iwamori *et al.*, 1985). Moreover, brains of weaver (*wv/wv*) and Purkinje cell degeneration (*pcd/pcd*) mutants, which lose granule and Purkinje cells, showed reduced concentrations of GD1a and GT1a, respectively, suggesting the localization of GD1a and GT1a in the

Table III
Monoclonal Antibodies Specific for Gangliosides Established by Tai *et al.* (1988)[a]

Ganglioside[b]	Designation	Class of Ig	Reference
GM3	GMR6	IgM (κ)	Kotani *et al.* (1992)
GM2	GMB28	IgM (κ)	"
GM1	GMR16	IgM (κ)	"
GD1a	GMR17	IgM (κ)	"
GT1a	GMR11	IgM (κ)	"
GD3	GMR19	IgM (κ)	Ozawa *et al.* (1992b)
GD2	GMR7	IgM (κ)	"
GD1b	GGR12	IgG2b (κ)	"
GT1b	GMR5	IgM (κ)	"
GQ1b	GMR13	IgM (κ)	"
GQ1bα	GGR41	IgG2a (κ)	Kusunoki *et al.* (1993)
GM4	AMR10	IgM (κ)	Ozawa *et al.* (1993)
O-Ac-disialoganglioside	GMR2	IgM (κ)	Ozawa *et al.* (1992a)
GM3(NeuGc-)	GMR8	IgM (κ)	Ozawa *et al.* (1992b)
GM2(NeuGc)	GMR14	IgM (κ)	Kawashima *et al.* (1993)
GD3(NeuGc-NeuGc-)	GMR3	IgM (κ)	Ozawa *et al.* (1992b)
GM1-L[c]	AMR38	IgM (κ)	Kawashima *et al.* (1994)
GD1a-L	AMR40	IgM (κ)	"
GD3-L	AMR19	IgM (κ)	"

[a]All of the monoclonal antibodies listed were established by Dr. Tadashi Tai and his collaborators (Tokyo Metropolitan Institute of Medical Science).
[b]Sialic acid moiety of gangliosides is *N*-acetylneuraminic acid (NeuAc) unless otherwise noted.
[c]L indicates ganglioside lactone.

granule and Purkinje cells, respectively (Seyfried *et al.*, 1982). Also, a characteristic reduction of GQ1b in the brain of the F1 mouse with a mutation in the T region of chromosome 17, which dies after 10–11 days during the prenatal period as a result of the retarded development of brain, indicates that GQ1b is an essential molecule for differentiation of neural cells or construction of brain tissues (Seyfried, 1987). Likewise, a significant increase in GD3 in the brains of mutant mice with retinal dystrophy or reactive gliosis suggests a possible localization of GD3 in proliferating glial cells (Seyfried *et al.*, 1982). Also, the properties of minor gangliosides as useful cell-specific markers of cholinergic neurons have clearly been shown for the α-series gangliosides, GT1aα (c-55 in Table I) and GQ1bα (c-56 in Table I), by preparation of monoclonal antibodies directed to particular gangliosides (Ando *et al.*, 1992; Hirabayashi *et al.*, 1992; Kusunoki *et al.*, 1993). In neurons and oligodendroglia, and in different anatomical regions of human brain, comparative studies on the concentration and composition of the major gangliosides were performed (Kracun *et al.*, 1984), but the results are based on the relative compositional rather than absolute differences. The relative compositions of GM4 in neurons and oligodendroglia were reported to be 0.9 and 5.9%, respectively, and that in the myelin, which originates from the plasma membrane of oligodendroglia, was 18.4%, suggesting that GM4 is a marker molecule for oligodendroglia and myelin (Yu and Iqbal, 1979). Furthermore, data on an enrichment of a- and b-series gangliosides in the temporal and occipital regions of human brains, respectively, were also obtained from comparison of the relative composition of gangliosides (Kracun *et al.*, 1984). However, a metabolic shift of ganglioside synthesis is known to be induced by temperature changes operating at the level of enzymatic activities (Rahmann, 1985) and pH (Iber *et al.*, 1990). When the pH of medium for the primary culture of cerebellar cells was changed from 7.4 to 6.2, the ganglioside composition shifted proportionally from the a-series to the b-series and the change was reversible. This phenomenon was largely ascribed to differences in optimum pH of sialyltransferase II and *N*-acetylgalactosaminyl transferase I, both of which are key enzymes at the initial steps for the syntheses of a- and b-series gangliosides, respectively. Thus, the expression of gangliosides in brain is regulated at the enzymatic as well as the genetic level.

4.2. Gangliosides in Extraneural Tissues

In contrast to the similarity in the core sialyl oligosaccharides in neural tissues, expression of gangliosides in extraneural tissues of vertebrates occurs in species-, organ-, tissue-, and cell type-specific modes, and some ganglioside oligosaccharides are characterized as being identical with the antigens defined immunologically as cell surface markers such as CDw60, CD76 (i-3), and CDw65 (i-21) (Table II). In mice, although the interstrain variation in ganglioside compositions of brain, thymus, kidney, lung, heart, spleen, and testis is quite low, gangliosides in erythrocytes and liver are expressed in a strain-specific manner and the enzymes responsible for their syntheses are genetically regulated according to Mendelian laws. Namely, the livers of C57BL/10 (B10) and SWR are clearly differentiated in terms of their ganglioside composition, the former and the latter being the GM2(NeuGc)- and the GM1(NeuGc)- plus GD1a(NeuGc)-expressing types, respectively. Because of the dominant character of the latter, liver of F_1 mice crossed with both strains contained GM1(NeuGc) plus GD1a (NeuGc). By back crossbreeding of F_1 and C57BL/10 (B10), the two different phenotypes were produced in a ratio of 1:1, indicating a difference in a single gene on the chromosome, and the analyses of glycosyltransferases and ganglioside phenotypes in the liver of H-2 congenic mice revealed that the gene (Ggm-1) is responsible for repression of the β1-3 galactosyltransferase needed to synthesize GM1(NeuGc) and that Ggm-1 is located 1 cM from the H2-k region. However, even in the mouse which is unable to synthesize GM1(NeuGc) in liver, GM1(NeuAc) is expressed in brain independently of Ggm-1 (Sakaizumi *et al.*, 1988), indicating that the Ggm-1 gene is only activated in the liver. By the same breeding experiments, Ggm-2, Gsl-4, Gsl-5, Gsl-6, and Gsl-7 (Hashimoto *et al.*, 1983; Sekine *et al.*, 1988, 1989; Nakamura *et al.*, 1990) were characterized to be the genes responsible for determination of ganglioside phenotypes in liver or erythrocytes, but it has not been clarified whether the genes encode the glycosyltransferases themselves. An analysis of the relationship between polymorphic variation of ganglioside expression in different tissues (Iwamori *et al.*, 1985) and a complete analysis of the above genes are required to understand the regulatory mechanism for tissue- and cell-specific expression of gangliosides. Although glycosphingolipids, including gangliosides, mostly are located in the outer leaflet of the lipid bilayer of the plasma membrane as well as in Golgi membranes, they also are distributed in, or associated with, cytoseletal structures.

The first report by Sakakibara *et al.* (1981a,b), who described immunocytochemical evidence for association of galactocerebroside with a colchicine-sensitive microtubule-like cytoskeletal structure of cultured cells and in the cytoplasm of epithelial cells in several tissues, was followed by the immunohistochemical observations that GM3 is associated with vimentin intermediate filaments of human umbilical vein endothelial cells and human fibroblasts and that globoside is associated with vimentin in human and mouse fibroblasts and in MDCK and HeLa cells, with desmin in smooth muscle cells, with keratin in keratinocytes and hepatoma cells, and with glial fibrillary acidic protein (GFAP) in glial cells (Gillard *et al.*, 1991, 1992). Calmodulin was also reported to be a Ca^{2+}-dependent ganglioside-binding protein (Higashi *et al.*, 1992) and a possible mechanism for involvement of gangliosides in the modulation of a calmodulin-dependent enzyme was brought forward (Higashi and Yamagata, 1992). At present, however, the functional role of cytoskeleton-associated glycosphingolipids remains conjectural. Direct binding of ganglioside GM2 to vimentin was also demonstrated in isolated vimentin from normal and Tay-Sachs disease human fibroblasts in tissues and in cells by immunofluorescence microscopy (Kotani *et al.*, 1994), and diagnostic application of this finding was carried out in cultured amniocytes (Sakuraba *et al.*, 1993).

5. EFFECTS OF EXOGENOUS GANGLIOSIDES ON DIFFERENTIATION OF NEURONS AND LEUKOCYTES

5.1. Effects of Gangliosides on Neuritogenesis

Nerve growth factor (NGF), a neurotrophic and neuritogenic peptide, is generally required for proliferation and differentiation including neurite outgrowth of neuronal cells, while the brain gangliosides have been shown not only to enhance the effect of NGF but also to exhibit NGF-like activities themselves, as reviewed often (Ledeen, 1984; Nagai and Tsuji, 1988; Schengrund, 1989; Skaper *et al.*, 1989; Bradley, 1990; Tettamanti and Riboni, 1993). The pioneering work showing the *in vivo* and *in vitro* effects of gangliosides on neural cells appeared in 1975 and then in 1977 as follows. Intraperiotoneal administration of bovine brain gangliosides to cats with experimental pre- or post-ganglionic anastomosis greatly influenced the regeneration and reinnervation process of both cholinergic and adrenergic nerve fibers (Ceccarelli *et al.*, 1976). Second, aberrant meganeurites, probably related to the accumulated gangliosides, were observed on similar neurons in individuals with GM2 as well as GM1 gangliosidosis (Purpura and Suzuki, 1976; Purpura and Baker, 1977). Enhancement of neuromuscular junctions between chick pectoral muscular cells and spinal cord-derived neuronal cells was observed by measurement of electrophysiological parameters after addition of gangliosides (Obata *et al.*, 1977). In the following decade, numerous reports on the *in vitro* effects of exogenous gangliosides on cultured neuroblastoma and pheochromocytoma cells, and on primary cultured neuronal cells appeared, and have been reviewed (Schengrund, 1989; Skaper *et al.*, 1989; Bradley, 1990). The relevant experiments were based on determination of the number and length of neurites, neuronal survival time, cell number, and enzymatic activities such as ornithine decarboxylase. However, since the neurite outgrowth of neuroblastoma cells is known to be induced or promoted also by addition of dimethylsulfoxide, hexamethylene bisacetamide, or cAMP-producing agents (Kimhi *et al.*, 1976), or even by depletion of serum, the effect of gangliosides still remains to be clarified relative to the mechanism of signal transduction. In fact, a broad specificity and the necessity for a relatively large amount of activating gangliosides and sialyl compounds as inducers of neuritogenesis were reported in murine neuroblastoma Neuro2a cells, which are insensitive to NGF. Although GM1 and mixed gangliosides have been reported to promote neurite outgrowth in Neuro2a (Roisen *et al.*, 1981), the c-series gangliosides, but not the a- and b-series gangliosides, were effective in sprouting neurites 3 hr after addition of gangliosides. Also, GD1b increased the length of neurites, while GD1a and GT1b stimulated both the length and number of neurites in Neuro2a (Matta *et al.*, 1986). Gangliosides GM1, GD1a, GT1a, and GQ1b, and several chemically synthetic sialyl compounds, such as α- and β-sialyl cholesterol, α- and β-sialyl alkyl glycerol ethers, and α-sialyl ceramide (but not β-sialyl ceramide), induced a long single neurite in Neuro2a, and the morphology was clearly different from that induced by 8-bromo cAMP, which gave a multipolar morphology (Tsuji *et al*,. 1988).

In contrast to the broad specificity and also the necessity of a micromolar level of sialyl compounds as the neuritogenic agent for Neuro2a cells, human neuroblastoma cell lines, GOTO and NB-1, strictly required GQ1b to sprout neurites and furthermore, the required amounts were at a level of a few nanomolar (Tsuji *et al.*, 1983). When GQ1b was added to a serum-free culture of those neuroblastoma cells, the number and length of neurites as well as the number of cells surviving in culture were increased. The optimum concentration of GQ1b was 5 ng/ml and the other gangliosides with different sialyl oligosaccharide moieties were completely inactive. Intact GQ1b was necessary for expression of the activity because neither the oligosaccharide nor the ceramide alone promoted neurite outgrowth. Also, since the neuritogenic activity of GQ1b was abolished in a dose-dependent manner by the oligosaccharide of GQ1b, a receptor molecule specific for GQ1b was presumed to be present on the cell surface to mediate the signal (Nakajima *et al.*, 1986; Nagai and Tsuji, 1988). One candidate reactive with GQ1b in the surface of GOTO cells is ecto-protein kinase(s), which is stimulated specifically by GQ1b at 5–10 ng/ml as optimum concentration (Tsuji *et al.*, 1988; Nagai and Tsuji, 1994). Evidence for direct coupling of GQ1b-dependent neuritogenesis with the GQ1b-specific ecto-protein kinase was provided by the use of a cell-impermeable ecto-protein kinase inhibitor, K-252b, which showed clearly the suppression of both neuritogenic and ecto-protein kinase activities in the cultured cells (Tsuji *et al.*, 1992; Nagai and Tsuji, 1994). The K-252b inhibitor also inhibited functional synapse formation between cultured neurons of rat cerebral cortex (Muramoto *et al.*, 1988). Importance of GQ1b in neural differentiation is also supported by the observation that T-locus mutant mice with defective neural differentiation do not contain GQ1b in the embryonic brain (Seyfried, 1987). However, primary cultured cells and cell lines other than GOTO and NB-1, in which neurite outgrowth is promoted in response to GQ1b, have not been reported as yet. The rat pheochromocytoma cell line PC12, which is sensitive to NGF, did not respond to exogenous gangliosides (Seifert, 1981), but the effect of NGF on the morphology, ornithine decarboxylase and tyrosine hydroxylase phosphorylation was enhanced in the presence of gangliosides, presumably by interaction of gangliosides with NGF, as shown in the process of nerve sprouting of dorsal root ganglia by inhibition with anti-GM1 antibody (Schwartz and Spirman, 1982; Ferrari *et al.*, 1983). NGF-promoted neuritogenesis in PC12 cells was also inhibited by K-252b, suggesting a possible coupling with ecto-protein phosphorylation (Nagashima *et al.*, 1991). Concerning the difference in the sensitivity of several cell lines to exogenous gangliosides, a possible relationship with the endogenous ganglioside composition in the target cells was suggested. The sensitive neuronal cell lines, B-103 and B-104, have a rather simple composition relative to the insensitive cell line, PC12 (Seifert, 1981), and the GQ1b-sensitive cell line, GOTO, does not contain endogenous GQ1b (Yamamoto *et al.*, 1990), suggesting the possibility that gangliosides that are not contained or are contained in a relatively low concentration in the target cells may be capable of inducing neuronal differentiation including neuritogenesis. However, contrary to this expectation, among the cell lines SH-SY5Y, Neuro2a, B-104, B-103, and NIA-103, the cells enriched with a relatively higher concentration of ganglio-series gangliosides, in particular, with GM1 at a concentration greater than 40 pmol/mg protein, neurite outgrowth was promoted in response to exogenous gangliosides, and the degree of neuritogenesis was higher in proportion to increase in the concentration of endogenous GM1.

Very recently cDNA encoding GD3 synthase (CMP-NeuAc: GM3 α2,8-sialyltransferase) was cloned (Haraguchi *et al.*, 1994; Nara *et al.*, 1994; Sasaki *et al.*, 1994). By transfection of this cDNA into Neuro 2a cells, the cells that expressed GD3 were cloned. The cells (N2a-GD3) expressed not only GD3 but also GQ1b, concomitantly decreased cell proliferation, and then spontaneously began to sprout neurites (Kojima *et al.*, 1994). Moreover, acetylcholine esterase as a marker of cholinergic neurons was induced. The transfection with the vector alone did not at all affect cell behavior or ganglioside composition of the parent nontransfected cells, which contained solely a- but not b-series gangliosides. The results definitely indicate the physiological significance of GQ1b as a causative factor for neuritogeneses as well as cholinergic differentiation. They also indicate that the mechanism of neurite sprouting in this system may be overlapped en route with that of exogenous GM1.

There are several working models to account for the action of exogenous gangliosides on neural cells (Skaper *et al.*, 1989), but the mechanism underlying the neurotrophic and neuritogenic activity of gangliosides remains to be fully explored. Differentiation of Neuro2a induced with monoclonal anti-GM3 antibody is associated with an elevation in adenylate cyclase, possibly by reducing Gi-protein inhibition (Chatterjee *et al.*, 1992). In PC12 cells, exogenous GM1 triggered Ca^{2+} entry through the dihydropyridine-sensitive Ca^{2+} channel (Hilbush and Levine, 1992), or by cell adhesion molecule-induced Ca^{2+} influx (Doherty *et al.*, 1992), and prevented the inhibition by K-252a, a cell membrane-permeable kinase inhibitor, of NGF-induced neurite regeneration and c-fos induction (Ferrari *et al.*, 1992). An involvement of microtubules in the distribution of exogenous gangliosides in the perikaryal and neurital surface of Neuro2a was demonstrated by electron microscopy with colcemid, taxol, and cytochalasin D (Fentie and Roisen, 1993). Thus, mechanisms important for the neurotrophic and neuritogenic activities of exogenous gangliosides appear to involve the cytoskeletal structure and membrane proteins such as GQ1b-specific ecto-protein kinase, ion channels, and signal-transducing systems including adenylate cyclase. The action of synthetic sialyl cholesterol (3-*O*-α- or β-NeuAc cholesterol) which is also capable of exhibiting a similar neuritogenic effect on Neuro 2a cells seems entirely different. The compounds were rapidly transported into cell nuclei without degradation and enhanced transcriptional activity of the nuclei distinctly (Yamashita *et al.*, 1991), suggesting the possibility that the compound in the nucleus plays a key role in neuritogenesis without using several of the more common second messengers.

On the basis of the *in vitro* effects of exogenous gangliosides, *in vivo* administration of gangliosides, mainly GM1, have been attempted in order to accelerate neural regeneration and repair in neurons injured mechanically and chemically in experimental animals and in human neurological diseases (Ledeen, 1984; Mahadik and Karpiak, 1988; Bradley, 1990). Exciting positive reports showing the peripheral nerve regeneration and the functional recovery of neurons of the central nervous tissue in experimental animals have appeared during the past 10 years, and restoration of muscle activity to a significant extent in trauma patients with spinal cord injury has been observed after intraveneous injection of GM1 (Geisler *et al.*, 1991). However, double-blind controlled studies of several human chronic neuromuscular diseases, such as amyotrophic lateral sclerosis, hereditary motor sensory neuropathy, spinocerebellar degenerations, and idiopathic polyneuropathy, in terms of the effect of exogenous gangliosides, have failed to confirm the results, giving a statistical significance at the 5% level (Bradley, 1990). Therefore, careful examination is required for application of gangliosides and the evaluation of their putative effects. Since the first experimental demonstration of autoimmune neurologic disorder (ganglioside syndrome) caused by the immunization of rabbits with brain gangliosides (Nagai *et al.*, 1976, 1978, 1980a,b), antiganglioside antibodies and the immunogenic property of ganglioside sialyl oligosaccharide moieties are known to be associated with several human neurological diseases, in particular, peripheral neuropathies (Chiba *et al.*, 1992; Yoshino *et al.*, 1992; Stevens *et al.*, 1993). An epitope identical with the GM1 structure was found in a lipopolysaccharide of *Campylobacter jejuni*, which elicits Guillain–Barré syndrome (Yuki *et al.*, 1992, 1993).

5.2. Effects of Gangliosides on Leukocyte Differentiation

Pluripotent hematopoietic stem cells have the potential to differentiate ultimately into T and B lymphocytes, granulocytes, monocytes, erythrocytes, or megakaryocytes through several stages during which their glycosphingolipid compositions are characteristically altered depending not only on differentiation stages but also on the directions of differentiation (Kiguchi *et al.*, 1990). The human myelogenous leukemia cell line, HL-60, is trapped at a stage in which the cells are bipotent: the cells are induced to differentiate into granulocytes in the presence of retinoic acid (Breitman *et al.*, 1980) or dimethylsulfoxide (Collins *et al.*, 1978), and into monocyte-macrophages in the presence of phorbol esters (Rovera *et al.*, 1979) or 1α,25-dihydroxy vitamin D_3 (Bar-Shavit *et al.*, 1983). As HL-60 cells, of which major glycolipids are LacCer, GM3, and IV^3NeuAc-nLc_4Cer, differentiated into monocytic cells, their content of GM3 increased remarkably (Nojiri *et al.*, 1984, 1986; Momoi and Yokota, 1983). Meanwhile, granulocytic HL-60 cells synthesize primarily neolacto-series gangliosides (Nojiri *et al.*, 1985). Changes in glycolipid composition as a result of differentiation are more significant than simply the display of markers of monocytic and granulocytic cells, because the addition of either GM3 or a neolacto-series ganglioside, IV^3NeuAc-nLc_4Cer, to the serum-free culture media of HL-60 cells can by itself induce differentiation of HL-60 cells into monocytic or granulocytic cells, respectively (Saito *et al.*, 1985; Nojiri *et al.*, 1986, 1988; Nakamura *et al.*, 1989). Growth of HL-60 was inhibited by GM3(NeuAc) in a dose-dependent manner and the inhibition was almost complete at a concentration of 50 μM. However, other gangliosides, such as GM1, GD1a, GD1b, GT1b, GM2, GM3(NeuGc), and a mixture of brain gangliosides, at the same concentrations, were much less active or showed no effect at all. Chemically synthesized GM3(αNeuAc) was much more potent in growth inhibition than the synthetic β-anomer, GM3(βNeuAc). Concurrently, striking changes in morphology, including a decrease in the nuclear–cytoplasmic ratio, a paler cytoplasm, lobulation of nuclei, and vacuolated cytoplasm with ruffled surface membranes, were observed, and monocytic differentiation by exogenous GM3(NeuAc) was demonstrated by increases in the number of α-naphthylbutyrate esterase-positive cells and in capacity for phagocytosis, as well as by the expression of cell surface antigens specific for mature monocytes and macrophages (Nojiri *et al.*, 1986). The bioactivity of exogenous GM3 as an inducer of differentiation was also demonstrated in the monocytic differentiation of U-937 cells (Nojiri *et al.*, 1986) and the megakaryocytic differentiation of K-562 cells (Nakamura *et al.*, 1991). Exogenous addition of IV^3NeuAc-nLc_4Cer to the medium of HL-60 cells triggered granulocytic differentiation (Nojiri *et al.*, 1988), and the activity was demonstrated even in retinoic acid-resistant HL-60 cells (Kitagawa *et al.*, 1989), suggesting that the influence of the exogenous IV^3NeuAc-nLc_4Cer is downstream from the retinoic acid effect. However, structural specificity of the sialyl compounds on granulocytic differentiation of HL-60 cells was rather broad, in comparison to GM3(NeuAc) on monocytic differentiation, because several synthetic sialyl compounds, such as sialyl cholesterol and sialyl alkyl glycerol ether, showed the same bioactivity (Saito *et al.*, 1990).

In the direction of differentiation of human myelogenous leukemic cells, HL-60, into either monocytic or granulocytic lineages, two glycosyltransferases might play an important role. HL-60 cells induced to differentiate with phorbol esters increased sialyltransferase I activity, yielding higher levels of GM3 in the monocytic lineage (Momoi *et al.*, 1986; Xia *et al.*, 1989), while retinoic acid increased UDP-GlcNAc:lactosylceramide β1-3-*N*-acetylglucosaminyltransferase activity, triggering the biosynthesis of IV^3NeuAc-nLc_4Cer in the granulocytic lineage (Nakamura *et al.*, 1992). Since exogenous GM3 was shown to be incorporated into the lysosomal compartment and catabolized rapidly (Nakamura *et al.*, 1989), its metabolites might function as an inducer of differentiation. In fact, a transitory but substantial increase in the concentration of ceramide by the activation of a neutral sphingomyelinase was observed in differentiation induced by 1α,25-dihydroxy vitamin D3 (Okazaki *et al.*, 1990). Exogenous addition of ceramide or increase of cellular ceramide by treatment with bacterial sphingomyelinase enhanced the differentiating ability of HL-60 cells, but the ceramide effect was not related to the inhibition of protein kinase C by the ceramide metabolite, sphinganine. Additionally, sphinganine, a known inhibitor of protein kinase C, was shown to prevent phorbol ester-induced differentiation of HL-60 cells (Merrill *et al.*, 1986), but had no effect on the induction with GM3 (Stevens *et al.*, 1989), indicating that exogenous GM3 is involved farther upstream in signal transduction than protein kinase C. The mechanism is as yet unidentified. However, since exogenous GM3 was effective in terms of growth inhibition and differentiation induction of fresh leukemia cells, it might be applicable to therapy of patients with leukemia.

6. GANGLIOSIDES AS RECEPTORS FOR BACTERIA, VIRUSES, AND BACTERIAL TOXINS

On the basis of observations that certain bacteria and viruses characteristically bind glycosphingolipids including gangliosides by a solid-phase overlay procedure, glycosphingolipids on the target cells are thought to be involved in the initial step of bacterial and viral infections (Karlsson, 1989), and lectins specific for the carbohydrate moiety on bacterial and viral surfaces are implicated in the binding with glycosphingolipids. Gangliosides and sialyl oligosaccharides in Table IV have been characterized to date to be the receptors for several bacteria, viruses, and bacterial toxins, and their structural specificity is quite

Table IV
Receptor Function of Gangliosides and Sialooligosaccharides for Bacteria, Viruses, and Bacterial Toxins

Receptors	Bacteria, viruses, and toxins	References
Bacteria		
GM3(NeuAcα2-3Gal)	*E. coli*, S-adhesin	Korhonen *et al.* (1984)
GM3, sulfatide	*H. pylori*	Saito *et al.* (1991)
GM3(NeuGc)	*E. coli*, K99-adhesin	Kyogashima *et al.* (1989)
NeuAcα2-3Galβ1-4 GlcNAc	*N. gonorrhoeae*	Stromberg *et al.* (1988)
GM1	*A. naeslundii*	Stromberg and Karlsson (1990)
Bacterial toxins		
GM2	*C. perfringens* δ	Jolivet-Reynaud and Alouf (1983)
GM1	*V. cholerae*	Cuatrecasas (1973)
	E. coli heat labile	Holmgren (1973)
GD1b, GT1b	*C. tetani*	Van Heyningen (1974)
GT1b, GD1b, GD1a	*C. botulinum* B,C,F	Ochanda *et al.* (1986)
GT1b	*V. parahaemolyticus* heat stable	Takeda *et al.* (1976)
GQ1b, GT1b, GD1a	*C. botulinum* A,E	Takamizawa *et al.* (1986)
II^3NeuAc-nLc_4Cer	*Staphylococcus* α	Kato *et al.* (1976)
Viruses		
NeuAcα2-3Gal		
NeuAcα2-6Gal	Influenza A	Suzuki *et al.* (1986)
	Influenza B	Suzuki *et al.* (1987)
Ac-*O*-9NeuAcα2-3Gal	Influenza C	Rogers *et al.* (1986)
NeuAcα2-8NeuAc	Sendai (HVJ)	Markwell *et al.* (1986)
NeuAcα2-3Gal	Newcastle (NDV)	Suzuki *et al.* (1985)
NeuAcα2-3Galβ1-4 GlcNAc	Hepatitis B (HVB)	Komai *et al.* (1988)

useful as a probe for detection of gangliosides at a femtomolar level and as a targeting agent for selective killing of ganglioside-bearing tumor cells. For example, *Clostridium perfringens* δ toxin lyses selectively those cells that express ganglioside GM2, such as melanoma, neuroblastoma, glioma, and retinoblastoma cells (Jolivet-Reynaud *et al.*, 1993). Although sensitive detection of gangliosides after extraction is made possible by utilizing several toxins (Takamizawa *et al.*, 1986), the specificity of binding should be carefully examined for application to tissues and cells. In fact, cholera toxin-binding glycoproteins were detected, but GM1 ganglioside, which has been well characterized as the receptor for cholera toxin, was absent in the rat intestinal epithelial cells which are the initial target cells in the pathogenesis of cholera toxin (Morita *et al.*, 1980). In the case of influenza virus, a structural requirement for an amino acid sequence in the hemagglutinin to form a pocket structure that accommodates a sialic acid moiety was clearly established (Weiss *et al.*, 1988), and type C strain was found to be able to bind to a wide spectrum of sialyl glycoconjugates including mucins, serum glycoproteins, and gangliosides containing 9-*O*-acetylated sialic acid (Zimmer *et al.*, 1992). These results indicate that the sialyl oligosaccharide moiety, but not the ceramide, is really involved in the receptor functions, although the ceramide possibly may affect the affinity of the oligosaccharide for the ligand in the binding of Propionibacterium to LacCer (Karlsson, 1989). In addition, a molecular unrelated to ganglioside is considered to be important in the binding of tetanus toxin to neuroblastoma cells based on the observation that neuraminidase- and β-galactosidase-treated cells retained binding ability for the toxin (Wiegandt, 1985). Besides the receptors for microbes, the existence of carbohydrate-recognizing lectin-type receptors has been presumed on the cell surface particularly in cell–cell interactions including cell adhesion. The isolation and identification of such receptor molecules from animal cells remain problematic in spite of the presumed cell biological significance.

7. TOOLS USEFUL FOR THE ANALYSIS OF GANGLIOSIDE FUNCTIONS

Many molecular probes such as monoclonal antibodies, microbial toxins, and lectins have been developed, which were found to be useful in making particular gangliosides and their carbohydrate changes detectable at an ultramicro scale and also to identify the type as well as lineage of cells *in vivo* and *in vitro*, particularly in the field of tumor and, recently, brain research. Ceramide glycanase (Li *et al.*, 1986) or endoglycoceramidase (Ito and Yamagata, 1990; Ito, 1991), which can split glycosphingolipids between the oligosaccharide and the ceramide, is useful to analyze the functional roles of glycosphingolipids at cellular levels (Ito *et al.*, 1993a,b; Ponce *et al.*, 1993; Muramoto *et al.*, 1994). Specific inhibitors for exo- and endoglycosidases and glycosyltransferases and recent gene transfection and targeting technologies also potentially are important. A synthetic ceramide analogue (1-phenyl-2-decanoylamino-3-morpholino-1-propanol, PDMP) (Inokuchi and Radin, 1987) which inhibits glucosylceramide biosynthesis was successfully applied to the depletion of glycosphingolipids including gangliosides at the cellular as well as organismic levels (Radin *et al.*, 1993). In a similar line of studies, establishment of a mutant cell line that is deficient in particular gangliosides and other glycosphingolipids should be taken into consideration. Analysis of a mouse B16 melanoma cell line, GM-95, deficient in GM3, showed that the deficiency is attributable to the first glycosylation step of ceramide and that the deficiency is associated with changes in cellular morphology and growth rate but not with cell growth *per se* (Ichikawa *et al.*, 1994).

Later (see Chapter 7), the roles of gangliosides in transmembrane signaling and in cell–cell recognition will be discussed in detail. Also, the quite recent concept that repressed sialylation of gangliosides may underlie the phenomenon of failed neuritogenesis in the central nervous system neuron afflicted with high-Ca^{2+} neuronodystrophy related to glutamate excititoxicity, as well as in the fetal alcohol syndrome, will be considered in Chapter 11.

REFERENCES

Acquotti, D., Fronza, G., Riboni, L., Sonnino, S., and Tettamanti, G., 1987, Ganglioside lactones: H-NMR determination of the inner ester position of GD1b-ganglioside lactone naturally occurring in human brain or produced by chemical synthesis, *Glycoconjugate J.* **4:**119–127.

Ando, S., Chang, N. C., and Yu, R. K., 1978, High performance thin-layer chromatography and densitometric determination of brain ganglioside compositions of several species, *Anal. Biochem.* **89:**437–450.

Ando, S., Yu, R. K., Scarsdale, J. N., Kusunoki, S., and Prestegard, J. H., 1989, High resolution proton NMR studies of gangliosides. Structure of two types of GD3 lactones and their reactivity with monoclonal antibody R24, *J. Biol. Chem.* **264:**3478–3483.

Ando, S., Hirabayashi, Y., Kon, K., Inagaki, F., Tate, S., and Whittaker, V. P., 1992, A trisialoganglioside containing a sialylα2-6-N-acetylgalactosamine residue is a cholinergic-specific antigen, *J. Biochem.* **111:**287–290.

Bar-Shavit, Z., Teitelbaum, S. L., Reitsman, P., Hall, A., Pegg, L. E., Trial, J., and Kahn, A., 1983, Induction of monocytic differentiation and bone resorption by 1α,25-dihydroxyvitamin D3, *Proc. Natl. Acad. Sci. USA* **80:**5907–5911.

Behr, J., and Lehn, J., 1973, The binding of divalent cations by purified gangliosides, *FEBS Lett.* **31:**297–300.

Blum, A. S., and Barnstable, C. J., 1987, O-Acetylation of a surface carbohydrate creates discrete molecular patterns during neural development, *Proc. Natl. Acad. Sci. USA* **84:**8716–8720.

Bouchon, B., Levery, S. B., Clausen, H., and Hakomori, S.-H., 1992, Production and characterization of a monoclonal antibody (BBH5) directed to ganglioside lactone, *Glycoconjugate J.* **9:**27–38.

Bouchours, J., Bouchours, D., and Hansson, G. C., 1987, Developmental changes of gangliosides of the rat stomach. Appearance of a blood group B-active ganglioside, *J. Biol. Chem.* **262:**16370–16375.

Bradley, W. G., 1990, Critical review of gangliosides and thyrotropin releasing hormone in peripheral neuromuscular diseases, *Muscle Nerve* **13:**833–842.

Braun, P. E., Morell, P., and Radin, N. S., 1970, Synthesis of C18 and C20-dihydrosphingosine, ketodihydrosphingosine and ceramides by microsomal preparations from mouse brain, *J. Biol. Chem.* **245:**335–341.

Breitman, T. R., Selonic, S. E., and Collins, S. J., 1980, Induction of differentiation of the human promyelocytic leukemia cell line (HL-60) by retinoic acid, *Proc. Natl. Acad. Sci. USA* **77:**2936–2940.

Callies, R., Schwarzmann, G., Radsak, K., Siegert, R., and Wiegandt, H., 1977, Characterization of the cellular binding of exogenous gangliosides, *Eur. J. Biochem.* **80:**425–432.

Ceccarelli, B., Aporti, F., and Finesso, M., 1976, Effects of brain gangliosides on functional recovery in experimental regeneration and reinnervation, *Adv. Exp. Med. Biol.* **71:**275–293.

Chatterjee, H., Chakraborty, M., and Anderson, G. M., 1992, Differentiation of Neuro2a neuroblastoma cells by an antibody to GM3 ganglioside, *Brain Res.* **583:**31–44.

Chege, N. W., and Pfeffer, S. R., 1990, Decompartmentation of the Golgi complex. Brefeldin A distinguishes trans-Golgi cisternae from the trans-Golgi network, *J. Cell Biol.* **111:**893–899.

Chiba, A., Kusunoki, S., Shimizu, T., and Kanazawa, I., 1992, Serum IgG antibody to ganglioside GQ1b is a possible marker of Miller Fisher syndrome, *Ann. Neurol.* **31**:677–679.

Collins, S. J., Rusetti, F. W., Gallagher, R. E., and Gallo, R. C., 1978, Terminal differentiation of human promyelocytic leukemia cells induced by dimethylsulfoxide and other polar compounds, *Proc. Natl. Acad. Sci. USA* **75**:2458–2462.

Corti, M., DeGiorgio, V., Ghidoni, R., Sonnino, S., and Tettamanti, G., 1980, Laser-light scattering investigation of the micellar properties of gangliosides, *Chem. Phys. Lipids* **26**:225–238.

Cuatrecasas, P., 1973, Gangliosides and membrane receptors for cholera toxin, *Biochemistry* **12**:3558–3566.

Czarniecki, M. F., and Thornton, E. R., 1977, C-NMR chemical shift titration of metal ion–carbohydrate complexes. An unexpected dichotomy for Ca^{2+}-binding between anomeric derivatives of N-acetylneuraminic acid, *Biochem. Biophys. Res. Commun.* **74**:553–558.

Dennis, R. D., Geyer, R., and Egge, H., 1985, Glycosphingolipids in insects. Chemical structure of ceramide tetra-, penta-, hexa- and heptasaccharides from Calliphora vicina pupae, *J. Biol. Chem.* **260**:5370–5375.

Doherty, P., Ashton, S. V., Skaper, S. D., Leon, A., and Walsh, F. S., 1992, Ganglioside modulation of neural cell adhesion molecule and N-cadherin-dependent neurite outgrowth, *J. Cell Biol.* **117**:1093–1099.

Felgner, P. L., Freire, E., Barenholz, Y., and Thompson, T. E., 1981, Asymmetric incorporation of trisialoganglioside into dipalmitoylphosphatidyl choline vesicles, *Biochemistry* **20**:2168–2172.

Fentie, I. H., and Roisen, F. J., 1993, The effects of cytoskeletal altering agents on the surface topography of GM1 in Neuro2a neuroblastoma cell membranes, *J. Neurochem.* **22**:498–506.

Ferrari, G., Fabris, M., and Gorio, A., 1983, Gangliosides enhance neurite outgrowth in PC12 cells, *Dev. Brain Res.* **8**:215–221.

Ferrari, G., Fabris, M., Fiori, M. G., Gabellini, N., and Volonte, C., 1992, Gangliosides prevent the inhibition by k-252a of NGF responses in PC 12 cells, *Dev. Brain Res.* **65**:35–42.

Formisano, S., Lee, G., Aloj, S. M., and Edelhoch, H. H., 1979, Critical micellar concentration of gangliosides, *Biochemistry* **18**:1119–1124.

Fredman, P., Mansson, J. E., Wikstrand, C. J., Vrionis, F. D., Rynmark, B. M., Bigner, D. D., and Svennerholm, L., 1989, A new ganglioside of the lactoseries, GalNAc-3′-isoLM1, detected in human meconium, *J. Biol. Chem.* **264**:12122–12125.

Furukawa, K., Chait, B. T., and Lloyd, K. O., 1988, Identification of N-glycolylneuraminic acid-containing gangliosides of cat and sheep erythrocytes, *J. Biol. Chem.* **263**:14939–14947.

Geisler, F. H., Dorsey, F. C., and Coleman, W. P., 1991, Recovery of motor function after spinal cord injury: A randomized, placebo-controlled trial with GM1 ganglioside, *N. Engl. J. Med.* **324**:1829–1838.

Gillard, B. K., Heath, J. P., Thurmon, L. T., and Marcus, D. M., 1991, Association of glycosphingolipids with intermediate filaments of human umbilical vein endothelial cells, *Exp. Cell Res.* **192**:433–444.

Gillard, B. K., Thurmon, L. T., and Marcus, D. M., 1992, Association of glycosphingolipids with intermediate filaments of mesenchymal, epithelial, glial and muscle cells, *Cell Motil. Cytoskel.* **21**:255–271.

Gross, S. K., Williams, M. A., and McCluer, R., 1980, Alkali labile, sodium borohydride-reducible ganglioside sialic acid residues in brain, *J. Neurochem.* **34**:1351–1361.

Hakomori, S., 1990, Bifunctional roles of glycosphingolipids. Modulators of transmembrane signalling and mediators for cellular interactions, *J. Biol. Chem.* **265**:18713–18716.

Haraguchi, M., Yamashiro, S., Yamamoto, A., Furukawa, K., Takamiya, K., Lloyd, K., Shiku, H., and Furukawa, K., 1944, Isolation of G_{D3} synthase gene by expression cloning of G_{M3} α2,8-sialyltransferase cDNA using anti-G_{D2} monoclonal antibody, *Proc. Natl. Acad. Sci. USA* **91**:10455–10459.

Hashimoto, Y., Otsuka, H., Sudo, K., Suzuki, K., Suzuki, A., and Yamakawa, T., 1983, Genetic regulation of GM2 expression in liver of mouse, *J. Biochem.* **93:**895–901.

Hayashi, K., and Katagiri, A., 1974, Studies on the interaction between gangliosides, protein and divalent cations, *Biochim. Biophys. Acta* **337:**107–117.

Hidari, K.I.-P., Irie, F., Suzuki, M., Kon, K., Ando, S., and Hirabayashi, Y., 1993, A novel ganglioside with a free amino group in bovine brain, *Biochem. J.* **296:**259–263.

Higashi, H., and Yamagata, T., 1992, Mechanism for ganglioside-mediated modulation of a calmodulin-dependent enzyme, *J. Biol. Chem.* **267:**9839–9843.

Higashi, H., Omori, A., and Yamagata, T., 1992, Calmodulin, a ganglioside-binding protein, *J. Biol. Chem.* **267:**9831–9838.

Hilbig, R., and Rahmann, H., 1980, Variability in brain gangliosides of fishes, *J. Neurochem.* **34:**236–240.

Hilbush, B. S., and Levine, J. M., 1992, Modulation of a Ca^{2+}-signalling pathway by GM1 ganglioside in PC12 cells, *J. Biol. Chem.* **267:**24789–24795.

Hirabayashi, Y., Hirota, M., Matsumoto, M., Tanaka, H., Obata, K., and Ando, S., 1988, Developmental changes of C-series polysialogangliosides in chick brains revealed by mouse monoclonal antibodies M6704 and M7163 with different epitope specificities, *J. Biochem.* **104:**973–979.

Hirabayashi, Y., Hirota, M., Suzuki, Y., Matsumoto, M., Obata, K., and Ando, S., 1989, Developmentally expressed O-acetylated ganglioside GT3 in fetal rat cerebral cortex, *Neurosci. Lett.* **106:**193–198.

Hirabayashi, Y., Hyogo, A., Nakao, T., Tsuchiya, K., Suzuki, Y., Matsumoto, M., Kon, K., and Ando, S., 1990, Isolation and characterization of extremely minor gangliosides GM1b and GD1α in adult bovine brains as developmentally regulated antigens, *J. Biol. Chem.* **265:**8144–8151.

Hirabayashi, Y., Nakao, T., Irie, F., Whittaker, V. P., Kon, K., and Ando, S., 1992, Structural characterization of a novel cholinergic neuron-specific ganglioside in bovine brain, *J. Biol. Chem.* **267:**12973–12978.

Holmgren, J., 1973, Comparison of the tissue receptors for Vibrio cholerae and Escherichia coli enterotoxins by means of gangliosides and natural cholerea toxoid, *Infect. Immun.* **8:**851–859.

Hoshi, M., and Nagai, Y., 1970, Biochemistry of mucolipids of sea urchin gametes and embryos: III. Mucolipids during early development, *Jpn. J. Exp. Med.* **40:**361–365.

Hoshi, M., and Nagai, Y., 1975, Novel sialosphingolipids from spermatozoa of the sea urchin, Arthocidaris crassispina, *Biochim. Biophys. Acta* **388:**152–162.

Iber, H., Echten, G. V., Klein, R. A., and Sandhoff, K., 1990, pH dependent changes of ganglioside biosynthesis in neuronal cell culture, *Eur. J. Cell Biol.* **52:**236–240.

Ichikawa, S., Nakajo, N., Sakiyama, H., and Hirabayashi, Y., 1994, A mouse B-16 melanoma mutant deficient in glycolipids, *Proc. Natl. Acad. Sci. USA* **91:**2703–2707.

Igarashi, M., Waki, H., Hirota, M., Hirabayashi, Y., Obata, K., and Ando, S., 1990, Differences in lipid composition between isolated growth cones from the forebrain and those from the brainstem in the fetal rat, *Dev. Brain Res.* **51:**1–9.

Inokuchi, J., and Radin, N., 1987, Preparation of active isomer of 1-phenyl-2-decanoylamino-3-morpholino-1-propanol, inhibitor of murine glucoceramide synthetase, *J. Lipid Res.* **28:**565–571.

Isono, Y., and Nagai, Y., 1966, Biochemistry of glycolipids of sea urchin gametes: I. Separation and characterization of new type of sulfolipid and sialoglycolipid, *Jpn. J. Exp. Med.* **36:**461–476.

Ito, M., and Yamagata, T., 1990, Endoglycoceramidase from *Rhodococcus* sp., *Methods Enzymol.* **179:**488–497.

Ito, M., Ikegami, Y., and Yamagata, T., 1991, Activator proteins for glycosphingolipid hydrolysis by endoglycoceramidase: Elucidation of biological functions of cell-surface glycosphingolipids in situ by endoglycoceramidase made possible using these activator proteins, *J. Biol. Chem.* **266:**7919–7926.

Ito, M., Ikegami, Y., Tai, T., and Yamagata, T., 1993a, Specific hydrolysis of intact erythrocyte cell-surface glycosphingolipids by endoglycoceramidase, *Eur. J. Biochem.* **218:**637–643.

Ito, M., Ikegami, Y., and Yamagata, T., 1993b, Kinetics of endoglycoceramidase action toward cell-surface glycosphingolipids of erythrocytes, *Eur. J. Biochem.* **218:**645–649.

Iwamori, M., and Nagai, Y., 1979, Ganglioside composition of brain in Tay–Sachs disease. Increased amounts of GD2 and N-acetylgalactosaminyl GD1a gangliosides, *J. Neurochem.* **32:**767–777.

Iwamori, M., and Nagai, Y., 1981, Comparative study on ganglioside compositions of various rabbit tissues. Tissue-specificity in ganglioside molecular species of rabbit thymus, *Biochim. Biophys. Acta* **665:**214–220.

Iwamori, M., Harpin, M. L., Lachapelle, F., and Baumann, N., 1985, Brain gangliosides of quaking and Shiverer mutants: Qualitative and quantitative changes of monosialogangliosides in quaking brain, *J. Neurochem.* **45:**73–78.

Iwamori, M., Noguchi, M., Yamamoto, T., Yago, M., Nozawa, S., and Nagai, Y., 1988, Selective terminal α2-3 and α2-6 sialylation of glycosphingolipids with lacto-series type 1 and 2 chains in human meconium, *FEBS Lett.* **233:**134–138.

Jacques, L. W., Riesco, B. F., and Weltner, W., 1980, NMR spectroscopy and calcium binding of sialic acids: N-glycolylneuraminic acid and periodate-oxidized N-acetylneuraminic acid, *Carbohydr. Res.* **83:**21–32.

Jolivet-Reynaud, C., and Alouf, J. E., 1983, Binding of *Clostridium perfringens* ^{125}I-labeled δ-toxin to erythrocytes, *J. Biol. Chem.* **258:**1871–1877.

Jolivet-Reynaud, C., Estrade, J., West, L. A., Alouf, J. E., and Chedid, L., 1993, Targeting of GM2-bearing tumor cells with the cytolytic *Clostridium perfringens* δ toxin, *Anticancer Drugs* **4:**65–75.

Kanda, S., Inoue, K., Nojima, S., Utsumi, H., and Wiegandt, H., 1982, Incorporation of spin-labeled ganglioside analogues into cell and liposomal membranes, *J. Biochem.* **91:**1707–1718.

Kannagi, R., Nudelman, E., and Hakomori, S., 1982, Possible role of ceramide in defining structure and function of membrane glycolipids, *Proc. Natl. Acad. Sci. USA* **79:**3470–3474.

Karlsson, K. A., 1970, Sphingolipid long chain bases, *Lipids* **5:**878–891.

Karlsson, K.-A., 1989, Animal glycosphingolipids as membrane attachment sites for bacteria, *Annu. Rev. Biochem.* **58:**309–350.

Kato, I., and Naiki, M., 1976, Ganglioside and rabbit erythrocyte membrane receptor for Staphylococcal alpha-toxin, *Infect. Immun.* **13:**289–291.

Kawashima, I., Ozawa, H., Kotani, M., Suzuki, M., Kawano, T., Gomibuchi, M., and Tai, T., 1993, Characterization of ganglioside expression in human melanoma cells: Immunological and biochemical analysis, *J. Biochem.* **114:**186–193.

Kawashima, I., Kotani, M., Ozawa, H., Suzuki, M., and Tai, T., 1994, Generation of monoclonal antibodies specific for ganglioside lactones: Evidence of the expression of lacatone on human melanoma cells, *Int. J. Cancer* **58:**263–268.

Kiguchi, K., Chubb, C.B.H., and Huberman, E., 1990, Glycosphingolipid patterns of peripheral blood lymphocytes, monocytes and granulocytes are cell specific, *J. Biochem.* **107:**8–16.

Kimhi, Y., Palfrey, C., Spector, I., Barak, Y., and Littauer, V. Z., 1976, Maturation of neuroblastoma cells in the presence of dimethylsulfoxide, *Proc. Natl. Acad. Sci. USA* **73:**462–466.

Kitagawa, S., Nojiri, H., Nakamura, M., Gallagher, R. E., and Saito, M., 1989, Human myelogenous leukemia cell line HL-60 cells resistant to differentiation induction by retinoic acid, *J. Biol. Chem.* **264:**16149–16154.

Kitajima, K., Inoue, Y., and Inoue, S., 1986, Polysialoglycoproteins of Salmonidae fish eggs:

Complete structure of 200-kDa polysialoglycoprotein from the unfertilized eggs of rainbow trout (Salmo gairdneri), *J. Biol. Chem.* **261**:5262–5269.

Kochetkov, N. K., and Smirnova, G. P., 1983, A disialoglycolipid with two sialic acid residues located in the inner part of the oligosaccharide chain from hepatopancreas of the starfish Patiria pectiria pectinifera, *Biochim. Biophys. Acta* **759**:192–198.

Kojima, S., Kurosawa, N., Nishi, T, Hanai, N., and Tsuji, S., 1994, Induction of cholinergic differentiation with neurite sprouting by *de novo* biosynthesis and expression of GD3 and b-series gangliosides in Neuro 2a cells, *J. Biol. Chem.* **269**:30451–30456.

Komai, K., Kaplan, M., and Peeples, M. E., 1988, The Vero cell receptor for the hepatitis B virus small S protein is a sialogylcoprotein, *Virology* **163**:629–634.

Korhonen, T. K., Baisanen-Rheu, V., Rhen, M., Pere, A., Parkkinen, A., and Finne, J., 1984, *Escherichia coli* fimbriae recognizing sialyl galactosides, *J. Bacteriol.* **159**:762–766.

Kotani, M., Ozawa, H., Kawashima, I., Ando, S., and Tai, T., 1992, Generation of one set of monoclonal antibodies specific for a-pathway ganglio-series gangliosides, *Biochim. Biophys. Acta* **1117**:97–103.

Kotani, M., Kawashima, I., Ozawa, H., Terashima, T., and Tai, T., 1993, Differential distribution of major gangliosides in rat central nervous system detected by specific monoclonal antibodies, *Glycobiology* **3**:137–146.

Kotani, M., Hosoya, H., Kubo, H., Itoh, K., Sakuraba, H., Kusubata, M., Inagaki, M., Yazaki, K., Suzuki, Y., and Tai, T., 1994, Evidence for direct binding of intracellularly distributed ganglioside GM2 to isolated vimentin intermediate filaments in normal and Tay–Sachs disease human fibroblasts, *Cell Struct. Funct.* **19**:81–87.

Kracun, I., Rosner, H., Cosovic, C., and Stavljenic, A., 1984, Topographical atlas of the gangliosides of the adult human brain, *J. Neurochem.* **43**:979–989.

Kusunoki, S., Chiba, A., Hirabayashi, Y., Irie, F., Kotani, M., Kawashima, I., Tai, T., and Nagai, Y., 1993, Generation of a monoclonal antibody specific for a new class of minor ganglioside antigens, GQ1bα and GT1aα: its binding to dorsal and lateral horn of human thoracic cord, *Brain Research* **623**:83–88.

Kyogashima, M., Ginsberg, V., and Krivan, H. C., 1989, *Escherichia coli* K99 binds to N-glycolylsialoparagloboside and N-glycolyl GM3 found in piglet small intestine, *Arch. Biochem. Biophys.* **270**:391–397.

Ledeen, R. W., 1984, Biology of gangliosides, neuritogenic and neurotrophic properties, *J. Neurosci. Res.* **12**:147–159.

Levery, S. B., Roberts, C. E., Salyan, M.E.K., Bouchon, B., and Hakomori, S., 1990, Strategies for characterization of ganglioside inner esters. II. Gas chromatography/mass spectrometry, *Biomed. Environ. Mass Spectrosc.* **19**:311–318.

Levine, J. M., Beasley, L., and Stallcup, W. B., 1984, The D1.1 antigen: A cell surface marker for germinal cells of the central nervous system, *J. Neurosci.* **4**:820–831.

Li, S.-C., DeGasperi, R., Muldrey, J. E., and Li, Y.-T., 1986, A unique glycosphingolipid-splitting enzyme (ceramide-glycanase from leech) cleaves the linkage between the oligosaccharide and the ceramide, *Biochem. Biophys. Res. Commun.* **141**:346–352.

Lipsky, N. G., and Pagano, R. E., 1985, Intracellular translocation of fluorescent sphingolipids in cultured fibroblasts: Endogenously synthesized sphingomyelin and glucocerebroside analogues pass through the Golgi en route to the plasma membrane, *J. Cell Biol.* **100**:27–34.

Mahadik, S. P., and Karpiak, S. K., 1988, Gangliosides in treatment of neural injury and disease, *Drug. Dev. Res.* **15**:337–360.

Markwell, M. A., Svennerholm, L., and Paulson, J. C., 1981, Specific gangliosides as host cell receptors for Sendai virus, *Proc. Natl. Acad. Sci. USA* **78**:5406–5410.

Markwell, M. A., Moss, J., Hom, B. E., Fishman, P. H., and Svennerholm, L., 1986, Expression of gangliosides as receptors at the cell surface controls. Infection of NCTC2071 cells by Sendai virus, *Virology* **155**:356–364.

Masserini, M., and Freire, E., 1987, Kinetics of ganglioside transfer between liposomal and synaptosomal membranes, *Biochemistry* **26**:237–242.

Matta, S. G., Yorke, G., and Roisen, F. J., 1986, Neuritogenic and metabolic effects of individual gangliosides and their interaction with nerve growth factor in cultures of neuroblastoma and pheochromocytoma, *Dev. Brain Res.* **27**:243–252.

Merrill, A. H., Sereni, A. M., Stevens, V. L., Hannun, Y. A., Bell, R. M., and Kinkade, J. M., 1986, Inhibition of phorobol ester-dependent differentiation of human promyelocytic leukemic (HL-60) cells by sphinganine and other long chain bases, *J. Biol. Chem.* **261**:12610–12615.

Merrill, A. H., Wang, E., and Mullins, R. E., 1988, Kinetics of long chain (sphingoid) base biosynthesis in intact LM cells: Effects of varying the extracellular concentration of serine and fatty acid precursors of this pathway, *Biochemistry* **27**:340–345.

Momoi, T., and Yokota, J., 1983, Alterations of glycolipid of human leukemia cell line HL-60 during differentiation, *J. Natl. Cancer Inst.* **70**:229–236.

Momoi, T., Shinmoto, M., Kasuya, J., Senoo, H., and Suzuki, Y., 1986, Activation of CMP-N-acetylneuraminic acid:lactosylceramide sialyltransferase during the differentiation of HL-60 cells induced by 12-O-tetradecanoyl phorbol-13-acetate, *J. Biol. Chem.* **261**:16270–16273.

Morell, A. G., Gregoriadis, G., and Scheinberg, I. H., 1971, The role of sialic acid in determining the survival of glycoproteins in the circulation, *J. Biol. Chem.* **246**:1461–1467.

Morita, A., Tsao, D., and Kim, Y. S., 1980, Identification of cholera toxin binding glycoprotein in rat intestinal microvillus membranes, *J. Biol. Chem.* **255**:2549–2553.

Mraz, M., Schwarzmann, G., Sattler, J., Seeman, B., Momoi, T., and Wiegandt, H., 1980, Aggregate formation of gangliosides at low concentrations in aqueous media, *Hoppe-Seyler's Z. Physiol. Chem.* **361**:177–185.

Muramoto, K., Kobayashi, K., Nakanishi, S., Matsuda, Y., and Kuroda, Y., 1988, Functional synapse formation between cultured neurons of rat cerebral cortex: Block by a protein kinase inhibitor which does not permeate the cell membrane, *Proc. Jpn. Acad. Ser. B* **64**:319–322.

Muramoto, K., Kawahara, M., Kobayashi, K., Ito, M., Yamagata, T., and Kuroda, Y., 1994, Endoglycoceramidase treatment inhibits synchronous oscillations of intercellular Ca^{2+} in cultured cortical neurons, *Biochem. Biophys. Res. Commun.* **202**:398–402.

Nagai, Y., and Tsuji, S., 1988, Cell biological significance of gangliosides in neural differentiation and development: Critique and proposals, in: *New Trends in Ganglioside Research: Neurochemical and Neuroregenerative Aspects, Fidia Research Series,* Vol. 14 (R. W. Ledeen, E. L. Hogan, G. Tettamanti, A. J. Yates, and R. K. Yu, eds.), Liviana Press/Springer-Verlag, Padova/Berlin, pp. 329–350.

Nagai, Y., and Tsuji, S., 1994, Significance of ganglioside-mediated glycosignal transduction in neuronal differentiation and development, *Prog. Brain Res.* **101**:119–126.

Nagai, Y., Momoi, T., Saito, M., Mitsuzawa, E., and Ohtani, S., 1976, Ganglioside syndrome, a new autoimmune neurologic disorder, experimentally induced with brain gangliosides, *Neurosci. Lett.* **2**:107–111.

Nagai, Y., Uchida, T., Takeda, S., and Ikuta, F., 1978, Restoration of activity for induction of experimental allergic peripheral neuritis by a combination of myelin basic protein P2 and gangliosides from peripheral nerve, *Neurosci. Lett.* **8**:247–254.

Nagai, Y., Ikuta, F., and Nagai, Y., 1980a, Neuropathological comparative studies on experimental allergic neuritis (EAN) induced in rabbits by P2 protein–ganglioside complexes, *Jpn. J. Exp. Med.* **50**:453–462.

Nagai, Y., Sakakibara, K., and Uchida, T., 1980b, Immunomodulatory roles of gangliosides in EAE and EAN, in: *Search for the Cancer of Multiple Sclerosis and Other Chronic Diseases of Central Nervous System* (A. Boese, ed.), Verlag Chemie, Weinheim, pp. 127–138.

Nagashima, K., Nakanishi, S., and Matsuda, Y., 1991, Inhibition of nerve growth factor-induced neurite outgrowth of PC12 cells by a protein kinase inhibitor which does not permeate the cell membrane, *FEBS Lett.* **293:**119–123.

Nakajima, J., Tsuji, S., and Nagai, Y., 1986, Bioactive gangliosides: Analysis of functional structures of the tetrasialoganglioside GQ1b which promotes neurite outgrowth, *Biochim. Biophys. Acta* **876:**65–71.

Nakamura, M., Ogino, H., Nojiri, H., Kitagawa, S., and Saito, M., 1989, Characteristic incorporation of ganglioside GM3, which induces monocytic differentiation in human myelogenous leukemia HL-60 cells, *Biochem. Biophys. Res. Commun.* **161:**782–789.

Nakamura, K., Hashimoto, Y., Moriwaki, K., Yamakawa, T., and Suzuki, A., 1990a, Genetic regulation of GM4 (NeuAc) expression in mouse erythrocytes, *J. Biochem.* **107:**3–7

Nakamura, M., Kirito, K., Yamamori, J., Nojiri, H., and Saito, M., 1990b, Gangliosides GM3 can induce megakaryocytoid differentiation of human leukemia cell, *Cancer Res.* **51:**1940–1945.

Nakamura, M., Tsunoda, A., Sakoe, K., Gu, J., Nishikawa, A., Taniguchi, N., and Saito, M., 1992, Total metabolic flow of glycosphingolipid biosynthesis is regulated by UDP-GlcNAc: lactosylceramide β1-3N-acetylglucosaminyltransferase in human hematopoietic cell line HL-60 during differentiation, *J. Biol. Chem.* **267:**23507–23514.

Nara, K., Watanabe, Y., Maruyama, K., Kasahara, K., Nagai, Y., and Sanai, Y., 1994, Expression cloning of a CMP-NeuAc:NeuAcα2-3Galβ1-4Glcβ1-4Glcβ1-1′Cer α2,8-sialyltransferase (GD3 synthase) from human melanoma cells, *Proc. Natl. Acad. Sci. USA* **91:**7952–7956.

Nohara, K., Suzuki, M., Inagaki, F., Ito, H., and Kaya, K., 1990, Identification of novel gangliosides containing lactosaminyl-GM1 structure from rat spleen, *J. Biol. Chem.* **265:**14335–14339.

Nohara, K., Suzuki, M., Inagaki, F., Sano, T., and Kaya, K., 1992, A novel disialoganglioside in rat spleen lymphocytes, *J. Biol. Chem.* **267:**14982–14986.

Nojiri, H., Takaku, H., Tetsuka, T., Motoyoshi, K., Miura, Y., and Saito, M., 1984, Characteristic expression of glycosphingolipid profiles in the bipotential cell differentiation of human promyelocytic leukemia cell line HL-60, *Blood* **64:**534–541.

Nojiri, H., Takaku, F., Ohta, M., Miura, Y., and Saito, M., 1985, Changes in glycosphingolipid composition during differentiation of human leukemic granulocytes in chronic myelogenous leukemia compared with in vitro granulocytic differentiation of human promyelocytic leukemia cell line HL-60, *Cancer Res.* **45:**6100–6106.

Nojiri, H., Takaku, F., Terui, Y., Miura, Y., and Saito, M., 1986, Ganglioside GM3: An acidic membrane component that increases during macrophage-like cell differentiation can induce monocytic differentiation of human myeloid and monocytoid leukemic cell lines HL-60 and U937, *Proc. Natl. Acad. Sci. USA* **83:**782–786.

Nojiri, H., Kitagawa, S., Nakamura, M., Kirito, K., Enomoto, Y., and Saito, M., 1988, Neolacto-series gangliosides induce granulocytic differentiation of human leukemic cell line HL-60, *J. Biol. Chem.* **263:**7443–7446.

Nudelman, E. D., Mandel, V., Levery, S. B., Kaizu, T., and Hakomori, S., 1987, A series of disialogangliosides with binary 2-3 sialosyl lactosamine structure defined by monoclonal antibody NVH2 are oncodevelopmentally regulated antigens, *J. Biol. Chem.* **264:**18719–18725.

Obata, K., and Tanaka, H., 1988, Molecular differentiation of the otic vesicle and neural tube in the chick embryo demonstrated by monoclonal antibodies, *Neurosci. Res.* **6:**131–142.

Obata, K., Oide, M., and Handa, S., 1977, Effects of glycolipids on in vitro development of neuromuscular junction, *Nature* **266:**369–371.

Ochanda, J. O., Syuto, B., Ohishi, I., Naiki, M., and Kubo, S., 1986, Binding of *Clostridium botulinum* neurotoxin to gangliosides, *J. Biochem.* **100:**27–33.

Ohsawa, T., and Nagai, Y., 1975, Immunological evidence for the localization of sialoglycosphingolipids at the cell surface of sea urchin spermatozoa, *Biochim. Biophys. Acta* **389:**69–83.

Okazaki, T., Bielawska, A., Bell, R. M., and Hannun, Y. A., 1990, Role of ceramides as a lipid mediator of 1α,25-dihydroxyvitamin D3-induced HL-60 cell differentiation, *J. Biol. Chem.* **265:**15823–15831.

Ozawa, H., Kawashima, I., and Tai, T., 1992a, Generation of murine monoclonal antibodies specific for N-glycolylneuraminic acid-containing gangliosides. *Arch. Biochem. Biophys.* **294:**423–433.

Ozawa, H., Kotani, M., Kawashima, I., Numata, M., Ogawa, T., Terashima, T., and Tai, T., 1993, Generation of a monoclonal antibody specific for ganglioside GM4: Evidence for GM4 expression on astrocytes in Chicken cerebellum, *J. Biochem.* **114:**5–8.

Pohlentz, G., Klein, D., Schwarzmann, G., Schmitz, D., and Sandhoff, K., 1988, Both GA2, GM2 and GD2 syntheses and GM1b, GD1a and GT1b syntheses are single enzymes in Golgi vesicles from rat liver, *Proc. Natl. Acad. Sci. USA* **85:**7044–7048.

Polley, M. J., Phillips, M. L., Wayer, E., Nudelman, A., Singhal, K., Hakomori, S., and Paulson, J. C., 1991, CD62 and endothelial cell leukocyte adhesion molecule 1(ELAM-1) recognize the same ligand, sialyl Lewis X, *Proc. Natl. Acad. Sci. USA* **88:**6224–6228.

Ponce, R. H., Yanagimachi, R., Urch, U. A., Yamagata, T., and Ito, M., 1993, Retention of hamster oolema fusibility with spermatozoa after various enzyme treatments: A search for the molecules involved in sperm–egg fusion, *Zygotes* **1:**163–171.

Prieto, P. A., and Smith, D. F., 1985, A new ganglioside in human meconium detected by antiserum against the human milk sialyl oligosaccharide, LS-tetrasaccharide b, *Arch. Biochem. Biophys.* **241:**281–289.

Purpura, D. P., and Suzuki, K., 1976, Distortion of neuronal geometry and formation of abberant synapses in neuronal storage disease, *Brain Res.* **116:**1–21.

Purpura, D. P., and Baker, H., 1977, Meganeurites and other aberrant processes of human neurons in feline GM1 gangliosidosis, *Brain Res.* **143:**13–26.

Radin, N., Shayman, J. E., and Inokuchi, J., 1993, Metabolic effects of inhibiting glycosylceramide synthesis with PDMP and other substances, *Adv. Lipid Res.* **26:**183–213.

Rahmann, H., 1985, Gedachtnisbilgung durch molekulare Bahnung in Synapsen mit Gangliosiden, *Funkt. Biol. Med.* **4:**249–261.

Riboni, L., Sonnino, S., Acquotti, D., Malesci, A., Ghidoni, R., Egge, H., Mingrino, S., and Tettamanti, G., 1986, Natural occurrence of gangliosides lactones: Isolation and characterization of GD1b inner ester from adult human brain, *J. Biol. Chem.* **261,** 8514–8519.

Rogers, G. N., Herrler, G., Paulson, J. C., and Klenk, H. D., 1986, Influenza C virus uses 9-O-acetyl-N-acetylneuraminic acid as a high affinity receptor determinant for attachment to cells, *J. Biol. Chem.* **261:**5947–5951.

Roisen, F. J., Bartfeld, H., Nagele, R., and Yorke, G., 1981, Ganglioside stimulation of axonal sprouting in vitro, *Science* **214:**577–578.

Rösner, H., Rahmann, H., Reuter, G., Schauer, R., Peter-Katalinic, J., and Egge, H., 1985a, Mass spectrometric identification of the pentasialoganglioside GP1c of embryonic chicken brain, *Biol. Chem. Hoppe-Seyler* **366:**1177–1181.

Rösner, H., Al-Aqtum, M., and Henke-Fahle, S., 1985b, Developmental expression of GD3 and polysialogangliosides in embryonic chicken nervous tissue reacting with monoclonal antiganglioside antibodies, *Dev. Brain Res.* **18:**85–95.

Roth, J., Kempf, A., Reuter, G., Schauer, R., and Gehring, W. J., 1992, Occurrence of sialic acids in *Drosophila melanogaster, Science* **256:**673–675.

Rovera, G., O'Brien, T. G., and Diamond, L., 1979, Induction of differentiation of human promyelocytic leukemia cells by tumor promoters, *Science* **204:**868–870.

Saito, M., Saito, M., and Rosenberg, A., 1984, Action of monensin, a monovalent cationophore, on cultured human fibroblast: Evidence that it induces high cellular accumulation of glucosyl- and lactosylceramide (gluco- and lactocerebroside), *Biochemistry* **23:**1043–1046.

Saito, M., Terui, Y., and Nojiri, H., 1985, An acidic glycosphingolipid, monosialoganglioside GM3 is a potent physiological inducer for monocytic differentiation of human promyelocytic leukemia cell line HL-60 cells, *Biochem. Biophys. Res. Commun.* **132:**223–231.

Saito, M., Nojiri, H., Ogino, H., Tao, A., Ogura, H., Itoh, M., Tomita, K., Ogawa, T., Nagai, Y., and Kitagawa, S., 1990, Synthetic sialyl glycolipids (sialo-cholesterol and sialo-diglyceride) induce granulocytic differentiation of human myelogenous leukemia cell line HL-60, *FEBS Lett.* **271:**85–88.

Saito, T., Natomi, H., Zhao, W., Okuzumi, K., Sugano, K., Iwamori, M., and Nagai, Y., 1991, Identification of glycolipid receptors for Helicobacter pylori by TLC-immunostaining, *FEBS Lett.* **282:**385–387.

Sakaizumi, M., Hashimoto, Y., Suzuki, A., Yamakawa, T., Kiuchi, Y., and Moriwaki, K., 1988, The locus controlling liver GM1 (NeuGc) expression is mapped 1 cM centromeric to H-2K, *Immunogenetics* **27:**57–60.

Sakakibara, K., Momoi, T., Uchida, T., and Nagai, Y., 1981a, Evidence for association of glycosphingolipid with a colchicine-sensitive microtubule-like cytoskeletal structure of cultured cells, *Nature* **239:**76–79.

Sakakibara, K., Iwamori, M., Uchida, T., and Nagai, Y., 1981b, Immunohistochemical localization of galactocerebroside in kidney, liver and lung of golden hamster, *Experientia* **37:**712–714.

Sakuraba, H., Itoh, K., Kotani, M., Tai, T., Yamada, H., Kurosawa, K., Kuroki, Y., Suzuki, H., Utsunomiya, T., Inoue, H., and Suzuki, Y., 1993, Prenatal diagnosis of GM2-gangliosidosis: Immunofluorescence analysis of GM2 in cultured amniocytes by confocal laser scanning microscopy, *Brain Dev.* **15:**278–282.

Saleh, M. N., Khazaeli, M. B., Wheeler, R. H., Dropcho, E., Liu, T. P., Urist, M., Miller, D. M., Lawson, S., Dixon, P., Russell, C. H., and Lobuglio, C., 1992, Phase 1-trial of the murine monoclonal anti-GD2 antibody 14G2a in metastatic melanoma, *Cancer Res.* **52:**4342–4347.

Sariola, H., Aufderheide, E., Bernhard, H., Henke-Fahle, S., Dippold, W., and Ekbolm, P., 1988, Antibodies to cell surface ganglioside GD3 perturb inductive epithelial–mesenchymal interactions, *Cell* **54:**235–245.

Sasaki, K., Kurata, K., Kojima, N., Kurosawa, N., Ohta, S., Hanai, N., Tsuji, S., and Nishi, T., 1994, Expression cloning of a G_{M3}-specific α2,8-sialyltransferase (G_{D3} synthase), *J. Biol. Chem.* **269(22):**15950–15956.

Schengrund, C., 1989, The role of gangliosides in neural differentiation and repair: A perspective, *Brain Res. Bull.* **24:**131–141.

Schwartz, M., and Spirman, N., 1982, Sprouting from chicken embryo dorsal root ganglia induced by nerve growth factor is specifically inhibited by affinity-purified antiganglioside antibodies, *Proc. Natl. Acad. Sci. USA* **79:**6080–6083.

Seifert, W., 1981, *Gangliosides in Nerve Cell Cultures,* Raven Press, New York, pp. 99–117.

Sekine, M., Nakamura, K., Suzuki, M., Inagaki, F., Yamakawa, T., and Suzuki, A., 1988, A single autosomal gene controlling the expression of the extended globoglycolipid carrying SSEA-1 determinant is responsible for the expression of two extended globogangliosides, *J. Biochem.* **103:**722–729.

Sekine, M., Sakaizumi, M., Moriwaki, K., Yamakawa, T., and Suzuki, A., 1989, Two genes controlling the expression of extended globoglycolipids in mouse kidney are closely linked to each other on chromosome 19, *J. Biochem.* **105:**680–683.

Seybold, U., and Rahmann, H., 1985, Brain gangliosides with different types of postnatal development (nidifugous and nidicolous type), *Dev. Brain Res.* **17:**201–208.

Seyfried, T., 1987, Ganglioside abnormalities associated with failed neural differentiation in a T-locus mutant mouse embryo, *Dev. Biol.* **123:**286–291.

Seyfried, T. N., Yu, R. K., and Miyazawa, N., 1982, Differential cellular enrichment of gangliosides in the mouse cerebellum: Analysis using neurological mutants, *J. Neurochem.* **38:**551–559.

Sillerud, L. O., Prestegard, J. H., Yu, R. K., Schafer, D. E., and Konigsberg, W. H., 1978, Assignment of the ^{13}C nuclear magnetic resonance spectrum of aqueous ganglioside GM1 micelles, *Biochemistry* **17:**2619–2628.

Sjoberg, E. R., Manzi, A. E., Khoo, K. H., Dell, A., and Varki, A., 1992, Structure that GD2 is an acceptor for ganglioside O-acetyltransferase in human melanoma cells, *J. Biol. Chem.* **267:**16200–16211.

Skaper, S. D., Leon, A., and Toffano, G., 1989, Ganglioside function in the development and repair of the nervous system, *Mol. Neurobiol.* **3:**173–199.

Smirnova, G. P., Kochetkov, N. K., and Sadovskaya, V. L., 1989, Gangliosides of the starfish Apelastevias japonica, evidence for a new linkage between two N-glycolylneuraminic acid residues through the hydroxyl group of the glycolic acid residue, *Biochim. Biophys. Acta* **920:**47–55.

Song, Y., Kitajima, K., Inoue, S., and Inoue, Y., 1991, Isolation and structural elucidation of a novel type of ganglioside. Deaminated neuraminic acid (KDN)-containing glycosphingolipids from rainbow trout sperm, *J. Biol. Chem.* **266:**21929–21935.

Stallcup, W. B., Beasley, L., and Levine, J., 1983, Cell surface molecules that characterize different stages in the development of cerebellar interneurons, *Cold Spring Harbor Symp. Quant. Biol.* **48:**761–774.

Stevens, A., Weller, M., and Wietholter, H., 1993, A characteristic ganglioside antibody pattern in the CSF of patients with amyotrophic lateral sclerosis, *J. Neurol. Neurosurg. Psychiat.* **56:**361–364.

Stevens, V. L., Winton, E. F., Smith, E. E., Owens, N. E., Kinkade, J. M., and Merrill, A. F., 1989, Differential effects of long chain (sphingoid) bases on the monocytic differentiation of human leukemia (HL-60) cells induced by phorbol esters, 1α,25-dihydroxyvitamin D3 or ganglioside GM3, *Cancer Res.* **49:**3229–3234.

Stromberg, N., and Karlsson, K.-A., 1990, Characterization of the binding of *Actinomyces naeslundii* (ATCC 12104) and *Actinomyces viscosuo* (ATCC 19246) to glycosphingolipids using a solid-phase overlay approach, *J. Biol. Chem.* **265:**11251–11258.

Stromberg, N., Deal, C., Nyberg, G., Normaek, S., So, M., and Karlsson, K.-A., 1988, Identification of carbohydrate structures that are possible receptors for *Neisseria gonorrhoeae, Proc. Natl. Acad. Sci. USA* **85:**4902–4906.

Stults, C. L., Sweeley, C. C., and Macher, B. A., 1989, Glycosphingolipids: Structure, biological source and properties, *Methods Enzymol.* **179:**167–214.

Sugita, M., 1979, Studies on the glycosphingolipids of the starfish, Asternia pectinifera. III. Isolation and structural studies of two novel gangliosides containing internal sialic acid residues, *J. Biochem.* **86:**765–772.

Sugita, M., Iwasaki, Y., and Hori, T., 1982, Studies on glycosphingolipids of larvae of the green bottle fly, Lusilia caeser. II. Isolation and structural studies on three glycosphingolipids with novel sugar sequences, *J. Biochem.* **92:**881–887.

Suzuki, Y., Suzuki, T., Matsunaga, M., and Matsumoto, M., 1985, Gangliosides as paramyxovirus receptor. Structural requirement of sialo-oligosaccharides in receptors for haemagglutinating virus of Japan (Sendai virus) and Newcastle disease virus, *J. Biochem.* **97:**1189–1199.

Suzuki, Y., Nagao, Y., Kato, H., Matsumoto, M., Nerome, K., Nakajima, K., and Nobusawa, E., 1986, Human influenza A virus hemagglutinin distinguishes sialyloligosaccharides in membrane-associated gangliosides as its receptor which mediates the adsorption and fusion processes of virus infection, *J. Biol. Chem.* **261:**17057–17061.

Suzuki, Y., Nagao, Y., Kato, H., Suzuki, T., Matsumoto, M., and Maruyama, J., 1987, The hemagglutinins of the human influenza viruses A and B recognize different receptor microdomains, *Biochim. Biophys. Acta* **903:**417–424.

Svennerholm, L., Bostrom, K., Fredman, P., Mansson, J. E., Rosengren, B., and Rynmark, B. M., 1989, Human brain gangliosides: Developmental changes from early fetal stage to advanced age, *Biochim. Biophys. Acta* **1005:**109–117.

Tai, T., Kawashima, I., Tada, N., and Ikegami, S., 1988, Different reactivities of monoclonal antibodies to ganglioside lactones, *Biochim. Biophys. Acta* **958:**134–138.

Takamizawa, K., Iwamori, M., Kozaki, S., Sakaguchi, G., Tanaka, R., Takayama, H., and Nagai, Y., 1986, TLC-immunostaining characterization of Clostridium botulinum type A neurotoxin binding to gangliosides and free fatty acids, *FEBS Lett.* **201:**229–232.

Takeda, Y., Takeda, T., Honda, T., and Miwatani, T., 1976, Inactivation of the biological activities of the thermostable direct hemolysin of Vivrio parahaemolyticus by ganglioside GT1, *Infect. Immun.* **14:**1–5.

Taki, T., Rokukawa, C., Kasama, T., Kon, K., Ando, S., Abe, T., and Handa, S., 1992a, Human meconium gangliosides. Characterization of a novel 1-type ganglioside with the NeuAcα2-6Gal, *J. Biol. Chem.* **267:**11811–11817.

Taki, T., Rokukawa, C., Kasama, T., and Handa, S., 1992b, Human hepatoma gangliosides. Occurrence of a novel type glycolipid with NeuAcα2-6Gal structure, *Cancer Res.* **52:**4805–4811.

Tettamanti, G., and Riboni, L., 1993, Gangliosides and modulation of the function of neural cells, *Adv. Lipid Res.* **25:**235–267.

Thorn, J. J., Levery, S. B., Salyan, M.E.K., Stroud, M. R., Cedergren, B., Nilsson, B., Hakomori, S., and Clausen, H., 1992, Structural characterization of X2glycosphingolipid. Its extended form and its sialosyl derivatives—Accumulation associated with the rare blood group phenotype, *Biochemistry* **31:**6509–6517.

Toffano, G., Benvegnu, D., Bonetti, A. C., Facci, L., Leon, A., Orlando, P., Ghidoni, R., and Tettamanti, G., 1980, Interactions of GM1 ganglioside with crude rat brain neuronal membranes, *J. Neurochem.* **35:**861–866.

Trinchera, M., and Ghidoni, R., 1989, Two glycosphingolipid sialyltransferases are localized in different sub-Golgi compartments in rat liver, *J. Biol. Chem.* **264:**15766–15769.

Tsuji, S., Arita, M., and Nagai, Y., 1983, GQ1b. A bioactive ganglioside that exhibits novel nerve growth factor (NGF)-like activities in two neuroblastoma cell lines, *J. Biochem.* **94:**303–306.

Tsuji, S., Yamashita, T., and Nagai, Y., 1988, A novel, carbohydrate signal-mediated cell surface protein phosphorylation: Ganglioside GQ1b stimulates ecto-protein kinase activity on the cell surface of a human neuroblastoma cell line, GOTO, *J. Biochem.* **104:**498–503.

Tsuji, S., Yamashita, T., Tanaka, M., and Nagai, Y., 1988, Synthetic sialylcompounds as well as natural gangliosides induce neuritogenesis in a mouse neuroblastoma cell line (Neuro 2a), *J. Neurochem.* **50:**414–423.

Tsuji, S., Yamashita, T., Matsuda, Y., and Nagai, Y., 1992, A novel glycosignaling system: GQ1b-dependent neuritogenesis of human neuroblastoma cell line, goto, is closely associated with GQ1b-dependent ecto-type protein phosphorylation, *Neurochem. Int.* **21(4):**549–554.

Ulrich-Bott, B., and Wiegandt, H., 1984, Micellar properties of glycosphingolipids in aqueous media, *J. Lipid Res.* **25:**1233–1245.

Van Echten, G., and Sandhoff, K., 1989, Modulation of ganglioside biosynthesis in primary cultured neurons, *J. Neurochem.* **52:**207–214.

Van Echten, G., Iber, H., Stotz, H., Takasuki, A., and Sandhoff, K., 1990, Uncoupling of ganglioside biosynthesis by brefeldin A, *Eur. J. Biochem.* **51:**135–139.

Van Heyningen, W. E., 1974, Gangliosides as membrane receptors for tetanus toxin, cholera and serotonin, *Nature* **249:**415–417.

Varki, A., Hooshmand, F., Diaz, S., Varki, N. M., and Hedrick, S. M., 1991, Developmental abnormalities in transgenic mice expressing a sialic acid-specific 9-O-acetyl esterase, *Cell* **65**:65–74.

Weiss, W., Brown, J. H., Cusack, S., Paulson, J. C., Skehel, J. J., and Wiley, D. C., 1988, Structure of the influenza virus haemagglutinin complexed with its receptor, sialic acid, *Nature* **333**:426–431.

Wiegandt, H., 1985, *Gangliosides*, Elsevier, Amsterdam, pp. 199–260.

Xia, X., Gu, X., Santorelli, A. C., and Yu, R. K., 1989, Effects of inducers of differentiation on protein kinase C and CMP-N-acetylneuraminic acid:lactosylceramide sialyltransferase activities of HL-60 leukemia cells, *J. Lipid Res.* **30**:181–188.

Yamamoto, H., Tsuji, S., and Nagai, Y., 1990, Tetrasialoganglioside GQ1b reactive monoclonal antibodies: Their characterization and application of GQ1b in some cell lines of neuronal and adrenal origin, *J. Neurochem.* **54**:513–517.

Yamashita, T., Tsuji, S., and Nagai, Y., 1991, Sialyl cholesterol is translocated into cell nuclei and it promotes neurite outgrowth in a mouse neuroblastoma cell line, *Glycobiology* **1**:149–154.

Yamato, K., and Yoshida, A., 1992, Biosynthesis of lactosylceramide and paragloboside by human lactose synthase A protein, *J. Biochem.* **92**:1123–1127.

Yohe, H. C., Roark, D. E., and Rosenberg, A., 1976, C20-sphingosine as a determining factor in aggregation of gangliosides, *J. Biol. Chem.* **251**:7083–7087.

Yoshino, H., Miyatani, N., Saito, M., Ariga, T., Lugaresi, A., Latov, N., Kushi, Y., Kasama, T., and Yu, R. K., 1992, Isolated bovine spinal motorneurons have specific ganglioside antigens recognized by sera from patients with motor neuron disease and motor neuropathy, *J. Neurochem.* **59**:1681–1691.

Young, W. W., Lutz, M. S., Mills, S. E., and Lechler-Osborn, E., 1990, Use of Brefeldin A to define sites of glycosphingolipid synthesis: GA2/GM2/GD2 synthase in trans to the Brefeldin A block, *Proc. Natl. Acad. Sci. USA* **87**:6838–6842.

Yu, R. K., and Iqbal, K., 1979, Sialosylgalactosyl ceramides as a specific marker for human myelin and oligodendroglial perikarya: Gangliosides of human myelin, oligodendroglia and neurons, *J. Neurochem.* **32**:293–300.

Yu, R. K., Macala, L. J., Taki, T., Weinfeld, H., and Yu, F. S., 1988, Developmental changes in ganglioside composition and synthesis in embryonic rat brain, *J. Neurochem.* **50**:1825–1829.

Yuki, N., Handa, S., Taki, T., Kasama, T., Takahashi, M., Saito, K., and Miyatake, T., 1992, Cross-reactive antigen between nervous tissue and a bacterium elicits Guillain–Barré syndrome. Molecular mimicry between ganglioside GM1 and liposaccharide from Penner's serotype 19 of Campylobacter jejuni, *Biomed. Res.* **13**:451–453.

Yuki, N., Taki, T., Kasama, T., Takahashi, M., Saito, K., Handa, S., and Miyatake, T., 1993, A bacterium lipopolysaccharide that elicits Guillain–Barré syndrome has a GM1 ganglioside-like structure, *J. Exp. Med.* **178**:1771–1775.

Zheng, M. Z., Fang, H., Tsuruoka, T., Tsuji, T., Sasaki, T., and Hakomori, S., 1993, Regulatory role of GM3 ganglioside in $\alpha5\beta1$ integrin receptor for fibronectin mediated adhesion of FUA169 cells, *J. Biol. Chem.* **268**:2217–2222.

Zhou, B., Li, X-C., Laine, R. A., Huang, R.T.C., and Li, Y.-T., 1989, Isolation and characterization of ceramide glycanase from the leech, *Macrobdella decora*, *J. Biol. Chem.* **264**:12272–12277.

Zimmer, G., Reuter, G., and Schauer, R., 1992, Use of influenza C-virus for detection of 9-O-acetylated sialic acid on immobilized glycoconjugates by esterase activity, *Eur. J. Biochem.* **204**:209–215.

Chapter 7

Role of Gangliosides in Transmembrane Signaling and Cell Recognition

Sen-itiroh Hakomori

1. INTRODUCTION

Sialic acid-containing glycosphingolipids (GSLs), collectively called "gangliosides," were discovered in the mid-1930s by Ernst Klenk (Cologne, Germany) (Klenk, 1942) and Gunnar Blix (Uppsala, Sweden), (Blix, 1936) (see Chapter 1). Since then, steadily increasing numbers of scientists have worked on isolation and characterization of gangliosides, determination of different molecular species, and their distribution in animal cells and tissues. Development of new separation technology (e.g., thin-layer and gas chromatography) and instrumental analysis (e.g., mass spectrometry, NMR spectroscopy), together with introduction of the monoclonal antibody (mAb) approach in immunochemistry, allowed identification of many previously unknown ganglioside species (especially those having complex lacto- or globo-series backbone structure) in the 1970s and 1980s.

Because high levels of ganglio-series gangliosides are found in neural tissues, ganglioside research has traditionally been linked with neurochemistry. However, aside from their role as cell type-specific antigens, the real function of gangliosides remains obscure (see Chapter 6). In comparison with studies of functional proteins, enzymes, antibodies, adhesion molecules, receptors, etc.,

Sen-itiroh Hakomori The Biomembrane Institute, and Departments of Pathobiology and Microbiology, University of Washington, Seattle, Washington 98195.

Biology of the Sialic Acids, edited by Abraham Rosenberg. Plenum Press, New York, 1995.

functional studies of gangliosides have not been very revealing, despite the great efforts of many researchers. Such studies are complicated by the large variety of gangliosides that exist, and their widespread occurrence in many types of cells and tissues.

Studies from this laboratory have been focused on GM3 and sialosylparagloboside (SPG), two fundamental gangliosides present in a great variety of extraneural tissues and cells. We have observed two distinct functions of these compounds and their derivatives: (1) control of transmembrane signaling or second messenger function and (2) modulation of cell–cell interaction or cell adhesion. This research area has been reviewed many times previously (Hakomori, 1984, 1987, 1990, 1993; Bremer and Hakomori, 1984; Hakomori *et al.*, 1990; Hakomori and Igarashi, 1993). The reader should refer to these earlier reviews in conjunction with this chapter.

2. ROLE OF GM3 GANGLIOSIDE AND ITS IMMEDIATE CATABOLITE AS MODULATORS OF TRANSMEMBRANE SIGNALING

Prompted by many startling discoveries concerning the structure and function of certain growth factor receptors, particularly the fact that the receptors are structurally as well as functionally associated with protein kinases (Cohen *et al.*, 1980; Ushiro and Cohen, 1980; Kasuga *et al.*, 1982c; Jacobs *et al.*, 1983; Heldin *et al.*, 1983; for a monograph see Bradshaw and Prentis, 1987), we began a search for possible effects of gangliosides and sphingolipids on growth factor receptors. A few examples of our findings will be summarized in this and subsequent sections.

GM3 is a ubiquitous component in many cell lines (e.g., BHK, 3T3, KB, and A431), and exogenous addition of anti-GM3 mAbs arrested the cell cycle at a defined stage (G1 phase) (Lingwood and Hakomori, 1977). Furthermore, fibroblast growth factor (FGF)-dependent BHK cell growth was inhibited by exogenous GM3 addition (Bremer and Hakomori, 1982). Thus, GM3 may be a general modulator of cell proliferation and the cell cycle.

These studies were subsequently extended to epidermal growth factor (EGF)-dependent cell growth, and effects of gangliosides on the EGF receptor. Growth of human ovarian epidermoid carcinoma KB and A431 cells was highly dependent on EGF, and was inhibited by exogenous addition of GM3. The following results (Bremer *et al.*, 1986) indicated that the effect of GM3 on cell growth was related to modulation of transmembrane signaling through EGF receptor Tyr kinase: (1) Neither GM3 nor GM1 had any effect on binding of [^{125}I]-EGF to its cell surface receptor. GM3 produced specific inhibition of EGF-stimulated Tyr phosphorylation of the EGF receptor in membrane preparations from both KB and A431 cells. (2) The effect of GM3 on EGF-dependent Tyr kinase was also observed in isolated EGF receptors after adsorption on antireceptor mAb/Sepharose complex. (3) Inhibition of Tyr phosphorylation of EGF receptor by GM3 was further confirmed by phospho-amino acid analysis. Tyr (but not Ser or Thr) phosphorylation was affected by GM3. (4) The inhibitory effect of GM3 on EGF-dependent receptor phosphorylation was reproduced in membranes isolated from GM3-fed A431 cells. Further elaborate studies on EGF-dependent growth of an epimerase-less mutant of Chinese hamster ovary cells were performed by Weis and Davis (1990). They found that the mutant, which also lacks EGF receptor, was incapable of synthesizing LacCer or GM3 unless Gal was added to culture medium. The cells were transfected to express EGF receptor, and then showed EGF-dependent growth in the absence of Gal. In fact, the EGF-dependent growth of these transfected cells was *inhibited* by the addition of Gal, which induces GM3 synthesis.

How does GM3 in cell membranes interact with growth factor receptors to modulate their function? Does GM3 directly and randomly interact with growth factor receptors, or does it act in some specific, organized fashion? Do GM3 derivatives exist in nature? If so, do they have distinct effects on transmembrane signaling, in comparison to native GM3?

We have shown that the inhibitory effect of GM3 on EGF receptor Tyr kinase of A431 cells is greatly enhanced in the presence of lyso-phosphatidylcholine (lyso-PC), but not lyso-phosphatidylethanolamine, lyso-phosphatidylserine, or lyso-phosphatidylinositol (Igarashi *et al.*, 1990). In an *in vitro* assay system, lyso-PC (but not the other compounds listed above) greatly stimulated EGF-dependent Tyr phosphorylation of EGF receptor. It is therefore assumed that lyso-PC promotes the inhibitory effect of GM3 in modulation of EGF receptor kinase (RK). Figure 1 illustrates the possible cooperative effect between lyso-PC and GM3, and specificities of ganglioside effects on various growth factor and hormone receptors.

Two derivatives of GM3 have been chemically detected, both displaying a significant effect on transmembrane signaling: (1) lyso-GM3, in which the *N*-fatty acyl residue is eliminated from GM3 to expose the free amino group of sphingosine (Sph), and (2) de-*N*-acetyl-GM3 (deNAcGM3), in which the *N*-acetyl group of sialic acid in GM3 is eliminated and the free amino group of the resulting neuraminic acid residue is exposed. Lyso-GM3 was detected in A431 cells (Hanai *et al.*, 1988b), and deNAcGM3 was detected by specific mAb DH5 in melanoma swiss 3T3 cells and A431 cells (Hanai *et al.*, 1988a), and in various human colonic cancers (E. D. Nudelman and S. Hakomori, unpublished). Lyso-GM3, in comparisons to GM3 (and other GSLs), showed the strongest inhibitory effect on protein kinase C (PKC) (Igarashi *et al.*, 1989).

We observed that deNAcGM3 had a stimulatory effect on EGF RK, particularly in the presence of detergent Triton X-100 (Hanai *et al.*, 1988a). This effect varied considerably depending on detergent quality, but was always stimulatory, in contrast to the inhibitory effect of parent GM3. A subsequent study by Song *et al.* (1991) showed that while deNAcGM3 had a stimulatory effect on EGF RK in the presence of Triton X-100 (in agreement with the findings of Hanai *et al.*), it had neither positive nor negative effect on EGF RK in the absence of detergent (i.e., using membrane vesicles subjected to repeated freezing/thawing which

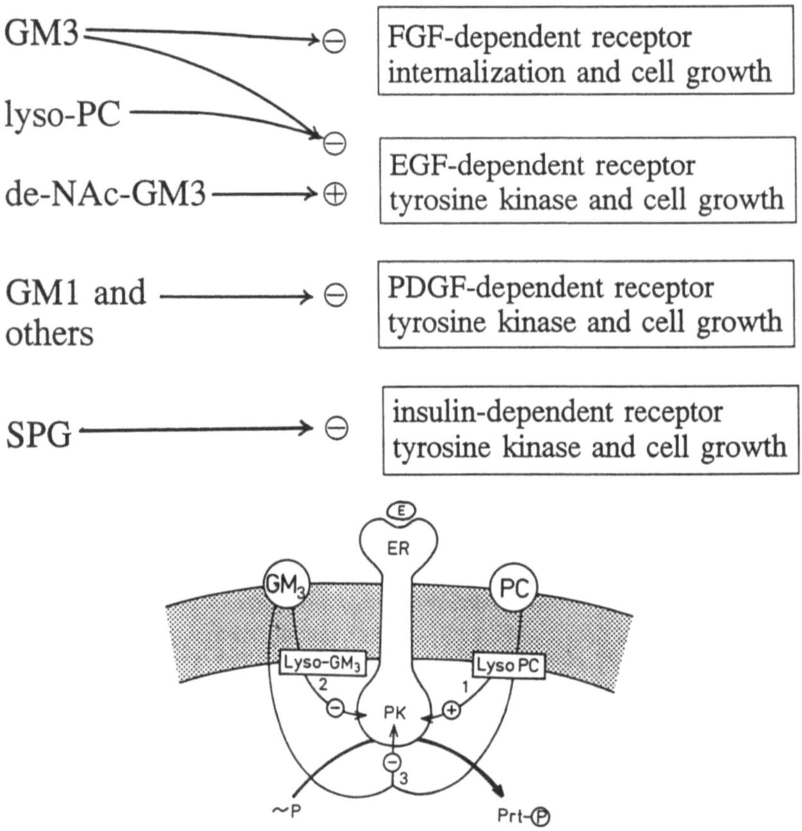

FIGURE 1. Cooperative effect of lyso-PC and GM3 on EGF RK activity. (Upper panel) Specificity of effect of gangliosides (shown on left) on various receptors for growth factors and hormones (shown on right). ⊖, inhibitory effect; ⊕, enhancing effect. Some of these effects occur via inhibition or promotion of receptor-associated tyrosine kinases (Lower panel). Cooperative effect of lyso-PC and GM3 in suppression of EGF RK activity. EGF (E), when bound to its receptor (ER), activates receptor-associated protein kinase (PK) (this mechanism is further illustrated in Figure 2). EGF-dependent activation of PK is promoted by lyso-PC (derived from PC) (route 1). GM3 inhibits PK, particularly in cooperation with lyso-PC (route 3). Lyso-GM3 (derived from GM3 by removal of fatty acyl group) also inhibits PK (route 2).

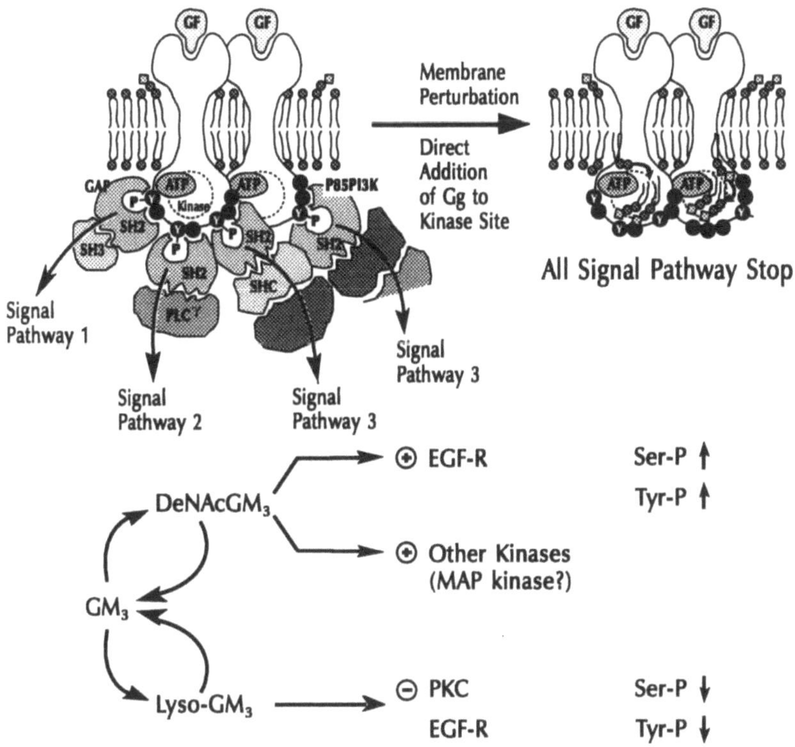

FIGURE 2. Role of gangliosides in transmembrane signaling (schematic). (Upper panel) Proposed mechanism of growth factor (GF)-dependent receptor–receptor interaction inducing activation of cytoplasmic tyrosine kinases (Schlessinger, 1988; Ullrich and Schlessinger, 1990). Activation of cytoplasmic kinase results in Tyr phosphorylation (Y-P), which is in turn recognized by a series of proteins that mediate the signaling. A few types of *src* homology protein 2 (SH2) immediately recognize Y-P structure; these include GTP-activating protein (GAP), phosphatidylinositol-3′-kinase binding protein (P85P13K), and other transducing proteins. Some of the activated SH2 protein is recognized by another *src* homology protein (SH3), or by phospholipase C-γ (PLCγ). The scheme of signaling pathways through SH2 and SH3 is adapted from Brugge (1993). Binding of growth factor can induce several signaling pathways, depending on the particular combination of SH2, SH3, and other signaling proteins. Gangliosides such as GM3 may directly bind to the kinase site, regardless of receptor–receptor interaction, and inhibit Tyr phosphorylation, thereby stopping all signaling pathways.

(Lower panel) Two immediate catabolites, deNAcGM3 and lyso-GM3, are created from GM3 and function as nonspecific second messengers in independent signaling pathways. Lyso-GM3 results from hydrolysis of fatty acid by ceramidase. It strongly inhibits PKC, and moderately inhibits EGF RK and possibly other kinases. deNAcGM3 results from hydrolysis of the *N*-acetyl group of GM3 by de-*N*-acetylase. It strongly enhances Ser phosphorylation, moderately enhances Tyr phosphorylation, and may mediate action of yet unknown kinases. Neither lyso-GM3 nor deNAcGM3 can be degraded further, but they may be converted back to GM3. These two cycles (GM3 $\rightleftharpoons$ lyso-GM3, GM3 $\rightleftharpoons$ deNAcGM3) may constitute important modulators acting antagonistically to each other in control of intracellular signaling.

increased ATP permeability). Song *et al.* observed no stimulation of cell growth by deNAcGM3.

We recently carried out more extensive studies on the effect of deNAcGM3 on EGF RK, and its distribution in various tissues (Zhou *et al.*, 1994). In the presence of deNAcGM3 and zero or low concentrations of Triton X-100, EGF RK was enhanced to twice control levels. At higher concentrations (>0.01%) of Triton X-100, enhancement of EGF RK declined. Ser phosphorylation rather than Tyr phosphorylation was enhanced in EGF receptors, suggesting that the enhancing effect of deNAcGM3 could occur through some unknown Ser or Tyr phosphokinase rather than through EGF RK, which is known to be directed exclusively toward a Tyr residue. We compared several cell lines for deNAcGM3 expression using specific mAb DH5. All lines tested contained deNAcGM3 as well as GM3. Surprisingly, some lines (e.g., KB, 3T6) contained higher levels of deNAcGM3 than of GM3 (Zhou *et al.*, 1994).

In view of the modulation by GM3, lyso-GM3, and deNAcGM3 of the signal transduction process, an obvious question is how binding of growth factors to their receptors induces activation of receptor-associated kinases. Schlessinger (1988) initially proposed that dimerization of EGF receptor occurs on addition of EGF at the cell surface, which induces activation of cytoplasmic RK. This hypothesis was subsequently extended to various other growth factor receptors and hormone receptors (Ullrich and Schlessinger, 1990). We examined the effect of GM3 on degree of dimerization of the EGF receptor. Unexpectedly, we found that GM3 does not affect degree of dimerization, but strongly inhibits EGF RK activity. This inhibitory effect may involve direct binding of GM3 to EGF RK, without dimerization taking place (Zhou *et al.*, 1994). The proposed modulatory role of GM3 and its derivatives on transmembrane signaling is illustrated in Figure 2.

3. ROLE OF OTHER GSLs AS MODULATORS OF RECEPTOR FUNCTION

3.1. GM1 as Modulator of PDGF Receptor and PDGF-Dependent Cell Growth

Growth of 3T3 cells in chemically defined medium requires PDGF, and this growth (both in terms of cell number increase and thymidine incorporation) was found to be inhibited by exogenous GM1, less so by GM3, and not at all by other gangliosides. Kinetic studies showed clearly that: (1) PDGF does not directly interact with gangliosides; (2) preincubation of cells with GM1 or GM3 alters the K_d of [^{125}I]-PDGF binding to cell surface receptors without alteration of receptor number; (3) ganglioside levels in membrane affect PDGF receptor Tyr phosphorylation, i.e., Tyr phosphorylation of PDGF receptor (170 kDa) associated with addition of PDGF was inhibited in a dose-dependent manner by GM1 and GM3, but not by other GSLs (Bremer *et al.*, 1984).

Effects of exogenous gangliosides added to culture media on PDGF-stimulated growth of intact 3T3 cells were studied more recently by Yates *et al.* (1993). PDGF-dependent increase of intracellular Ca^{2+} concentration was inhibited by gangliosides in the order GM1 $\geq$ GT1b $>$ GM2 $>$ GM3, whereas PDGF-stimulated Tyr phosphorylation of PDGF receptor was inhibited in the order GD1a = GT1b $>$ GM1 $>$ GM2 $>$ GM3. None of these gangliosides bound to PDGF directly. Thus, ganglioside-dependent modulation of PDGF receptor function has been clearly demonstrated *in situ*.

3.2. Sialosylparagloboside (SPG) as a Specific Modulator of the Insulin Receptor

Insulin-dependent cell growth, through the insulin receptor, is well characterized. The receptor consists of two 135-kDa insulin-binding subunits and two 95-kDa subunits showing Tyr kinase activity. Tyr phosphorylation of the 95-kDa subunits occurs when insulin binds to its receptor (Kasuga *et al.*, 1982a,b; Czech, 1985). Insulin-dependent phosphorylation of the 95-kDa subunits was specifically inhibited by 2→3SPG, but not by other gangliosides. However, exogenous 2→3SPG did not alter binding of insulin to cellular receptors. When IM9 cells (which exhibit insulin-dependent growth) were preincubated with 2→3SPG and subsequently treated with insulin, *in situ* receptor phosphorylation as detected by immunoprecipitation with antireceptor mAb was greatly reduced. Insulin-dependent growth of IM9, HL60, and K562 cells was inhibited by 2→3SPG, and 2→3SPG-treated HL60 cells differentiated into myelomonocytes, as evidenced by morphological and surface marker changes. We concluded that addition of 2→3SPG inhibits insulin-dependent cell growth, and triggers differentiation in HL60 cells (Nojiri *et al.*, 1991).

4. MODULATION OF CELL ADHESION BY GANGLIOSIDES: POSSIBLE MODULATION OF INTEGRIN AND OTHER RECEPTOR FUNCTIONS

In an early study of membrane receptors able to recognize pericellular fibronectin (FN), it was observed that FN-dependent cell adhesion was inhibited by various gangliosides, particularly polysialogangliosides (Kleinman *et al.*, 1979). This finding suggested that a ganglioside could be the receptor for FN. Many related studies followed. We observed that GT1b ganglioside produced equal inhibition of cell adhesion not only to FN but also to gelatin and other substrates, indicating that this ganglioside may not be a specific receptor for FN

(Rauvala *et al.*, 1981). The general idea of "integrin receptors" was not well established at the time of these studies.

Integrin receptors for various adhesive proteins (initially FN receptor, later vitronectin and many others) were identified during the years from 1979 to 1985. The idea that gangliosides may modulate cell adhesion has been almost forgotten in the current literature. However, two findings should be noted: (1) GM3 has been found as a component of the "detergent-insoluble cell adhesion matrix," which represents a specific cell adhesion site morphologically known as "adhesion plaque." Thus, GM3 may play an important role in control of a functional protein present at the adhesion plaque, as evidenced not only by its presence in the plaque, but also by the ability of GM3 to block cell-to-substratum attachment and cell spreading in the presence of Ca^{2+} and Mg^{2+}. The association of GM3 with adhesion plaques in normal cells is much higher than in transformed cells, indicating that GM3 may have a specific function in control of cell adhesion (Okada *et al.*, 1984). (2) Integrin receptors isolated from human melanoma cells by affinity chromatography on an Arg-Gly-Asp-Ser-containing peptide/Sepharose column contained GD3 ganglioside, and a close association was demonstrated between GD3 and the FN-dependent adhesion site in melanoma cells, suggesting that GD3 is a specific cofactor for the integrin receptor (Cheresh *et al.*, 1987). However, GD3 is not widely present, i.e., some cells showing FN-dependent adhesion do not contain GD3 or GD2 which also can show activity in this latter system.

In more recent studies, mouse mammary carcinoma mutant cell line FUA169, characterized by high GM3 content, was established from a parent cell line FM3A/F28-7, which has high LacCer content but no GM3. In contrast to F28-7 cells, FUA169 cells showed clear adhesion to FN. Several lines of evidence indicate that FUA169 adhesion to FN requires the presence of GM3, which supports the functionality of the integrin receptor: (1) Both FUA169 and F28-7 cells express the same quantity of FN integrin receptor, which consists of $\alpha5\beta1$ (sensitive to RGDS peptide) and $\alpha4\beta1$ (sensitive to CS1 peptide). However, adhesion to FN-coated plates, regardless of type of FN, was much higher for FUA169 than for F28-7 cells. (2) F28-7 cells, which normally lack GM3 and adhere only weakly to FN, acquired GM3 during incubation in GM3-containing medium, and subsequently adhered strongly to FN. (3) Cholesterol–lecithin liposomes (cholesterol was ^{14}C-labeled) incorporating $\alpha5\beta1$ receptor isolated from human placenta showed clear adhesion to FN-coated plates, and this adhesion was completely inhibited by RGDS peptide and by anti-β_1 mAb ZH1. When liposomes included a moderate quantity of GM3 (0.2–0.4 nmol per 55 μg phosphatidylcholine, 33 μg cholesterol, 5 μg $\alpha5\beta1$ in the liposome), adhesion was enhanced significantly. In contrast, adhesion was greatly reduced below control level for $\alpha5\beta1$ liposomes containing a higher quantity (>2 nM) of GM3. Adhesion to FN was also inhibited, but never enhanced, for $\alpha5\beta1$ liposomes with

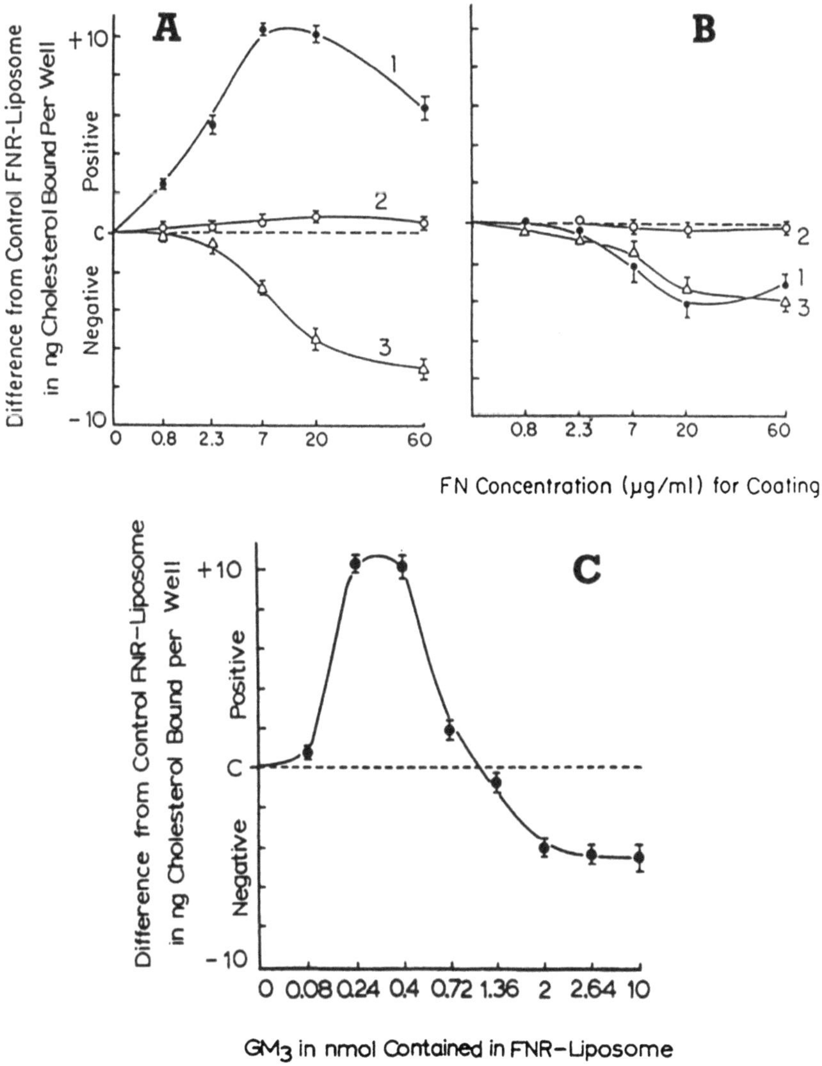

FIGURE 3. Effects of varying concentrations of GSLs on integrin-dependent adhesion to FN. (A) Purified α5β1 integrin receptor was incorporated in liposomes. Liposomes of constant composition in terms of PC, [^{14}C]cholesterol, and α5β1, with various quantities of GM3, LacCer, or GlcCer were prepared, and aliquots of ~1.25 μg cholesterol per well were added to Pro-bind plates previously coated by 1:3 sequence dilution of FN (abscissa). Specific binding was quantitated by scintillation counting, and expressed as ng cholesterol bound. α5β1 liposomes (55 μg PC, 33 μg [^{14}C]cholesterol, 5 μg α5β1 receptor) containing GM3 2.2 μg (2 nM) (△) (curve 3), 0.44 μg (0.4 nM) (●) (curve 1), and 0.088 μg (0.08 nM) (○) (curve 2) were compared with control FN receptor liposomes containing no GM3 (dashed line "C") in terms of binding activity. (B) All experimental conditions and symbols as in (A) except using LacCer instead of GM3. (C) Each well of Pro-bind assay plates was coated with 12.5 μg/ml FN in PBS, and saturated with 1% BSA in PBS. Binding assays were performed using PC/[^{14}C]cholesterol/FN receptor liposomes containing various concentrations of GM3. Abscissa: GM3 quantity (nmol) contained in liposomes. Ordinate: difference of adhesion (expressed as ng cholesterol bound per well) of GM3-containing liposome versus control liposome. Value for control (i.e., FN receptor liposome without GM3) was 26 ng cholesterol bound per well. Reprinted with permission from Zheng *et al.* (1993).

similar composition but containing 0.4 nmol (or other quantities) of LacCer or GlcCer instead of GM3 (Zheng *et al.*, 1992, 1993). Effects of GM3 and other GSLs on FN-dependent adhesion of α5-β1 liposomes are illustrated in Figure 3. In summary, GM3 and other gangliosides may modulate not only membrane-associated RKs and PKC, but also functions of integrin receptors, some of which are involved in transmembrane signaling.

5. DIRECT INVOLVEMENT OF GANGLIOSIDES IN CELL–CELL ADHESION

5.1. Are Fucogangliosides Containing Sialosyl-Le^x or Sialosyl-Le^a the Natural Ligands for E- and P-Selectins?

Considerable evidence has been gathered in recent years showing that certain gangliosides are directly involved in cell–cell adhesion. For example, lacto-series fucogangliosides [sialosyl-Le^x (SLe^x) and sialosyl-Le^a (SLe^a)] have been claimed to be target structures recognized by the lectin domain of E- and P-selectin. SLe^x-expressing cells (HL60, Chinese hamster ovary) are capable of binding to activated endothelial cells (ECs) or platelets, and this binding was inhibited by SLe^x-liposomes or anti-SLe^x mAbs (Phillips *et al.*, 1990; Polley *et al.*, 1991). Fluorescent microsphere beads coated with SLe^x or SLe^a bind to activated platelets but not to paraformaldehyde-fixed platelets (Handa *et al.*, 1991). These data from our laboratory (based on use of fucogangliosides), as well as data from other studies (Lowe *et al.*, 1990; Berg *et al.*, 1991; Takada *et al.*, 1991), strongly indicate that the epitope structure recognized by both E- and P-selectin is related to SLe^x or SLe^a (for review see Paulson, 1992; and Chapters 2 and 6 for further details).

However, the exact identity of naturally occurring ligands bound by these selectins is still not clear, particularly under dynamic flow conditions (Kojima *et al.*, 1992a).

The functional role of E- and P-selectin in inflammatory response of neutrophils has been well documented in recent years, leading to great interest in the pharmaceutical application of SLe^x oligosaccharides for blocking this inflammatory response (Mulligan *et al.*, 1993a,b). It is possible that ganglioside liposomes containing SLe^x or SLe^a could be useful for this purpose. Nevertheless, the physiological role of SLe^x and SLe^a as natural ligands for these selectins (particularly P-selectin) remains ambiguous and controversial. While gangliosides carrying SLe^x or SLe^a epitope bind directly to activated platelets (Handa *et al.*, 1991), they do not bind to P-selectin–Ig fusion protein, which binds strongly to an unidentified glycoprotein (K. Handa, K. Ito, and S. Hakomori, unpublished). Specific glycoproteins carrying *O*-glycosidically linked SLe^x or SLe^a are capable of binding to P- or E-selectin, as evidenced by the fact that selectin-dependent cell binding was strongly inhibited by an inhibitor of *O*-glycosylation extension (Kojima *et al.*, 1992b). Therefore, gangliosides may not be the natural ligands of P-selectin. The role of gangliosides as natural ligands of E-selectin also remains to be clarified.

5.2. Direct Involvement of GM3 in Cell Adhesion

It has become clear that GM3 ganglioside functions as a cell surface adhesion molecule in GSL–GSL interaction. GM3 interacts strongly with Gg3Cer (asialo-GM2) and moderately with LacCer or globoside, but shows negative interaction (repulsion) with GM3 (Figure 4). Cells expressing high levels of GM3 (e.g., B16 mouse melanoma) interact strongly with cells expressing high Gg3Cer (e.g., mouse T-cell lymphoma L6158) (Kojima and Hakomori, 1989, 1991). B16 cells also bind strongly to human or mouse ECs, which express LacCer or Gg3Cer. These interactions are inhibited by liposomes containing GM3, Gg3Cer, or LacCer.

In general, multiple adhesion systems are involved in cell–cell adhesion. For example, adhesion of neutrophils to activated platelets or ECs involves E- or P-selectins, intercellular adhesion molecule 1 (ICAM-1), and β_2

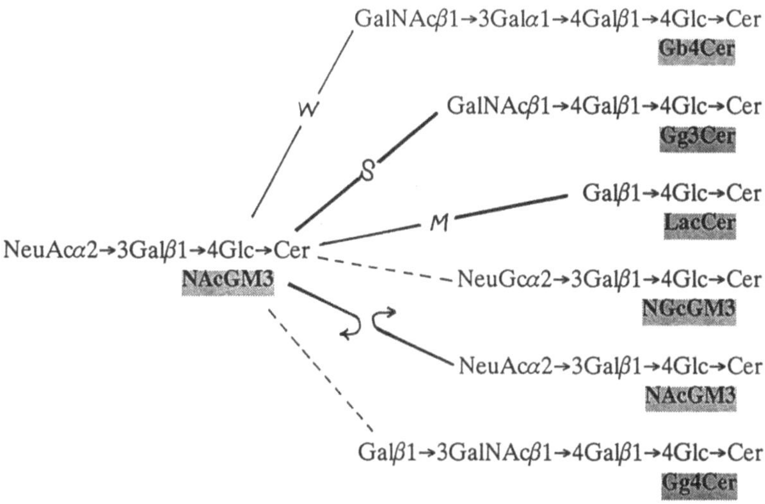

FIGURE 4. Interaction of various GSLs with *N*-acetyl-GM3. "S" indicates a strong interaction; "M," a moderately strong interaction; and "W," a weak interaction. The dashed lines indicate no interaction, and the bidirectional arrows a negative or repellent interaction. Based on data from adhesion of GSL-liposome to solid-phase GSL, or multivalent GSL oligosaccharide to GSL affixed in a column (Kojima and Hakomori, 1989, 1991).

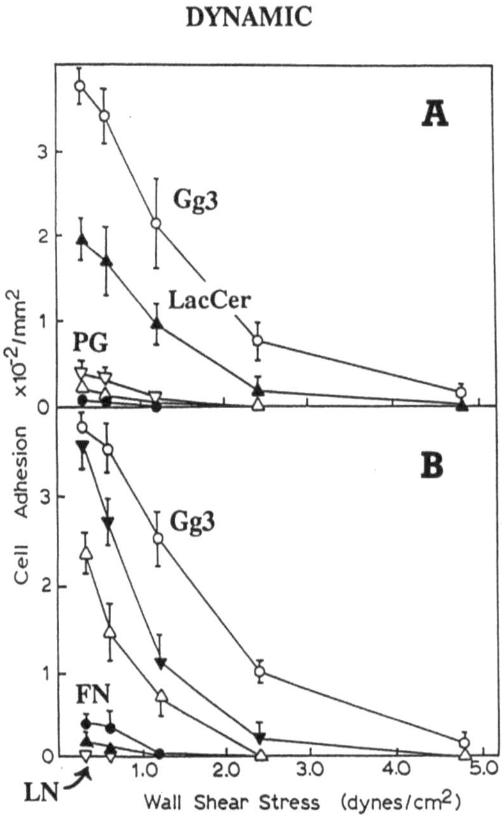

FIGURE 5. Adhesion of BL6 cells to mouse ECs is based on interaction of GM3 (expressed on BL6 cells) with Gg3 or LacCer (expressed on ECs). (A, B) Laminar flow dynamic adhesion system. Wall shear stress was calculated as described by Lawrence *et al.* (1990). One of the parallel plates was coated with Gg3-liposome, LacCer-liposome, FN, or laminin (LN), and BL6 cells suspended in medium were passed through the laminar flow chamber. See Kojima *et al.* (1992c) for experimental details. Adhesion based on Gg3 or LacCer predominated over that based on FN or LN, regardless of shear stress. (C) Static adhesion system. FN- or LN-dependent adhesion became obvious only after ~30 min of incubation. In contrast, Gg3- or LacCer-dependent adhesion were obvious at 20 min. These results suggest that there is a longer "lag time" for integrin-based cell adhesion compared to adhesion based on carbohydrate–carbohydrate interaction, in a static system.

integrin. It appears that a weak but rapid reaction between carbohydrate and selectin takes place initially, followed by a stronger but slower reaction between ICAM-1 and β_2 integrin (Lawrence and Springer, 1991). We expect, therefore, that cellular interactions based on carbohydrate–carbohydrate interaction take place faster than those involving integrin receptors. Indeed, interaction between GM3 and Gg3Cer or LacCer occurs very rapidly compared to interaction of

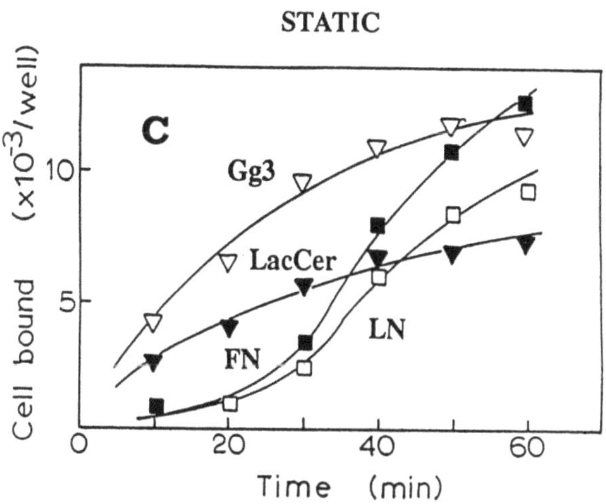

FIGURE 5. *(Continued)*

lectins or integrin receptors, particularly in a dynamic flow system (Kojima *et al.*, 1992c).

Comparative adhesion of B16 cells based on integrin-dependent, lectin-dependent, and carbohydrate–carbohydrate mechanisms, under dynamic flow conditions, is illustrated schematically in Figure 5. At certain shear stress values, carbohydrate-based adhesion is much stronger than integrin-dependent adhesion.

6. CONCLUSIONS

Gangliosides play two major roles in membrane-mediated physiological processes: (1) control of transmembrane signaling and (2) mediation of cell–cell or cell–substratum interaction.

6.1. Control of Transmembrane Signaling

(1) GM3 by itself or together with lyso-PC directly inhibits receptor-associated kinases such as EGF receptor-associated Tyr kinase. This type of inhibition is independent of receptor–receptor dimerization. Through such inhibition, many physiological channels initiated by Tyr phosphorylation are blocked or modified. (2) GM1 or higher gangliosides inhibit PDGF RK through some yet unexplored mechanism. (3) SPG specifically inhibits insulin RK. Other related gangliosides had no effect whatsoever on insulin RK. (4) Lyso-GM3 and de-NAcGM3, two immediate degradation products of GM3, modulate transmembrane signaling or act as second messengers. Lyso-GM3 strongly inhibits PKC

activity, whereas deNAcGM3 promotes Ser kinases of unknown origin and thereby enhances Ser phosphorylation by the EGF receptor. Both lyso-GM3 and deNAcGM3 are present in all types of cells so far examined.

6.2. Mediation of Cell–Cell or Cell–Substratum Interaction

(1) Liposomes containing SLe^x or SLe^a, or mAbs directed to these epitopes, inhibit cell adhesion mediated by E- or P-selectin. However, the naturally occurring, real epitope of these selectins has not been clearly identified. Possible E-selectin epitopes include gangliosides bearing SLe^x or SLe^a. The real P-selectin epitope is probably carried by mucin-type glycoproteins rather than gangliosides. (2) GM3 modulates FN receptor function. $\alpha 5\beta 1$ function is maximized at a certain optimal concentration of GM3, but is inhibited at higher GM3 concentrations. (3) GM3 at the cell surface is involved in direct binding of Gg3Cer- or LacCer-expressing cells, through GM3-Gg3Cer or GM3-LacCer interaction in the presence of bivalent cation. This type of interaction occurs more quickly than integrin-dependent interaction, and presumably plays an essential role in initiation of certain types of cell adhesion.

ACKNOWLEDGMENT. I thank Stephen Anderson, Ph.D., for valuable help in preparation of this manuscript.

REFERENCES

Berg, E. L., Robinson, M. K., Mansson, O., Butcher, E. C., and Magnani, J. L., 1991, A carbohydrate domain common to both sialyl Le^a and sialyl Le^x is recognized by the endothelial cell leukocyte adhesion molecule ELAM-1, *J. Biol. Chem.* **266:**14869–14872.

Blix, G., 1936, Über die Kohlenhydratgruppen des Submaxillarismucins, *Hoppe-Seyler's Z. Physiol. Chem.* **240:**43–54.

Bradshaw, R. A., and Prentis, S., 1987, *Oncogenes and Growth Factors,* Elsevier, Amsterdam.

Bremer, E. G., and Hakomori, S., 1982, GM_3 ganglioside induces hamster fibroblast growth inhibition in chemically-defined medium: Ganglioside may regulate growth factor receptor function, *Biochem. Biophys. Res. Commun.* **106:**711–718.

Bremer, E. G., and Hakomori, S., 1984, Gangliosides as receptor modulators, *Adv. Exp. Med. Biol.* **174:**381–394.

Bremer, E. G., Hakomori, S., Bowen-Pope, D. F., Raines, E., and Ross, R., 1984, Ganglioside-mediated modulation of cell growth, growth factor binding, and receptor phosphorylation, *J. Biol. Chem.* **259:**6818–6825.

Bremer, E. G., Schlessinger, J., and Hakomori, S., 1986, Ganglioside-mediated modulation of cell growth: Specific effects of GM_3 on tyrosine phosphorylation of the epidermal growth factor receptor, *J. Biol. Chem.* **261:**2434–2440.

Brugge, J. S., 1993, New intracellular targets for therapeutic drug design, *Science* **260:**918–919.

Cheresh, D. A., Pytela, R., Pierschbacher, M. D., Klier, F. G., Ruoslahti, E., and Reisfeld, R. A., 1987, An Arg-Gly-Asp-directed receptor on the surface of human melanoma cells exists in a divalent cation-dependent functional complex with the disialoganglioside GD2, *J. Cell Biol.* **105:**1163–1173.

Cohen, S., Carpenter, G., and King, L., 1980, Epidermal growth factor receptor–protein kinase interactions: Co-purification of receptor and epidermal growth factor-enhanced phosphorylation activity, *J. Biol. Chem.* **255:**4834–4842.

Czech, M. P., 1985, The nature and regulation of the insulin receptor: Structure and function, *Annu. Rev. Physiol.* **47:**357–381.

Hakomori, S., 1984, Glycosphingolipids as differentiation-dependent, tumor-associated markers and as regulators of cell proliferation, *Trends Biochem. Sci.* **9:**453–455.

Hakomori, S., 1987, Ganglioside-mediated modulation of growth factor receptor function and cell adhesion, in: *Gangliosides and Modulation of Neuronal Functions* (H. Rahmann, eds.), Springer-Verlag, Berlin, pp. 465–479.

Hakomori, S., 1990, Bifunctional role of glycosphingolipids: Modulators for transmembrane signaling and mediators for cellular interactions, *J. Biol. Chem.* **265:**18713–18716.

Hakomori, S., 1993, Structure and function of sphingoglycolipids in transmembrane signalling and cell–cell interactions, *Biochem. Soc. Trans.* **21:**583–595.

Hakomori, S., and Igarashi, Y., 1993, Gangliosides and glycosphingolipids as modulators of cell growth, adhesion, and transmembrane signaling, *Adv. Lipid Res.* **25:**147–162.

Hakomori, S., Igarashi, Y., Nojiri, H., Bremer, E. G., Hanai, N., and Nores, G. A., 1990, Bioactive gangliosides modulating transmembrane signaling, in: *Trophic Factors and the Nervous System* (L. A. Horrocks, N. H. Neff, A. J. Yates, and M. Hadjiconstantinou, eds.), Raven Press, New York, pp. 135–158.

Hanai, N., Dohi, T., Nores, G. A., and Hakomori, S., 1988a, A novel ganglioside, de-*N*-acetyl-GM_3 ($II^3NeuNH_2LacCer$), acting as a strong promoter for epidermal growth factor receptor kinase and as a stimulator for cell growth, *J. Biol. Chem.* **263:**6296–6301.

Hanai, N., Nores, G. A., MacLeod, C., Torres-Mendez, C.-R., and Hakomori, S., 1988b, Ganglioside-mediated modulation of cell growth: Specific effects of GM_3 and lyso-GM_3 in tyrosine phosphorylation of the epidermal growth factor receptor, *J. Biol. Chem.* **263:**10915–10921.

Handa, K., Nudelman, E. D., Stroud, M. R., Shiozawa, T., and Hakomori, S., 1991, Selectin GMP-140 (CD62; PADGEM) binds to sialosyl-Le^a and sialosyl-Le^x, and sulfated glycans modulate this binding, *Biochem. Biophys. Res. Commun.* **181:**1223–1230.

Heldin, C.-H., Ek, B., and Ronnstrand, L., 1983, Characterization of the receptor for platelet-derived growth factor on human fibroblasts: Demonstration of an intimate relationship with a 185,000-dalton substrate for the plate-derived growth factor-stimulated kinase, *J. Biol. Chem.* **258:**10054–10061.

Igarashi, Y., Nojiri, H., Hanai, N., and Hakomori, S., 1989, Gangliosides that modulate membrane protein function, *Methods Enzymol.* **179:**521–541.

Igarashi, Y., Kitamura, K., Zhou, Q., and Hakomori, S., 1990, A role of lyso-phosphatidylcholine in GM3-dependent inhibition of epidermal growth factor receptor autophosphorylation in A431 plasma membranes, *Biochem. Biophys. Res. Commun.* **172:**77–84.

Jacobs, S., Kull, F. C., Jr., Earp, H. S., Svoboda, M. E., Van Wyk, J. J., and Cuatrecasas, P., 1983, Somatomedin-C stimulates the phosphorylation of the beta-subunit of its own receptor, *J. Biol. Chem.* **258:**9581–9584.

Kasuga, M., Hedo, J. A., Yamada, K. M., and Kahn, C. R., 1982a, The structure of insulin receptor and its subunits: Evidence for multiple non-reduced forms and a 210,000 possible proreceptor, *J. Biol. Chem.* **257:**10392–10399.

Kasuga, M., Karlsson, F. A., and Kahn, C. R., 1982b, Insulin stimulates the phosphorylation of the 95,000-dalton subunit of its own receptor, *Science* **215:**185–187.

Kasuga, M., Zick, Y., Blithe, D. L., Crettaz, M., and Kahn, C. R., 1982c, Insulin stimulates tyrosine phosphorylation of the insulin receptor in a cell-free system, *Nature* **298:**667–669.

Kleinman, H. K., Martin, G. R., and Fishman, P. H., 1979, Ganglioside inhibition of fibronectin-mediated cell adhesion to collagen, *Proc. Natl. Acad. Sci. USA* **76:**3367–3371.

Klenk, E., 1942, Über die Ganglioside, eine neue Gruppe von zuckerhaltigen Gehirnlipoiden, *Hoppe-Seyler's Z. Physiol. Chem.* **273:**76–86.

Kojima, N., and Hakomori, S., 1989, Specific interaction between gangliotriaosylceramide (Gg3) and sialosyllactosylceramide (GM3) as a basis for specific cellular recognition between lymphoma and melanoma cells, *J. Biol. Chem.* **264:**20159–20162.

Kojima, N., and Hakomori, S., 1991, Cell adhesion, spreading, and motility of G_{M3}-expressing cells based on glycolipid–glycolipid interaction, *J. Biol. Chem.* **266:**17552–17558.

Kojima, N., Handa, K., Newman, W., and Hakomori, S., 1992a, Multi-recognition capability of E-selectin in a dynamic flow system, as evidenced by differential effects of sialidases and anti-carbohydrate antibodies on selectin-mediated cell adhesion at low vs. high wall shear stress: A preliminary note, *Biochem. Biophys. Res. Commun.* **189:**1686–1694.

Kojima, N., Handa, K., Newman, W., and Hakomori, S., 1992b, Inhibition of selectin-dependent tumor cell adhesion to endothelial cells and platelets by blocking O-glycosylation of these cells, *Biochem. Biophys. Res. Commun.* **182:**1288–1295.

Kojima, N., Shiota, M., Sadahira, Y., Handa, K., and Hakomori, S., 1992c, Cell adhesion in a dynamic flow system as compared to static system: Glycosphingolipid–glycosphingolipid interaction in the dynamic system predominates over lectin- or integrin-based mechanisms in adhesion of B16 melanoma cells to non-activated endothelial cells, *J. Biol. Chem.* **267:**17264–17270.

Lawrence, M. B., and Springer, T. A., 1991, Leukocytes roll on a selectin at physiologic flow rates: Distinction from and prerequisite for adhesion through integrins, *Cell* **65:**859–873.

Lawrence, M. B., Smith, C. W., Eskin, S. G., and McIntire, L. V., 1990, Effect of venous shear stress on CD18-mediated neutrophil adhesion to cultured endothelium, *Blood* **75:**227–237.

Lingwood, C., and Hakomori, S., 1977, Selective inhibition of cell growth and associated changes in glycolipid metabolism induced by monovalent antibodies to glycolipids, *Exp. Cell Res.* **108:**385–391.

Lowe, J. B., Stoolman, L. M., Nair, R. P., Larsen, R. D., Berhend, T. L., and Marks, R. M., 1990, ELAM-1-dependent cell adhesion to vascular endothelium determined by a transfected human fucosyltransferase cDNA, *Cell* **63:**475–484.

Mulligan, M. S., Lowe, J. B., Larsen, R. D., Paulson, J. C., Zheng, Z.-L., DeFrees, S., Maemura, K., Fukuda, M., and Ward, P. A., 1993a, Protective effects of sialylated oligosaccharides in immune complex-induced acute lung injury, *J. Exp. Med.* **178:**623–631.

Mulligan, M. S., Paulson, J. C., DeFrees, S., Zheng, Z.-L., Lowe, J. B., and Ward, P. A., 1993b, Protective effects of oligosaccharides in P-selectin-dependent lung injury, *Nature* **364:**149–151.

Nojiri, H., Stroud, M. R., and Hakomori, S., 1991, A specific type of ganglioside as a modulator of insulin-dependent cell growth and insulin receptor tyrosine kinase activity: Possible association of ganglioside-induced inhibition of insulin receptor function and monocytic differentiation induction in HL60 cells, *J. Biol. Chem.* **266:**4531–4537.

Okada, Y., Mugnai, G., Bremer, E. G., and Hakomori, S., 1984, Glycosphingolipids in detergent-insoluble substrate attachment matrix (DISAM) prepared from substrate attachment material (SAM): Their possible role in regulating cell adhesion, *Exp. Cell Res.* **155:**448–456.

Paulson, J. C., 1992, Selectin/carbohydrate-mediated adhesion of leukocytes, in: *Adhesion: Its Role in Inflammatory Disease* (J. M. Harlan and D. Y. Liu, eds.), Freeman, San Francisco, pp. 19–42.

Phillips, M. L., Nudelman, E. D., Gaeta, F.C.A., Perez, M., Singhal, A. K., Hakomori, S., and Paulson, J. C., 1990, ELAM-1 mediates cell adhesion by recognition of a carbohydrate ligand, sialyl-Le^x, *Science* **250:**1130–1132.

Polley, M. J., Phillips, M. L., Wayner, E. A., Nudelman, E. D., Singhal, A. K., Hakomori, S., and

Paulson, J. C., 1991, CD62 and endothelial cell–leukocyte adhesion molecule 1 (ELAM-1) recognize the same carbohydrate ligand, sialyl-Lewis x, *Proc. Natl. Acad. Sci. USA* **88:**6224–6228.

Rauvala, H., Carter, W. G., and Hakomori, S., 1981, Studies on cell adhesion and recognition: I. Extent and specificity of cell adhesion triggered by carbohydrate-reactive proteins (glycosidases and lectins) and by fibronectin, *J. Cell Biol.* **88:**127–137.

Schlessinger, J., 1988, Signal transduction by allosteric receptor oligomerization, *Trends Biochem. Sci.* **13:**443–447.

Song, W., Vacca, M. F., Welti, R., and Rintoul, D. A., 1991, Effects of gangliosides G_{M3} and de-*N*-acetyl G_{M3} on epidermal growth factor receptor kinase activity and cell growth, *J. Biol. Chem.* **266:**10174–10181.

Takada, A., Ohmori, K., Takahashi, N., Tsuyuoka, K., Yago, A., Zenita, K., Hasegawa, A., and Kannagi, R., 1991, Adhesion of human cancer cells to vascular endothelium mediated by a carbohydrate antigen, sialyl Lewis A, *Biochem. Biophys. Res. Commun.* **179:**713–719.

Ullrich, A., and Schlessinger, J., 1990, Signal transduction by receptors with tyrosine kinase activity, *Cell* **61:**203–212.

Ushiro, H., and Cohen, S., 1980, Identification of phosphotyrosine as a product of epidermal growth factor-activated protein kinase in A431 cell membranes, *J. Biol. Chem.* **255:**8363–8365.

Weis, F.M.B., and Davis, R. J., 1990, Regulation of epidermal growth factor receptor signal transduction: Role of gangliosides, *J. Biol. Chem.* **265:**12059–12066.

Yates, A. J., VanBrocklyn, J., Saqr, H. E., Guan, Z., Stokes, B. T., and O'Dorisio, M. S., 1993, Mechanisms through which gangliosides inhibit PDGF-stimulated mitogenesis in intact Swiss 3T3 cells: Receptor tyrosine phosphorylation, intracellular calcium, and receptor binding, *Exp. Cell Res.* **204:**38–45.

Zheng, M., Tsuruoka, T., Tsuji, T., and Hakomori, S., 1992, Regulatory role of GM3 ganglioside in integrin function, as evidenced by its effect on function of α5β1-liposomes: A preliminary note, *Biochem. Biophys. Res. Commun.* **186:**1397–1402.

Zheng, M., Fang, H., Tsuruoka, T., Tsuji, T., Sasaki, T., and Hakomori, S., 1993, Regulatory role of G_{M3} ganglioside in α5β1 integrin receptor for fibronectin-mediated adhesion of FUA169 cells, *J. Biol. Chem.* **268:**2217–2222.

Zhou, Q., Hakomori, S., Kitamura, K., and Igarashi, Y., 1994, G_{M3} directly inhibits tyrosine phosphorylation and de-*N*-acetyl-G_{M3} directly enhances serine phosphorylation of epidermal growth factor receptor, independently of receptor–receptor interaction, *J. Biol. Chem.* **269:** 1959–1965.

Chapter 8

Biochemistry and Function of Sialidases

Megumi Saito and Robert K. Yu

1. INTRODUCTION

Sialidases (EC 3.2.1.18; *N*-acylneuraminosyl glycohydrolase) are a family of exoglycosidases that catalyze the cleavage of nonreducing sialic acid residues ketosidically linked to mono- or oligosaccharide chains of glycoconjugates. They are widely distributed in viruses, bacteria, fungi, mycoplasma, and protozoa as well as avian and mammalian species (Rosenberg and Schengrund, 1976; Corfield *et al.*, 1981a; Corfield and Schauer, 1982; Conzelmann and Sandhoff, 1987; Corfield, 1992). Among the various sialidase species, viral and bacterial enzymes have been studied extensively; a number of them have been purified to homogeneity and characterized for their properties and structures. Mammalian sialidases are more labile and often are bound tightly to membranes, hindering successful purification of these enzymes. Much attention, however, has been directed toward these enzymes as interest in the metabolism and biological function of sialoglycoconjugates in mammalian cells has grown in recent years (Schauer, 1982, 1985, 1991; Ledeen, 1989; Schengrund, 1990; Varki, 1992). The term "sialidase" was first proposed by Heimer and Meyer (1956), and "neuraminidase" was introduced a year later (Gottschalk, 1957). Both names have been used interchangeably in the literature. Since the enzyme does not usually apply to neuraminic acid itself, but to its derivatives, sialic acid, the term "sialidase" is deemed more appropriate (Rosenberg and Schengrund, 1976).

Megumi Saito and Robert K. Yu Department of Biochemistry and Molecular Biophysics, Medical College of Virginia, Virginia Commonwealth University, Richmond, Virginia 23298.

Biology of the Sialic Acids, edited by Abraham Rosenberg. Plenum Press, New York, 1995.

1.1. Substrates

While a variety of sialoglycoconjugates such as glycoproteins, poly- and oligosaccharides, gangliosides, and synthetic glycosides can serve as the substrates for sialidases, their susceptibility to the action of a specific sialidase differs. Such differences are derived mainly from the specific structure of the substrates, i.e., the structure and linkage type of the sialic acid residues and nature of the aglycones. Earlier studies on this subject have been reviewed thoroughly in several monographs (Rosenberg and Schengrund, 1976; Corfield *et al.*, 1981a).

1.1.1. Anomeric Configuration

Naturally occurring sialic acid has nine carbons in the carbohydrate backbone; the configurational carbon is positioned at C-7 which has the D configuration. The ketosidic bond of naturally occurring sialidase-susceptible sialic acid residues has the α-D configuration and is equatorial to the pyranoid ring (Yu and Ledeen, 1969). It is this configuration that is recognized by most sialidases. The anomeric structure of the sialic acid released by sialidase was determined as the α-D-anomer, which rapidly mutarotates to the β-D form in aqueous solution (Friebolin *et al.*, 1980; Chong *et al.*, 1992).

1.1.2. Linkage

The susceptibility of a sialic acid residue to sialidase action is greatly influenced by the type of ketosidic linkage to the adjacent sugar moiety which is usually a galactose, *N*-acetylgalactosamine, *N*-acetylglucosamine, or another sialic acid (Schauer, 1982). The *O*-ketosidic linkage of the sialic acid residue is mandatory for sialidase action. Since the enzyme cannot cleave synthetic α-*N*- or α-*S*-glycosides, these compounds can serve as inhibitors for the enzyme (Khorlin *et al.*, 1970). In most sialoglycoconjugates, the sialic acid residues are attached to the neighboring hexose through an α2-3 or α2-6 linkage, or to another sialic acid residue through an α2-8 linkage. The susceptibility of each linkage to sialidase action varies from one enzyme species to another, constituting one of the important characteristics of the enzyme. While the existence of α2-4 and α2-9 linkages in natural compounds has been reported, little information is available concerning their properties as the substrates for sialidases (Huang and Orlich, 1972).

1.1.3. Carboxyl Group

The presence of the free carboxyl group in the sialic acid residue is essential for sialidase action (Yu and Ledeen, 1969). The methyl esters of sialic acids in submandibular gland mucin and II³Neu5AcLac have been shown to be resistant to the action of *Vibrio cholerae* sialidase (Corfield *et al.*, 1981a). The carboxyl group of sialic acid can react with the C-9 or C-7 hydroxyl group of the same or adjacent sialic acid molecule, or with the C-3 hydroxy group of the adjacent galactose residue, forming lactones (Yu *et al.*, 1985; Ando *et al.*, 1989). Sialidases are generally unable to cleave these lactones in colominic acid (Aminoff, 1961; Cassidy *et al.*, 1965) or GD1b (Riboni *et al.*, 1986).

1.1.4. *O*-Substitution

Substitution of one or more of the hydroxyl groups in a sialic acid residue with a substituent affects its susceptibility to sialidase action (Schauer, 1982, 1991; Corfield *et al.*, 1981a, 1992). 4-*O*-Acetylated sialic acid is resistant to bacterial (Corfield *et al.*, 1986; Kleineidam *et al.*, 1990) and mammalian sialidases (Schauer *et al.*, 1980; Schauer and Wember, 1984). The 4-*O*-acetylated derivative, however, can be hydrolyzed at slow rates by sialidases from Newcastle disease virus, influenza A virus, and fowl plague virus (Corfield *et al.*, 1986; Kleineidam *et al.*, 1990). It is known that the hydroxyl group at C-4 in sialic acid interacts with the carbonyl group of the *N*-acyl group through hydrogen bonding (Czarniecki and Thornton, 1977). Thus, the reduced susceptibility of the 4-*O*-acetylated derivatives to sialidase action is attributed to specific conformational changes resulting from the blockage of this interaction in the molecules (Schauer *et al.*, 1980). It has also been reported that the 4-*O*-methyl derivative is resistant to *V. cholerae* sialidase, but can be hydrolyzed by fowl plague sialidase at a rate even greater than the nonsubstituted compound (Beau and Schauer, 1980). *O*-Acetylation of the C-7–C-9 side chain of sialic acid makes the sialic acid residue less susceptible to the enzyme (Corfield *et al.*, 1986; Kleineidam *et al.*, 1990).

1.1.5. *N*-Substitution

Two major types of *N*-substituted sialic acids are found in natural sialoglycoconjugates: *N*-acetyl and *N*-glycolylneuraminic acids. The former is usually more susceptible to sialidase than the *N*-glycolyl derivative (Corfield *et al.*, 1981a,b). Synthetic derivatives with *N*-formyl, *N*-propionyl, or other groups with larger sizes have reduced or abolished susceptibility to sialidase action (Brossmer and Nebelin, 1969).

1.1.6. C-7–C-9 Chain

The shortening of the C-7–C-9 side chain in a sialic acid residue generally renders it less susceptible to the sialidase action (Suttajit and Winzler, 1971; Veh *et al.*, 1977). This modification is achieved by oxidation with periodate and successive reduction with tritiated borohydride for the preparation of radiolabeled derivatives (van Lenten and Ashwell, 1971). It should be noted that this modification converts the susceptible ketosidic bond from an α-D to a β-L configuration because the configurational carbon is shifted to C-6 which is of L-configuration (Yu and Ledeen, 1969).

1.1.7. Branched Sialic Acid Residues

Sialic acid residues are attached not only to the terminal galactose moiety of an oligosaccharide, but also to an internal sugar, forming a branched structure. Sialic acids linked to internal GalNAc or GlcNAc through an α2-6 linkage are usually cleaved by sialidase at rates similar to those for α2-6-linked residues linked to the terminal galactose, as shown with glycoprotein oligosaccharides (Corfield *et al.*, 1981b) and gangliosides such as $IV^3Neu5AcIII^6Neu5AcGg_4Cer$ (GD1α) (Hirabayashi *et al.*, 1990). Another branched structure is found in GM1 ($II^3Neu5AcGg_4Cer$) or GM2 (II^35AcGg_3Cer), where the sialic acid residue is linked to the internal galactose residue. While these gangliosides and their oligosaccharides exhibit resistance to most viral and bacterial sialidases (Schauer *et al.*, 1980), they can be hydrolyzed, though at slower rates, by sialidase from *Arthrobacter ureafaciens* (Saito *et al.*, 1979) and some mammalian membrane-bound sialidases (Tallman and Brady, 1973; Miyagi and Tsuiki, 1986; Saito and Yu, 1986; Miyagi *et al.*, 1990a; Sagawa *et al.*, 1990). The lower susceptibility of these compounds to the enzyme action is attributed to the steric hindrance of GalNAc, based on the interaction of the carbonyl oxygen of the *N*-acetyl group of GalNAc and the sialic acid carboxyl group (Ledeen and Salsman, 1965; Schauer *et al.*, 1980; Koerner *et al.*, 1983; Scarsdale *et al.*, 1990). As compared with GM2, GM1 is usually hydrolyzed by sialidase at a much slower rate, suggesting a long-range effect from the terminal galactose residue in the GM1 molecule (Saito and Yu, 1992). Interestingly, GM1 becomes more susceptible to sialidase when it is developed on thin-layer plates and treated *in situ* by the enzyme; GM1 can be efficiently hydrolyzed by *A. ureafaciens* or *V. cholerae* sialidases in the absence of detergent (Saito *et al.*, 1985; Portoukalian and Bouchon, 1986). Although the reason for this is unknown, it may suggest that the gangliosides developed on TLC plates take specific conformation favorable to sialidase action relative to those in aqueous solutions.

1.1.8. Aglycones

The susceptibility of a sialic acid residue to sialidase is influenced not only by its structure and linkage, but also by the remainder of the glycoconjugate structure, i.e., the composition and length of the oligosaccharide chain and nature of the aglycone. This is particularly conspicuous for complex sialoglycoconjugates; significant differences are observed between the susceptibility of glycoproteins (or gangliosides) and their free oligosaccharides to the enzyme. On the other hand, hydrolysis of synthetic substrates with simple structures is less affected by the nature of the aglycone, as shown by the lack of appreciable differences in the K_m values for methyl, pentyl, decyl, benzyl, phenyl, and other α-ketosidic derivatives (Corfield *et al.*, 1981a).

1.2. Sialidase Assay

To date, various assay methods for sialidase activity have been developed. In principle, the hydrolytic reaction catalyzed by sialidase can be assayed by quantitating the released sialic acid or aglycone. The thiobarbituric acid procedure described by Warren (1959) or its modification by Aminoff (1961) is used frequently for the colorimetric determination of released, free sialic acid.

Highly sensitive sialidase assay can be achieved using radiolabeled substrates. The sialic acid residues of glycoconjugates are radiolabeled by mild oxidation with periodate and subsequent reduction with $NaB[^3H]_4$ (van Lenten and Ashwell, 1971; Bernacki and Bosmann, 1973). Radiolabeled substrates can also be prepared by radiolabeling the aglycones. Two commonly used radiolabeled substrates are tritiated *N*-acetylneuramin-lactit[^{3}H]ol, which is prepared by reduction of Neu5AcLac with $NaB[^3H]_4$ (Bhavanandan *et al.*, 1975), and a radiolabeled ganglioside (e.g., GM3), which is prepared by $NaB[^3H]_4$ reduction of the double bond in the sphingosine base (Schwarzmann, 1978; Ghidoni *et al.*, 1981; Yohe *et al.*, 1983). The sialidase activity is determined by measuring the radioactivity of released lactitol or lactosylceramide, respectively.

Sialidase activity can also be measured using an artificial fluorogenic substrate, 4-methylumbelliferyl-α-D-*N*-acetylneuraminic acid (4MU-Neu5Ac) (Thomas *et al.*, 1978; Potier *et al.*, 1979; Warner and O'Brien, 1979; Myers *et al.*, 1980) or chromogenic substrates including 3-methoxyphenyl-*a*-D-*N*-acetylneuraminic acid (Tuppy and Palese, 1969) and *p*-nitrophenyl-α-D-*N*-acetylneuraminic acid (Khorlin *et al.*, 1970). Recently, two linkage-specific chromogeneic substrates have been synthesized enzymatically by linking *N*-acetylneuraminic acid to the hydroxyl group of 4-nitrophenyl-β-D-galactopyranoside at position 3 or 6 (Kodama *et al.*, 1991). After termination of the reaction with Neu5Ac2en, a potent sialidase inhibitor, the reaction product, 4-nitrophenyl galactopyranoside, is converted to chromogenic 4-nitrophenol using β-galactosidase.

In addition to these methods, new techniques based on the principle of enzyme-linked immunosorbent assay (ELISA) have recently been introduced for sialidase assay. A glycoprotein or ganglioside substrate is coated in plastic microtiter wells and incubated with an enzyme preparation, followed by detection of the desialylated product using an appropriate ligand. Thus, the activity of *Clos-*

tridium perfringens sialidase directed toward neolacto-series gangliosides has been successfully assayed using monoclonal antibody H-11, which specifically reacts with nLc_4Cer, and a peroxidase-conjugated second antibody in this manner (Taki *et al.*, 1990). Similarly, the reaction product, GM1, from GD1a or GD1b has been detected using peroxidase-conjugated cholera toxin B subunit (Ogura *et al.*, 1992). Desialylated products from natural substrates such as human erythrocytes, fetuin, and gangliosides have also been quantitated using peanut agglutinin (PNA) that reacts with the Galβ1-3GalNAc structure exposed after the terminal sialic acid residues have been removed (Lambre *et al.*, 1991). The linkage specificity of sialidase has been studied using various lectins including those from *Sambucus nigra* (SNA) and *Maackia amurensis* (MAA), which recognize specifically α2-6- and α2-3-bound sialic acid residues, respectively (Rogerieux *et al.*, 1993). Catalytically active sialidase bands on polyacrylamide gel can be stained with 4MU-Neu5Ac following electrophoresis (Berg *et al.*, 1985).

2. VIRAL SIALIDASES

Sialidase is an intrinsic membrane glycoprotein of ortho- and paramyxoviruses including influenza A and B viruses, fowl plague virus, Newcastle disease virus, parainfluenza virus, Sendai virus, measles virus, mumps virus, and murine pneumonia virus (Rosenberg and Schengrund, 1976; Corfield *et al.*, 1981a) (see Chapter 9). Among these sialidases, the influenza virus enzymes have been studied most extensively (Air and Laver, 1989; Colman, 1989). Influenza A virus sialidase is classified into nine subtypes (N1–N9) with no distinct serological cross-reactivity. Among these subtypes, N9 is distinct in that it also possesses the hemagglutination activity (Laver *et al.*, 1984). The hemagglutinin activity of N9 sialidase appears to be localized at a site separate from the catalytic site (Nuss and Air, 1991). Influenza B virus contains a single type of sialidase which has less than 25% sequence identity compared with either N2 or N9 sialidase (Shaw *et al.*, 1982). In Newcastle disease virus and other paramyxoviruses, the sialidase activity resides in a hemagglutinin-sialidase glycoprotein, which catalyzes both hemagglutinin and sialidase activities at separate molecular sites (Scheid *et al.*, 1972; Scheid and Choppin, 1974; Portner *et al.*, 1987; Garcia-Sastre *et al.*, 1991).

2.1. Purification

Viral membrane sialidases have been solubilized with the aid of detergents or proteolytic enzymes such as pronase or trypsin (Drzeniek *et al.*, 1966; Laver, 1978). The intact influenza virus sialidase (about 240 kDa) can be solubilized with detergents. Treatment of the viruses with proteases produces a soluble sialidase "head" with a molecular mass of approximately 200 kDa which retains both antigenic and enzymatic activities of the intact molecule (Wrigley *et al.*, 1973; Laver *et al.*, 1982). The molecular mass of the hemagglutinin-sialidase glycoprotein in paramyxoviruses is larger than that of the influenza viral sialidase; the monomer of Newcastle disease virus sialidase has a molecular mass of 74 kDa which is a glycosylated product of a 67-kDa polypeptide (Nishikawa *et al.*, 1986).

2.2. Properties

The kinetic properties and substrate specificity of viral sialidases have been investigated by many researchers (Drzeniek *et al.*, 1966; Drzeniek, 1972, 1973; Huang and Orlich, 1972; Kessler *et al.*, 1977; Thomas *et al.*, 1978; Corfield *et al.*, 1981b, 1982, 1983; Cabezas *et al.*, 1989; Kleineidam *et al.*, 1990; Kodama *et al.*, 1991; Xu *et al.*, 1993). The pH optima of these enzymes range between 5.5 and 6.5 with Neu5AcLac as the substrate, and the K_m values are 1–2 mM.

Most viral sialidases have strict specificity for α2-3-linked sialic acid residues (Drzeniek, 1973; Corfield *et al.*, 1982). The hydrolytic rates of II^3Neu5Ac-Lac relative to the α2-6 isomer are reported to be 470-, 230-, and 56-fold for Newcastle disease virus, influenza A virus, and fowl plague virus sialidases, respectively. GM3 is less susceptible than the oligosaccharide NeuAcα2-3Lac. II^3NeuAcGgOse$_4$Cer (GM1) and its oligosaccharide are totally resistant to the sialidase action. The α2-8-linked sialic acid residue in disialyllactose is cleaved by Newcastle disease virus sialidase at a rate comparable to NeuAcα2-3Lac, but is poorly hydrolyzed by the influenza and fowl plague virus enzymes (Corfield *et al.*, 1981a, 1982). The influenza B virus sialidase is also more active toward II^3NeuAcLac, II^3NeuAcLacCer, and IV^3NeuAcLcOse$_4$Cer than the corresponding α2-6 isomers (Xu *et al.*, 1993). The enzyme shows lower V_{max} and K_m values than does influenza A virus sialidase (Cabezas *et al.*, 1989).

Recent studies have demonstrated that the substrate specificity of viral sialidases, especially those from influenza viruses, can undergo changes during viral evolution. Among the various subtypes of sialidases from influenza A virus (the N2 subtypes) isolated between 1957 and 1987, the early subtypes exhibit strict specificity for α2-3Gal-linked sialic acids of a derivatized α-1 acid protein, whereas later subtypes isolated after 1972 cleave both α2-3- and α2-6-linked sialic acids at comparable rates (Baum and Paulson, 1991). A similar tendency has also been observed for influenza B virus sialidase with various substrates (Xu *et al.*, 1993). The drift in substrate specificity of viral sialidase is probably related to evolutionary changes in the gene structure of the enzyme (Martinez *et al.*, 1983; Air *et al.*, 1990; Nerome *et al.*, 1991; Henrickson and Savatski, 1992; Saito *et al.*, 1993). However, the structural basis for these changes has not been fully elucidated. This acquired substrate specificity may be complementary to the specificity of the hemagglutinin receptor (Baum and Paulson, 1991; Xu *et al.*, 1993).

Previous studies have suggested that viral sialidases require Ca^{2+} for full activity (Dimmock, 1971; Carroll and Paulson, 1982) and stability (Baker and Gandhi, 1976). A recent study on steady-state kinetics of influenza A virus sialidase reaction has demonstrated that Ca^{2+} increases the initial reaction rate by accelerating the enzyme–substrate interaction (Chong *et al.*, 1991). Based on sequence analysis of viral sialidases of various subtypes, it has been proposed that Ca^{2+} binds at the subunit interface and transmits a conformational change in the active site of the enzyme (Chong *et al.*, 1991). The existence of Ca^{2+}-binding sites in influenza virus sialidase has previously been postulated from its binding capacity to samarium (Varghese *et al.*, 1983; Chong *et al.*, 1991). More recently, two Ca^{2+}-binding sites with moderate and high affinity have been demonstrated (Burmeister *et al.*, 1992). The binding site with the lower affinity is located between the active site and a large surface antigen loop and is assumed to be present in all influenza A and B virus sialidases.

2.3. Structure

The nucleotide sequences of many viral sialidases have been determined and the amino acid sequences deduced. They include the enzymes from influenza A virus (Fields *et al.*, 1981; Hiti and Nayak, 1982; Ward *et al.*, 1982; Blok and Air, 1982; Blok *et al.*, 1982; Martinez *et al.*, 1983; Lentz *et al.*, 1984; Air *et al.*, 1985; Dale *et al.*, 1986; Saito *et al.*, 1993), influenza B virus (Shaw *et al.*, 1982; Air *et al.*, 1990; Nerome *et al.*, 1991; Burmeister *et al.*, 1991, 1993), Newcastle disease virus (Jorgensen *et al.*, 1987; Nagy *et al.*, 1990), Sendai virus (Blumberg *et al.*, 1985; Miura *et al.*, 1985; Takahashi *et al.*, 1992), and parainfluenza virus (Hiebert *et al.*, 1985; Henrickson and Savatski, 1992). Influenza virus sialidase is a tetramer of four identical subunits, each of which is composed of a box-shaped head, and a centrally attached stalk containing a hydrophobic region that is embedded in the viral membrane. While the globular head portion contains the full enzymatic and antigenic activities and can be released by protease treatment, the N-terminus of the polypeptide is located inside the viral membrane and consists of six polar amino acids. This sequence is totally conserved in each of the nine different influenza A subtypes (Blok and Air, 1982; Air *et al.*, 1985), but not in influenza B virus (Shaw *et al.*, 1982). The N-terminus is followed by about 30 uncharged hydrophobic amino acids, which form a membrane-anchor and signal sequence for translocation into the endoplasmic reticulum (Fields *et al.*, 1981; Bos *et al.*, 1984). This anchor sequence is not conserved. Between the head and anchor regions is a stretch of about 40 amino acids which forms the stalk. There is no conservation of sequence in this region. Evidence has been provided suggesting that changes in the stalk length affect the accessibility of substrates to sialidase on the viral surface (Els *et al.*, 1985) and the efficiency of elution, replication, and virulence of the virus (Castrucci and Kawaoka, 1993). There is no posttranslational cleavage of the polypeptides and no splitting of the signal peptide (Blok *et al.*, 1982). Potential glycosylation sites have been shown in the polypeptides from different subtypes (Fields *et al.*, 1981; Hiti and Nayak, 1982; Dale *et al.*, 1986).

The hemagglutinin-sialidase glycoproteins of paramyxoviruses are also linked to viral membranes via the hydrophobic amino acid sequences at the N-termini of the polypeptides. Differing from the influenza virus sialidases, the primary translation product of hemagglutinin-sialidases from some strains are inactive precursors which become active only after processing by proteolytic cleavage (Scheid and Choppin, 1974; Nagai *et al.*, 1976). Similarity in the secondary structure has been observed between the paramyxovirus and influenza virus sialidases (Jorgensen *et al.*, 1987).

The three-dimensional structures of some viral sialidases have been studied by X-ray diffraction of the crystallized head regions of the enzymes or their complexes with antibodies or sialic acid, providing vital information about the antigenic and catalytic sites of the enzymes (Varghese *et al.*, 1983, 1992; Colman *et al.*, 1983, 1987; Baker *et al.*, 1987; Laver *et al.*, 1987, 1989; Varghese and Colman, 1991; Tulip *et al.*, 1991, 1992a,b; Burmeister *et al.*, 1992; Taylor *et al.*, 1993). For example, the catalytic site of influenza virus N2 sialidase is located in the deep pocket on the upper corners of the box-shaped head of the molecule, and the antigenic determinants form a nearly continuous surface across the top of the monomer encircling the catalytic site (Colman *et al.*, 1983). The binding of influenza sialidase with monoclonal antibodies has recently been studied in detail (Tulip *et al.*, 1991, 1992a,b; Malby *et al.*, 1993). Knowledge of the structure of the enzyme is clearly critical in understanding its mode of action and in facilitating the design of antiviral agents (von Itzstein *et al.*, 1993).

The catalytic site of viral sialidase has also been vigorously explored, although the precise structure and mechanism of catalysis remain to be clarified (Lentz *et al.*, 1987; Jorgensen *et al.*, 1987; Gorman *et al.*, 1991; Sheehan and Iorio, 1992; Varghese *et al.*, 1992; Burmeister *et al.*, 1992; Takahashi *et al.*, 1992; Colman *et al.*, 1993). Analysis of 14 isolates with site-specific mutated sialidases suggests the involvement of His-274 and Glu-276 in catalysis (Lentz *et al.*, 1987). An X-ray diffraction study of sialidase–sialic acid complex demonstrated that three arginine residues in influenza A virus sialidase interact with the carboxyl group of the sugar which is equatorial to the saccharide ring as a consequence of its distorted geometry. The ketosidic oxygen is positioned within a hydrogen-bonding distance of Asp-151, implying this residue is involved in catalysis (Varghese *et al.*, 1992). Another crystallographic study on influenza B virus sialidase–sialic acid complexes showed the structure of a putative catalytic site in which arginine residues play important roles, and demonstrated that this structure is stabilized by a second shell of ten highly conserved residues principally by an extensive network of hydrogen bonds (Burmeister *et al.*, 1992).

A kinetic study on the biosynthesis of influenza A virus sialidase has been reported recently (Hogue and Nayak, 1992). The enzyme protein is first synthesized as a monomer and, within 10 min, a significant fraction of the monomer is assembled into dimers or tetramers in the endoplasmic reticulum. The mature enzyme is detected at the host cell surface 30 min after its synthesis. Its concentration is increased to the maximum level after 1 hr, and decreases later, presumably because of its incorporation into the virion.

2.4. Biological Function

Viral sialidases have been assumed to play a key role in viral infection. The enzyme may facilitate the transport of the viral particles to and from the site of infection by degrading sialylated mucins (Burnet and Stone, 1947) and help the elution of progeny viruses from the host cells by destroying the hemagglutinin receptors on the cell surface (Palese *et al.*, 1974; Palese and Compans, 1976). The removal of sialic acid residues from newly synthesized hemagglutinin and sialidase is also necessary to prevent self-aggregation of the virus (Palese *et al.*, 1974; Palese and Compans, 1976). For certain paramyxoviruses, the enzyme is thought to be involved in fusion of the virion with the infected cell (Huang *et al.*, 1980, 1985; Shibata *et al.*, 1981; Merz and Wolinsky, 1983; Iorio and Glickman, 1992).

3. BACTERIAL AND FUNGAL SIALIDASES

Sialidase is found in a variety of bacterial and fungal strains, including *Clostridium, Vibrio, Corynebacterium, Bacteroides, Streptococcus, Pasteurella, Salmonella, Arthrobacter, Actinomyces,* and *Streptomyces* species (Rosenberg and Schengrund, 1976; Corfield *et al.*, 1981a; Corfield, 1992). Most sialidase-producing bacteria are symbionts or pathogens for animals, suggesting that the enzyme may be a pathogenic factor in microbial infection (Müller, 1974; Schauer, 1983; Corfield, 1992). While sialidase is usually released into the culture medium, it also exists in cell-bound forms (Fraser and Brown, 1981; Teufel *et al.*, 1989; Guzman *et al.*, 1990; Corfield *et al.*, 1992). The sialidase activity in some bacterial strains can be induced by culturing the microorganisms in the presence of sialic acid, sialic acid glycosides, or *N*-acetylmannosamine (Drzeniek, 1972, 1973; Müller, 1974). This property has been utilized for preparing large amounts of highly active enzymes.

3.1. Purification

Sialidases have been purified from many different bacterial strains using conventional purification methods, e.g., ammonium sulfate precipitation, gel filtration, and ion exchange and absorption column chromatography. They include *Clostridium perfringens* (Cassidy *et al.*, 1965; Nees *et al.*, 1975; Roggentin *et al.*, 1988a), *Clostridium chauvoei* (Heuermann *et al.*, 1991), *Clostridium sordellii* (Roggentin *et al.*, 1987), *Vibrio cholerae* (Galen *et al.*, 1992), *Bacteroides fragilis* (Berg *et al.*, 1983; Tanaka *et al.*, 1992, 1994), *Streptococcus pneumoniae* (Davis *et al.*, 1979; Lock *et al.*, 1988; Camara *et al.*, 1991), *Salmonella typhimurium* (Hoyer *et al.*, 1991, 1992), *Pasteurella haemolytica* (Straus *et al.*, 1993), *Arthrobacter ureafaciens* (Uchida *et al.*, 1977), *Arthrobacter sialophius* (Kessler *et al.*, 1982), *Actinomyces viscosus* (Teufel *et al.*, 1989), and *Micromonospora viridifaciens* (Aisaka *et al.*, 1991; Sakurada *et al.*, 1992).

3.2. Properties

The properties of sialidases from some bacterial strains have been reviewed (Rosenberg and Schengrund, 1976; Corfield *et al.*, 1981a). The pH optimum of most bacterial sialidases typically falls between 5 and 7. The kinetic properties can be affected significantly by the assay conditions (Corfield *et al.*, 1981a). An increase in the ionic strength can cause the pH optimum to shift to more acidic ranges and broaden the pH-dependency curves for some bacterial sialidases (Lipovac *et al.*, 1973; Barton *et al.*, 1975; Saito *et al.*, 1979). The K_m values with Neu5AcLac as the substrate range between 0.5 and 5 mM for most bacterial enzymes.

Bacterial and fungal sialidases show diverse substrate specificity (Table I). *Vibrio cholerae* and *Clostridium perfringens* sialidases hydrolyze glycoprotein substrates at rates higher than oligosaccharide substrates. Differing from many other bacterial sialidases, the *Arthrobacter ureafaciens* enzyme effectively hydrolyzes ganglioside GM1 in the presence of bile acids; the V_{max} value for the enzyme is 130-fold higher than that for *C. perfringens* (Saito *et al.*, 1979).

Substrate linkage specificity also differs among bacterial and fungal sialidases. In general, Neu5Acα2-3Lac is a better substrate than the α2-6 isomer. Especially, *Salmonella typhimurium* sialidase is highly specific to the α2-3 linkage, as compared with the viral enzymes. On the other hand, *Arthrobacter ureafaciens* and *Actinomyces viscosus* sialidases preferentially hydrolyze the α2-6 linkage. *Bacteroides fragilis* and *Actinomyces viscosus* sialidases cleave α2-8 sialic acid residues of colominic acid more efficiently than α2-3 or α2-6 residues.

The molecular masses of native sialidases have been determined by gel filtration and SDS-PAGE, or deduced from the nucleotide sequences coding for

Table I
Substrate Specificity of Bacterial and Fungal Sialidases

Substrates	Sialidase activity[a]							
					Arthrobacter ureafaciens[e]			
	Vibrio cholerae[b]	*Clostridium perfringens*[b]	*Clostridium chauvoei*[c]	*Bacteroides fragilis*[d]	I	II	*Salmonella typhimurium*[f]	*Actinomyces viscosus*[g]
Oligo- and polysaccharides								
II^3Neu5AcLac	100	100	100	100	100	100	100	100
II^6Neu5AcLac	53	44	62	70	157	157	0.4	462
Colominic acid (α2-8)	30	33	7	152	63	69	0.1	300
Glycoproteins								
Fetuin (α2-3>α2-6)	340	272	—	75	59	24	17	—
α1-Acid glycoprotein (α2-6>α2-3)	1000	555	—	—	—	—	—	761
Submandibular gland mucin (α2-6)	400	139	61	10	—	—	—	123
Submaxillary gland mucin (α2-6)	—	—	—	2	56	32	—	—
Gangliosides								
Ganglioside mixtures	(360)[h]	(350)[h]	27	—	78	0.7	34	285
Synthetic								
4MU-Neu5Ac	1580	605	—	87	—	—	1050	—

[a]Each value represents a relative sialidase activity when the activity directed toward II^3Neu5AcLac is regarded as 100.
[b]Corfield *et al.* (1981).
[c]Heuermann *et al.* (1991).
[d]Tanaka *et al.* (1992).
[e]Uchida *et al.* (1979).
[f]Hoyer *et al.* (1991).
[g]Teufel *et al.* (1989).
[h]Values obtained in the absence of bile acids.

Table II
Molecular Masses and Ca^{2+} Requirement of Bacterial and Fungal Sialidases

Strains	Ca^{2+} requirement	Molecular mass[a]	Reference
Clostridium perfringens	−	69 kDa	Nees *et al.* (1975)
		72 kDa	van der Horst *et al.* (1990a)
		*42,770 Da	Roggentin *et al.* (1988a)
Clostridium chauvoei	+	30 kDa (2 subunits)	Heuermann *et al.* (1991)
Clostridium septicum		*111 kDa	Rothe *et al.* (1991)
Clostridium sordellii	−	*44,735 Da (0 subunits)	Roggentin *et al.* (1987)
Vibrio cholerae	+	*85.6 kDA	Galen *et al.* (1992)
Bacteroides fragilis	+	92 kDa	Berg *et al.* (1983)
		165 kDa (3 subunits)	Tanaka *et al.* (1992)
Streptococcus pyrogenes	+	90 kDA	Davis *et al.* (1979)
Streptococcus peneumoniae	−	107 kDA	Lock *et al.* (1988)
Salmonella typhimurium	−	*41,300 Da	Hoyer *et al.* (1992)
Actinomyces viscosus	−	150 kDa	Teufel *et al.* (1989)
		*113 kDa	Henningsen *et al.* (1991)
		*92,871 Da	Yeung (1993)
Micromonospora viridifaciens	−	*68,840 Da	Aisaka *et al.* (1991)
Pasteurella haemolytica	+	170 kDa	Straus *et al.* (1993)
Arthrobacter ureafaciens	−	39 kDa (subtype I)	Uchida *et al.* (1979)
		51 kDa (subtyle II)	Uchida *et al.* (1979)

[a]Asterisks indicate molecular masses of enzyme polypeptides, calculated based on the nucleotide sequences.

the enzymes (Table II). Recently, it has been suggested that bacterial sialidases can be classified into two families depending on the molecular size and Ca^{2+} requirement (Taylor *et al.*, 1992a). Many sialidases that do not require the cation have molecular masses around 42 kDa and belong to a "small" sialidase family. They include enzymes from *Clostridium perfringens, C. sordellii, Salmonella typhimurium,* and *Micromonospora viridifaciens*. Significant sequence homology has been observed among these enzymes. Analysis of the primary structure of these enzymes suggests that the "small" sialidases share a similar tertiary structure (Hoyer *et al.*, 1992). On the other hand, *Vibrio cholerae* sialidase, which requires Ca^{2+} for full activity, has a molecular mass of 86 kDa and belongs to a "large" sialidase family. Sequence analysis reveals little homology with viral or "small" bacterial sialidases (Galen *et al.*, 1992).

3.3. Structure

Recently, sialidase genes have been cloned from several bacterial and fungal strains, including *Clostridium perfringens* (Roggentin *et al.*, 1988a, 1992), *Clos-*

tridium septicum (Rothe *et al.*, 1991), *Clostridium sordellii* (Rothe *et al.*, 1989), *Vibrio cholerae* (Vimr *et al.*, 1988; Galen *et al.*, 1992), *Bacteroides fragilis* (Russo *et al.*, 1990), *Streptococcus pneumoniae* (Berry *et al.*, 1988; Camara *et al.*, 1991), *Salmonella typhimurium* (Hoyer *et al.*, 1992), *Actinomyces viscosis* (Henningsen *et al.*, 1991; Yeung and Fernandez, 1991; Yeung, 1993), and *Micromonospora viridifaciens* (Sakurada *et al.*, 1992). The genes of bacterial sialidases have no significant sequence homology to influenza virus sialidase genes (Roggentin *et al.*, 1989; Rothe *et al.*, 1991). A comparison of the deduced amino acid sequences of bacterial sialidases also gives no indication of conserved histidine or glutamic acid residues, as demonstrated in certain viral enzymes (Roggentin *et al.*, 1992).

It has been demonstrated that there are short conserved nucleotide sequences in bacterial sialidases (Roggentin *et al.*, 1989). The enzymes from *Clostridium perfringens, C. sordellii, Vibrio cholerae,* and *Salmonella typhimurium* contain a sequence of 12 amino acids, which are usually repeated at four positions at distances that are also conserved. Of the 12 amino acids, 5 are highly conserved at defined positions, showing a sequence of Ser-X-Asp-X-Gly-X-Thr-Try. This sequence is often referred to as the "Asp" box. Subsequent studies have shown that this sequence is conserved in other bacterial enzymes (Russo *et al.*, 1990; Rothe *et al.*, 1991; Henningsen *et al.*, 1991; Sakurada *et al.*, 1992; Yeung, 1993). Interestingly, these conserved sequences are not restricted to bacterial sialidases, but are also found in sialidases from influenza A virus (Roggentin *et al.*, 1989). The "Asp" box has also been demonstrated in *Trypanosoma cruzi* sialidase (Pereira *et al.*, 1991) and a cytosolic sialidase from rat skeletal muscle (Miyagi *et al.*, 1993).

Despite the conservation of these sequences in many sialidase species, their function is still unknown. Previous studies with chemical modification of amino acid residues have demonstrated that the tryptophan or arginine residues may be involved in catalysis by sialidases from *Clostridium perfringens* or *Streptomyces griseus*, respectively, although the location of these amino acid residues has not been identified (Bachmayer, 1972; Kabayo and Hutchinson, 1975). Recently, the role of these sequences has been examined using *C. perfringens* sialidases containing one of three different site-specific mutations in the two conserved nucleotide regions (Roggentin *et al.*, 1992). Among these mutations, substitution of Arg-37 by Lys in a short conserved region upstream from the four repeated sequences causes significant changes in K_m, V_{max}, and K_i values with the sialidase inhibitor Neu5Ac2en; this suggests the involvement of this amino acid residue in substrate binding. Temperature- and pH-dependency of enzyme activity or reactivity of the enzyme to an antibody raised against the wild-type enzyme are also changed as a result of this mutation. The other mutations, located in the second repeated region, have less influence on the activity or immunoreactivity of the enzyme.

Recently, evidence has been presented suggesting that some bacterial sialidases may share similar secondary structures with the viral enzymes (Hoyer *et al.*, 1992). Photoaffinity labeling of the *Salmonella typhimurium* enzyme with an azido derivative of Neu5Ac2en reveals a secondary structural similarity of the labeled peptide to the active sites of influenza A and Sendai virus sialidases (Warner *et al.*, 1992). Preliminary crystallographic studies on bacterial sialidases have been reported for the *Vibrio cholerae, S. typhimurium* LT2 (Taylor *et al.*, 1992a), and *Micromonospora viridifaciens* enzymes (Taylor *et al.*, 1992b).

3.4. Biological Function

It has been postulated that sialidase may play important roles in the function of bacteria (for review see Corfield, 1992). The sialidase in enteric bacteria living in the large intestine hydrolyzes mucin oligosaccharides (and probably gangliosides), providing an energy source for the microorganisms. Sialic acid permease and acylneuraminate pyruvate lyase are assumed to serve for the uptake and utilization of the released sialic acid as a carbon source (Hoskins *et al.*, 1992; Vimr and Hoyer, 1992). Hydrolysis of sialoglycoconjugates by sialidase may be controlled, at least in part, by the content of *O*-acetylated sialic acid residues in mucin glycoproteins. The presence of the *O*-acetyl group in sialic acid is known to retard the rate of bacterial sialidase action (Corfield *et al.*, 1986, 1992).

The finding that a large number of pathogenic bacteria possess sialidase activity suggests the possible importance of this enzyme in bacterial infection (Müller, 1974; Corfield *et al.*, 1981a; Schauer, 1983). Nevertheless, previous studies have failed to establish a positive correlation between sialidase activity and the infectivity of bacteria including group B streptococcal (Mattingly *et al.*, 1980; Orefici *et al.*, 1984) and pneumococcal species (O'Toole and Stahl, 1975). Recently, evidence has been provided suggesting the possible involvement of *Pseudomonas aeruginosa* sialidase in respiratory diseases during the course of cystic fibrosis (Cacalano *et al.*, 1992).

Immunoassay of bacterial sialidases in biological materials using specific antibodies against the enzymes has recently been developed and successfully applied for the diagnosis of clostridial infection including gas gangrene (Roggentin *et al.*, 1988b, 1991, 1993).

4. SIALIDASES OF PROTOZOA AND INVERTEBRATES

Sialidases have been found in several protozoan species including *Trypanozoma cruzi* (Pereira, 1983; Pereira and Hoff, 1986), *T. vivax* (Esievo, 1983), *T. brucei* (Engstler *et al.*, 1992), *Naegleria fowleri* (Eisen and Franson, 1987), and *Acanthamoeba* (Pellegrin *et al.*, 1991). Among them, the enzyme in *T. cruzi*, the causative agent for Chagas' disease, has been characterized most extensively. The expression of sialidase activity in *T. cruzi* is regulated developmentally; the enzyme activity is highest in the highly infective, nonmultiplicative trypomastigotes, whereas the multiplicative epimastigote form within mammalian host cells contains only moderate levels of activity. No sialidase activity is detected in amastigotes, the multiplicative form in the insect vector (Pereira, 1983; Pereira and Hoff, 1986; Sauto-Padron *et al.*, 1990). The enzyme is anchored to the surface of the parasite via a glycosylphosphatidylinositol (GPI) anchor (Priori *et al.*, 1991; Rosenberg *et al.*, 1991). It has an optimal pH between 6.0 and 6.5 and actively hydrolyzes α1-acid glycoproteins, fetuin, and Neu5AcLac (Pereira, 1983).

T. cruzi sialidase is highly polymorphic. The enzyme from trypomastigotes contains a set of polypeptides that differ in size and number according to the strains and clones (Cavallesco and Pereira, 1988; Priori *et al.*, 1990). Although the reason for this diversity is unknown, it may arise from distinct enzyme genes or cotranslational modifications, or occur in the initiation step of protein synthesis at different methionine codons (Pereira *et al.*, 1991). Recently, the primary structure of *T. cruzi* sialidase was determined (Pereira *et al.*, 1991). The polypeptide consists of 1162 amino acids with four distinct domains. The cysteine-rich domain at the N-terminus contains a stretch of 332 amino acids nearly 30% identical to the *Clostridium perfringens* sialidase, three repeated motifs highly conserved in bacterial and viral sialidases ("Asp" box), and two segments with similarity to the YWTD repeats found in the low-density lipoprotein (LDL) receptor. This domain is connected by a structure characteristic of the type III modules of fibronectin to a long terminal repeat that contains a number of potential phosphorylation sites. The hydrophobic sequence at the C-terminus is assumed to anchor the enzyme to the cell membrane.

It is known that *T. cruzi* possesses a unique enzyme, trans-sialidase, which catalyzes the direct transfer of sialic acid among a variety of molecules without utilizing CMP-sialic acid (Previato *et al.*, 1985; Zingales *et al.*, 1987; Schenkman *et al.*, 1992; Frevert *et al.*, 1992; Vandekerckhove *et al.*, 1992; Ferrero-Garcia *et al.*, 1993; Scudder *et al.*, 1993). Recent evidence has suggested that both sialidase and trans-sialidase activities may be catalyzed by a single protein (Schenkman *et al.*, 1992; Parodi *et al.*, 1992; Uemura *et al.*, 1992). The two enzyme activities share similar properties in many respects, including inhibition by antibodies, anchorage by GPI, chromatographic behaviors, pH optimum, and heat stability. Furthermore, the sialidase and trans-sialidase reactions are coupled; increases in the concentration of acceptor molecules accelerate the trans-sialidase reaction concomitant with inhibition of the sialidase reaction (Schenkman *et al.*, 1992; Parodi *et al.*, 1992). There is also evidence suggesting that the shed-acute-phase-antigen molecule of *T. cruzi* carries both enzyme activities (Parodi *et al.*, 1992).

T. cruzi sialidase may play an important role in the infectivity of the parasite. Trypomastigotes are able to desialylate myocardial and vascular endothelial cells (Libby *et al.*, 1986), erythrocytes (Pereira, 1983), and lysosomal membrane glycoproteins (Hall *et al.*, 1992). Sialidase activity is correlated inversely with the virulence of the parasite; inhibitors of the enzyme enhance the *in vitro* infection of host cells by the parasite (Pereira and Hoff, 1986; Cavallesco and Pereira, 1988; Priori *et al.*, 1990).

Sialidase in *T. brucei,* the pathogen for African nagana, has been partially purified and characterized (Engstler *et al.*, 1992). Like *T. cruzi,* this enzyme is linked to the surface of the parasite via a GPI anchor. The enzyme hydrolyzes 4MU-Neu5Ac, Neu5AcLac, and glycoproteins more actively than gangliosides at neutral pH, showing preferential hydrolysis of α2-3-linked as compared with α2-6-linked sialic acids. The molecular mass is estimated to be 67 kDa.

Sialidase activity has been detected in certain invertebrate species. A recent study has demonstrated that the leech *Macrobdella decora* contains not only an ordinary sialidase, but also a novel sialidase which releases 2,7-anhydro-α-*N*-acetylneuraminic acid from various sialoglycoconjugates (Li *et al.*, 1990). The purified enzyme has a molecular mass of 84 kDa and shows strict specificity for the α2-3 linkage in Neu5AcLac and gangliosides; the enzyme does not cleave the α2-6 and α2-8 linkages (Chou *et al.*, 1992). The mechanism of action of this enzyme is unknown.

A soluble sialidase in the starfish *Asterias rubens* has been purified to homogeneity (Schauer and Wember, 1989). This enzyme has a molecular mass of 230 kDa with four identical subunits and preferentially cleaves α2-3-linked sialic acids in various substrates. Another enzyme in the hepatopancreas from the shrimp *Penaeus japonicus* has been shown to be a 32-kDa protein which binds β-galactosidase *in vitro.* No protective protein has been detected for this enzyme (Chuang and Yang, 1990) (see Section 5.1 for protective protein).

5. MAMMALIAN SIALIDASES

Compared to viral and bacterial sialidases, information about the properties and molecular structures of mammalian sialidases is still limited reflecting in part the difficulty encountered in the purification of the enzyme. In general, mammalian sialidases are unstable and, in the case of membrane-bound enzymes, they appear to become more labile when dissociated from the membrane. Since different isoenzymes with different subcellular localizations usually exist in a single tissue, the isolation of a pure subcellular organelle is required for the purification of a particular enzyme species. In addition, recent studies have suggested the existence of at least two different sialidase activities in individual subcellular organelles, e.g., cytosol (Venerando *et al.*, 1975), lysosomes (Miyagi and Tsuiki, 1984; Michalski *et al.*, 1986; Miyagi *et al.*, 1990a,b), and plasma membranes (Zeigler *et al.*, 1989; Miyagi *et al.*, 1990a,b; Chiarini *et al.*, 1990; Pitto *et al.*, 1992). No conclusive evidence, however, has been provided to show that the different activities represent distinct enzyme entities.

Sialidase activity of mammalian origin was first detected in commercial preparations of bovine and human glycoproteins (Warren and Spearing, 1960) and in the supernatants of tissue homogenates from different rat organs (Carubelli, 1962). Subsequently, the enzyme activity was shown in soluble and particulate fractions of various mammalian tissues (Taha and Carubelli, 1967; Mahadevan *et al.*, 1967; Leibovitz and Gatt, 1968; Horvat and Touster, 1968; Tettamanti and Zambotti, 1968). The activity associated with the particulate fraction was initially assumed to be of lysosomal origin (Taha and Carubelli, 1967; Mahadevan *et al.*, 1967; Horvat and Touster, 1968; Aronson and de Duve, 1968). Thereafter, sialidase activity was demonstrated to be present not only in the cytosol and lysosomes, but also in the plasma membranes of cells, such as brain synaptosomal membranes (Schengrund and Rosenberg, 1970; Tettamanti *et al.*, 1972) and liver plasma membranes (Schengrund *et al.*, 1972). The sialidase activities in the different subcellular fractions have been discriminated from each other based on their differences in pH dependency, effects of ions, kinetic parameters, substrate specificity, and stability (Visser and Emmelot, 1973). The existence of sialidase in the Golgi apparatus has also been suggested although additional confirmation has not been forthcoming (Kishore *et al.*, 1975).

5.1. Lysosomal Sialidases

Lysosomal sialidases in various animal and human tissues have been investigated. They include liver (Meyer *et al.*, 1981; Michalski *et al.*, 1982, 1986; Miyagi and Tsuiki, 1984), brain (Sato *et al.*, 1989; Fiorilli *et al.*, 1989, 1991; Miyagi *et al.*, 1990a,b; Pitto *et al.*, 1992), placenta (Verheijen *et al.*, 1985, 1987; Hiraiwa *et al.*, 1987, 1988; van der Horst *et al.*, 1989), mammary gland (Tulsiani and Carubelli, 1971), thyroid (van Dessel *et al.*, 1984), spermatozoal acrosomes (Srivastava and Abou-Issa, 1977), kidney (Sato *et al.*, 1989), testis (Verheijen *et al.*, 1982), salivary glands (Sato *et al.*, 1989), pulmonary parenchyma (Terdizis-Trabelsi *et al.*, 1991), fibroblasts (Hong *et al.*, 1980; Zeigler and Bach, 1981; Chigorno *et al.*, 1986; Lieser *et al.*, 1989; Zeigler *et al.*, 1989), leukocytes (Verheijen *et al.*, 1983; Schauer and Wember, 1984), lymphocytes (Yamada *et al.*, 1983), and macrophages (Pilatte *et al.*, 1987).

5.1.1. Sialidase–β-Galactosidase–Carboxypeptidase Complex

Recent studies have shown that the lysosomal sialidase exists as a complex with β-galactosidase, protective protein, and other proteins including *N*-acetyl-α-galactosaminidase (Verheijen *et al.*, 1982, 1985, 1987; Hoogeveen *et al.*, 1983; Tsuji *et al.*, 1989). The lysosomal enzyme has been identified as a 66-kDa protomer which consists of a 38-kDa core protein and 7 to 14 *N*-linked oligosaccharide chains (van der Horst *et al.*, 1989). The β-galactosidase protomer has a molecular mass of 63 kDa and exists in an aggregated form in the complex. The protective protein, which has recently been identified as carboxypeptidase, is a 54-kDa heterodimer of 32- and 20-kDa protomers linked together by a disulfide bond (Galjart *et al.*, 1988; van der Horst *et al.*, 1989). This protein is required for the multimerization of β-galactosidase as well as the expression and stabilization of sialidase activity; lack of this protein causes galactosialidosis, a lysosomal storage disease with a combined deficiency of sialidase and β-galactosidase activities (Hoogeveen *et al.*, 1981, 1983; d'Azzo *et al.*, 1982; Verheijen *et al.*, 1985; see Chapter 9). While the sialidase protomer dissociated from the complex is inactive, the activity can be restored by association with β-galactosidase and carboxypeptidase (van der Horst *et al.*, 1989; Scheibe *et al.*, 1990). The complex shows three different molecular forms with molecular masses of 679, 235, and 147 kDa, which are in equilibrium and are interchangeable; the sialidase activity is detected only in the largest form (Potier *et al.*, 1990a). A molecular structure was proposed for the sialidase–β-galactosidase–carboxypeptidase complex (Potier *et al.*, 1990a).

5.1.2. Multiforms of Lysosomal Enzymes

Recent studies have suggested that lysosomes may contain more than one sialidase species. Sialidase activity directed toward oligosaccharides or 4MU-Neu5Ac can be released from rat liver lysosomes by hypotonic shock, implying that this enzyme activity is probably located in the intralysosomal space, i.e., lysosomal matrix (Miyagi and Tsuiki, 1984). The enzyme from lysosomes of bovine testis (Verheijen *et al.*, 1982) and human placenta (Verheijen *et al.*, 1987; Hiraiwa *et al.*, 1987, 1988) can also be liberated by similar procedures, which imply an intra-lysosomal origin of the enzyme. On the other hand, much of the sialidase activity in human liver lysosomes remains in the lysosomal membranes after hypotonic treatment and requires detergents for solubilization of the activity, suggesting that this enzyme is more tightly bound to the lysosomal membranes (Michalski *et al.*, 1982, 1986). The membrane-bound form of lysosomal sialidase has been demonstrated in bovine thyroid (van Dessel *et al.*, 1984) and rat brain (Miyagi *et al.*, 1990a). Lysosomal membranes from human liver have been shown to contain two distinct sialidase activities which can be distinguished by their responses to the inhibitors, Neu5Ac2en and *N*-(4-nitrophenyl)-oxamic acid, as well as by substrate specificity. One preferentially hydrolyzes gangliosides and Neu5AcLac, while the other is more active toward fetuin and sialylhexasaccharides (Michalski *et al.*, 1986). Similarly, rat liver lysosomes also contain two sialidase activities which can be separated by AH-Sepharose column chromatography (Miyagi *et al.*, 1990b).

5.1.3. Sialidase Activity in Light and Heavy Lysosomal Fractions

Sialidase activities in light and heavy lysosomal fractions have been characterized in rat mammary gland (Tulsiani and Carubelli, 1971), human liver (Meyer *et al.*, 1981), and mouse brain (Fiorilli *et al.*, 1991). The heavy lysosomal fraction shows higher specific enzyme activity than the light fraction, but there is no difference in pH optimum, K_m value, or substrate specificity between the two fractions. In mouse brain, the sialidase activities in the heavy and light lysosomal fractions exhibit different developmental profiles, suggesting that the expression of these activities may be regulated independently (Fiorilli *et al.*, 1991).

5.1.4. Purification

Lysosomal sialidases have been purified from various tissues and cells, including rabbit spermatozoal acrosomes (Srivastava and Abou-Issa, 1977), bovine testis (Verheijen *et al.*, 1982), human liver (Michalski *et al.*, 1982), human leukocytes (Schauer and Wember, 1984), bovine thyroid (van Dessel *et al.*, 1984), rat liver (Miyagi and Tsuiki, 1984), human placenta (Verheijen *et al.*, 1985; Hiraiwa *et al.*, 1987, 1988), and rat brain (Miyagi *et al.*, 1990a). Rabbit spermatozoal sialidase has been extracted from inner acrosomal membranes and purified by about 50-fold to near homogeneity (Srivastava and Abou-Issa, 1977). Sialidase in human leukocytes has also been solubilized and purified to homogeneity by gel filtration and affinity chromatography (Schauer and Wember, 1984). The purified enzyme appears as a single band on SDS-PAGE. Although the subcellular localization of this enzyme has not been determined, it is most likely of lysosomal origin based on its kinetic properties and substrate specificity.

The membrane-bound lysosomal enzyme in rat brain has been solubilized from the particulate fraction and partially purified (Miyagi *et al.*, 1990a). The sialidase activity can be separated into two peaks, corresponding to sialidases I and II, by AH-Sepharose column chromatography. Using specific antibodies against the enzymes, it has been shown that sialidase I is localized in synaptosomal membranes, and sialidase II in both synaptosomal and lysosomal membranes. The latter species has been further purified by 2150-fold with a recovery of 0.7%.

Lysosomal sialidases are always copurified with β-galactosidase when isolated from tissues such as bovine testis (Verheijen *et al.*, 1982), porcine testis (Yamamoto and Nishimura, 1987), and human placenta (Verheijen *et al.*, 1985, 1987; Hiraiwa *et al.*, 1987, 1988). This sialidase–β-galactosidase complex can be isolated by affinity chromatography with β-galactosidase-specific columns, e.g., one linked with *p*-aminophenylthiogalactoside, resulting in significant increases in the specific activity of sialidase. Using this technique, the enzyme from human placenta has been purified as much as 54,600-fold (Verheijen *et al.*, 1985; Yamamoto and Nishimura, 1987; Hiraiwa *et al.*, 1988).

5.1.5. Molecular Mass

Different molecular masses have been reported for lysosomal sialidases from various tissues and animal species: 48.5 kDa for human leukocytes (Schauer and Wember, 1984), 56 to 60 kDa for rat liver (Beauregard and Potier, 1982; Miyagi and Tsuiki, 1984), and 70 kDa for human liver (Michalski *et al.*, 1982) and rat brain (Miyagi *et al.*, 1990a). Analysis of human placental sialidase using a photoaffinity labeling technique gave a molecular mass of 61 kDa (Warner *et al.*, 1990), which is comparable to the value of 66 kDa obtained with SDS-PAGE (van der Horst *et al.*, 1989). Using the same techniques, the molecular mass of the bovine testis enzyme was determined to be 55 kDa (van der Horst, 1990b). Higher values (240 and 203 kDa) for human fibroblast enzymes as determined by the radiation inactivation method may imply that the enzymes exist in the form of complexes under certain conditions (Hong *et al.*, 1980).

5.1.6. Kinetic Properties

Lysosomal sialidase is optimally active at acidic pH. In most cases, the pH optimum determined with Neu5AcLac or 4MU-Neu5Ac is between 4.0 and 4.5 (Tulsiani and Carubelli, 1971; Srivastava and Abou-Issa, 1977; Meyer *et al.*, 1981; Verheijen *et al.*, 1982, 1985; Michalski *et al.*, 1982, 1986; van Dessel *et al.*, 1984; Hiraiwa *et al.*, 1987, 1988; Sato *et al.*, 1989). This may vary slightly depending on assay conditions (Miyagi and Tsuiki, 1984; Schauer and Wember, 1984; Terzidis-Trabelsi *et al.*, 1991). The K_m values with Neu5AcLac as the substrate usually range between 0.4 and 2.0 mM (Tulsiani and Carubelli, 1971; Srivastava and Abou-Issa, 1977; Miyagi and Tsuiki, 1984; van Dessel *et al.*, 1984), although higher values have been reported for human liver (17 mM) (Meyer *et al.*, 1981) and leukocytes (7.9 mM) (Schauer and Wember, 1984). The K_m values obtained with 4MU-Neu5Ac are lower, ranging from 0.1 to 0.4 mM (van Dessel *et al.*, 1984; Pilatte *et al.*, 1987; Terzidis-Trabelsi *et al.*, 1991).

Regarding the effect of Ca^{2+} on the activity of lysosomal enzymes, conflicting results have been reported in the literature. Ca^{2+} has been reported to have no effect on the activity of the enzyme from rat mammary gland (Tulsiani and Carubelli, 1971), rat liver (Miyagi and Tsuiki, 1984), or rat brain (Miyagi *et al.*, 1990a), but, at a concentration of 1 mM, slightly stimulates sialidase activity in human leukocytes (Schauer and Wember, 1984) and human liver (Michalski *et*

Table III
Substrate Specificity of Lysosomal Sialidases

Substrates	Sialidase activity[a]					
	Rat liver[b]	Human liver[c]	Porcine testis[d]	Human placenta[e]	Rat brain[f]	Human leucocyte[g]
Oligo- and polysaccharides						
Neu5AcLac (α2-3,α2-6)	—	—	100	—	—	—
II^3Neu5AcLac	100	100	—	100	100	100
II^3Neu5GcLac	—	86	—	—	—	52
II^6Neu5AcLac	66	33	—	36	41	18
IV^3Neu5AcLc_4	—	119	—	—	—	—
IV^6Neu5AcLc_4	—	86	—	—	—	—
Sialylhexasaccharide I (α2-3)	—	138	—	—	—	—
Sialylhexasaccharide II (α2-6)	—	95	—	—	—	—
Colominic acid (α2-8)	39	100	108	57	—	—

Glycoproteins						
Fetuin (α2-3>α2-6)	2	310	41	20	64	90
α1-Acid glycoprotein (α2-6>α2-3)	1	—	—	7	21	—
Transferrin (α2-6)	1	—	—	7	—	—
Submaxillary mucin (α2-6)	0	90	5	0	31	—
Gangliosides						
Ganglioside mixtures	0	271	3	—	—	9
II^3Neu5AcLacCer (GM3, *N*-acetyl)	—	476	—	441	164	—
II^3Neu5GcLacCer (GM3, *N*-glycolyl)	—	252	—	—	97	—
II^3Neu5AcGg$_3$Cer (GM2)	—	—	—	57	84	—
II^3Neu5AcGg$_4$Cer (GM1)	—	33	—	59	16	—
IV^3Neu5AcII^3Neu5AcGg$_4$Cer (GD1a)	—	—	—	197	179	—
II^3(Neu5Ac)$_2$Gg$_4$Cer (GD1b)	—	—	—	147	—	—
Synthetic						
4MU-Neu5Ac	201	—	214	—	95	200

[a]Each value represents a relative sialidase activity when the activity directed toward II^3Neu5AcLac or Neu5AcLac is regarded as 100.
[b]Miyagi and Tsuiki (1984).
[c]Michalski *et al.* (1986).
[d]Yamamoto and Nishimura (1987).
[e]Hiraiwa *et al.* (1987).
[f]Miyagi *et al.* (1990a).
[g]Schauer and Wember (1984).

al., 1986). The effect of Cu^{2+} also varies. While Cu^{2+} strongly inhibits enzyme activity in bovine thyroid (van Dessel *et al.*, 1984) or human liver (Michalski *et al.*, 1986), it has almost no effect on sialidases from rat mammalian gland (Tulsiani and Carubelli, 1971), human leukocytes (Schauer and Wember, 1984), and rat brain (Miyagi *et al.*, 1990a).

5.1.7. Substrate Specificity

Earlier studies indicated that lysosomal sialidase preferentially hydrolyzes oligosaccharide and glycoprotein substrates, and is less active toward ganglioside substrates (Rosenberg and Schengrund, 1976). Subsequent studies with partially purified enzyme preparations have demonstrated the diverse substrate specificity of the enzyme (Table III). Both Neu5AcLac and 4MU-Neu5Ac are effectively hydrolyzed by most lysosomal enzyme species. However, the susceptibility of glycoproteins to the enzyme is variable. For example, the enzyme from rat liver can hardly hydrolyze fetuin or submaxillary mucin, whereas the counterpart from human liver is active toward both substrates. The susceptibility of gangliosides to lysosomal sialidases also differs among the various enzyme species (Chigorno *et al.*, 1986; Fiorilli *et al.*, 1989; Lieser *et al.*, 1989; Zeigler *et al.*, 1989; Pitto *et al.*, 1992). The activity of ganglioside sialidase in lysosomes from rat brain and cultured rat granule cells can be distinguished from the activity in plasma membranes based on differences in pH optimum, kinetic parameters, responses to detergents, and stability (Fiorilli *et al.*, 1989; Pitto *et al.*, 1992). Thus, treatment of cultured human fibroblasts with Cu^{2+} at millimolar concentrations totally inactivates the activity of ganglioside sialidase in plasma membranes, but has no effect on the activity in the lysosomal fraction (Lieser *et al.*, 1989). In human fibroblasts, the plasma membrane sialidase activity toward GM3 is stimulated by Triton X-100, whereas the activity of the lysosomal enzyme is enhanced by sodium cholate (Lieser *et al.*, 1989) or sodium glycodeoxycholate (Schneider-Jakob and Cantz, 1991). The sialidase activities directed toward ganglioside as well as oligosaccharide and glycoprotein substrates are drastically reduced in fibroblasts from patients with sialidosis or galactosialidosis, suggesting that these activities may be contained within a single enzyme entity (Lieser *et al.*, 1989; Zeigler *et al.*, 1989; Schneider-Jakob and Cantz, 1991).

Recently, the lysosomal sialidase was shown to require an activator protein for the hydrolysis of gangliosides; the degradation of various gangliosides is accelerated by the presence of the sulfatide activator protein, a physiological lipid solubilizer. This effect is different from that rendered by detergents such as taurodeoxycholate (Fingerhut *et al.*, 1992). This activator protein appears to be different from another ganglioside-sialidase-activating protein which does not accelerate the hydrolysis of GM1 by β-galactosidase (Ben-Yoseph *et al.*, 1991).

The existence of an activator protein for stimulating the activity of oligosaccharide sialidase has also been described (Chabas *et al.*, 1987).

Most lysosomal sialidases preferentially catalyze the hydrolysis of the α2-3-linked sialic acids compared to the α2-6 residues. Exceptions to this are the enzymes from rabbit spermatozoal acrosomes (Srivastava and Abou-Issa, 1977) and human liver (Michalski *et al.*, 1982), which are more active on α2-6-linked sialic acids on glycoproteins, i.e., Cowper's gland mucin or submandibular gland mucin. The enzymes from human liver and porcine testis are more active on α2-8-linked sialic acids in colominic acid than other lysosomal enzymes (Table III).

5.1.8. Structure

No information is available concerning the molecular structure of lysosomal sialidase. It was recently suggested that prosaposin, a 70-kDa precursor of activator proteins, may also be the precursor for lysosomal sialidase. Significant structural similarities have been observed between activator protein 1 and influenza virus sialidase (Potier, 1988). An antiprosaposin immunoglobulin preparation cross-reacted with a 60-kDa protein, identified as sialidase in a human placenta sialidase preparation (Potier *et al.*, 1990b). However, this claim has been challenged by several groups. The 60-kDa protein was found to react with an anti-IgG antibody present in the antiprosaposin antibody preparation. This antibody was identified as IgG heavy chain protein Tie (Hiraiwa *et al.*, 1991). Furthermore, a normal sialidase level has been observed in fibroblasts from patients with a deficiency of activator proteins 1 and 2 (Paton *et al.*, 1992).

5.2. Plasma Membrane-Bound Sialidase

The activity of plasma membrane-bound sialidase has been characterized using membrane preparations from various tissues and cells, including liver (Schengrund *et al.*, 1972; Visser and Emmelot, 1973; Miyagi and Tsuiki, 1986; Miyagi *et al.*, 1990b; Sagawa *et al.*, 1990), thyroid gland (van Dessel *et al.*, 1984), fibroblasts (Zeigler and Bach, 1981; Chigorno *et al.*, 1986; Lieser *et al.*, 1989; Zeigler *et al.*, 1989; Schneider-Jakob and Cantz, 1991), synaptosomal plasma membranes (Schengrund and Rosenberg, 1970; Tettamanti *et al.*, 1972; Schengrund and Nelson, 1975; Yohe and Rosenberg, 1977, 1978; Cruz and Gurd, 1978, 1983; Moran *et al.*, 1986; Chiarini *et al.*, 1990; Miyagi *et al.*, 1990a; Saito *et al.*, 1994), myelin (Yohe *et al.*, 1983, 1986; Saito and Yu, 1986, 1992, 1993), cultured cerebellar granule cells (Pitto *et al.*, 1992), erythrocytes (Sagawa *et al.*, 1990; Chiarini *et al.*, 1993), and lymphocytes (Yamada *et al.*, 1983).

5.2.1. Localization of Plasma Membrane-Bound Enzymes

There is ample evidence suggesting that the plasma membrane-bound sialidase may exist as an ectoenzyme on the cell surface. Treatment of transformed hamster embryo fibroblasts with the diazonium salt of sulfanilic acid, a non-penetrating inhibitor for various ectoenzymes, resulted in a 90% loss of 5′-nucleotidase activity and 40% loss of sialidase activity directed toward gangliosides with no concomitant decrease in the activities of lactate dehydrogenase and succinate dehydrogenase (Schengrund *et al.*, 1976). Similarly, treatment of human fibroblasts with Cu^{2+} produced a total loss in the sialidase activity in plasma membranes, but not in lysosomes (Lieser *et al.*, 1989). It has also been shown that the sialidase on the surface of a cultured fibroblast cell can cleave the sialic acid residues of sialoglycoconjugates on the surface of an adjacent cell, resulting in changes in cell-to-cell contact (packing of the cells). The hydrolytic reaction in the packed cells is totally inhibited by pretreatment of the cells with Hg^{2+} (Rosenberg, 1981).

Recent studies have suggested that certain plasma membrane-bound sialidases are linked to the membranes via a GPI anchor, as observed with the sialidase from *T. cruzi* (Rosenberg *et al.*, 1991). Phosphatidylinositol-specific phospholipase C (PIPLC) released sialidase activity (28% of total) from purified synaptosomal membranes of pig brain, but not from lysosomal membranes (Chiarini *et al.*, 1990). The presence of GPI-anchored sialidase was also shown in human erythrocyte membranes (Chiarini *et al.*, 1993). On the other hand, the sialidase activity in rat brain myelin could not be solubilized with PIPLC despite a concomitant release of alkaline phosphatase activity (about 40% of total), suggesting a different mode of association of the enzyme with the membranes (Saito and Yu, unpublished data).

5.2.2. Reactions with Ganglioside Substrates

Differing from ordinary enzyme reactions with soluble enzymes and hydrophilic substrates, sialidase reactions with membrane preparations and lipid substrates show unique kinetic features. The ganglioside substrates are thought first to be absorbed to the membranes, then incorporated into the membranes, and finally to gain access to the enzyme through lateral diffusion (Sandhoff and Pallmann, 1978; Scheel *et al.*, 1985). Thus, membrane-bound sialidase-catalyzed hydrolysis of gangliosides, but not of other hydrophilic substrates, can be accelerated by various membrane-fluidizing agents that enhance lateral diffusion of the substrate molecules (Sandhoff *et al.*, 1976; Sandhoff and Pallmann, 1978; Scheel *et al.*, 1982, 1985). Increases in pH and ionic strength inhibit the hydrolysis of gangliosides by interfering with the absorption process of the substrates (Scheel *et al.*, 1982, 1985).

5.2.3. Multiforms of Plasma-Membrane Enzymes

Recent studies have suggested the existence of more than one type of sialidase activity in plasma membranes of various cell types. The sialidase activity in the plasma membranes of human fibroblasts has been separated into two different activities using Percoll gradient centrifugation; one was activated by cholate and the other partially inhibited by the same detergent (Zeigler *et al.*, 1989). The pH–activity curve for the sialidase activity in purified plasma membranes from rat cultured granule cells gave two separate peaks with pH optima of 3.9 and 6.0 with GD1a and of 3.9 and 5.4 with 4MU-Neu5Ac (Pitto *et al.*, 1992). Similarly, two sialidase activities with different pH optima (pH 4.2 and 6.6) have been observed in an enzyme preparation solubilized from pig brain membranes using PIPLC (Chiarini *et al.*, 1990). A rat synaptosomal membrane fraction contained two immunologically distinct sialidase activities (Miyagi *et al.*, 1990a). It is, however, still unclear whether these sialidase activities represent distinct enzyme entities.

5.2.4. Uneven Distribution of Sialidase Activity

There is evidence suggesting an uneven distribution of sialidase activity in plasma membranes. The enzyme activity in synaptosomal membranes is significantly higher than the activity in neuronal perikarya, suggesting an enrichment of the enzyme activity in nerve ending plasma membranes (Schengrund and Nelson, 1976; Saito *et al.*, 1995). The sialidase activity in myelin subfractions from rat brain is also unevenly distributed, and its distribution is apparently regulated developmentally; the activity is enriched in the compact myelin fraction during the early phase of myelinogenesis and, thereafter, becomes concentrated in the heavy myelin fraction (Saito and Yu, 1992).

5.2.5. Purification

To date, none of the plasma membrane-bound sialidases has been purified to homogeneity. Partial purification has been achieved for the rat brain (Miyagi *et al.*, 1990a) and erythrocyte enzymes (Sagawa *et al.*, 1990). The sialidase activity solubilized from the particulate fraction of rat brain has been separated into two species, sialidases I and II, by AH-Sepharose column chromatography. Sialidase I has been further purified using two hydrophobic columns and gel filtration to obtain an overall purification of 450-fold. The enzyme has a molecular mass of 70 kDa as determined by gel filtration (Miyagi *et al.*, 1990a). The sialidase in human erythrocyte membranes has been purified by 30-fold (Sagawa *et al.*, 1990).

5.2.6. Kinetic Properties

The pH optimum of plasma membrane-bound sialidases with ganglioside substrates generally is between 4.5 and 5.5, as reported for rat liver (Miyagi and Tsuiki, 1986), mouse and rat synaptosomal membranes (Fiorilli *et al.*, 1989; Miyagi *et al.*, 1990a), rat brain myelin (Yohe *et al.*, 1986), and rat erythrocyte ghost (Sagawa *et al.*, 1990). Slightly lower values ranging from pH 4.0 to 4.5 have been reported for the enzyme activities in mouse synaptosomal membranes (Saito *et al.*, 1995), rat liver plasma membranes (Visser and Emmelot, 1973), and human erythrocyte enzyme solubilized with PIPLC (pH 4.2) (Chiarini *et al.*, 1993). The observed K_m values with ganglioside substrates are generally within 0.1 to 0.3 mM for the enzyme activities from mouse and rat synaptosomal membranes (Fiorilli *et al.*, 1989; Miyagi *et al.*, 1990a) and erythrocyte membranes (Chiarini *et al.*, 1993). However, much lower values (around 10 μM) have also been reported for the activities in rat brain myelin (Yohe *et al.*, 1986) and mouse synaptosomal membranes (Saito *et al.*, 1994). The effects of Ca^{2+} on the activity of plasma membrane-bound enzymes are varied, depending on the tissue source (Schengrund and Nelson, 1976; Yohe *et al.*, 1986; Miyagi *et al.*, 1990a).

5.2.7. Substrate Specificity

It has generally been assumed that gangliosides are the best substrates for the plasma membrane-bound sialidase. This, however, is not always the case (Table IV). Sialidases in the plasma membranes of rat liver, erythrocytes, or synaptosomes have been shown to be active exclusively on ganglioside substrates; the hydrolytic rates of Neu5AcLac by these enzyme species are less than 10% of the corresponding values for GM3. On the other hand, myelin-associated sialidase in rat brain hydrolyzed neuramin-lactitol and GM3 at comparable rates. Evidence has been provided suggesting that the two myelin-associated sialidase activities are likely catalyzed by a single enzyme entity (Saito and Yu, 1986). The plasma membrane enzyme in human lymphocytes is more active toward neuramin-lactitol and 4MU-Neu5Ac than GM3 (Yamada *et al.*, 1983). The synaptosomal enzyme appears to be more active toward α2-3-linked sialic acids than the α2-6-linked isomer (Miyagi *et al.*, 1990a).

Sialidase activities directed toward ganglioside substrates in plasma membrane and lysosomal fractions can be distinguished based on their responses to detergents and inhibitors. The requirement of detergents significantly differs between both enzyme activities (see Section 5.1). Heparin or heparin sulfate almost completely inhibits the GM3 sialidase activity in the plasma membrane preparation, but only slightly affects the lysosomal enzyme activity. The inhibition of the plasma membrane enzyme, but not the lysosomal enzyme, can be

Table IV
Substrate Specificity of Plasma Membrane-Bound Sialidases

	Sialidase activity[a]			
Substrates	Rat liver[b]	Human brain synaptosomes[c]	Rat brain myelin[d]	Human erythrocytes[b]
Gangliosides				
II^3Neu5AcLacCer (GM3, *N*-acetyl)	100	100	100	100
II^3Neu5GcLacCer (GM3, *N*-glycolyl)	—	58	—	—
II^3Neu5AcGg$_3$Cer (GM2)	—	17	40	—
II^3Neu5AcGg$_4$Cer (GM1)	11	7	4	14
IV^3Neu5AcII^3Neu5AcGg$_4$Cer (GD1a)	99	97	—	97
II^3(Neu5Ac)$_2$LacCer (GD3)	—	92	—	—
Glycoproteins				
Fetuin (α2-3>α2-6)	0	5	46	0
α1-Acid glycoprotein (α2-6>α2-3)	3	3	—	2
Submaxillary mucin (α2-6)	—	2	—	—
Glycophorin	—	20	—	—
Oligosaccharides				
Neu5AcLac (α2-3,α2-6)	3	—	—	3
II^3Neu5AcLac	—	7	95[e]	—
Synthetic				
4MU-Neu5Ac	8	9	—	4

[a]Each value represents a relative sialidase activity when the activity directed toward II^3Neu5AcLacCer is regarded as 100.
[b]Sagawa *et al.* (1990).
[c]Miyagi *et al.* (1990a).
[d]Saito and Yu (1986).
[e]*N*-acetylneuraminyl(α2-3)lactitol was used instead of Neu5AcLac.

reversed by dilution (Schneider-Jakob and Cantz, 1991). In human skin fibroblasts and rat brain, Cu^{2+} (1 mM) strongly inhibits the activity of plasma membrane-bound enzymes toward ganglioside substrates, but only slightly affects the activity of lysosomal enzymes (Lieser *et al.*, 1989; Miyagi *et al.*, 1990a). Cu^{2+}, however, only mildly inhibits the sialidase activity in rat brain myelin (Yohe *et al.*, 1986).

5.3. Cytosolic Sialidase

Cytosolic sialidase has been investigated in various tissues, including brain (Ohmann *et al.*, 1970; Venerando *et al.*, 1975, 1978, 1982, 1983; Miyagi *et al.*, 1990b), liver (Taha and Carubelli, 1967; Meyer *et al.*, 1981; Miyagi and Tsuiki, 1984, 1985; Dairaku *et al.*, 1986; Miyagi *et al.*, 1990b), skeletal muscle (Dairaku *et al.*, 1986; Miyagi *et al.*, 1990b, 1993), kidney (Kuratowska and Kubicka, 1967; Tettamanti and Zambotti, 1968; Dairaku *et al.*, 1986), mammary gland

(Tulsiani and Carubelli, 1971), fibroblasts (Zeigler and Bach, 1981), alveolar macrophages (Pilatte *et al.*, 1987), salivary glands (Sato *et al.*, 1989), and pulmonary parenchyma (Terzidis-Trabelsi *et al.*, 1991). The distribution of the cytosolic sialidase activity in various rat tissues has been examined in detail (Dairaku *et al.*, 1986; Sato *et al.*, 1989). The sialidase activity (units/mg protein) in individual tissues is reported to be 1.1 for heart and testis, 1.9 for liver and stomach, 2.3 for intestine, 6.0 for brain, 21.9 for kidney, and 31.2 for skeletal muscle (Dairaku *et al.*, 1986). In a separate study, a higher level of the activity is found in liver than in brain tissues (Sato *et al.*, 1989). No activity has been detected in lung, spleen (Dairaku *et al.*, 1986), or erythrocytes (Sagawa *et al.*, 1990).

Recently, a specific antibody against the purified cytosolic sialidase from rat skeletal muscle was prepared and its immunoreactivity with other sialidases examined (Miyagi *et al.*, 1990b). The antibody completely immunoprecipitates the cytosolic enzyme in rat liver and skeletal muscle, but not the synaptosomal and lysosomal enzymes in rat brain and liver. These results suggest that a single cytosolic sialidase species may exist in rat tissues which is immunologically distinct from the lysosomal and plasma membrane enzymes.

5.3.1. Intracellular Distribution

In brain, the activity of cytosolic sialidase appears to be enriched in nerve endings (Venerando *et al.*, 1978). Two soluble fractions have been prepared from pig brain, one by osmotic shock of the isolated synaptosomes (nerve ending cytosol) and the other by mild homogenization of brain tissues (neuronal perikaryal/glial cell cytosol). The specific activity of the sialidase in nerve ending cytosol is 15 to 20 times higher than that in neuronal perikarya/glial cell cytosol. While both enzyme activities have similar pH dependency and K_m values, their stability at 4°C or on freezing and thawing differs (Venerando *et al.*, 1987). In rat brain, the activities of the two cytosolic sialidases exhibit different developmental profiles (Venerando *et al.*, 1982, 1983).

It has been reported that a neutral sialidase activity is associated with rat liver microsomal membranes and this activity can be released from the membranes with 0.5 M NaCl. The solubilized enzyme is indistinguishable from the cytosolic enzyme with respect to pH optima and K_m values. Isotonic sucrose solution does not solubilize the enzyme. It appears that this enzyme may represent cytosolic enzyme which is adsorbed onto the membranes (Miyagi and Tsuiki, 1985).

5.3.2. Multiforms of Cytosolic Enzymes

Previous studies have demonstrated the presence of two different forms of sialidase in the cytosol of pig brain (Venerando *et al.*, 1975). The sialidase activity in the supernatant of tissue homogenates can be separated into two peaks, corresponding to sialidases A and B, by hydroxylapatite and cellulose gel chromatography. These two sialidase activities differ in pH optima (4.7 and 4.9 for sialidases A and B, respectively), effects of ions, and K_m values (2.2 and 0.46 mM, respectively, with Neu5AcLac as the substrate). The existence of different cytosolic sialidases in other tissues or species has not been reported.

5.3.3. Formation of Enzyme–Ganglioside Complex

Partially purified cytosolic sialidase from pig brain interacts with micellar GT1b on incubation at 37°C and pH 4.8, forming two molecular sizes of noncatalytic enzyme–ganglioside complexes (Venerando *et al.*, 1985, 1987). The smaller complex has a molecular mass of 420 kDa and consists of one ganglioside micelle and two enzyme proteins. The larger complex appears to be a dimer of two smaller ones. These two forms are in equilibrium with each other. Formation of an enzyme–ganglioside complex has also been observed with GD1a, GD1b, and GM1. The ability of the sialidase to bind gangliosides is corroborated by the recent finding that a specific interaction exists between myelin-associated sialidase and GM1 (Saito and Yu, 1993; see Section 5.4.).

5.3.4. Purification

The cytosolic sialidase from rat liver and skeletal muscle has been purified to homogeneity. The rat liver enzyme is purified 83,000-fold by procedures including Blue-Sepharose and heparin-Sepharose column chromatography (Miyagi and Tsuiki, 1985). The skeletal muscle enzyme has been purified 22,700-fold (Miyagi *et al.*, 1990b). In both cases, the purified enzyme appears as a single protein band on SDS-PAGE analysis. The brain enzyme (sialidase A) has been purified 2000-fold from pig brain by procedures including Affi-Gel Blue chromatography (Venerando *et al.*, 1985), and the kidney enzyme purified 3700-fold by ammonium sulfate fractionation and gel filtration (Kuratowska and Kubicka, 1967).

5.3.5. Molecular Mass

The reported molecular mass of cytosolic sialidase is smaller than those of the lysosomal and plasma membrane enzymes, ranging from 40 kDa for the rat brain (Venerando *et al.*, 1985) to 43 kDa for the rat liver enzyme (Miyagi and Tsuiki, 1985). A slightly higher value (56 kDa) has also been reported for the latter using the radiation inactivation method (Beauregard and Potier, 1982). The molecular mass of the cytosolic sialidase from rat skeletal muscle, previously reported as 43 kDa based on SDS-PAGE (Miyagi *et al.*, 1990b), has recently been calculated as 42,381 Da from the deduced amino acid sequence (Miyagi *et al.*, 1993).

5.3.6. Kinetic Properties

The optimal pH for the cytosolic sialidase with Neu5AcLac or 4MU-Neu5Ac generally ranges between 5.0 and 6.0 (Tulsiani and Carubelli, 1971; Miyagi and Tsuiki, 1985; Dairaku *et al.*, 1986; Pilatte *et al.*, 1987; Sato *et al.*, 1989; Miyagi *et al.*, 1990b). Lower values have been reported for the enzyme from rat brain (pH 4.7–4.9) (Venerando *et al.*, 1975), rat kidney (pH 4.7) (Tettamanti and Zambotti, 1968), pulmonary parenchyma (pH 4.6) (Terzidis-Trabelsi *et al.*, 1991), and human liver (pH 4.0) (Meyer *et al.*, 1981). The K_m values with Neu5AcLac as the substrate vary widely from as low as 0.46 to 8 mM, depending on the source of the enzyme (Tulsiani and Carubelli, 1971; Venerando *et al.*, 1975; Meyer *et al.*, 1981; Miyagi and Tsuiki, 1985). The K_m values with 4MU-Neu-5Ac as the substrate range from 0.1 to 0.7 mM (Miyagi and Tsuiki, 1985; Pilatte *et al.*, 1987; Saito *et al.*, 1989; Terzidis-Trabelsi *et al.*, 1991).

Ca^{2+} has no effect on the enzymes from the mammary gland (Tulsiani and Carubelli, 1971), brain (sialidase A) (Venerando *et al.*, 1975), or liver (Miyagi and Tsuiki, 1985). Brain sialidase B, however, is slightly activated by the cation (Venerando *et al.*, 1975). Cu^{2+} at 1 mM strongly inhibits the activity of most mammalian enzymes (Tulsiani and Carubelli, 1971; Miyagi and Tsuiki, 1985).

5.3.7. Substrate Specificity

Substrate specificity of the cytosolic sialidase from rat liver, kidney, and skeletal muscle is summarized in Table V. The enzyme from rat liver is active toward 4MU-Neu5Ac, sialyloligosaccharides, GM3, and, to a lesser degree, glycoprotein substrates. The enzyme preferentially hydrolyzes the α2-3-linked sialic acid residues as compared with the α2-6- and α2-8-linked residues.

5.3.8. Structure

Recently, a cDNA clone encoding the cytosolic sialidase of rat skeletal muscle was prepared and expressed in *Escherichia coli* (Miyagi *et al.*, 1993). The nucleotide sequence encodes 379 amino acid residues with a calculated molecular mass of 42,381 Da. The deduced amino acid sequence does not resemble any of the viral, bacterial, or parasitic sialidases, although it contains two "Asp" blocks, which are conserved in these enzymes (Roggentin *et al.*, 1989).

Table V
Substrate Specificity of Cytosolic Sialidases

Substrates	Sialidase activity[a]		
	Rat liver[b]	Rat muscle[c]	Rat kidney[c]
Oligo- and polysaccharides			
Neu5AcLac (α2-3,α2-6)	—	100	100
II³Neu5AcLac	100	—	—
II³Neu5GcLac	15	—	—
Fetuin glycopeptides	25	22	49
Colominic acid (α2-8)	18	—	—
Glycoproteins			
Fetuin (α2-3>α2-6)	33	50	0
α1-Acid glycoprotein (α2-6>α2-3)	48	—	—
Submaxillary mucin (α2-6)	0	—	—
Gangliosides			
Ganglioside mixtures	62	58	0
II³Neu5AcLacCer (GM3, *N*-acetyl)	180	—	—
II³Neu5GcLacCer (GM3, *N*-glycolyl)	—	—	—
II³Neu5AcGg$_3$Cer (GM2)	0	—	—
II³Neu5AcGg$_4$Cer (GM1)	0	—	—
IV³Neu5AcII³Neu5AcGg$_4$Cer (GD1a)	71	—	—
II³(Neu5Ac)$_2$Gg$_4$Cer (GD1b)	54	—	—
Synthetic			
4MU-Neu5Ac	190	190	435

[a]Each value represents a relative sialidase activity when the activity directed toward II³Neu5AcLac or Neu5AcLac is regarded as 100.
[b]Miyagi and Tsuiki (1985).
[c]Dairaki *et al.* (1986).

5.4. Biological Function

A large body of evidence suggests that sialidase participates in diverse functions of the cell. One of its most important functions is the catabolism of sialoglycoconjugates. Lysosomal sialidase is assumed to play a central role in the degradation of sialocompounds in cells, as shown by sialidosis (sialidase deficiency) or galactosialidosis (combined deficiency of sialidase and β-galactosidase activities), both of which are characterized by the accumulation and excretion of excess sialyloligosaccharides (for review see Suzuki, 1993, and Chapter 9). Interestingly, the ganglioside levels in brain and other tissues are not elevated in these disorders. Although it was previously postulated that a deficiency of lysosomal ganglioside sialidase might be the primary defect in mucolipidosis IV (Bach *et al.*, 1979; Ben-Yoseph *et al.*, 1982), recent studies seem to refute this concept as a normal level of sialidase activity was found in lysosomes and plasma membranes of fibroblasts from patients with this disorder (Lieser *et al.*, 1989; Schneider-Jakob and Cantz, 1991; see Chapter 9 for further details). Thus, the exact role of lysosomal sialidase in the catabolism of gangliosides is still unclear.

The plasma membrane-bound enzyme plays an important role in the degradation of membrane gangliosides. *In vitro* incubation of plasma membrane preparations leads to hydrolysis of membrane gangliosides by endogenous sialidase (Heijlman and Roukema, 1972; Tettamanti *et al.*, 1975; Yohe and Rosenberg, 1977; Saito and Yu, 1992). When exogenous gangliosides are incorporated into the plasma membranes of cerebellar granule cells, sialidase-sensitive gangliosides such as GD1a and GD1b are rapidly degraded by plasma membrane-bound sialidase, leading to the accumulation of GM1 which is not attacked by the enzyme (Riboni *et al.*, 1991). Evidence was also presented suggesting a crucial role of myelin-associated sialidase in the developmental changes of ganglioside composition in brain myelin (Saito and Yu, 1992). While plasma membrane-bound sialidase can hydrolyze membrane sialoglycoproteins under *in vitro* conditions (Cruz and Gurd, 1978, 1983), its role in the catabolism of sialoglycoproteins in membranes still remains to be clarified.

It has been reported that treatment of cells with purified preparation of sialidase can affect a variety of cellular functions. The ensuing functional changes may result from the removal of terminal sialic acid residues or exposure of penultimate galactose residues. Thus, sialidase can modify receptor function (van Aswegen *et al.*, 1990; Hechler *et al.*, 1990; Haddad and Gies, 1992), cell growth and proliferation (Vaheri *et al.*, 1972; Nilsson *et al.*, 1982), cell differentiation (Wu and Ledeen, 1991), cell-to-cell binding and interactions (Schauer *et al.*, 1990; Guzman *et al.*, 1990; True *et al.*, 1990; Fischer *et al.*, 1991; Stenberg *et al.*, 1991; Domingues *et al.*, 1992), immunocyte function (Taira and Nariuchi, 1988; Houde and Arora, 1990; Nakamura *et al.*, 1991; Karasuno *et al.*, 1992), membrane fusion (Igarashi and Bando, 1990; Puri *et al.*, 1992), and membrane fluidity (Ando *et al.*, 1986; Ohbayashi *et al.*, 1987). It is also known that the growth of fibroblasts in culture may be controlled, at least partially, by sialidase that is released from the cells into the culture medium (Usuki *et al.*, 1988a,b; Ogura and Sweeley, 1992). Recently, it has been demonstrated that sialidase treatment modulates membrane permeability to Na^+ (Recio-Pinto *et al.*, 1990) and Ca^{2+} (Werner *et al.*, 1991; Wu and Ledeen, 1991; Yee *et al.*, 1991; Post, 1992). The effect of sialidase on ion permeability may be involved in the underlying mechanisms for some of the cellular events induced by sialidase.

Information about the function of sialidase has also been provided by examining sialidase activity during various cellular processes. Sialidase activity is increased in confluent and postconfluent S20Y murine neuroblastoma cells and may account for the reduction of total ganglioside sialic acid observed with increasing cell density (Schengrund and Repman, 1982). Sialidase activity is also stimulated during differentiation of various cell types, including human HL-60 promyelocytic leukemic cells (Nojiri *et al.*, 1982), mouse T lymphocytes (Landolfi *et al.*, 1985), and human monocytes (Lambre *et al.*, 1989/1990). In addition, sialidase activity toward GD1a in cultured rat cerebellar granule cells increased 100-fold during *in vitro* differentiation (Pitto *et al.*, 1989). A recent study demonstrated that the activation of human neutrophils with phorbol myristate acetate or other agents was accompanied by the redistribution of sialidase activity from intracellular stores to the cell surface (Cross and Wright, 1991). Changes in sialidase activity with neoplastic transformation have also been reported (Schengrund *et al.*, 1973; Sagawa *et al.*, 1988; Miyagi *et al.*, 1984, 1990c,d). Numerous studies have demonstrated a developmental change of the sialidase activity in a variety of tissues, cells, and subcellular organelles (Carubelli and Tulsiani, 1971; Schiller *et al.*, 1979; Segler-Stahl *et al.*, 1981; Wille and Trenkner, 1981; Cruz and Gurd, 1981, 1983; Venerando *et al.*, 1982, 1983; Moran *et al.*, 1986; Yohe *et al.*, 1986; Saito and Yu, 1992; Saito *et al.*, 1992, 1995). Of particular interest is that the developmental profiles of sialidase activity in oligodendroglial cells and myelin suggest the possible involvement of myelin-associated sialidase in the formation and maintenance of the myelin sheath (Saito and Yu, 1992; Saito *et al.*, 1992). No information is available about the function of cytosolc sialidase though it may play some role in the catabolism of sialoglycoproteins and gangliosides in the cytoplasm.

Recently, a novel function of sialidase as a "dynamic membrane adhesion molecule" has been proposed for myelin-associated sialidase (Saito and Yu, 1993). Myelin-associated sialidase degrades the complex ganglio-series gangliosides synthesized in oligodendroglial perikarya and converts them to GM1, which is a poor substrate of the enzyme. Moreover, rat oligodendroglial cells bind specifically with GM1 and this binding can be abolished by Neu5Ac2en, a specific sialidase inhibitor, suggesting that the binding is mediated by oligodendroglial plasma membrane surface sialidase. Following the slow hydrolysis of GM1, the product asialo-GM1 loses affinity for the enzyme, resulting in dissociation of two adjacent membranes. This observation is particularly interesting in that it not only accounts for the greatly simplified ganglioside pattern in myelin in which GM1 predominates, but also offers a mechanism for the formation of the multilamellar structure of myelin during myelinogenesis. Interactions with GM1 through a similar mechanism have been observed with the cytosolic sialidase in brain tissues (Venerando *et al.*, 1985, 1987) and *Clostridium perfringens* sialidase (Perillo *et al.*, 1994). There is an intriguing possibility that this type of membrane "association–dissociation" mechanism mediated by enzyme–substrate complex formation may underlie a number of dynamic cellular processes, such as cellular aggregation and dissociation. Future research is needed to elucidate the generality of this novel functional role of sialidases.

ACKNOWLEDGMENTS. We wish to acknowledge Dr. Cara-Lynne Schengrund for her helpful suggestions. The work from the authors' laboratories has been supported by USPHS grants NS 11853 to R.K.Y. and NS 27865 to M.S.

REFERENCES

Air, G. M., and Laver, W. G., 1989, The neuraminidase of influenza virus, *Proteins* **6**:341–356.

Air, G. M., Ritchie, L. R., Laver, W. G., and Colman, P. M., 1985, Gene and protein sequence of an influenza neuraminidase with hemagglutinin activity, *Virology* **145**:117–122.

Air, G. M., Laver, W. G., Luo, M., Stray, S. J., Legrone, G., and Webster, R. G., 1990, Antigenic, sequence, and crystal variation in influenza B neuraminidase, *Virology* **177**:578–587.

Aisaka, K., Igarashi, A., and Uwajima, T., 1991, Purification crystallization and characterization of neuraminidase from Micromonospora viridifaciens, *Agric. Biol. Chem.* **55**:997–1004.

Aminoff, D., 1961, Methods for the quantitative estimation of N-acetylneuraminic acid and their application to hydrolysates of sialomucoids, *Biochem. J.* **81**:384–392.

Ando, S., Tanaka, Y., and Kon, K., 1986, Membrane aging of the brain synaptosomes with special reference to gangliosides, in: *Gangliosides and Neuroplasticity* (G. Tettamanti, R. W. Ledeen, K. Sandhoff, Y. Nagai, and G. Taffano, eds.), Liviana Press, Podova, pp. 105–112.

Ando, S., Yu, R. K., Scarsdale, J. N., Kusunoki, S., and Prestegard, J. H., 1989, High-resolution proton NMR studies of gangliosides. Structures of two types of GD3 lactones and their reactivities with monoclonal antibody R24, *J. Biol. Chem.* **264**:3478–3483.

Aronson, N.N., and de Duve, C., 1968, Digestive activity of lysosomes. II. The digestion of macromolecular carbohydrates by extracts of rat liver lysosomes, *J. Biol. Chem.* **243**:4564–4573.

Bach, G., Zeigler, M., Shaap, T., and Kohn, G., 1979, Mucolipidosis: Ganglioside sialidase deficiency, *Biochem. Biophys. Res. Commun.* **90**:1341–1347.

Bachmayer, H., 1972, Effect of tryptophan modification on the activity of bacterial and viral neuraminidases, *FEBS Lett.* **23**:217–219.

Baker, N. J., and Gandhi, S. S., 1976, Effect of Ca^{++} on the stability of influenza virus neuraminidase, *Arch. Virol.* **52**:7–18.

Baker, A. T., Varghese, J. N., Laver, W. G., Air, G. M., and Colman, P. M., 1987, Three-dimensional structure of neuraminidase of subtype N9 from an avian influenza virus, *Proteins* **2**:111–117.

Barton, N. W., Lipovac, V., and Rosenberg, A., 1975, Effects of strong electrolyte upon the activity of *Clostridium perfringens* sialidase towards sialyllactose and sialoglycolipids, *J. Biol. Chem.* **250**:8462–8466.

Baum, L. G., and Paulson, J. C., 1991, The N2 neuraminidase of human influenza virus has acquired a substrate specificity complementary to the hemagglutinin receptor specificity, *Virology* **180**:10–15.

Beau, J. M., and Schauer, R., 1980, Metabolism of 4-O-methyl-N-acetylneuraminic acid a synthetic sialic acid, *Eur. J. Biochem.* **106**:531–540.

Beauregard, G., and Potier, M., 1982, Radiation inactivation of enzymes at low dose rates: Identical molecular weights of rat liver cytosolic and lysosomal neuraminidases, *Anal. Biochem.* **122**:379–384.

Ben-Yoseph, Y., Momoi, T., Hahn, L. C., and Nadler, H. L., 1982, Catalytically defective ganglioside neuraminidase in mucolipidosis IV, *Clin. Genet.* **21**:374–381.

Ben-Yoseph, Y., Mitchell, D. A., Yager, R. M., and Pretzlaff, R. K., 1991, Stimulation of GM3 ganglioside activity by an activator protein in patient with mucolipidosis IV and controls, *Enzyme* **45:**23–29.

Berg, J. O., Lindqvist, L., Andersson, G., and Nord, C. E., 1983, Neuraminidase in Bacteroides fragilis, *Appl. Environ. Microbiol.* **46:**75–80.

Berg, W., Gutschker-Gdaniec, G., and Schauer, R., 1985, Fluorescent staining of sialidases in polyacrylamide gel electrophoresis and ultrathin-layer isoelectric focusing, *Anal. Biochem.* **145:**339–342.

Bernacki, R. J., and Bosmann, H. B., 1973, Rat-liver-sialidase activity utilizing a tritium-labeled sialic acid derivative of glycoprotein substrates, *Eur. J. Biochem.* **34:**425–433.

Berry, A. M., Paton, J. C., Glare, E. M., Hansman, D., and Catcheside, D.E.A., 1988, Cloning and expression of the pneumococcal neuraminidase gene in Escherichia coli, *Gene* **71:**299–305.

Bhavanandan, V. P., Yeh, A. K., and Carubelli, R., 1975, Neuraminidase assay utilizing sialyl-oligosaccharide substrates with tritium-labeled aglycone, *Anal. Biochem.* **69:**385–394.

Blok, J., and Air, G. M., 1982, Sequence variation at the 3′ end of the neuraminidase gene from 39 influenza type A viruses, *Virology* **121:**211–229.

Blok, J., Air, G. M., Laver, W. G., Ward, C. W., Lilley, G. G., Woods, E. F., Roxburgh, C. M., and Inglis, A. S., 1982, Studies on the size, chemical composition, and partial sequence of the neuraminidase (NA) from type A influenza viruses show that the N-terminal region of the NA is not processed and serves to anchor the NA in the viral membrane, *Virology* **119:**109–121.

Blumberg, W. P., Giorgi, C., Roux, L., Raju, R., Dowling, P., Chollet, A., and Kolakofsky, D., 1985, Sequence determination of the Sendai virus HN gene and its comparison to the influenza virus glycoproteins, *Cell* **41:**269–278.

Bos, T. J., Davis, A. R., and Nayak, D. P., 1984, NH_2-terminal hydrophobic region of influenza virus neuraminidase provides the signal function in translocation, *Proc. Natl. Acad. Sci. USA* **81:**2327–2331.

Bossart, P., Babu, Y. S., Cook, W. J., Air, G. M., and Laver, W. G., 1988, Crystallization and preliminary X-ray analyses of two neuraminidases from influenza B virus B/Hong Kong/8/73 and B/Lee/40, *J. Biol. Chem.* **263:**6421–6423.

Brossmer, R., and Nebelin, E., 1969, Synthesis of N-formyl- and N-succinyl-D-neuraminic acid on the specificity of neuraminidase, *FEBS Lett.* **4:**335–336.

Burmeister, W. P., Daniels, R. S., Dayan, S., Gagnon, J., Cusack, S., and Ruigrok, R.W.H., 1991, Sequence and crystallization of influenza virus B/Beijing/1/87 neuraminidase, *Virology* **180:**266–272.

Burmeister, W. P., Ruigrok, R.W.H., and Cusack, S., 1992, The 2.2 Å resolution crystal structure of influenza B neuraminidase and its complex with sialic acid, *EMBO J.* **11:**49–56.

Burmeister, W. P., Baudin, F., Cusack, S., and Ruigrok, W. H., 1993, Comparison of structure and sequence of influenza B/Yamagata and B/Beijing neuraminidases shows a conserved "head" but much greater variability in the "stalk" and NB protein, *Virology* **192:**683–686.

Burnet, F. M., and Stone, J. D., 1947, The receptor-destroying enzyme of V. cholerae, *Aust. J. Exp. Biol. Med.* **25:**227–233.

Cabezas, J. A., Milicua, M., Bernal, C. S., Villar, E., Pres, N., and Hannoun, C., 1989, Kinetic studies on the sialidase of three influenza B and three influenza A virus strains, *Glycoconj. J.* **6:**219–227.

Cacalano, G., Kays, M., Saiman, L., and Prince, A., 1992, Production of the Pseudomonas aeruginosa neuraminidase is increased under hyperosmolar conditions and is regulated by genes involved in alginate expression, *J. Clin. Invest.* **89:**1866–1874.

Camara, M., Mitchell, T. J., Andrew, P. W., and Boulnois, G. J., 1991, Streptococcus pneumoniae produces at least two distinct enzymes with neuraminidase activity: Cloning and expression of a second neuraminidase gene in Escherichia coli, *Infect. Immun.* **59:**2856–2858.

Carroll, S. M., and Paulson, J. C., 1982, Complete metal ion requirement of influenza virus N1 neuraminidases, *Arch. Virol.* **71**:273–277.
Carubelli, E., 1962, Neuraminidase activity in mammalian organs, *Biochim. Biophys. Acta* **60**:196–197.
Carubelli, E., and Tulsiani, D.R.P., 1971, Neuraminidase activity in brain and liver of rats during development, *Biochim. Biophys. Acta* **237**:78–87.
Cassidy, J. T., Jourdian, G. W., and Roseman, S., 1965, The sialic acid. VI. Purification and properties of sialidase from Clostridium perfringens, *J. Biol. Chem.* **240**:3501–3506.
Castrucci, M. R., and Kawaoka, Y., 1993, Biologic importance of neuraminidase stalk length in influenza A virus, *J. Virol.* **67**:759–764.
Cavallesco, R., and Pereira, M.E.A., 1988, Antibody to Trypanosoma cruzi neuraminidase enhances infection in vitro and identifies a subpopulation of trypomastigotes, *J. Immunol.* **140**:617–625.
Chabas, A., Guardiola, A., and Burguera, J. M., 1987, An activator protein of oligosaccharide sialidase, *Biochem. Int.* **15**:449–457.
Chiarini, A., Fiorilli, A., Siniscalco, C., Tettamanti, G., and Venerando, V., 1990, Solubilization of the membrane-bound sialidase from pig brain by treatment with bacterial phosphatidylinositol phospholipase C, *J. Neurochem.* **55**:1576–1584.
Chiarini, A., Fiorilli, A., di Francesco, L., Venerando, B., and Tettamanti, G., 1993, Human erythrocyte sialidase is linked to the plasma membrane by a glycosylphosphatidylinositol anchor and partly located on the outer surface, *Glycoconj. J.* **10**:64–71.
Chigorno, V., Cardace, G., Pitto, M., Sonnino, S., Ghidoni, R., and Tettamanti, G., 1986, A radiometric assay for ganglioside sialidase to the determination of the enzyme subcellular location in cultured human fibriblasts, *Anal. Biochem.* **153**:283–294.
Chong, A.K.J., Pegg, M. S., and von Itzstein, M., 1991, Influenza virus sialidase: Effect of calcium on steady-state kinetic parameters, *Biochim. Biophys. Acta* **1077**:65–71.
Chong, A.K.J., Pegg, M. S., Taylor, N. R., and von Itzstein, M., 1992, Evidence for a sialosyl cation transition-state complex in the reaction of sialidase from influenza virus, *Eur. J. Biochem.* **207**;335–343.
Chou, M.-Y., Li, Y.-T., and Li, S.-C., 1992, α2-3 specific sialidase, sialidase L, *Glycobiology* **5**:495.
Chuang, N.-N., and Yang, B. C., 1990, A sialidase from hepatopancreas of the shrimp Penaeus japonicus (Crustacea: Decapoda): Reversible binding with the acidic β-galactosidase, *Comp. Biochem. Physiol.* **97C**:353–356.
Colman, P. M., 1989, Neuraminidase: Enzyme and antigen, in: *The Influenza viruses* (R. M. Krug, ed.), Plenum Press, New York, pp. 175–210.
Colman, P. M., Varghese, J. N., and Laver, W. G., 1983, Structure of the catalytic and antigenic sites in influenza virus neuraminidase, *Nature* **303**:41–44.
Colman, P. M., Laver, W. G., Varghese, J. N., Baker, A. T., Tulloch, P. A., Air, G. M., and Webster, R. G., 1987, Three-dimensional structure of a complex of antibody with influenza virus neuraminidse, *Nature* **326**:358–363.
Colman, P. M., Hoyne, P. A., and Lawrence, M. C., 1993, Sequence and structure alignment of paramyxovirus hemagglutinin-neuraminidase with influenza virus neuraminidase, *J. Virol.* **67**:2972–2980.
Conzelmann, E., and Sandhoff, K., 1987, Glycolipid and glycoprotein degradation, *Adv. Enzymol.* **60**:89–216.
Corfield, A. P., and Schauer, R., 1982, Metabolism of sialic acid, in: *Sialic Acids—Chemistry, Metabolism, and Function* (R. Schauer, ed.), Springer-Verlag, Berlin, pp. 195–261.
Corfield, A. P., Michalski, J. C., and Schauer, R., 1981a, The substrate specificity of sialidases from microorganisms and mammals, in: *Perspectives in Inherited Metabolic Diseases* (G. Tettamanti, P. Durand, and S. di Donate, eds.), Vol. 4, Edi Ermes, Milan, Italy, pp. 3–70.
Corfield, A. P., Veh, R. W., Wember, M., Michalski, J. C., and Schauer, R., 1981b, The release of

N-acetyl and N-glycolyl-neuraminic acid from soluble complex carbohydrates and erythrocytes by bacterial, viral and mammalian sialidases, *Biochem. J.* **197:**293–299.

Corfield, A. P., 1992, Bacterial sialidases—Roles in pathogenesis and nutrition, *Glycobiology* **2:**509–521.

Corfield, A. P., Wember, M., Schauer, R., and Rott, R., 1982, The specificity of viral sialidases: The use of oligosaccharide substrates to probe enzyme characteristics and strain-specific differences, *Eur. J. Biochem.* **124:**521–525.

Corfield, A. P., Higa, H., Paulson, J. C., and Schauer, R., 1983, The specificity of viral and bacterial sialidases for α(2-3)- and α(2-6)-linked sialic acids in glycoproteins, *Biochim. Biophys. Acta* **744:**121–126.

Corfield, A. P., Sander-Wewer, M., Veh, R. W., Wember, M., and Schauer, R., 1986, The action of sialidases on substrates containing O-acetylsialic acids, *Biol. Chem. Hoppe-Seyler* **367:**433–439.

Corfield, A. P., Wagner, S. A., Clamp, J. R., Kriaris, M. S., and Hoskins, L. C., 1992, Mucin degradation in the human colon: Production of sialidase, sialate O-acetylesterase, N-acetylneuraminate lyase, arylesterase, and glycosulfatase activities by strains of fecal bacteria, *Infect. Immun.* **60:**3971–3978.

Cross, A. S., and Wright, D. G., 1991, Mobilization of sialidase from intracellular stores to the surface of human neutrophils and its role in stimulated adhesion responses of these cells, *J. Clin. Invest.* **88:**2067–2076.

Cruz, T. F., and Gurd, J. W., 1978, Reaction of synaptic plasma membrane sialoglycoproteins with intrinsic sialidase and wheat germ agglutinin, *J. Biol. Chem.* **253:**7314–7318.

Cruz, T. F., and Gurd, J. W., 1981, The effects of development on activity, specificity and endogenous substrates of synaptic membrane sialidase, *Biochim. Biophys. Acta* **675:**201–208.

Cruz, T. F., and Gurd, J. W., 1983, Identification of intrinsic and sialoglycoprotein substrates in rat brain synaptic junctions, *J. Neurochem.* **40:**1599–1604.

Czarniecki, M. F., and Thornton, E. R., 1977, Carbon-13 NMR spin lattice relaxation in the N-acetylneuraminic acids, *J. Am. Chem. Soc.* **99:**8273–8279.

Dairaku, K., Miyagi, T., Wakui, A., and Tsuiki, S., 1986, Cytosolic sialidases of rat tissues with special reference to skeletal muscle enzyme, *Biochem. Int.* **13:**741–748.

Dale, B., Brown, R., Miller, J., White, R. T., Air, G. M., and Cordell, B., 1986, Nucleotide and deduced amino acid sequence of the influenza neuraminidase genes of two equine serotypes, *Virology* **155:**460–468.

Davis, L., Baig, M. M., and Atoub, E. M., 1979, Properties of extracellular neuraminidase produced by group A streptococcus, *Infect. Immun.* **24:**780–786.

d'Azzo, A., Hoogeveen, A. T., Reuser, A.J.J., Robinson, D., and Galjaard, H., 1982, Molecular defect in combined β-galactosidase and neuraminidase deficiency in man, *Proc. Natl. Acad. Sci. USA* **79:**4535–4539.

Dimmock, N. J., 1971, Dependence of the activity of an influenza virus neuraminidase upon Ca^{2+}, *J. Gen. Virol.* **13:**481–483.

Domingues, R. M., Cavalcanti, S. M., Andrale, A. F., and Ferreira, M. C., 1992, Sialic acid as receptor of Bacteroides fragilis lectin-like adhesion, *Int. J. Microbiol. Virol. Parasitol.* **277:** 340–344.

Drzeniek, R., 1972, Viral and bacterial neuraminidases, *Curr. Top. Microbiol. Immunol.* **59:**35–74.

Drzeniek, R., 1973, Substrate specificity of neuraminidase, *Histochem. J.* **5:**271–290.

Drzeniek, R., Seto, J. T., and Rott, R., 1966, Characterization of neuraminidases from myxoviruses, *Biochim. Biophys. Acta* **128:**547–558.

Eisen, D., and Franson, R. C., 1987, Acid-active neuraminidases in growth media from cultures of pathogenic Naegleria fowleri and in sonicates of rabbit alveolar macrophages, *Biochim. Biophys. Acta* **924:**369–372.

Els, M. C., Air, G. M., Murti, K. G., Webster, R. G., and Laver, W. G., 1985, An 18-amino acid deletion in an influenza neuraminidase, *Virology* **142:**241–247.

Engstler, M., Reuter, G., and Schauer, R., 1992, Purification and characterization of a novel sialidase found in procyclic culture forms of Trypanosoma brucei, *Mol. Biochem. Parasitol.* **54**:21–30.
Esievo, K.A.N., 1983, Trypanosoma vivax, stock V953: Inhibitory effect of type A influenza virus anti-HAV8 serum on in vitro neuraminidase (sialidase) activity, *J. Parasitol.* **69**:491–495.
Ferrero-Garcia, M. A., Trombetta, S. E., Sanchez, D. O., Reglero, A., Frasch, A. C., and Parodi, A. J., 1993, The action of Trypanosoma cruzi trans-sialidase on glycolipids and glycoproteins, *Eur. J. Biochem.* **213**:765–771.
Fields, S., Winter, G., and Brownlee, G. G., 1981, Structure of the neuraminidase gene in human influenza virus A/PR/8/34, *Nature* **290**:213–217.
Fingerhut, R., van der Horst, G.T.J., Verheijen, F. W., and Conzelmann, E., 1992, Degradation of gangliosides by the lysosomal sialidase requires an activator protein, *Eur. J. Biochem.* **208**:623–629.
Fiorilli, A., Venerando, B., Siniscalco, C., Monti, E., Bresciani, R., Caimi, L., Preti, A., and Tettamanti, G., 1989, Occurrence in brain lysosomes of a sialidase active on ganglioside, *J. Neurochem.* **53**:672–680.
Fiorilli, A., Siniscalco, C., Chiarini, A., di Francesco, L., Venerando, B., and Tettamanti, G., 1991, Occurrence of sialidase activity in two distinct and highly homogeneous populations of lysosomes prepared from the brain of developing mouse, *FEBS Lett.* **282**:235–238.
Fischer, C., Kelm, S., Ruch, B., and Schauer, R., 1991, Reversible binding of sialidase-treated rat lymphocytes by homologous peritoneal macrophages, *Carbohydr. Res.* **213**:263–273.
Fraser, A. G., and Brown, R., 1981, Neuraminidase production by Bacteroidacea, *J. Med. Microbiol.* **14**:63–73.
Frevert, U., Schenkman, S. D., and Nussenzweig, V., 1992, Stage-specific expression and intracellular shedding of the cell surface trans-sialidase of Trypanosoma cruzi, *Infect. Immun.* **60**:2349–2360.
Friebolin, H., Brossmer, R., Keilich, G., Ziegler, D., and Supp, M., 1980, H-NMR-Spektroskopischer Nachweis der N-Acetyl-α-D-Neuraminsaure als primares Spaltprodukt der Neuraminidasen, *Hoppe-Seyler's Z. Physiol. Chem.* **361**:697–702.
Galen, J. E., Ketley, J. M., Fasano, A., Richardson, S. H., Wasserman, S. S., and Kaper, J. B., 1992, Role of Vibrio cholerae neuraminidase in the function of cholera toxin, *Infect. Immun.* **60**:406–415.
Galjart, N., Gillemans, N., Harris, A., van der Horst, G.T.J., Verheijen, F. W., Galjaard, H., and d'Azzo, A., 1988, Expression of cDNA encoding the human protective protein associated with lysosomal β-galactosidase and neuraminidase: Homology to yeast proteases, *Cell* **54**:755–764.
Garcia-Sastre, A., Cobaleda, C., Cabezas, J. A., and Villar, E., 1991, On the inhibition mechanism of the sialidase activity from Newcastle disease virus, *Biol. Chem. Hoppe-Seyler* **372**:923–927.
Ghidoni, R., Sonnino, S., Masserini, M., Orlando, P., and Tettamanti, G., 1981, Specific tritium labeling of gangliosides at the 3-position of sphingosine, *J. Lipid Res.* **22**:1286–1295.
Gorman, W. L., Takahashi, T., Scroggs, R. A., and Portner, A., 1991, Identification of amino acid positions associated with neuraminidase activity of the hemagglutinin-neuraminidase glycoprotein of Sendai virus, *Virology* **180**:803–806.
Gottschalk, A., 1957, Neuraminidase: The specific enzyme of influenza virus and Vibrio cholerae, *Biochim. Biophys. Acta* **23**:645–646.
Guzman, C. A., Plate, M., and Pruzzo, C., 1990, Role of neuraminidase-dependent adherence in Bacterioides fragilis attachment to human epithelial cells, *FEMS Microbiol. Lett.* **59**:187–192.
Haddad, E. B., and Gies, J. P., 1992, Neuraminidase reduced the number of super-high-affinity (3H)oxotremorin-M binding sites in lung, *Eur. J. Pharmacol.* **211**:273–276.
Hall, F. B., Webster, P., Ma, A. K., Joiner, K. A., and Andrews, N. W., 1992, Desialylation of lysosomal membrane glycoproteins by Trypanosoma cruzi: A role for the surface neuraminidase in facilitating parasite entry into the host cell cytoplasm, *J. Exp. Med.* **176**:313–325.

Hechler, V., Mersel, M., Dreyfus, H., and Maitre, M., 1990, Effects of phospholipases, proteases and neuraminidase on gamma-hydroxybutyrate binding sites, *Mol. Cell. Biochem.* **93:**87–94.

Heijlman, J., and Roukema, P. A., 1972, The action of calf brain sialidase on gangliosides, sialoglycoproteins and sialoglycopeptides, *J. Neurochem.* **19:**2567–2575.

Heimer, R., and Meyer, K., 1956, Studies on sialic acid of submaxillary mucoid, *Proc. Natl. Acad. Sci. USA* **42:**728–734.

Henningsen, M., Roggentin, P., and Schauer, R., 1991, Cloning, sequencing, expression of the sialidase gene from Actinomyces viscosus DSM 43798, *Biol. Chem. Hoppe-Seyler* **372:**1065–1072.

Henrickson, K. J., and Savatski, L. L., 1992, Genetic variation and evolution of human parainfluenza virus type 1 hemagglutinin neuraminidase: Analysis of 12 clinical isolates, *J. Infect. Dis.* **166:**995–1005.

Heuermann, D., Roggentin, P., Kleineidam, R., and Schauer, R., 1991, Purification and characterization of a sialidase from Clostridium chauvoei NC08596, *Glycoconj. J.* **8:**95–101.

Hiebert, S. W., Peterson, R. G., and Lamb, R. A., 1985, Hemagglutinin-neuraminidase protein of the paramyxovirus simian virus 5: Nucleotide sequence of the mRNA predicts an N-terminal membrane anchor, *J. Virol.* **54:**1–6.

Hirabayashi, Y., Hyogo, A., Nakao, T., Tsuchiya, K., Suzuki, Y., Matsumoto, M., Kon, K., and Ando, S., 1990, Isolation and characterization of extremely minor gangliosides, GM1b and GD1α, in adult bovine brains as developmentally regulated antigens, *J. Biol. Chem.* **265:**8144–8151.

Hiraiwa, M., Uda, Y., Nishizawa, M., and Miyatake, T., 1987, Human placental sialidase partial purification and characterization, *J. Biochem.* **101:**1273–1279.

Hiraiwa, M., Nishizawa, M., Uda, Y., Nakajima, T., and Miyatake, T., 1988, Human placental sialidase: Further purification and characterization, *J. Biochem.* **103:**86–90.

Hiraiwa, M., Uda, Y., Tsuji, S., Miyatake, T., Martin, B. M., Tayama, M., O'Brien, J. S., and Kishimoto, Y., 1991, Human placental sialidase complex: Characterization of the 60 KDA protein that cross-reacts with anti-saposin antibodies, *Biochem. Biophys. Res. Commun.* **177:** 1211–1216.

Hiti, A. L., and Nayak, D. P., 1982, Complete nucleotide sequence of the neuraminidase gene of human influenza virus A/WSN/33, *J. Virol.* **41:**730–734.

Hogue, B. G., and Nayak, D. P., 1992, Synthesis and processing of the influenza virus neuraminidase, a type II transmembrane glycoprotein, *Virology* **188:**510–517.

Hong, V. N., Beauregard, G., Potier, M., Belisle, M., Mameli, L., Gatti, R., and Durand, P., 1980, Studies on the properties of human leucocyte neuraminidases, *Biochim. Biophys. Acta* **616:**259–270.

Hoogeveen, A. T., d'Azzo, A., Brossmer, R., and Galjaard, H., 1981, Correction of combined β-galactosidase/neuraminidase deficiency in human fibroblasts, *Biochem. Biophys. Res. Commun.* **103:**292–300.

Hoogeveen, A. T., Verheijen, F. W., and Galjaard, H., 1983, The relation between human lysosomal β-galactosidase and its protective protein, *J. Biol. Chem.* **258:**12143–12146.

Horvat, A., and Touster, O., 1968, On the lysosomal occurrence and the properties of the neuraminidase of rat liver and of the neuraminidase of rat liver and of Ehrlich ascites tumor cells, *J. Biol. Chem.* **243:**4380–4390.

Hoskins, L. C., Boulding, E. T., Gerken, T. A., Harouny, V. R., and Krisris, M. S., 1992, Mucin glycoprotein degradation by mucin-oligosaccharide-degrading strains of human fecal bacteria, *Microbiol. Ecol. Health Dis.* **5:**193–207.

Houde, M., and Arora, D. J., 1990, Stimulation of tumor necrosis factor secretion by purified influenza virus neuraminidase, *Cell. Immunol.* **129:**104–111.

Hoyer, L. L., Roggentin, P., Schauer, R., and Vimr, E. R., 1991, Purification and properties of cloned

Salmonella typhimurium LT2 sialidase with virus-typical kinetic preference for sialyl α2-3 linkages, *J. Biochem.* **110**:462–467.

Hoyer, L. L., Hamilton, A. C., Steenbergen, S. M., and Vimr, E. R., 1992, Cloning, sequencing and distribution of the Salmonella typhimurium LT2 sialidase gene, nanH, provides evidence for interspecies gene transfer, *Mol. Microbiol.* **6**:873–884.

Huang, R.T.C., and Orlich, M., 1972, Substrate specificities of the neuraminidases of Newcastle disease and fowl plague viruses, *Hoppe-Seyler's Z. Physiol. Chem.* **353**:318–322.

Huang, R. T., Rott, R., Wahn, K., Klenk, H. D., and Kohama, T., 1980, The function of the neuraminidase in membrane fusion induced by myxoviruses, *Virology* **107**:313–319.

Huang, R. T. C., Dietsch, E., and Rott, R., 1985, Further studies on the role of neuraminidase and the mechanism of low pH dependence in influenza virus-induced membrane fusion, *J. Gen. Virol.* **66**:295–301.

Igarashi, M., and Bando, Y., 1990, Enhanced efficiency of cell hybridization by neuraminidase treatment, *J. Immunol. Methods* **135**:91–93.

Iorio, R. M., and Glickman, R. L., 1992, Fusion mutants of Newcastle disease virus selected with monoclonal antibodies to the hemagglutinin-neuraminidase, *J. Virol.* **66**:6626–6633.

Jorgensen, E. D., Collins, P. L., and Iomedico, P. T., 1987, Cloning and nucleotide sequence of Newcastle disease virus hemagglutinin-neuraminidase mRNA: Identification of a putative sialic acid binding site, *Virology* **156**:12–24.

Kabayo, J. P., and Hutchinson, D. W., 1975, Studies on a neuraminidase from Streptomyces griseus, *FEBS Lett.* **78**:221–224.

Karasuno, T., Kanayama, Y., Nishiura, T., Nakao, H., Yonezawa, T., and Tarui, S., 1992, Glycosidase inhibitors (castanospermine and swainsonine) and neuraminidase inhibit pokeweed mitogen-induced B cell maturation, *Eur. J. Immunol.* **22**:2003–2008.

Kessler, J., Heck, J., Tannenbaum, S. W., and Flashner, M., 1982, Substrate and product specificity of Arthrobacter sialophilus neuraminidase, *J. Biol. Chem.* **257**:5056–5060.

Kessler, N., Bardeletti, G., and Aymard, M., 1977, The neuraminidase of human parainfluenza 1 virus (HA2 virus), *J. Gen. Virol.* **37**:547–556.

Khorlin, A. Y., Privalova, I. M., Zakstelskaya, L. Y., Molibog, E. V., and Evstigneeva, N. A., 1970, Synthetic inhibitors of Vibrio cholerae neuraminidase and neuraminidases of some influenza virus strains, *FEBS Lett.* **8**:17–19.

Kishore, G. S., Tulsiani, D.R.P., Bhavanandan, V. P., and Carubelli, R., 1975, Membrane-bound neuraminidase of rat liver: Neuraminidase activity in Golgi apparatus, *J. Biol. Chem.* **250**: 2655–2659.

Kleineidam, R. G., Furuhata, K., Ogura, H., and Schauer, R., 1990, 4-Methylumbelliferyl-α-glycosides of partially O-acetylated N-acetylneuraminic acids as substrates of bacterial and viral sialidases, *Biol. Chem. Hoppe-Seyler* **371**:715–719.

Kodama, H., Baum, L. G., and Paulson, J. C., 1991, Synthesis of linkage-specific sialoside substrates for colorimetric analysis of neuraminidases, *Carbohydr. Res.* **218**:111–119.

Koerner, T.A.W., Prestegard, J. H., Demou, P. C., and Yu, R. K., 1983, High-resolution proton NMR studies of gangliosides. Use of homonuclear two-dimensional spin–echo J-correlated spectroscopy for determination of residue composition and anomeric configuration, *Biochemistry* **22**:2676–2687.

Kuratowska, Z., and Kubicka, T., 1967, Purification and some properties of the neuraminidae from rabbit kidney, *Acta Biochim. Pol.* **14**:255–259.

Lambre, C. R., Greffard, A., Gattegno, L., and Saffar, L., 1989/1990, Modification of sialidase activity during the monocyte–macrophage differentiation in vitro, *Immunol. Lett.* **23**:179–182.

Lambre, C. R., Terzidis, H., Greffard, A., and Webster, R. G., 1991, An enzyme-linked lectin assay for sialidase, *Clin. Chim. Acta* **198**:183–194.

Landolfi, N. F., Leone, J., Womack, J. E., and Cook, R. G., 1985, Activation of T lymphocytes results in an increase in H2 encoded neuraminidase, *Immunogenetics* **22:**159–167.

Laver, W. G., 1978, Crystallization and peptide maps of neuraminidase "heads" from H2N2 and H3N3 influenza virus strains, *Virology* **86:**78–87.

Laver, W. G., Air, G., Webster, R. G., and Markoff, L. J., 1982, Amino acid sequence changes in antigenic variants of type A influenza virus 2 neuraminidase, *Virology* **122:**450–460.

Laver, W. G., Colman, P. M., Webster, R. G., Hinsha, V. S., and Air, G. M., 1984, Influenza virus neuraminidase with hemagglutinin activity, *Virology* **137:**314–323.

Laver, W. G., Webster, R. G., and Colman, P. M., 1987, Crystals of antibodies complexed with influenza virus neuraminidase show isosteric binding of antibody to wild-type and variant antigens, *Virology* **156:**181–184.

Laver, W. G., Thompson, S. D., Murti, K. G., and Portner, A., 1989, Crystallization of Sendai virus HN protein complexed with monoclonal antibody Fab fragments, *Virology* **171:**291–293.

Ledeen, R. W., 1989, Biosynthesis, metabolism, and biological effects of gangliosides, in: *Neurobiology of Glycoconjugates* (R. U. Margolis and R. K. Margolis, eds.), Plenum Press, New York, pp. 43–83.

Ledeen, R., and Salsman, K., 1965, Structure of Tay-Sachs ganglioside, *Biochemistry* **4:**2225–2232.

Leibovitz, Z., and Gatt, S., 1968, Enzymatic hydrolysis of sphingolipids. VII. Hydrolysis of gangliosides by a neuraminidase from calf brain, *Biochim. Biophys. Acta* **152:**136–143.

Lentz, M. R., Air, G. M., Laver, W. G., and Webster, R. G., 1984, Sequence of the neuraminidase gene of influenza virus A/Tokyo/3/67 and previously uncharacterized monoclonal variants, *Virology* **135:**257–265.

Lentz, M. R., Webster, R. T., and Air, G. M., 1987, Site-directed mutation of the active site of influenza neuraminidase and implications for the catalytic mechanism, *Biochemistry* **26:**5351–5358.

Li, Y. T., Nakagawa, H., Ross, S. A., Hansson, G. C., and Li, S. C., 1990, A novel sialidase which releases 2,7-anhydro-α-N-acetylneuraminic acid from sialoglycoconjugates, *J. Biol. Chem.* **265:**21629–21633.

Libby, P., Alroy, J., and Pereira, M.E.A., 1986, A neuraminidase from Trypanosoma cruzi removes sialic acid from the surface of mammalian myocardial and endothelial cells, *J. Clin. Invest.* **77:**127–135.

Lieser, M., Harms, E., Kern, H., Bach, G., and Cantz, M., 1989, Ganglioside GM3 sialidase activity in fibroblasts of normal individuals and of patients with sialidosis and mucolipidosis IV, Biochem. J. **260:**69–74.

Lipovac, V., Barton, N., and Rosenberg, A., 1973, Control of the action of *Vibrio cholerae* sialidase on mammalian brain gangliosides by ionic strength, *Biochemistry* **12:**1858–1861.

Lock, R. A., Paton, J. C., and Hansman, D., 1988, Purification and immunological characterization of neuraminidase produced by Streptococcus pneumoniae, *Microbiol. Pathog.* **4:**33–43.

Mahadevan, S., Nduaguba, J. C., and Tappel, A. L., 1967, Sialidase of rat liver and kidney, *J. Biol. Chem.* **242:**4409–4413.

Malby, R. L., Caldwell, J. B., Gruen, L. C., Harley, V. R., Ivancic, N., Kortt, A. A., Lilley, G. G., Power, B. E., Webster, R. G., and Colman, P. M., 1993, Recombinant antineuraminidase single chain antibody: Expression, characterization, and crystallization in complex with antigen, *Proteins* **16:**57–73.

Martinez, C., Del Rio, L., Portela, A., Domingo, E., and Ortin, J., 1983, Evolution of the influenza virus neuraminidase gene during drift of the N2 subtype, *Virology* **130:**539–545.

Mattingly, S. J., Milligan, T. W., Pierpont, A. A., and Straus, D. C., 1980, Extracellular neuraminidase production by clinical isolates of group B streptococci from infected neonates, *J. Clin. Microbiol.* **12:**633–635.

Merz, D. C., and Wolinsky, J. S., 1983, Conversion of nonfusing mumps virus infections to fusing infections by selective proteolysis of the HN glycoprotein, *Virology* **131**:328–340.
Meyer, D. M., Lemonnier, M., and Bourrillon, R., 1981, Human liver neuraminidase, *Biochem. Biophys. Res. Commun.* **130**:1302–1309.
Michalski, J. C., Corfield, A. P., and Schauer, R., 1982, Solubilization and affinity chromatography of a sialidase from human liver, *Hoppe-Seyler's Z. Physiol. Chem.* **363**:1097–1102.
Michalski, J. C., Corfield, A. P., and Schauer, R., 1986, Properties of human liver lysosomal sialidase, *Hoppe-Seyler's Z. Physiol. Chem.* **367**:715–722.
Miura, N., Nakatani, Y., Ishiura, M., Uchida, T., and Okada, Y., 1985, Molecular cloning of a full-length cDNA encoding the hemagglutinin-neuraminidase glycoprotein of Sendai virus, *FEBS Lett.* **188**:112–116.
Miyagi, T., and Tsuiki, S., 1984, Rat-liver lysosomal sialidase. Solubilization, substrate specificity and comparison with the cytosolic sialidase, *Eur. J. Biochem.* **141**:75–81.
Miyagi, T., and Tsuiki, S., 1985, Purification and characterization of cytosolic sialidase from rat liver, *J. Biol. Chem.* **260**:6710–6716.
Miyagi, T., and Tsuiki, S., 1986, Evidence for sialidase hydrolyzing gangliosides GM2 and GM1 in rat liver plasma membrane, *FEBS Lett.* **206**:223–228.
Miyagi, T., Goto, T., and Tsuiki, S., 1984, Sialidase of rat hepatomas: Quantitative and qualitative comparison with rat liver sialidase, *Gann* **75**:1076–1082.
Miyagi, T., Sagawa, J., Konno, K., Handa, S., and Tsuiki, S., 1990a, Biochemical and immunological studies on two distinct ganglioside-hydrolyzing sialidases from the particulate fraction of rat brain, *J. Biochem.* **107**:787–793.
Miyagi, T., Sagawa, J., Konno, K., Handa, S., and Tsuiki, S., 1990b, Immunological discrimination of intralysosomal, cytosolic, and two membrane sialidases present in rat tissues, *J. Biochem.* **107**:794–798.
Miyagi, T., Konno, K., Sagawa, J., and Tsuiki, S., 1990c, Neoplastic alteration of a membrane-associated sialidase of rat liver, *Jpn. J. Cancer Res.* **81**:915–919.
Miyagi, T., Sagawa, J., Kuroki, T., Matsuya, Y., and Tsuiki, S., 1990d, Tumor-promoting phorbol ester induced alterations of sialidases and sialyltransferase activities of JB6 cells, *Jpn. J. Cancer Res.* **81**:1286–1292.
Miyagi, T., Konno, K., Emori, Y., Kawasaki, H., Suzuki, K., Yasui, A., and Tsuiki, S., 1993, Molecular cloning and expression of cDNA encoding rat skeletal muscle cytosolic sialidase, *J. Biol. Chem.* **268**:26435–26440.
Moran, N. M., Breen, K. C., and Regan, C. M., 1986, Characterization and cellular localization of a developmentally regulated rat neural sialidase, *J. Neurochem.* **47**:18–22.
Müller, H. E., 1974, Neuraminidase of bacteria and protozoa and their pathogenic role, *Behring Inst. Mitt.* **55**:34–56.
Myers, R. W., Lee, R. T., Lee, Y. C., Thomas, G. H., Reynolds, L. W., and Uchida, Y., 1980, The synthesis of 4-methylumbelliferyl α-ketoside of N-acetylneuraminic acid and its use in a fluorometric assay for neuraminidase, *Anal. Biochem.* **101**:166–174.
Nagai, T., Klenk, H.-D., and Rott, R., 1976, Proteolytic cleavage of viral glycoproteins and the significance of the virulence of Newcastle disease virus, *Virology* **72**:494–508.
Nagy, E., Derbyshire, J. B., Dobos, P., and Krell, P. J., 1990, Cloning and expression of NDV hemagglutinin-neuraminidase cDNA in a Baculovirus expression vector system, *Virology* **176**:426–438.
Nakamura, M., Yoshida, T., Isobe, K. I., Iwamoto, T., Jamschedur Rahman, S. M., Zhang, Y. H., Hasegawa, T., Ichihara, M., and Nakashima, I., 1991, Modulation of the secondary antibody response of murine lymphocytes to sheep red blood cells in vitro by neuraminidase and exoglycosidases, *Immunol. Lett.* **29**:235–240.

Nees, S., Veh, R. W., and Schauer, R., 1975, Purification and characterization of neuraminidase from Clostridium perfringens, *Hoppe-Seyler's Z. Physiol. Chem.* **356:**1027–1042.

Nerome, K., Kanegae, Y., Yoshioka, Y., Itamura, S., Ishida, M., Gojobori, T., and Oya, A., 1991, Evolutionary pathway of N2 neuraminidases of swine and human influenza A viruses: Origin of the neuraminidase genes of two reassortance (H1N2) isolated from pigs, *J. Gen. Virol.* **72:**693–698.

Nilsson, J., Ksiazek, T., and Thyberg, J., 1982, Effects of neuraminidase on DNA synthesis, proliferation and endocytosis of cultivated arterial smooth muscle cells, *Exp. Cell Res.* **142:**333–339.

Nishikawa, K., Morishita, T., Toyoda, T., Miyadai, T., Yoshida, T., and Nagai, Y., 1986, Topological and operational delineation of antigenic sites on the HN glycoprotein of Newcastle disease virus and their structural requirements, *J. Virol.* **60:**987–993.

Nojiri, H., Takaku, F., Tetsuka, T., and Saito, M., 1982, Stimulation of sialidase activity during cell differentiation of human promyelocytic leukemic cell line HL-60, *Biochem. Biophys. Res. Commun.* **104:**1239–1246.

Nuss, J. M., and Air, G. M., 1991, Transfer of the hemagglutinin activity of influenza virus neuraminidase subtype N9 into an N2 neuraminidase background, *Virology* **183:**496–504.

Ogura, K., and Sweeley, C. C., 1992, Mitogenic effects of bacterial neuraminidase and lactosylceramide on human cultured fibroblasts, *Exp. Cell Res.* **199:**169–173.

Ogura, K., Ogura, M., Anderson, R. L., and Sweeley, C. C., 1992, Peroxidase-amplified assay of sialidase activity toward gangliosides, *Anal. Biochem.* **200:**52–57.

Ohbayashi, T., Taka, M., and Mohri, T., 1987, Effect of neuraminidase treatment on the lipid fluidity of the intestinal brush-border membranes, *Biochim. Biophys. Acta* **905:**57–64.

Ohmann, R., Rosenberg, A., and Svennerholm, L., 1970, Human brain sialidase, *Biochemistry* **9:**3774–3782.

Orefici, G., de Stasio, A., Guarino, M., Martini, A., and Orsi, N., 1984, Group b streptococci: Extracellular Neuraminidase production and virulence in mouse, *Microbiologica* **7:**75–78.

O'Toole, R. D., and Stahl, W. L., 1975, Experimental pneumococcal meningitis: Effects of neuraminidase and other pneumococcal constituents on cerebrospinal fluid in the intact dog, *J. Gen. Virol.* **33:**159–163.

Palese, P., and Compans, R. W., 1976, Inhibition of influenza virus replication in tissue culture by 2-deoxy-2,3-dehydro-N-trifluoroacetylneuraminic acid (FANA): Mechanism of action, *J. Gen. Virol.* **33:**159–163.

Palese, P., Tobita, K., and Ueda, M., 1974, Characterization of temperature sensitive influenza virus mutants defective in neuraminidase, *Virology* **61:**397–410.

Parodi, A. J., Pollevick, G. D., Mautner, M., Buschiazzo, A., Sanchez, D. O., and Frasch, A.C.C., 1992, Identification of the gene(s) coding for the trans-sialidase of Trypanosoma cruzi, *EMBO J.* **11:**1705–1710.

Paton, B. C., Schmid, B., Kustermann-Kuhn, B., Poulos, A., and Harzer, K., 1992, Additional biochemical findings in a patient and fetal sibling with a genetic defect in the sphingolipid activator protein (SAP) precursor, prosaposin, *Biochem. J.* **285:**481–488.

Pellegrin, J. J., Ortega-Barria, E., Barza, M., Baum, J., and Pereira, M.E.A., 1991, Neuraminidase activity in Acanthamoeba species trophozoites and cysts, *Invest. Ophthalmol. Vis. Sci.* **32:**3061–3066.

Pereira, M.E.A., 1983, A developmentally regulated neuraminidase activity in Trypanosoma cruzi, *Science* **219:**1444–1446.

Pereira, M.E.A., and Hoff, R., 1986, Heterogeneous distribution of neuraminidase activity in strains and clones of Trypanosoma cruzi and its possible association with parasitic myotropism, *Mol. Biochem. Parasitol.* **20:**183–189.

Pereira, M.E.A., Meija, S., Ortega-Barria, E., Matzilevitch, D., and Prioli, R. P., 1991, The Trypanosoma cruzi neuraminidase contains sequences similar to bacterial neuraminidases, YWTD repeats of the low density lipoprotein receptor, and type III modules of fibronectin, *J. Exp. Med.* **175:**179–191.
Perillo, M., Yu, R. K., and Maggio, B., 1994, Modulation of the activity of *Clostridium perfringens* neuraminidase by the molecular organization of gangliosides in monolayers, *Biochim. Biophys. Acta* **1193:**155–164.
Pilatte, Y., Bignon, J., and Lambre, C. R., 1987, Lysosomal and cytosolic sialidases in rabbit alveolar macrophages: Demonstration of increased lysosomal activity after in vivo activation with bacillus Calmette-Guerin, *Biochim. Biophys. Acta* **923:**150–155.
Pitto, M., Chigorno, V., Giglioni, A., Valsecchi, M., and Tettamanti, G., 1989, Sialidase in cerebellar granule cells differentiating in culture, *J. Neurochem.* **53:**1464–1470.
Pitto, M., Giglioni, A., and Tettamanti, G., 1992, Dual subcellular localization of sialidase in cultured granule cells differentiated in culture, *Neurochem. Int.* **21:**367–374.
Portner, A., Scroggs, R. A., and Metzger, D. W., 1987, Distinct functions of antigenic sites of the HN glycoprotein of Sendai virus, *Virology* **158:**61–68.
Portoukalian, J., and Bouchon, B., 1986, Hydrolysis of all gangliosides, including GM1 and GM2, on thin-layer plates by Vibrio cholerae neuraminidase, *J. Chromatogr.* **380:**386–392.
Post, J. A., 1992, Removal of sarcolemmal sialic acid residues results in a loss of sarcolemmal functioning and integrity, *Am. J. Physiol.* **263:**H147–H152.
Potier, M., 1988, Structure of the lysosomal sphingolipid activator protein 1 by homology with influenza virus neuraminidase, *Biochem. Biophys. Res. Commun.* **155:**32–37.
Potier, M., Mameli, L., Belisle, M., Dallaire, L., and Melancon, S. B., 1979, Fluorometric assay of neuraminidase with a sodium (4-methylumbelliferyl-α-D-N-acetylneuraminate) substrate, *Anal. Biochem.* **94:**287–296.
Potier, M., Michaud, L., Tranchemontagne, J., and Thauvette, L., 1990a, Structure of the lysosomal neuraminidase–β-galactosidase–carboxypeptidase multienzymic complex, *Biochem. J.* **267:** 197–202.
Potier, M., Lamotagne, S., Michaud, L., and Tranchemontagne, J., 1990b, Human neuraminidase is a 60-kDa processing product of prosaposin, *Biochem. Biophys. Res. Commun.* **173:**449–456.
Previato, J. O., Andrade, A. F., Pessolani, M. C., and Mendoca-Previato, L., 1985, Incorporation of sialic acid into Trypanosoma cruzi macromolecules. A proposal for a new metabolic route, *Mol. Biochem. Parasitol.* **16:**85–96.
Priori, R. P., Mejia, J. S., and Pereira, M.E.A., 1990, Monoclonal antibodies against Trypanosoma cruzi neuraminidase reveal enzymes polymorphism, recognize a subset of trypomastigotes, and enhance infection in vitro, *J. Immunol.* **144:**4384–4391.
Priori, R. P., Mejia, J. S., and Pereira, M.E.A., 1991, Trypanosoma cruzi: Localization of neuraminidase on the surface of trypomastigotes, *Trop. Med. Parasitol.* **42:**146–150.
Puri, A., Grimaldi, S., and Blumenthal, R., 1992, Role of viral envelope sialic acid membrane fusion mediated by the vesicular stomatitis virus envelope glycoprotein, *Biochemistry* **31:** 10108–10113.
Recio-Pinto, E., Thornhill, W. B., Duch, D. S., Levinson, S. R., and Urban, B. W., 1990, Neuraminidase treatment modifies the function of electroplax sodium channels in planar lipid bilayers, *Neuron* **5:**675–684.
Riboni, L., Sonnino, S., Acquotti, D., Malesci, A., Ghidoni, R., Egge, H., Mingrino, S., and Tettamanti, G., 1986, Natural occurrence of ganglioside lactones. Isolation of GD1b inner ester from adult human brain, *J. Biol. Chem.* **261:**8514–8519.
Riboni, L., Prinetti, A., Bassi, R., and Tettamanti, G., 1991, Cerebellar granule cells in culture exhibit a ganglioside-sialidase presumably linked to the plasma membrane, *FEBS* **287:**42–46.
Rogerieux, F., Belaise, M., Terzidis-Trabelsi, H., Greffard, A., Pilatte, Y., and Lambre, C. R., 1993,

Determination of the sialic acid linkage specificity of sialidase using lectins in a solid phase assay, *Anal. Biochem.* **211**:200–204.

Roggentin, P., Berg, W., and Schauer, R., 1987, Purification and characterization of sialidase from Clostridium sordeii G12, *Glycoconj. J.* **4**:349–359.

Roggentin, P., Rothe, B., Lottspeich, F., and Schauer, R., 1988a, Cloning and sequencing of a Clostridium perfringens sialidase gene, *FEBS Lett.* **238**:31–34.

Roggentin, P., Gutschker-Gdaniec, G.H.M., Hobrecht, R., and Schauer, R., 1988b, Early diagnosis of clostridial gas gangrene using sialidase antibodies, *Clin. Chim. Acta* **173**:251–262.

Roggentin, P., Rothe, B., Kaper, J. B., Galen, J., Lawrisuk, L., Vimr, E. R., and Schauer, R., 1989, Conserved sequences in bacterial and viral sialidases, *Glycoconj. J.* **6**:349–353.

Roggentin, P., Hobrecht, R., Tirpitz, D., Rothe, B., and Schauer, R., 1991, Application of sialidase antibodies for the diagnosis of clostridial infections, *Clin. Chim. Acta* **196**:97–106.

Roggentin, T., Kleineidam, R. G., Schauer, R., and Roggentin, P., 1992, Effects of site-specific mutations on the enzymatic properties of a sialidase from Clostridium perfringens, *Glycoconj. J.* **9**:235–240.

Roggentin, T., Kleineidam, R. G., Majewski, D. M., Tirpitz, D., Roggentin, P., and Schauer, R., 1993, An immunoassay for the rapid and specific detection of three sialidase-producing clostridia causing gas gangrene, *J. Immunol. Methods* **157**:125–133.

Rosenberg, A., 1981, Cell surface sialidase, in: *Perspectives in Inherited Metabolic Diseases*, Vol. IV (G. Tettamanti, P. Duran, and S. di Donato, eds.), Edi. Emeres, Milan, pp. 111–124.

Rosenberg, A., and Schengrund, C. L., 1976, Sialidases, in: *Biological Roles of Sialic Acid* (A. Rosenberg and C. L. Schengrund, eds.), Plenum Press, New York, pp. 295–359.

Rosenberg, I. A., Prioli, R. P., Ortega-Barria, E., and Pereira, M. E. A., 1991, Stage-specific phospholipase C-mediated release of Trypanosoma cruzi neuraminidase, *Mol. Biochem. Parasitol.* **46**:303–306.

Rothe, B., Roggentin, P., Frank, R., Blocker, H., and Schauer, R., 1989, Cloning, sequencing and expression of a sialidase gene from Clostridium sordellii G12, *J. Gen. Microbiol.* **135**:3087–3096.

Rothe, B., Rothe, B., Roggentin, P., and Schauer, R., 1991, The sialidase gene from Clostridium septicum: Cloning, sequencing, expression in Escherichia coli and identification of conserved sequences in sialidase and other proteins, *Mol. Gen. Genet.* **226**:190–197.

Russo, T. A., Thompson, J. S., Godoy, V. G., and Malamy, M. H., 1990, Cloning and expression of the Bacteroides fragilis TAL2480 neuraminidase gene, nanH, in Escherichia coli, *J. Bacteriol.* **172**:2594–2600.

Sagawa, J., Miyagi, T., and Tsuiki, S., 1988, Membrane-associated sialidase of rat liver and its decrease in hepatomas, *Jpn. J. Cancer Res. (Gann)* **79**:69–73.

Sagawa, J., Miyagi, T., and Tsuiki, S., 1990, Characterization of the major sialidases of various types of rats blood cells: Their comparison with rat liver sialidases, *J. Biochem.* **107**:452–456.

Saito, M., and Yu, R. K., 1986, Further characterization of a myelin-associated neuraminidase: Properties and substrate specificity, *J. Neurochem.* **47**:632–641.

Saito, M., and Yu, R. K., 1992, Role of myelin-associated neuraminidase in the ganglioside metabolism of rat brain myelin, *J. Neurochem.* **58**:83–87.

Saito, M., and Yu, R. K., 1993, Possible role of myelin-associated neuraminidase in membrane adhesion, *J. Neurosci. Res.* **36**:127–132.

Saito, M., Sugano, K., and Nagai, Y., 1979, Action of Arthrobacter ureafaciens sialidase on sialoglycolipid substrates, *J. Biol. Chem.* **254**:7845–7854.

Saito, M., Kasai, N., and Yu, R. K., 1985, In situ immunological determination of basic carbohydrate structures of gangliosides on thin-layer plates, *Anal. Biochem.* **148**:54–58.

Saito, M., Sato-Bibbee, C., and Yu, R. K., 1992, Neuraminidase activities in oligodendroglia cells of the rat brain, *J. Neurochem.* **58**:78–82.

Saito, M., Tanaka, Y., Tang, C.-P., Yu, R. K., and Ando, S., 1995, Characterization of sialidase activity in mouse synaptic plasma membranes and its age-related changes, *J. Neurosci. Res.* **40:**401–406.

Saito, T., Kawaoka, Y., and Webster, R. G., 1993, Phylogenetic analysis of the N8 neuraminidase gene of influenza A viruses, *Virology* **193:**868–876.

Sakurada, K., Ohta, T., and Hasegawa, M., 1992, Cloning, expression, and characterization of the Micromonospora viridifaciens neuraminidase gene in Streptomyces lividans, *J. Bacteriol.* **174:**6896–6903.

Sandhoff, K., and Pallmann, B., 1978, Membrane-bound neuraminidae from calf brain: Regulation of oligosialoganglioside degradation by membrane fluidity and membrane components, *Proc. Natl. Acad. Sci. USA* **75:**122–126.

Sandhoff, K., Schraven, J., and Nowoczek, G., 1976, Effect of xenon, nitrous oxide and halothane on membrane-bound sialidase from calf brain, *FEBS Lett.* **62:**284–287.

Sato, A., Hiramatsu, M., Kashimata, M., Murayama, M., Minami, N., and Minami, N., 1989, Characterization of sialidase in the rat salivary glands, *Enzymes* **41:**200–208.

Sauto-Padron, T., Harth, G., and de Souza, W., 1990, Immunocytochemical localization of neuraminidase in Trypanosoma cruzi, *Infect. Immun.* **58:**586–592.

Scarsdale, J. N., Prestegard, J. H., and Yu, R. K., 1990, NMR computational studies of interactions between remote residues in gangliosides, *Biochemistry* **29:**9843–9855.

Schauer, R., 1982, Chemistry, metabolism, and biological functions of sialic acid, *Adv. Carbohydr. Chem. Biochem.* **40:**131–234.

Schauer, R., 1983, Glycosidases with special reference to the pathophysiological role of sialidase, in: *Structural Carbohydrates of the Liver* (H. Popper, W. Reutter, F. Gudat, and E. Kottgen, eds.), MTP Press, pp. 83–97.

Schauer, R., 1985, Sialic acid and their role as biological masks, *Trends Biochem. Sci.* **9:**357–360.

Schauer, R., 1991, Biosynthesis and function of N- and O-substituted sialic acids, *Glycobiology* **1:**449–452.

Schauer, R., and Wember, M., 1984, Isolation and characterization of an oligosaccharide- and glycoprotein-specific sialidase from human leucocytes, *Hoppe-Seyler's Z. Physiol. Chem.* **365:**419–426.

Schauer, R., and Wember, M., 1989, Isolation and characterization of a sialidase from the starfish Asterias rubens, *Biol. Chem. Hoppe-Seyler* **370:**183–190.

Schauer, R., Veh, R. W., Sander, M., Corfield, A. P., and Wiegandt, H., 1980, Neuraminidase-resistant sialic acid residues of gangliosides, *Adv. Exp. Med. Biol.* **125:**283–294.

Schauer, R., Fischer, C., Kluge, A., Lee, H., and Ruch, B., 1990, Mechanism of binding and uptake of sialidase-treated blood cells and glycoproteins by the galactose-specific receptor of rat peritoneal macrophages, *Biomed. Biochim. Acta* **49:**S230–S235.

Scheel, G., Acevedo, E., Conzelmann, E., Nehrkorn, H., and Sandhoff, K., 1982, Model for the interaction of membrane-bound substrates and enzymes: Hydrolysis of gangliosides GD1a by sialidase of neural membranes isolated from calf brain, *Eur. J. Biochem.* **127:**245–253.

Scheel, G., Schwarzmann, G., Hoffmann-Bleihauer, P., and Sandhoff, K., 1985, The influence of ganglioside insertion into brain membranes on the rate of ganglioside degradation by membrane-bound sialidase, *Eur. J. Biochem.* **153:**29–35.

Scheibe, R., Hein, K., and Wenzel, K. W., 1990, Lysosomal β-galactosides from rat liver: Purification, molecular forms and association with neuraminidase, *Biomed. Biochim. Acta* **49:**547–556.

Scheid, A., and Choppin, P. W., 1974, Identification of biological activities of paramyxovirus glycoproteins. Activation of cell fusion, hemolysis, and infectivity by proteolytic cleavage of an inactive precursor protein of Sendai virus, *Virology* **57:**475–490.

Scheid, A., Caliguiri, L. A., Compans, R. W., and Choppin, P. W., 1972, Isolation of paramyxovirus glycoprotein: Association of both hemagglutinin and neuraminidase activities with the larger SV5 glycoprotein, *Virology* **50:**640–652.

Schengrund, C.-L., 1990, The role(s) of gangliosides in neural differentiation and repair: A prospective, *Brain Res. Bull.* **24:**131–141.

Schengrund, C.-L., and Nelson, J. T., 1975, Influence of cation concentration of the sialidase activity of neuronal synaptic membranes, *Biochem. Biophys. Res. Commun.* **63:**217–223.

Schengrund, C.-L., and Nelson, J. T., 1976, Ganglioside sialidase activity in bovine neuronal perikarya, *Neurochem. Res.* **1:**171–180.

Schengrund, C.-L., and Repman, M. A., 1982, Density-dependent changes in gangliosides and sialidase activity of murine neuroblastoma cells, *J. Neurochem.* **39:**940–947.

Schengrund, C.-L., and Rosenberg, A., 1970, Intracellular location and properties of bovine brain sialidase, *J. Biol. Chem.* **245:**6196–6200.

Schengrund, C.-L., Jensen, D. S., and Rosenberg, A., 1972, localization of sialidase in the plasma membrane of rat liver cells, *J. Biol. Chem.* **247:**2742–2746.

Schengrund, C.-L., Lausch, R. N., and Rosenberg, A., 1973, Sialidase activity in transformed cells, *J. Biol. Chem.* **248:**4424–4428.

Schengrund, C.-L., Rosenberg, A., and Repman, M. A., 1976, Ecto-ganglioside-sialidase activity of herpes simpex virus transformed hamster embryo fibroblasts, *J. Cell Biol.* **70:**555–561.

Schenkman, S., de Varvalho, L. P., and Nussenzweig, V., 1992, Trypanosoma cruzi trans-sialidase and neuraminidase activities can be mediated by the same enzyme, *J. Exp. Med.* **175:**567–575.

Schiller, H., Segler, K., Jeserich, G., Rosner, H., and Rahmann, H., 1979, Properties of membrane-bound neuraminidase in the developing trout brain, *Life Sci.* **25:**2029–2033.

Schneider-Jakob, H. R., and Cantz, M., 1991, Lysosomal and plasma membrane gangliosides GM3 sialidases of cultured human fibroblasts, *Biol. Chem. Hoppe-Seyler* **372:**443–450.

Schwarzmann, G., 1978, A simple and novel method for tritium labelling of gangliosides and other sphingolipids, *Biochim. Biophys. Acta* **529:**105–114.

Scudder, P., Doom, J. P., Chuenkova, M., Manger, I. D., and Pereira, M. E., 1993, Enzymatic characterization of β-D-galactoside α2,3-trans-sialidase from Trypanosoma cruzi, *J. Biol. Chem.* **268:**9886–9891.

Segler-Stahl, K., Rosner, H., and Rahmann, H., 1981, Membrane-bound neuraminidase in the developing mouse brain, *Life Sci.* **29:**1363–1367.

Shaw, M. W., Lamb, R. A., Erickson, B. W., Briedis, D. J., and Choppin, P. W., 1982, Complete nucleotide sequence of the neuraminidase gene of influenze B virus, *Proc. Natl. Acad. Sci. USA* **79:**6817–6821.

Sheehan, J. P., and Iorio, R. M., 1992, A single acid substitution in the hemagglutinin-neuraminidase of Newcastle disease virus results in a protein deficient in both functions, *Virology* **189:**778–781.

Shibata, H., Kanda, T., Hazama, A., Adachi, A., and Matsumoto, M., 1981, Parainfluenza 3 virus: Plaque-type variants lacking neuraminidase activity, *Infect. Immun.* **34:**262–267.

Srivastava, P. N., and Abou-issa, H., 1977, Purification and properties of rabbit spermatozoal acrosomal neuraminidase, *Biochem. J.* **161:**193–200.

Stenberg, P., Levin, J., Baker, G., Mok, Y., and Corash, L., 1991, Neuraminidase-induced thrombocytopenia in mice: Effects on thrombopoiesis, *J. Cell. Physiol.* **147:**7–16.

Straus, D. C., Unbehagen, P. J., and Purdy, C. W., 1993, Neuraminidase production by a Pasteurella haemolytica A1: Strain associated with bovine pneumonia, *Infect. Immun.* **61:**253–259.

Suttajit, M., and Winzler, R., 1971, Effect of modification of N-acetylneuraminic acid on the binding of glycoproteins to influenza virus and on susceptibility to cleavage by neuraminidase, *J. Biol. Chem.* **246:**3398–3404.

Suzuki, K., 1993, Genetic disorders of lipid, glycoprotein, and mucopolysaccharide metabolism, in: *Basic Neurochemistry,* 5th ed. (G. J. Siegel, B. W. Argaroff, R. W. Albers, and P. B. Molinoff, eds.), Raven Press, New York, pp. 793–812.

Taha, B., and Carubelli, R., 1967, Mammalian neuraminidase: Intracellular distribution and changes of enzyme activity during location, *Arch. Biochem. Biophys.* **119:**55–61.

Taira, S., and Nariuchi, H., 1988, Possible role of neuraminidase in activated T cells in the recognition of allogeneic Ia, *J. Immunol.* **141**:440–446.

Takahashi, T., Ryan, K. W., and Portner, A., 1992, Expression of cDNA encoding the Sendai virus hemagglutinin-neuraminidase gene: Characterization of wild-type mutant gene products, *Virology* **187**:837–840.

Taki, T., Nishiwaki, S., Ishii, K., and Handa, S., 1990, A simple and specific assay of glycosyltransferase and glycosidase activities by an enzyme linked immunosorbent assay method, and its application to assay of galactosyltransferase activity in sera from patients with cancer, *J. Biochem.* **107**:493–498.

Tallman, J. F., and Brady, R. O., 1973, The purification and properties of a mammalian neuraminidase (sialidase), *Biochim. Biophys. Acta* **293**:434–443.

Tanaka, H., Ito, F., and Iwasaki, T., 1992, Purification and characterization of sialidase from Bacteroides fragilis SBT3182, *Biochem. Biophys. Res. Commun.* **189**:524–529.

Tanaka, H., Ito, F., and Iwasaki, T., 1994, Two sialidases which preferentially hydrolyze sialyl α2-8 linkage from Bacteroides fragilis SBT3182, *J. Biochem.* **115**:318–321.

Taylor, G., Vimr, E., Garman, E., and Laver, G., 1992a, Purification, crystallization and preliminary crystallographic study of neuraminidase from Vibrio cholerae and Salmonella typhimurium LT2, *J. Mol. Biol.* **226**:1287–1290.

Taylor, G., Dineley, L., Glowka, M., and Laver, G., 1992b, Crystallization and preliminary crystallographic study of neuraminidase from Micromonospora viridifaciens, *J. Mol. Biol.* **225**:1135–1136.

Taylor, G., Garman, E., Webster, R., Saito, T., and Laver, G., 1993, Crystallization and preliminary X-ray studies of influenza A virus neuraminidase of subtype N5, N6, N8, and N9, *J. Mol. Biol.* **230**:345–348.

Terzidis-Trabelsi, H., Pilatte, Y., Greffard, A., Bignon, J., and Lambre, C., 1991, Sialidase in the guinea pig pulmonary parenchyma: Increased activity in the cytosolic and microsomal subcellular fractions after stimulation with bacillus Calmette Guerin, *Biol. Chem. Hoppe-Seyler* **372**:437–442.

Tettamanti, G., and Zambotti, V., 1968, Purification of neuraminidase from pig brain and its action on different gangliosides, *Enzymologia* **35**:61–74.

Tettamanti, G., Preti, A., Lombardo, A., Gasparini, M., and Zambotti, V., 1972, Assay of brain particulate neuraminidase III. Preparation of the enzyme devoid of endogenous substrates, *Biochim. Biophys. Acta* **258**:228–237.

Tettamanti, G., Preti, A., Lombardo, A., Suman, T., and Zambotti, V., 1975, Membrane-bound neuraminidase in the brain of different animals: Behavior of the enzymes on endogenous sialo derivatives and rationale for its assay, *J. Neurochem.* **25**:451–456.

Teufel, M., Roggentin, P., and Schauer, R., 1989, Properties of sialidase isolated from Actinomyces viscosus DSM 43798, *Biol. Chem. Hoppe-Seyler* **370**:435–443.

Thomas, J. J., Folgler, E. C., Nist, D. L., Thomas, B. J., and Jones, R. H., 1978, Km values of influenza virus neuraminidases for a new fluorogenic substrate, 4-methylumbelliferone N-acetylneuraminic acid ketoside, *Anal. Biochem.* **88**:461–467.

True, D., Singer, M. S., Lasky, L. A., and Rosen, S. D., 1990, Requirement for sialic acid on the endothelial ligand of a lymphocyte homing receptor, *J. Cell Biol.* **111**:2757–2764.

Tsuji, S., Yamauchi, T., Hiraiwa, M., Isobe, T., Okuyama, T., Sakimura, K., Takahashi, Y., Nishizuka, M., Uda, Y., and Miyatake, T., 1989, Molecular cloning of a full-length cDNA for human α-N-acetylgalactosaminidase (α-galactosidase B), *Biochem. Biophys. Res. Commun.* **163**:1498–1504.

Tulip, W. R., Varghese, J. N., Baker, A. T., Van Donkelaar, A., Laver, W. G., Webster, R. G., and Colman, P. M., 1991, Refined atomic structures of N9 subtype influenza virus neuraminidase and escape mutants, *J. Mol. Biol.* **221**:487–497.

Tulip, W. R., Varghese, J. N., Laver, W. G., Webster, R. G., and Colman, P. M., 1992a, Refined crystal structure of the influenza virus N9 neuraminidase-NC41 Fab complex, *J. Mol. Biol.* **227:**122–148.

Tulip, W. R., Varghese, J. N., Webster, R. G., Laver, W. G., and Colman, P. M., 1992b, Crystal structures of two mutant neuraminidase–antibody complexes with amino acid substrates in the interface, *J. Mol. Biol.* **227:**149–159.

Tulsiani, D.R.P., and Carubelli, R., 1971, Studies on the soluble and lysosomal neuraminidase of rat mammary glands, *Biochim. Biophys. Acta* **227:**139–153.

Tuppy, H., and Palese, P., 1969, A chromogenic substrate for the investigation of neuraminidase, *FEBS Lett.* **3:**72–75.

Uchida, Y., Tsukada, Y., and Sugimori, T., 1977, Distribution of neuraminidase in Arthrobacter and its purification by affinity chromatography, *J. Biochem.* **82:**1425–1433.

Uchida, Y., Tsukada, Y., and Sugimori, T., 1979, Enzymatic properties of neuraminidase from Arthrobacter ureafaciens, *J. Biochem.* **86:**1573–1585.

Uemura, H., Schenkman, S., Nussenzweig, V., and Eichinger, D., 1992, Only some members of a gene family in Trypanosoma cruzi encode proteins that express both trans-sialidase and neuraminidase activities, *EMBO J.* **11:**3837–3844.

Usuki, S., Lyu, S.-C., and Sweeley, C. C., 1988a, Sialidase activities of cultured human fibroblasts and the metabolism of GM3 ganglioside, *J. Biol. Chem.* **263:**6847–6853.

Usuki, S., Hoops, P., and Sweeley, C. C., 1988b, Growth control of human foreskin fibroblasts and inhibition of extracellular sialidase activity by 2-deoxy-2,3-dehydro-N-acetylneuraminic acid, *J. Biol. Chem.* **263:**10595–10599.

Vaheri, A., Ruoslahti, E., and Nordling, S., 1972, Neuraminidase stimulates division and sugar uptake in density-inhibited cell cultures, *Nature New Biol.* **238:**211–212.

van Aswegen, C. H., Van Resburg, H.G.J., Becker, P. J., Wittliff, J. L., and Du Plessis, D. J., 1990, Influence of sialic acid on the binding activity of estrogen receptors, *Clin. Physiol. Biochem.* **8:**169–178.

van der Horst, G.T.J., Galjart, N. T., d'Azzo, A., Galjaard, H., and Verheijen, F. W., 1989, Identification and in vitro reconstitution of lysosomal neuraminidase from human placenta, *J. Biol. Chem.* **264:**1317–1322.

van der Horst, G.T.J., Mancini, G.M.S., Brossmer, R., Rose, U., and Verheijen, F. W., 1990a, Photoaffinity labeling of a bacterial sialidase with an aryl azide derivative of sialic acid, *J. Biol. Chem.* **265:**10801–10804.

Vandekerckhove, F., Schenkman, S., de Carvalho, P. L., Tomlinson, S., Kiso, M., Yoshida, M., Hasegawa, A., and Nussenzweig, V., 1992, Substrate specificity of the Trypanosoma cruzi trans-sialidase, *Glycobiology* **2:**541–548.

van der Horst, G.T.J., Rose, U., Brossmer, R., and Verheijen, F. W., 1990b, Photoaffinity labeling of the lysosomal neuraminidase from bovine testis, *FEBS Lett.* **277:**42–44.

van Dessel, G., De Wolf, M., Lagrou, A., Hilderson, H., and Dierick, W., 1984, Characterization, purification, and subcellular localization of bovine thyroid sialidases, *J. Biochem.* **96:**937–947.

van Lenten, L., and Ashwell, G., 1971, Studies on the chemical and enzymatic modification of glycoproteins, *J. Biol. Chem.* **246:**1889–1894.

Varghese, J. N., and Colman, P. M., 1991, Three-dimensional structure of the neuraminidase of influenza virus A/Tokyo/3/67 at 2.2 A resolution, *J. Mol. Biol.* **221:**473–486.

Varghese, J. N., Laver, W. G., and Colman, P. M., 1983, Structure of the influenza virus glycoprotein antigen neuraminidase at 2.9 A resolution, *Nature* **303:**35–40.

Varghese, J. N., McKimm-Breschkin, J. L., Caldwell, J. B., Kortt, A. A., and Colman, P. M., 1992, The structure of the complex between influenza virus neuraminidase and sialic acid, the viral receptor, *Proteins* **14:**327–332.

Varki, A., 1992, Diversity in the sialic acids, *Glycobiology* **2:**25–40.

Veh, R. W., Corfield, A. P., Sander, M., and Schauer, R., 1977, Neuraminic acid-specific modification and tritium labeling of gangliosides, *Biochim. Biophys. Acta* **486:**145–160.

Venerando, B., Tettamanti, G., Cestaro, B., and Zambotti, V., 1975, Studies on brain cytosol neuraminidase. I. Isolation and partial characterization of two forms of the enzyme from pig brain, *Biochim. Biophys. Acta* **403:**461–472.

Venerando, B., Preti, A., Lombardo, A., Cestaro, B., and Tettamanti, G., 1978, Studies on brain cytosol neuraminidase. II. Extractability, solubility and intraneuronal distribution of the enzyme in pig brain, *Biochim. Biophys. Acta* **527:**17–30.

Venerando, B., Goi, G. C., Preti, A., Fiorilli, A., Lombardo, A., and Tettamanti, G., 1982, Cytosolic sialidase in developing rat forebrain, *Neurochem. Int.* **4:**313–320.

Venerando, B., Fiorilli, A., Malesci, A., Goi, G. C., Lombardo, A., Preti, A., and Tettamanti, G., 1983, Cytosolic sialidase from the nerve ending of developing rat forebrain, *Neurochem. Int.* **5:**619–624.

Venerando, B., Fiorilli, A., Masserini, M., Giuliani, A., and Tettamanti, G., 1985, Interaction of pig brain cytosolic sialidase with gangliosides. Formation of catalytically inactive enzyme–ganglioside complexes, *Biochim. Biophys. Acta* **833:**82–92.

Venerando, B., Fiorilli, A., Caimi, L., and Tettamanti, G., 1987, Interaction of pig brain cytosolic sialidase with gangliosides. The formation of catalytically inactive enzyme–ganglioside complexes required homogenous ganglioside micelles and is a resensible phenomenon, *J. Biochem.* **102:**1167–1176.

Verheijen, F., Brossmer, R., and Galjaard, H., 1982, Purification of β-galactosidase and acid neuraminidase from bovine testis: Evidence for an enzyme complex, *Biochem. Biophys. Res. Commun.* **108:**868–875.

Verheijen, F. W., Janse, H. C., Van Diggelen, O. P., Bakker, H. D., Loonen, M.B.C., Durand, P., and Galjaard, H., 1983, Two genetically different MU-NANA neuraminidases in human leucocytes, *Biochem. Biophys. Res. Commun.* **117:**470–478.

Verheijen, F. W., Palmeri, S., Hoogeveen, T., and Galjaard, H., 1985, Human placental neuraminidase: Activation, stabilization and association with β-galactosidase and its protective protein, *Eur. J. Biochem.* **149:**315–321.

Verheijen, F. W., Palmeri, S., and Galjaard, H., 1987, Purification and partial characterization of lysosomal neuraminidase from human placenta, *Eur. J. Biochem.* **162:**63–67.

Vimr, E. R., Lawrisuk, L., Galen, J., and Kaper, J. B., 1988, Cloning and expression of the Vibrio cholerae neuraminidase gene nanH in Escherichia coli, *J. Bacteriol.* **170:**1495–1504.

Visser, A., and Emmelot, P., 1973, Studies on plasma membranes, *J. Membr. Biol.* **14:**73–84.

von Itzstein, M., Wu, W-Y., Kok, G. B., Pegg, M. S., Dyason, J. C., Jin, B., Phan, T. V., Smythe, M. L., White, H. F., Oliver, S. W., Colman, P. M., Vargese, J. N., Ryan, D. M., Woods, J. M., Bethell, R. C., Hotham, V. J., Cameron, J. M., and Penn, C. R., 1993, Rational design of potent sialidase-based inhibitors of influenza virus replication, Science, **363:**418–423.

Ward, C. W., Elleman, T. C., and Azad, A. A., 1982, Amino acid sequence of the pronase-released heads of neuraminidase subtype N2 from the Asian strain A/Tokyo/3/67 of influenza virus, *Biochem. J.* **207:**91–95.

Warner, T. G., and O'Brien, J. S., 1979, Synthesis of 2′-(4-methylumbelliferyl)-α-D-N-acetylneuraminic acid and detection of skin fibroblast neuraminidase in normal humans and in sialidosis, *Biochemistry* **18:**2783–2786.

Warner, T. G., Louie, A., and Potier, M., 1990, Photolabeling of the α-neuraminidase/β-galactosidase complex from human placenta with a photoreactive neuraminidase inhibitor, *Biochem. Biophys. Res. Commun.* **173:**13–19.

Warner, T. G., Harris, R., McDowell, R., and Vimr, E. R., 1992, Photolabeling of Salmonella typhimurium LT2 sialidase, *Biochem. J.* **285:**957–964.

Warren, L., 1959, The thiobarbituric acid assay of sialic acids, *J. Biol. Chem.* **234:**1971–1975.

Warren, L., and Spearing, C. W., 1960, Mammalian sialidase (neuraminidase), *Biochem. Biophys. Res. Commun.* **3**:489–492.
Werner, G., Addick, K., Fricke, U., Klaus, W., Sarram, M., and Gielen, W., 1991, Sialic acid removal modulates the myocardial and vascular activity of calcium channel ligands, *Biochem. Pharmacol.* **42**:S77–S87.
Wille, W., and Trenkner, E., 1981, Changes in particulate neuraminidase activity during normal and staggerer mutant mouse development, *J. Neurochem.* **37**:443–446.
Wrigley, N. G., Skehel, J. J., Charlwood, P. A., and Brand, C. M., 1973, The size and shape of influenza virus neuraminidase, *Virology* **51**:525–529.
Wu, G., and Ledeen, R. W., 1991, Stimulation of neurite outgrowth in neuroblastoma cells by neuraminidase: Putative role of GM1 ganglioside in differentiation, *J. Neurochem.* **56**:95–103.
Xu, G., Suzuki, T., Hanagata, G., Deya, E., Kiso, M., Hasegawa, A., and Suzuki, Y., 1993, Drift of the sialyl-linkage specific recognition of the sialidase of influenza B virus isolates, *J. Biochem.* **113**:304–307.
Yamada, T., Tsuji, S., and Miyatake, T., 1983, Lysosomal sialidase deficiency in sialidosis with partial β-galactosidase deficiency, *Biochim. Biophys. Acta* **755**:106–111.
Yamamoto, Y., and Nishimura, K., 1987, Copurification and separation of β-galactosidase and sialidase from porcine testis, *Int. J. Biochem.* **19**:435–442.
Yee, H. F., Kuwata, J. H., and Langer, G. A., 1991, Effects of neuraminidase on cellular calcium and contraction in cultured cardiac myocytes, *J. Mol. Cell. Cardiol.* **23**:175–185.
Yeung, M. K., 1993, Complete nucleotide sequence of the Actinomyces viscosus T14V sialidase gene: Presence of a conserved repeating sequencing among strain of Actinomyces spp., *Infect. Immun.* **61**:109–116.
Yeung, M., and Fernandez, S. R., 1991, Isolation of a neuraminidase gene from Actinomyces viscosus T14V, *Appl. Environ. Microbiol.* **57**:3062–3069.
Yohe, H. C., and Rosenberg, A., 1977, Action of intrinsic sialidase of rat brain synaptic membranes on membrane sialolipid and sialoprotein components in situ, *J. Biol. Chem.* **252**:2412–2418.
Yohe, H. C., and Rosenberg, A., 1978, Effect of neurotoxic divalent cations on the activity of the intrinsic nerve ending membrane associated sialidase of bovine brain, *Neurochem. Res.* **3**:101–113.
Yohe, H. C., Jacobson, R. I., Yu, R. K., 1983, Ganglioside–basic protein interaction: Protection of gangliosides against neuraminidase action, *J. Neurosci.* **9**:401–412.
Yohe, H. C., Saito, M., Ledeen, R. W., Kunishita, T., Sclafani, J. R., and Yu, R. K., 1986, Further evidence for an intrinsic neuraminidase in CNS myelin, *J. Neurochem.* **46**:623–629.
Yu, R. K., and Ledeen, R. W., 1969, Configuration of the ketosidic bond of sialic acid, *J. Biol. Chem.* **244**:1306–1313.
Yu, R. K., Koerner, T.A.W., Ando, S., and Prestegard, J. H., 1985, High-resolution proton NMR studies of gangliosides. 3. Elucidation of the structure of ganglioside GM3-lactone, *J. Biochem.* **98**:1367–1373.
Zeigler, M., and Bach, G., 1981, Cellular localization of neuraminidase in cultured human fibroblasts, *Biochem. J.* **198**:505–508.
Zeigler, M., Sury, V., and Bach, G., 1989, The identification of lysosomal ganglioside sialidase in human cells, *Eur. J. Biochem.* **183**:455–458.
Zingales, B., Carniol, R., de Lederkremer, R. M., and Colli, W., 1987, Direct sialic acid transfer from a protein donor to glycolipids of trypomastigote forms of Trypanosoma cruzi, *Biochem. Parasitol.* **26**:35–144.

Chapter 9

Sialic Acid as Receptor Determinant of Ortho- and Paramyxoviruses

Georg Herrler, Jürgen Hausmann, and Hans-Dieter Klenk

1. INTRODUCTION

The designation myxoviruses has been chosen historically for a group of viruses comprising influenza viruses, Newcastle disease virus, and mumps virus, because they were able to interact with mucins (see Chapter 5). A characteristic feature of these viruses was the presence of two activities that appeared to counteract each other. On the one hand, the viruses bind to receptors present on erythrocytes, resulting in a hemagglutination reaction. On the other hand, they contain a receptor-destroying enzyme rendering erythrocytes resistant to the viral agglutinating activity. The receptor determinant recognized by this group of viruses turned out to be sialic acid, and the receptor-destroying enzyme has been characterized as a sialidase and a sialate *O*-acetylesterase. Despite the similarities, members of the myxovirus group differ in several fundamental aspects, e.g., the presence of a segmented or a nonsegmented genome. Therefore, they have been grouped into two taxonomic families. Influenza A, B, and C viruses belong to the family Orthomyxoviridae. Viruses such as mumps virus, Newcastle disease virus, Sendai virus, and other parainfluenza viruses are members of the family Paramyxoviridae. In addition to the receptor-binding and the receptor-destroying activity, these viruses have a fusion activity. They differ, however, in

---

Georg Herrler, Jürgen Hausmann, and Hans-Dieter Klenk Virology Institute, University of Marburg, Marburg, Germany.

Biology of the Sialic Acids, edited by Abraham Rosenberg. Plenum Press, New York, 1995.

the distribution of the three activities on the viral surface glycoproteins. With influenza A and B viruses, receptor-binding and fusion activity are functions of the hemagglutinin (HA). A second glycoprotein (NA) is responsible for the sialidase activity. In the case of paramyxoviruses, the receptor-binding and the sialidase activity are combined on one type of glycoprotein designated HN, whereas the fusion activity is localized on a separate glycoprotein (F). Influenza C viruses have only a single glycoprotein that is responsible for all three activities. They also differ from the other viruses in that they recognize *N*-acetyl-9-*O*-acetylneuraminic acid (Neu5,9Ac_2) rather than *N*-acetylneuraminic acid (Neu5Ac) as receptor determinant. Furthermore, the receptor-destroying enzyme of influenza C viruses is a sialate 9-*O*-acetylesterase. In this chapter, the interaction of viral proteins with sialic acid and the biological significance of the receptor-binding and the receptor-destroying activity are discussed.

2. RECEPTOR-BINDING PROTEINS

2.1. The HA Protein of Influenza A and B Viruses

HA is the receptor-binding protein of influenza A viruses. The HA protein is a type I integral membrane protein that protrudes from the viral envelope (reviewed by Lamb, 1989). The HA protein is modified during its transport from the rough endoplasmic reticulum to the host cell's plasma membrane, where the virus acquires its lipid envelope by a budding process. The posttranslational modifications of HA include the attachment of *N*-linked oligosaccharides and fatty acid molecules. Neither glycosylation nor acylation has been unambiguously correlated with a defined function of HA. Another posttranslational modification, however, is indispensable for the virus to acquire its fusion activity: the proteolytic cleavage of HA into the subunits HA_1 and HA_2 (reviewed by Klenk and Garten, 1994). HAs that contain the motif Arg-Arg/Lys-Xxx-Arg are cleaved intracellularly by a ubiquitous protease that has been identified as furin. For those influenza strains, including the human ones, that contain only a single Arg at the cleavage site, proteolytic activation of HA occurs at the cell surface or after release of the virus from the infected cell. In natural infections, the activating protease is expected to be present in the respiratory epithelium, the target tissue of influenza viruses. A corresponding enzyme has been isolated recently from Clara cells derived from the bronchial epithelium of rats and mice. Influenza viruses grown in embryonated eggs are activated by a protease that is related to blood-clotting factor X. The two products generated by the proteolytic cleavage (HA_1 and HA_2) are held together by a single disulfide bond. The membrane-anchored subunit HA_2 contains at the N-terminus a conserved hydrophobic sequence rich in glycine residues. This region is important for the fusion activity of influenza A viruses.

HA is a trimer. The monomers are not covalently linked and have a length of about 560 amino acids. At the C-terminus, a short cytoplasmic tail of 10 amino acids is followed by a stretch of about 27 hydrophobic residues that anchors HA in the membrane. The ectodomain of the HA of strain A/Aichi/68 can be released by bromelain treatment of the virus and was used to determine the three-dimensional structure by X-ray crystallography to 3 Å resolution (Wilson *et al.*, 1981). These studies indicated that the influenza HA extends 135 Å from the viral lipid bilayer and consists of two major elements. The membrane proximal portion has been described as an elongated stem comprising the HA_2 and part of the HA_1 subunit. On top of this stem, there is a globular domain

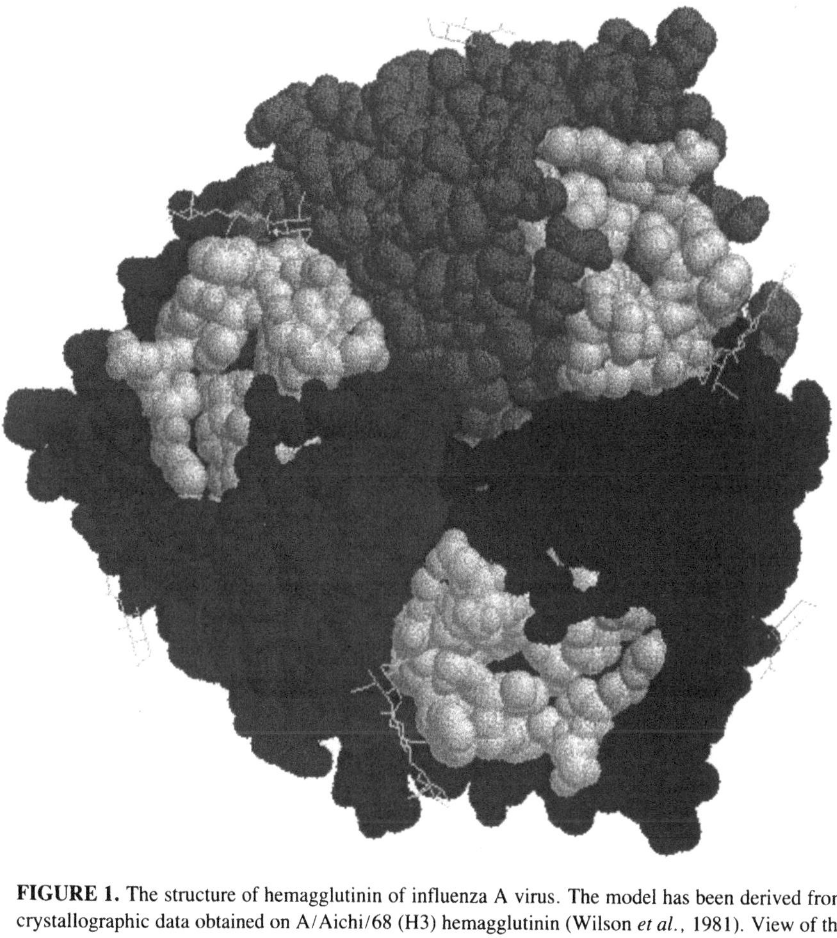

FIGURE 1. The structure of hemagglutinin of influenza A virus. The model has been derived from crystallographic data obtained on A/Aichi/68 (H3) hemagglutinin (Wilson *et al.*, 1981). View of the hemagglutinin trimer is down the threefold symmetry axis. The three monomers (black, dark gray, and light gray) and the receptor binding sites (white) are shown. For details see text.

derived from residues of the HA_1 subunit (Figure 1). The globular head region contains most of the antigenic epitopes as well as the receptor-binding site. The latter domain is a depression at the tip of the molecule (Weis *et al.*, 1988). Several of the amino acids that form the receptor pocket are conserved among the 14 serological subtypes of HA (Nobusawa *et al.*, 1991). Therefore, the overall shape of the binding site appears to be very similar in the HAs of different strains of influenza A viruses. The conserved residues include Tyr98 and Trp153 at the bottom of the cavity. At the rear side, His183 is completely and Glu190 and Leu194 are highly conserved, being replaced by related amino acids (Asp190 and Ile194) in the H1 subtype of influenza A viruses. The right side of the receptor-binding site is formed by amino acids 134–138, with Gly134 being conserved and residues 136 (Thr or Ser) and Ala138 being highly conserved. The amino acids 224–228 at the left side appear to be less important for maintaining the basic structure of the receptor-binding site. They are less conserved than the residues mentioned above and a deletion mutant lacking amino acids 224–230 was still able to bind sialic acid-containing receptors (Daniels *et al.*, 1987). However, as discussed below (Section 5.3), residues located at the left side of the cavity may modulate the receptor-binding specificity.

At the N-terminus of HA_2, there is a stretch of 25 hydrophobic amino acids that is rich in glycine residues and plays a key role in the fusion activity of influenza viruses (reviewed by Wharton *et al.*, 1989). It has a helical structure and is located within the stem region of HA about 3 nm from the viral membrane. The N-terminal glycine is buried in the trimeric interface of the molecule. In an acidic environment, e.g., within an endosome, HA undergoes a conformational change resulting in the exposure of the hydrophobic peptide and the subsequent fusion of the viral envelope with the target membrane. Recently it has been proposed that a coiled coil structure of 80 Å in the native state is extended in the fusogenic form to a 135-Å coiled coil, relocating the fusion peptide by 100 Å toward the target membrane (Carr and Kim, 1993).

HA of influenza B viruses has not been characterized to the same extent as the corresponding protein of influenza A viruses. The amino acid sequences deduced from the nucleotide suquence of different strains indicate that the amino acid identity between HAs of type A and B influenza viruses is about 30% with most of the cysteine residues being conserved (Lamb, 1989). It is expected that the main structural features are very similar in both proteins.

2.2. The HEF Protein of Influenza C Virus

Influenza C viruses differ from influenza A and B viruses by containing only a single type of surface glycoprotein, the HEF protein (reviewed by Herrler and Klenk, 1991). The three-letter designation HEF indicates that the protein has three activities: receptor-binding (*h*emagglutination), receptor-inactivation (*es*-

terase), and *fusion* activity (Herrler *et al.*, 1988a). The spike proteins are arranged in a regular pattern consisting mainly of hexagons. The glycoprotein is inserted into the viral lipid envelope as a homotrimer. The monomers comprise about 640 amino acids. Like HA of influenza A and B viruses, the HEF protein is a type I membrane glycoprotein. A hydrophobic amino acid sequence at the C-terminus is assumed to function as a membrane anchor that is followed by a cytoplasmic tail of only three amino acids. Another hydrophobic sequence located around amino acid 450 is probably involved in the fusion activity. Similar to influenza A and B viruses, the fusion activity of influenza C virus requires proteolytic cleavage into the subunits HEF1 and HEF2 that are held together by disulfide bonds (Ohuchi *et al.*, 1982; Kitame *et al.*, 1982) Cleavage occurs between a single arginine and the hydrophobic sequence mentioned above. HEF2 is the membrane-anchored portion of the influenza C glycoprotein containing the "fusion peptide" at its N-terminus. Proteolytic cleavage is not required for the hemagglutinating and the receptor-destroying activity.

Additional posttranslational modifications are the addition of seven Asn-linked oligosaccharides and acylation with fatty acids attached presumably to the cysteine residue preceding the three amino acids of the cytoplasmic tail. In contrast to most other glycoproteins that contain palmitic acid, stearic acid was found to be the prevailing fatty acid attached to the HEF protein (Veit *et al.*, 1990). Another unusual feature of the influenza C glycoprotein is the low electrophoretic mobility under nonreducing conditions, suggesting a molecular weight of about 100,000 in contrast to the expected size of about 80,000. The slow electrophoretic migration is the result of a conformational change that takes place about 5–10 min after synthesis and involves the formation of intramolecular disulfide bonds (Szepanski *et al.*, 1994).

Because the three-dimensional structure of the HEF protein has not been elucidated, the receptor-binding site is not exactly known. Mutants of influenza C virus have been described that differed from the parent virus in the receptor-binding activity. The variants were obtained either by growing the parent virus in different host cells (Szepanski *et al.*, 1992; Umetsu *et al.*, 1992) or by selection with monoclonal antibodies for neutralization-resistant mutants (Matsuzaki *et al.*, 1992). With some of the variants the receptor-binding activity was improved, whereas others were less efficient than the parent virus in the recognition of cellular receptors. Sequence analysis revealed that the change of the receptor-binding activity was the result of point mutations in a domain of the HEF protein comprising amino acids 178–284. Mutations within amino acids 178 and 226 in general had a negative effect on the recognition of receptors. However, mutations within amino acids 245 and 284 improved the receptor-binding activity. Several mutants had an amino acid exchange in positions 283 and 284. These mutations are located next to the sequence 279-Gly-Asn-Ser-Gly-282, that in similar form (GQSG) is part of the receptor-binding pocket in HA of several influenza A subtypes. Therefore, the sequence around amino acid 280 may be involved in the receptor-binding pocket of the influenza C glycoprotein. An exact localization of this functional domain awaits elucidation of the three-dimensional structure by X-ray crystallography.

2.3. The HN Protein of Paramyxoviruses

Viruses of the genus *Paramyxovirus* within the family Paramyxoviridae contain two surface proteins protruding from the lipid envelope: the F protein that is responsible for the fusion activity of these viruses and the HN protein that has receptor-binding as well as receptor-destroying activity. The mature HN protein is a homotetramer the monomers of which (depending on the strain) may be held together by disulfide bonds. The monomers have a size of about 570–580 amino acids (reviewed by Morrison and Portner, 1991). Though the overall sequence similarity between the HN proteins of different strains is low, structurally important cysteine and glycine residues have been reported to be conserved. In contrast to the HA and HEF proteins described above, the HN protein is a type II integral membrane protein. It contains only a single long stretch of hydrophobic amino acids. This sequence is located about 20–30 amino acids from the N-terminus and probably serves both as a signal peptide and as a membrane anchor.

Posttranslational modifications of all HN proteins are the formation of intramolecular disulfide bonds and the attachment of *N*-linked oligosaccharides. Acylation with fatty acid has been observed only with some strains of paramyxoviruses. In order to be biologically active, the HN protein of a few strains of Newcastle disease virus has to be converted by proteolytic cleavage from a precursor (HN_0) to the mature form (HN). Only the latter form is able to agglutinate erythrocytes (Nagai and Klenk, 1977). The functional domain of the hemagglutinating activity, i.e., the sialic acid-binding site, is unknown. Sequence analysis suggested some amino acids that may be involved in the interactions with receptors. However, as the receptor-binding site is expected to be a conformation-dependent domain, the structure and location of this functional site will be known only after the three-dimensional structure has been determined by X-ray crystallography.

3. RECEPTOR-DESTROYING ENZYMES

3.1. Sialidases

3.1.1. The NA Protein of Influenza A and B Viruses

The sialidase (NA) is the receptor-inactivating enzyme of the influenza A and B viruses which cleaves off the terminal sialic acids of oligosaccharides. It exists in nine serologically distinct subtypes and represents a type II membrane glycoprotein of approximately 60 kDa. In viral membranes, NA is found as a tetramer consisting of two disulfide-linked dimers. The tetramer has a circular fourfold symmetry and an average size of 240 kDa, and is composed of a large box-shaped head domain of 100 × 100 × 60 Å (Varghese *et al.*, 1983) which is anchored in the membrane via a slender stalk of length up to 100 Å and diameter 15 Å. This morphologically defined stalk can be further divided into three domains, the cytoplasmic domain of 6 amino acids, the N-terminal hydrophobic domain, and the actual stalk domain. The cytoplasmic domain containing 6 amino acids is completely conserved in the A type sialidases and has recently been shown to be essential for incorporation of NA into virions and also for the assembly of influenza A virus (Bilsel *et al.*, 1993). The N-terminal hydrophobic domain of 29 amino acids shows only low sequence homology, provides the signal function for translocation of the NA polypeptide into the ER (Bos *et al.*, 1984), and anchors the NA glycoprotein in the membrane (Ward *et al.*, 1982; Blok *et al.*, 1982). The stalk region connecting the head domain of approximately 390 amino acids to the membrane anchor is highly variable in length, varying from 19 to 57 amino acids, and shows considerable sequence variation. The significance of the stalk length has been studied recently by the reverse genetics technique described by Luytjes *et al.* (1989). Shortening of the stalk to about 15 amino acids or even complete removal did not affect the replication of the respective recombinant influenza viruses in tissue culture, but the stalkless mutant showed reduced virulence and an attenuated phenotype in mice (Castrucci and Kawaoka, 1993). In contrast, insertions of 14 to 41 amino acids into the stalk region were tolerated and had no effect on replication of the respective recombinant viruses (Castrucci and Kawaoka, 1993; Luo *et al.*, 1993).

The number of potential glycosylation sites in the different NAs varies from 3 to 7, and NA contains mannose-rich as well as complex oligosaccharides. Glycosylation at a specific site has been correlated with the unique growth properties and virulence of the WSN strain of influenza virus (Schulman and Palese,1977; Li *et al.*, 1993). In contrast to HA, influenza NA does not undergo any proteolytic processing.

The three-dimensional structures of the head domains of N2, N9, and type B sialidases have been determined with high resolution (Varghese and Colman, 1991; Tulip *et al.*, 1991; Burmeister *et al.*, 1992). The results confirmed that these sialidases have essentially the same structure and suggest that the structure of the sialidases of other subtypes will be very similar. The head domain of a monomer is composed of six β-sheets arranged like the blades of a propeller. Each of the β-sheets is made up of four antiparallel β-strands forming a "W" topology and connected by short peptide loops.

Each monomer has a functional site on the upper surface of the head domain (Figure 2). The enzyme catalyzes the hydrolysis of the α-ketosidic linkage be-

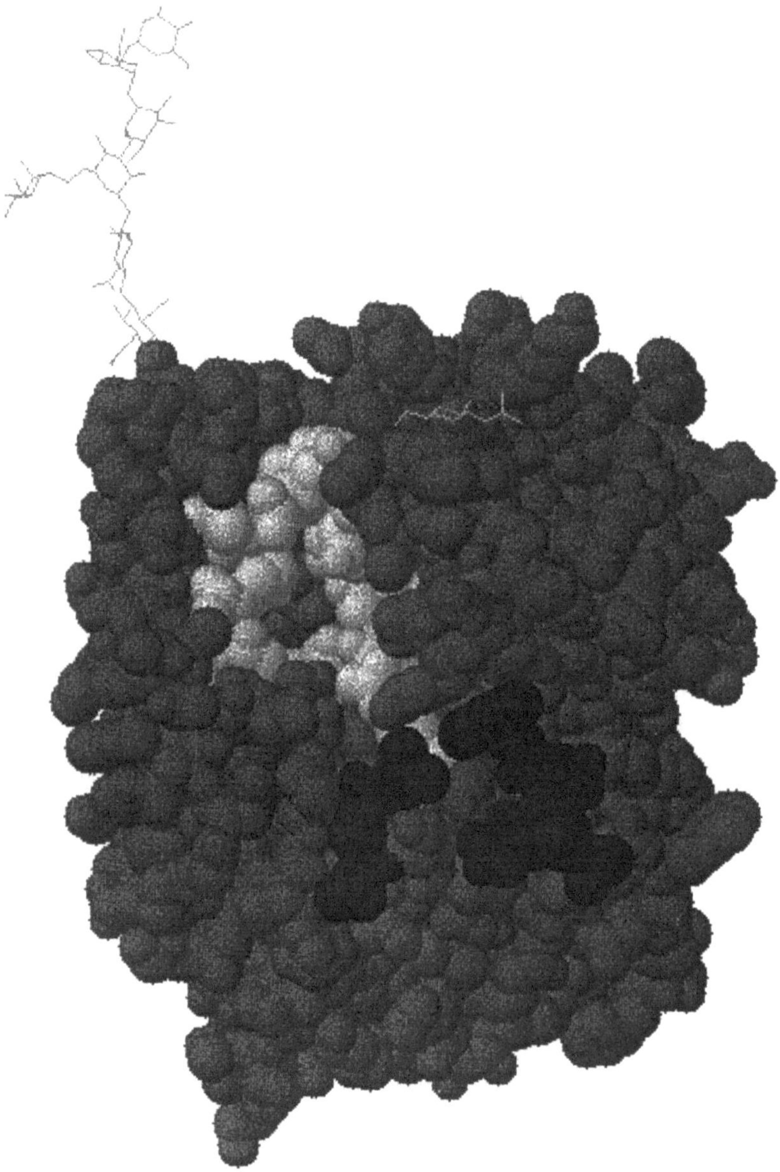

Figure 2. The structure of the sialidase of influenza A virus. The model has been derived from the crystallographic data obtained on A/tern/Australia/G70C/75 sialadase (Baker *et al.*, 1987). View of the monomer is down the fourfold symmetry axis. The catalytic site (white) is indicated. The receptor binding site (black) observed on some sialidase molecules is also shown. For details see text.

tween a terminal sialic acid and the penultimate sugar (Klenk *et al.*, 1955; Gottschalk, 1957). The α2-3 linkage is preferably cleaved by the influenza virus sialidases, but most of these sialidases also hydrolyze the α2-6 linkage (Corfield and Schauer, 1982; Corfield *et al.*, 1983) (see Chapter 8). An interesting observation was the drift of the enzyme specificity of N2 sialidases from human influenza viruses isolated over 30 years toward a higher activity against the α2-6 linkage, which is also preferably recognized by the H2 and H3 HAs of these viruses (Baum and Paulson, 1991). The sialidase of the early virus isolates exclusively hydrolyzed the α2-3 linkage, whereas the activity against the α2-6 linkage increased in the later isolates and was finally comparable to the activity toward the α2-3 linkage. Other aspects of the substrate specificity concern the nature of the aglycon portion of the sialic acid-containing glycoconjugate and the substituents of the hydroxyl groups of the sialic acid, especially those at C-7–C-9. For example, 9-*O*-acetylation of sialic acid has been shown to decrease the activity of the sialidase against this substrate (Munoz-Barroso *et al.*, 1992). Sialidase activity can be increased by addition of the bivalent cations calcium and magnesium (Wilson *et al.*, 1967; Dimmock, 1971), but there is no absolute requirement for these metal ions (Wilson *et al.*, 1967).

The active site (Figure 2) has been studied by mutational analysis of conserved residues (Lentz *et al.*, 1987) and X-ray analysis of NA crystals soaked with sialic acid (Colman *et al.*, 1983; Varghese *et al.*, 1992; Burmeister *et al.*, 1992). It is located in a large depression on the top of the molecule and contains a considerable number of charged residues. The active center is lined entirely by 18–20 residues that are invariant in all types and subtypes of NA analyzed so far (Colman, 1989). In the case of the N9 subtype, 12 residues make contact with the substrate and 6 are responsible for maintaining the structural integrity of the active site (Bossart-Whitaker *et al.*, 1993). The local chemical environment in the active site of the sialidase is predominantly basic. The carboxyl group of the substrate sialic acid is bound by three invariant arginine residues (Arg119, Arg294, and Arg372). Trp180, Ile224, and Arg153 are thought to interact with the *N*-acetyl group of sialic acid, the latter hydrogen-bonding to the carbonyl atom of the N-acetyl moiety. This residue may be dispensable for binding of the substrate in spite of its high conservation (Colman *et al.*, 1993). The aspartic acid residue at position 152 (N9 numbering) is most likely involved in catalyzing the hydrolysis of the glycosidic bond via a sialosyl cation transition state complex (Varghese *et al.*, 1992; Chong *et al.*, 1992). A water molecule probably serves as a proton donor involving Glu278 (N9 numbering) in the transfer of this proton (Chong *et al.*, 1991; Varghese *et al.*, 1992). Other residues of the active site together with the ones mentioned above form an intricate network of hydrogen bonds between each other and the enzyme and the substrate. Recently, an empty side pocket of the substrate binding site of the enzyme adjacent to the hydroxyl group at C-4 (O4) of the bound substrate sialic acid was identified, and a derivative of the long-known NA inhibitor DANA (2-deoxy-2,3-didehydro-D-*N*-acetylneuraminic acid) was synthesized with a guanidino side chain at the O4 position of the sialic acid. This side group exactly filled this empty pocket, interacting with two glutamic acid residues (von Itzstein *et al.*, 1993). The resulting molecule was a 1000-fold more potent NA inhibitor than DANA and was specific for the influenza A and B sialidases. The active site of influenza B sialidases is very similar to that of the A type sialidases with the same conserved residues arranged in an almost identical way, although the overall sequence identity between the compared A and B sialidases was only around 30%. Sialidases of the N9 subtype and an N1 sialidase have been found to possess hemagglutinating activity (Laver *et al.*, 1984; Hausmann *et al.*, in preparation). The structural basis for this activity has been elucidated by investigation of escape mutants (Webster *et al.*, 1987). The two short stretches of amino acids responsible for the HA activity are located on two loops on the surface of the NA molecule and form a shallow pocket separate from the large depression of the active site of the enzyme (Figure 2). The significance of these residues in the N9 NA for the HA activity of the sialidase was demonstrated by introduction of these amino acids at the respective positions in an N2 sialidase by site-directed mutagenesis of the N2 gene (Nuss and Air, 1991). By this approach, it has been possible to transfer HA activity to an N2 sialidase that previously had no such activity by itself. The significance of the HA activity of the sialidase for virus replication is still unclear but it seems to be conserved, at least in some sialidases.

3.1.2. The HN Protein of Paramyxoviruses

The HN protein has already been described in Section 2.3 as a receptor-binding protein. The two-letter designation indicates that the protein is not only a hemagglutinin, but also a sialidase, i.e., it acts as an exoglycosidase hydrolytically cleaving terminal sialic acid from oligosaccharides. The sialidases of paramyxoviruses are more restricted in their linkage specificity than bacterial sialidases. Under conditions where the enzyme completely releases sialic acid attached in an α2-3 linkage to galactose, it is virtually inactive toward substrates containing α2-6-linked sialic acid (Drzeniek, 1972). A difference between sialidases from influenza viruses and paramyxoviruses was observed when the enzymes of fowl plague virus and Newcastle disease virus were compared. Both enzymes effectively released sialic acid from α2-3 linkage to galactose, but only the paramyxovirus sialidase was able to cleave sialic acid attached in an α2-8 linkage to another sialic acid molecule.

Because the three-dimensional structure of HN has not been determined, the exact structure and location of the active site are unknown. Sequence analysis indicated, however, that there are several invariant residues in the HN protein of different paramyxoviruses that have counterparts in the enzyme-active center of influenza virus sialidase. Therefore, a model for the three-dimensional structure of the HN protein has been proposed that is similar to the structure determined for the influenza sialidases (Colman *et al.*, 1993).

3.2. Sialate 9-*O*-Acetylesterase

The HEF protein of influenza C virus has been described in Section 2.2 as an HA recognizing *N*-acetyl-9-*O*-acetylneuraminic acid. In addition to the receptor-binding activity, it also has receptor-destroying activity resulting from its ability to catalyze the release of the acetyl group from C-9 of *N*-acetyl-9-*O*-acetylneuraminic acid. The acetylesterase of strain Johannesburg/1/66 was found to hydrolyze the *O*-acetyl group from C-4 of sialic acid at a 30-fold reduced rate and to be inactive toward 7-*O*-acetyl residues (Herrler *et al.*, 1985). The enzyme activity is inhibited by diisopropyl fluorophosphate (DFP)and some isocoumarins, indicating that the influenza C esterase belongs to the class of serine hydrolases (Muchmore and Varki, 1987; Vlasak *et al.*, 1989). The active site of these enzymes involves a catalytic triad composed of the amino acids serine, histidine, and aspartate that are crucial for the cleavage of the substrate. Covalent binding of DFP to amino acid 71 indicated that this residue represents the active-site serine of the influenza C acetylesterase (Herrler *et al.*, 1988b; Vlasak *et al.*, 1989). This residue is part of the motif Gly-Asp-Ser that is present in the active site of several serine proteases. Sequence comparison with other serine hydrolases suggested that His368 or His369 and Asp261 may be the other residues of the catalytic triad of the influenza C esterase. The importance of these amino acids has been confirmed by analysis of variant proteins obtained by site-directed mutagenesis (Pleschka, Klenk, and Herrler, submitted). A detailed characterization of the functional domains awaits elucidation of the three-dimensional structure.

4. VIRUS RECEPTORS

Virus receptors are defined as surface structures that allow viruses to attach to the cell and to initiate an infection. Distinct surface proteins have been identified as virus receptors, e.g., CD4 for HIV and ICAM-1 for rhinoviruses. The actual domain of the protein that is recognized by the virus is designated receptor determinant. Applying this terminology to viruses that attach to any derivatives of neuraminic acid on the cell surface, sialic acid is designated as receptor determinant. As sialic acids are present on many glycoproteins and glycolipids, different surface molecules may serve as receptors for these viruses.

4.1. Receptor Determinant Sialic Acid

The importance of sialic acid for the binding of influenza viruses to cells already had become obvious almost 40 years ago, when the receptor-destroying enzyme was shown to be a sialidase, releasing terminal sialic acid from glycoconjugates (Klenk *et al.*, 1955; Rosenberg *et al.*, 1956). Since then the ability of influenza viruses to use sialic acid as a receptor determinant for binding to cells has been studied in great detail. Some of the methods applied for this purpose involved the reconstitution of receptors on the cell surface after removal of endogenous sialic acid by sialidase treatment. Receptors were generated, for example, by incubation of asialo-cells with sialyltransferase and CMP-activated sialic acid (Markwell and Paulson, 1980; Rogers *et al.*, 1983b). Depending on the specificity of the transferase and the choice of CMP-sialic acid, cells were obtained that contained a defined type of sialic acid bound to a surface glycoconjugate in a defined type of linkage. Analyzing such cells for susceptibility to infection or—in the case of erythrocytes—agglutinability by the respective virus provided information about the preference of different virus strains for a certain type of sialic acid and for a certain type of linkage. In another approach, receptors were generated by incubation of asialo-cells with purified gangliosides (Markwell *et al.*, 1981; Suzuki *et al.*, 1985). The glycolipids attach to the cell surface and in part are incorporated into the plasma membrane. The ability of influenza viruses to bind to such cells and/or to induce fusion of the viral envelope with the cell membrane indicates whether or not the gangliosides are suitable as receptors. From the composition of the glycolipids, information about the specificity of the sialic acid binding activity can be deduced. The interaction between influenza viruses and sialic acid has also been analyzed by using NMR spectroscopy to calculate the dissociation constant and by determining the crystal structure of HA complexed with different types of sialic acids (Sauter *et al.*, 1992).

Using these methods it has been shown that influenza A and B viruses preferentially bind to *N*-acetylneuraminic acid, though some strains may additionally recognize *N*-glycoloylneuraminic acid or *N*-acetyl-9-*O*-acetylneuraminic acid (Higa *et al.*, 1985). This is in contradistinction to influenza C viruses which only use 9-*O*-acetylated sialic acid for binding to cells and do not accept Neu5Ac or Neu5Gc as receptor determinant (Rogers *et al.*, 1986). Apart from the type of sialic acid, the binding of influenza viruses is also dependent on the linkage that connects the sialic acid residue with the oligosaccharide of the receptor molecule. Strains of influenza viruses may have a preference for sialic acid attached to galactose in an α2-6 linkage or for sialic acid attached in an α2-3 linkage. Some strains recognize both linkage types efficiently (Rogers and Paulson, 1983). Even sugar residues not directly connected to sialic acid may affect the binding of influenza viruses. Several strains of influenza virus have been shown to differentiate between Neu5Acα2,3GalGlcNAc and Neu5Acα2,3GalGalNAc, i.e., carbohydrate structures characteristic for *N*- and *O*-linked oligosaccharides, respectively (see Chapter 5). In the case of paramyxoviruses, most information is available for Sendai virus and Newcastle disease virus. Both viruses can use *N*-acetylneuraminic acid α2-3 linked to galactose as a receptor determinant for attachment to cells (Markwell and Paulson, 1980). The disialo group Neu5Ac-α2,8Neu5Ac attached in α2-3 linkage to galactose is recognized more efficiently than is a monosialo group as has been shown with gangliosides containing the respective carbohydrate structures (Markwell *et al.*, 1981). Apart from the type of sialic acid and the type of linkage, the binding of viruses to sialic acid-containing receptors depends also on the valency of the interaction. Free sialic acid is a poor inhibitor of the hemagglutinating activity of influenza viruses. Glycoproteins containing several oligosaccharide side chains, or gangliosides in the form of micelles or liposomes, are potent inhibitors. The inhibitory activity of glycoproteins has been shown to increase after chemical cross-linking or aggregation. The valency of the interaction between viruses and receptors depends not only on the number of oligosaccharides with terminal sialic acid available for binding of the trimeric HA spike, but also on the spatial arrangement of the oligosaccharides. Studies with macroglobulins from different species have shown that glycoproteins of similar size and similar number of sialic acid residues may show dramatic differences in their ability to inhibit the hemagglutinating activity of influenza viruses (Pritchett and Paulson, 1989). The difference is explained by a higher valency of interaction between the virus and some inhibitors because of a more favorable distribution of the carbohydrate side chains and thus of the sialic acid residues. Taken together, the recognition of sialic acid-containing receptors depends on several factors: (1) the type of sialic acid, (2) the type of linkage connecting sialic acid with the oligosaccharide, (3) the structure of the oligosaccharide carrying the terminal sialic acid, (4) the number of sialic acid residues present on the receptor, and (5) the spatial arrangement of the sialic acid residues.

4.2. Receptor Molecules (Glycoproteins, Gangliosides)

Though the role of *N*-acetylneuraminic acid as a receptor determinant for influenza virus has been known for 40 years, the actual receptor used by this virus to infect cells has not been identified unequivocally. As sialic acid is a frequent sugar on glycoproteins and glycolipids of vertebrate cells, most surface components contain *N*-acetylneuraminic acid or a derivative of it. All sialylated molecules present on the cell surface are potential receptors for influenza viruses or paramyxoviruses provided they contain sialic acid according to the requirements discussed above (Section 4.1). In the case of erythrocytes, the majority of the surface-bound sialic acid is present on glycophorin. Following isolation from the plasma membrane, this glycoprotein is a potent inhibitor of hemagglutination and, therefore, represents a major erythrocyte receptor for influenza virus (Marchesi and Andrews, 1971). The glycoprotein GP-2, isolated from bovine erythrocytes, is a strong inhibitor of the hemagglutinating activity of Sendai virus, and thus has to be considered as a potential receptor for this virus (Suzuki *et al.*, 1983).

In the case of cultured cells, a defined receptor molecule has not been identified. Among the different surface components containing sialic acid, it remains to be shown which of them are actually used by the virus for infecting the cell. From studies with Sendai virus, it has been concluded that glycolipids rather than glycoproteins may serve as virus receptors. Several gangliosides of the ganglio series, GD1a, GT1b, and GQ1b, have been shown to be potential receptors for Sendai virus with the latter one being functional already at very low concentrations (Markwell *et al.*, 1981). Gangliosides containing lacto-series oligosaccharide chains were also found to be potential receptors for Sendai virus and Newcastle disease virus (NDV) (Suzuki *et al.*, 1985). Both viruses efficiently recognized *N*-acetylneuraminosylparagloboside with high affinity, while the active ganglioside was a functional receptor only for Sendai virus, but not for NDV. Gangliosides of the ganglio series have been shown to be present in cells susceptible to Sendai virus infection in an amount sufficient to function as virus receptor (Markwell *et al.*, 1984). The role of glycolipids as receptors for Sendai virus is well documented. It has, however, not been excluded that glycoproteins are also involved in the initiation of infection. They may function either as alternative receptors or in cooperation with gangliosides. Because of their size, glycoproteins are more exposed on the cell surface than are glycolipids. Therefore, it is very possible that glycoproteins containing appropriate sialic acids serve as primary attachment sites. From there the virion may, possibly with the help of the receptor-destroying enzyme, proceed to glycolipid receptors. Compared to glycoprotein receptors, the binding to gangliosides would reduce the distance between the viral and the cellular membrane and might thus facilitate the subsequent fusion reaction that is required for the viral genome to get into the cell.

Glycolipids are potential receptors also for influenza viruses (Suzuki *et al.*, 1986; Herrler and Klenk, 1987). As discussed above, glycoproteins may function as alternative receptors. Cells that are resistant to infection by influenza C virus because of a lack of receptors, have been shown to become susceptible to infection if they are resialylated with 9-*O*-acetylated sialic acid (Szepanski *et al.*, 1992). As the α2,6-sialyltransferase used attaches sialic acid specifically to oligosaccharides present on glycoproteins, the virus obviously used glycoprotein receptors to infect these cells. The same approach has been applied to influenza A virus. In this case, cells were pretreated with sialidase to inactivate endogenous receptors. The resistance of the asialo-cells was then overcome by resialylation of the cells with *N*-acetylneuraminic acid (Carroll and Paulson, 1985).

Recently, sensitive binding assays with proteins immobilized on nitrocellulose have been described to identify sialylated proteins that are recognized by viruses (Schultze *et al.*, 1993). Applying this method to cell surface proteins, it should be possible to determine the cellular proteins that are likely to function as virus receptors.

5. ROLE OF SIALIC ACID IN VIRUS REPLICATION

5.1. Virus Entry

It is clear that the HA of influenza viruses and paramyxoviruses plays a prime role in virus attachment to the cell surface. The molecular details of receptor binding have been described above. Although the biological role of the sialidase is not fully understood, there is evidence that the enzyme is also involved in virus entry. It has been assumed that the sialidase permits transport of the virus through mucin in the respiratory tract, thus allowing virus access to the target epithelial cells. Similarly, in systemic infection the enzyme may prevent the virus from being trapped by serum inhibitors. Studies with reconstituted viral envelopes suggested that the sialidase may also be involved in the fusion process, permitting virus penetration (Huang *et al.*, 1980). However, studies with cloned HA (White *et al.*, 1982) indicated no requirement for sialidase, at least for low-pH-dependent fusion of cells. Subsequent studies by Huang *et al.* (1985) included the observation that antisialidase antisera can inhibit fowl plague virus-induced fusion of erythrocytes and that fusion can be restored on addition of *Vibrio cholerae* sialidase. Partial inhibition of hemolysis was also caused by the same antisera. Huang *et al.* (1985) went on to show that the surface properties of erythrocytes were the determining factors in low-pH-dependent fusion and hemolysis by influenza virus. The influence of target membrane composition on fusion has also been demonstrated by Haywood and Boyer (1985). It therefore appears to be important to consider target membrane viability in assessing factors influencing virus-mediated membrane fusion.

5.2. Virus Maturation

Newly synthesized virus particles bud from the plasma membrane of infected cells, and these particles are free of sialic acid (Klenk *et al.*, 1970). Removal of sialic acid from the carbohydrate side chains of the viral glycoproteins and from cellular glycoconjugates in the plasma membrane appears to be a prerequisite for elution of virus from infected cells (Palese *et al.*, 1974; Palese and Compans, 1976).

It has been shown that, in a special case, the sialidase determines whether cleavage of HA occurs (Schulman and Palese, 1977). These authors showed that A/WSN/3-A/HK/68 or WSN-A/FM/1/47 reassortants that derived only the NA gene of the WSN virus did produce infectious virus with cleaved HA in MDCK cells, whereas reassortants that were identical to WSN except for its NA gene, did not. Similar results were obtained among reassortants for A/turkey/Ontario/7732/66 (H5N9) and WSN virus. Again, WSN sialidase determined whether the infectious virus was produced in chicken embryo cells. Thus, the role of sialidase for HA cleavage is restricted to an interaction between a specific HA–NA combination. The mechanism by which a viral NA in some instances may facilitate cleavage of HA is not understood.

5.3. Spread of Infection

Comparison of influenza viruses isolated from humans and other species showed differences in their ability to agglutinate erythrocytes and to combine with glycoprotein inhibitors of hemagglutination. These properties were attributed to functional differences among the viral HA and led to the suggestion that viruses may interact with more than one type of receptor on cells. Thus, the antigenically similar variants RI/5^+ and RI/5^- differed in their capacities to elute from erythrocytes and to bind to hemagglutination-inhibiting soluble glycoproteins (Choppin and Tamm, 1960). These variations were related to differential receptor specificities of the HA to α2-3- and α2,6-linked sialic acid, respectively. In contrast, the sialidases of the two viruses had similar specificities, hydrolyzing only the α2-3 linkage, which explains the failure of the RI/5^+ virus to elute from erythrocytes (Carroll *et al.*, 1981). These studies provide an explanation for the receptor gradient whereby viruses could be ranked by their ability to remove their own receptors from erythrocytes and those for viruses higher up the gradient, while leaving the receptors for other viruses intact.

Such variations in receptor specificity may also be important for the adaptation of a virus from one host system to another. Early studies by Burnet and Bull (1943) documented a receptor shift on adaptation of human influenza viruses (H1N1) to growth in chicken embryos. The original isolate (O virus) could only be grown in the amnion but after several passages could be propagated in the allantoic cavity as the derived (D) virus. Concomitant with adaptation, the binding properties of the virus changed from preferential agglutination of guinea pig erythrocytes by O virus to agglutination of chick erythrocytes by D virus. More recently, human influenza B viruses isolated and propagated by growth in MDCK cells have been found to undergo dramatic selection for host variants when adapted to growth in chicken embryos (Schild *et al.*, 1983). Sequence analyses of the HA genes of theses viruses before and after egg adaptation revealed an amino acid substitution resulting in the loss of a carbohydrate attachment site located on the top of the HA molecule adjacent to the receptor-binding site. Similarly, during egg adaptation of influenza A viruses (H1N1) a variety of amino acid exchanges could be detected, located mostly on the periphery of the receptor-binding site (Robertson *et al.*, 1985).

Genetic dimorphism has also been observed in swine influenza viruses whereby cloned viruses exhibited either high yields (H virus) or low yields (L virus) in eggs. Genetic reassortment analysis revealed that the phenotypic properties of the two viruses were carried by the RNA segment coding for HA. Although both H and L forms of the virus were found in influenza virus isolates from swine, experimental infections of swine suggested that the H form is more virulent (Kilbourne *et al.*, 1978). The primary structural difference between the H and L HAs appears to be a single amino acid change of glutamic acid (H) to glycine (L) at residue 155 (Both *et al.*, 1983). It is not clear whether this exchange is associated with quantifiable differences in receptor-binding properties, but it is remarkable that it is located in a region near the receptor-binding site.

It is generally accepted that the influenza pandemic in 1968 arose from genetic reassortment of the previously circulating human virus and an avian virus similar to A/duck/Ukraine/1/63 which contributed the H3 HA to the new Hongkong virus (Scholtissek *et al.*, 1978). However, it has also been clearly shown that the avian H3 viruses bind preferentially to α2-3-linked neuraminic acid, whereas the human viruses of the same serotype bind to α2-6-linked neuraminic acid. Comparative sequence analyses have revealed that the difference in receptor specificity is related in large part to a point mutation at the receptor binding site involving the exchange of glutamine residue 226 with the avian viruses for leucine with the human strains (Rogers and Paulson, 1983; Rogers *et al.*, 1983a,b). It is therefore reasonable to assume that generation of the Hongkong virus involved selection of a receptor-binding variant that allowed propagation in human tissue.

These observations taken together clearly indicate that the receptor specificities of HA and sialidase are important determinants for the tissue tropism and host range of influenza viruses. It is likely that receptor specificity also controls pathogenicity, but this concept has to be proven by further experimental data.

REFERENCES

Baker, A. T., Varghese, J. N., Laver, W. G., Air, G. M., and Colman, P. M., 1987, Three-dimensional structure of neuraminidase of subtype N9 from an avian influenza virus, *Proteins* **2:**111–117.

Baum, L. G., and Paulson, J. C., 1991, The N2 neuraminidase of human influenza virus has acquired a substrate specificity complementary to the hemagglutinin receptor specificity, *Virology* **180:**10–15.

Bilsel, P., Castrucci, M., and Kawaoka, Y., 1993, Mutations in the cytoplasmatic tail of influenza A virus neuraminidase affect incorporation into virions, *J. Virol.* **67:**6762–6767.

Blok, J., Air, G. M., Laver, W. G., Ward, C. W., Lilley, G. G., Woods, E. F., Roxburgh, C. M., and Inglis, E. S., 1982, Studies on the size, chemical composition and partial sequence of the neuraminidase (NA) from type A influenza virus show that the N-terminal region of the NA is not processed and serves to anchor the NA in the viral membrane, *Virology* **119:**109–121.

Bos, T. J., Davis, A. R., and Nayak, D. P., 1984, NH_2-terminal hydrophobic region of influenza virus neuraminidase provides the signal function in translocation, *Proc. Natl. Acad. Sci. USA* **81:**2327–2331.

Bossart-Whitaker, P., Carson, M., Babu, J. S., Smith, C. D., Laver, W. G., and Air, G. M., 1993, Three-dimensional structure of influenza A N9 neuraminidase and its complex with the inhibitor 2-deoxy 2,3-dehydro-N-acetyl neuraminic acid, *J. Mol. Biol.* **232:**1069–1083.

Both, G. M., Shi, C. H., and Kilbourne, E. D., 1983, The hemagglutinin of swine influenza virus: A single amino acid change pleiotropically affects viral antigenicity and replication, *Proc. Natl. Acad. Sci. USA* **80:**6996–7000.

Burmeister, W. P., Ruigrok, R.W.H., and Cusack, S., 1992, The 2.2Å resolution crystal structure of influenza B neuraminidase and its complex with sialic acid, *EMBO J.* **11:**49–56.

Burnet, F. M., and Bull, D. R., 1943, Changes in influenza virus associated with adaptation to passage in chick embryo, *Aust. J. Exp. Med. Sci.* **21:**55–69.

Carr, C. M., and Kim, P. S., 1993, A spring-loaded mechanism for the conformational change of influenza hemagglutinin, *Cell* **73:**823–832.

Carroll, S. M., and Paulson, J. C., 1985, Differential infection of receptor-modified host cells by receptor-specific influenza viruses, *Virus Res.* **3:**165–179.

Carroll, S. M., Higa, H. H., Paulson, J. C., 1981, Different cell-surface receptor determinants of antigenically similar influenza virus hemagglutinins, *J. Biol. Chem.*, **256:**8357–8363.

Castrucci, M. R., and Kawaoka, Y., 1993, Biologic importance of neuraminidase stalk length in influenza A virus, *J. Virol.* **67:**759–764.

Chong, A. K., Pegg, M. S., and von Itzstein, M., 1991, Characterisation of an ionisable group involved in binding and catalysis by sialidase from influenza virus, *Biochem. Int.* **24:**165–171.

Chong, A. K., Pegg, M. S., Taylor, N. R., and von Itzstein, M., 1992, Evidence for sialosyl cation transition-state complex in the reaction of sialidase from influenza virus, *Eur. J. Biochem.* **207:**335–343.

Choppin, P. W., and Tamm, I., 1960, Studies of two kinds of virus particles which comprise influenza A2 virus strains. I. Characterization of stable homogenous substrains in reactions with specific antibody, mucoprotein inhibitors, and erythrocytes, *J. Exp. Med.* **112:**895–920.

Colman, P. M., 1989, Neuraminidase: Enzyme and antigen, in: *The Influenza Viruses* (R. M. Krug, ed.), Plenum Press, New York, pp. 175–218.

Colman, P. M., Varghese, J. N., and Laver, W. G., 1983, Structure of the catalytic and antigenic sites in influenza virus neuraminidase, *Nature* **303:**41–44.

Colman, P. M., Hoyne, P. A., and Lawrence, M. C., 1993, Sequence and structure alignment of paramyxovirus hemagglutinin-neuraminidase with influenza neuraminidase, *J. Virol.* **67:**2972–2980.

Corfield, A. P., and Schauer, R., 1982, Metabolism of sialic acids, in: *Sialic Acids—Chemistry, Metabolism and Function* (R. Schauer, ed.), Springer-Verlag, Berlin, pp. 195–261.

Corfield, A. P., Higa, H., Paulson, J. C., and Schauer, R., 1983, The specificity of viral and bacterial sialidases for α2-3 and α2-6 linked sialic acids in glycoproteins, *Biochim. Biophys. Acta* **744:**121–126.

Daniels, R. S., Jeffries, S., Yates, P., Schild, G. C., Rogers, G. N., Paulson, J. C., Wharton, S. A.,

Douglas, A. R., Skehel, J. J., and Wiley, D. C., 1987, The receptor binding and membrane fusion properties of influenza virus variants selected using anti-haemagglutinin monoclonal antibodies, *EMBO J.* **6:**1459–1465.

Dimmock, N. J., 1971, Dependence of the activity of an influenza virus neuraminidase upon Ca^{2+}, *J. Gen. Virol.* **13:**481–483.

Drzeniek, R., 1972, Substrate specificities of neuraminidases, *Histochem. J.* **5:**271–290.

Gottschalk, A., 1957, Neuraminidase: The specific enzyme of influenza virus and Vibrio cholerae, *Biochim. Biophys. Acta* **23:**645–646.

Haussman, J., Kretzshcmear, E., Garten, W., and Klenk, H. D., 1995, N1 neuraminidase of influenza virus A1FPV1Rostock134 has hem-adsorbing activity, *J. Gen. Virol.*, in press.

Haywood, A. M., and Boyer, A. P., 1985, Fusion of influenza virus membranes with liposomes of pH 7.5, *Proc. Natl. Acad. Sci. USA* **82:**4611–4615.

Herrler, G., and Klenk, H.-D., 1987, The surface receptor is a major determinant of the cell tropism of influenza C virus, *Virology* **159:**102–108.

Herrler, G., and Klenk, H.-D., 1991, Structure and function of the HEF glycoprotein of influenza C virus, *Adv. Virus Res.* **40:**213–234.

Herrler, G., Rott, R., Klenk, H.-D., Müller, H.-P., Shukla, A. K., and Schauer, R., 1985, The receptor-destroying enzyme of influenza C virus is neuraminate-O-acetylesterase, *EMBO J.* **4:**1503–1506.

Herrler, G., Dürkop, I., Becht, H., and Klenk, H.-D., 1988a, The glycoprotein of influenza C virus is the haemagglutinin, esterase and fusion factor, *J. Gen. Virol.* **69:**839–846.

Herrler, G., Multhaup, G., Beyreuther, K., and Klenk, H.-D., 1988b, Serine 71 of the glycoprotein HEF is located at the active site of the acetylesterase of influenza C virus, *Arch. Virol.* **102:**269–274.

Higa, H. H., Rogers, G. N., and Paulson, J. C., 1985, Influenza virus hemagglutinins differentiate between receptor determinants bearing N-acetyl-, N-glycolyl-, and N,O-diacetylneuraminic acids, *Virology* **144:**279–282.

Huang, R.T.C., Rott, R., Wahn, K., Klenk, H.-D., and Kohama, T., 1980, The function of the neuraminidase in membrane fusion induced by myxoviruses, *Virology* **107:**313–319.

Huang, R.T.C., Dietsch, E., and Rott, R., 1985, Further studies on the role of neuraminidase and the mechanism of low pH dependence in influenza virus-induced membrane fusion, *J. Gen. Virol.* **66:**295–301.

Kilbourne, E. D., 1978, Genetic dimorphism in influenza viruses: Characterization of stably associated hemagglutinin mutants differing in antigenicity and biological properties, *Proc. Natl. Acad. Sci. USA* **75:**6258–6262.

Kitame, F., Sugawara, K., Ohwada, K., and Homma, M., 1982, Proteolytic activation of hemolysis and fusion by influenza C virus, *Arch. Virol.* **73:**357–361.

Klenk, E., Fallard, H., and Lempfrid, H., 1955, Über die enzymatische Wirkung von Influenzaviren, *Hoppe-Seyler's Z. Physiol. Chem.* **301:**235–246.

Klenk, H.-D., and Garten, W., 1994, Host cell proteases controlling virus pathogenicity, *Trends Microbiol.* **2:**39–42.

Klenk, H.-D., Compans, R. W., and Choppin, P. W., 1970, An electron microscopic study of the presence or absence of neuraminic acid in enveloped viruses, *Virology* **42:**1158–1162.

Lamb, R. A., 1989, Genes and proteins of the influenza viruses, in: *The Influenza Viruses* (R. M. Krug, ed.), Plenum Press, New York, pp. 1–87.

Laver, W. G., Colman, P. M., Webster, R. G., Hinshaw, V. S., and Air, G. M., 1984, Influenza virus neuraminidase with hemagglutinin activity, *Virology* **137:**314–323.

Lentz, M. R., Air, G. M., and Webster, R. G., 1987, Site-directed mutation of the active site of influenza neuraminidase and implications for the catalytic mechanism, *Biochemistry* **26:**5351–5358.

Li, S., Schulman, J. L., Itamura, S., and Palese, P., 1993, Glycosylation of neuraminidase determines the neurovirulence of influenza A/WSN/33 virus, *J. Virol.* **67:**6667–6673.

Luo, G., Chung, J., and Palese, P., 1993, Alterations of the stalk of the influenza virus neuraminidase: Deletions and insertions, *Virus Res.* **29:**141–153.

Luytjes, W., Krystal, M., Enami, M., Parvin, J. D., and Palese, P., 1989, Amplification, expression, and packaging of a foreign gene by influenza virus, *Cell* **59:**1107–1113.

Marchesi, V. T., and Andrews, E. P., 1971, Glycoproteins: Isolation from cell membranes with lithium diiodosalicylate, *Science* **174:**1247–1248.

Markwell, M.A.K., and Paulson, J. C., 1980, Sendai virus utilizes specific sialyloligosaccharides as host cell receptor determinants, *Proc. Natl. Acad. Sci. USA* **77:**5693–5697.

Markwell, M.A.K., Svennerholm, L., and Paulson, J. C., 1981, Specific gangliosides function as host cell receptors for Sendai virus, *Proc. Natl. Acad. Sci. USA* **78:**5406–5410.

Markwell, M.A.K., Fredman, P., and Svennerholm, L., 1984, Receptor ganglioside content of three hosts for Sendai virus MDBK, HeLa, and MDCK cells, *Biochim. Biophys. Acta* **775:**7–16.

Matsuzaki, M., Sugawara, K., Adachi, K., Hongo, S., Nishimura, H., Kitame, F., and Nakamura, K., 1992, Location of neutralizing epitopes on the hemagglutinin-esterase protein of influenza C virus, *Virology* **189:**79–89.

Morrison, T., and Portner, A., 1991, Structure, function, and intracellular processing of the glycoproteins of Paramyxoviridae, in: *The Paramyxoviruses* (D. W. Kingsbury, ed.), Plenum Press, New York, pp. 347–382.

Muchmore, E. A., and Varki, A., 1987, Selective inactivation of influenza C esterase: A probe for detecting 9-O-acetylated sialic acids, *Science* **230:**1293–1295.

Munoz-Barroso, I., Garcia-Sastre, A., Villar, E., Manuguerra, J. C., Hannoun, C., and Cabezas, J. A., 1992, Increased influenza A virus sialidase activity with N-acetyl-9-O-acetylneuraminic acid-containing substrates resulting from influenza C virus O-acetylesterase action, *Virus Res.* **25:**145–153.

Nagai, Y., and Klenk, H.-D., 1977, Activation of precursors to both glycoproteins of Newcastle disease virus by proteolytic cleavage, *Virology* **77:**125–134.

Nobusawa, E., Aoyama, T., Kato, H., Suzuki, Y., Tateno, Y., and Nakajima, K., 1991, Comparison of complete amino acid sequences and receptor-binding properties among 13 serotypes of hemagglutinins of influenza A viruses, *Virology* **182:**475–485.

Nuss, J. M., and Air, G. M., 1991, Transfer of the hemagglutinin activity of influenza virus neuraminidase subtype N9 into an N2 neuraminidase background, *Virology* **183:**496–504.

Ohuchi, M., Ohuchi, R., and Mifune, K., 1982, Demonstration of hemolytic and fusion activities of influenza C virus, *J. Virol.* **42:**1076–1079.

Palese, P., and Compans, R. W., 1976, Inhibition of influenza virus replication in tissue culture by 2-deoxy-2,3-dehydro-N-trifulone-acetyl-neuraminic acid (FANA): Mechanism of action, *J. Gen. Virol.* **33:**159–163.

Palese, P., Tobita, K., Ueda, M., and Compans, R. W., 1974, Characterization of temperature-sensitive influenza virus mutants defective in neuraminidase, *Virology* **61:**397–410.

Pritchett, T. J., and Paulson, J. C., 1989, Basis for the potent inhibition of influenza virus infection by equine and guinea pig α_2-macroglobulin, *J. Biol. Chem.* **264:**9850–9858.

Robertson, J. S., Naeve, C. W., Webster, R. G., Bootman, J. S., Newman, R., and Schild, G. C., 1985, Alterations in the hemagglutinin associated with adaptation of influenza B virus to growth in eggs, *Virology* **143:**166–174.

Rogers, G. N., and Paulson, J. C., 1983, Receptor determinants of human and animal influenza virus isolates: Differences in receptor specificity of the H3 hemagglutinin based on species of origin, *Virology* **127:**361–373.

Rogers, G. N., Paulson, J. C., Daniels, R. S., Skehel, J. J., and Wilson, D. C., 1983a, Single amino acid substitutions in influenza haemagglutinin change receptor binding specificity, *Nature* **304**:76–78.

Rogers, G. N., Pritchett, T. J., Lane, J. L., and Paulson, J. C., 1983b, Differential sensitivity of human, avian, and equine influenza A viruses to a glycoprotein inhibitor of infection: Selection of receptor specific variants, *Virology* **131**:394–408.

Rogers, G. N., Herrler, G., Paulson, J. C., and Klenk, H.-D., 1986, Influenza C virus uses 9-O-acetyl-N-acetylneuraminic acid as a high affinity receptor determinant for attachment to cells, *J. Biol. Chem.* **261**:5947–5951.

Rosenberg, A., Howe, C., and Chargaff, E., 1956, Inhibition of influenza virus hemagglutination by a brain lipid fraction, *Nature* **177**:234–235.

Sauter, N. K., Hanson, J. E., Glick, G. D., Brown, J. H., Crowther, R. L., Park, S. J., Skehel, J. J., and Wiley, D. C., 1992, Binding of influenza virus hemagglutinin to analogs of its cell surface receptor, sialic acid: Analysis by proton nuclear magnetic resonance spectroscopy and X-ray crystallography, *Biochemistry* **31**:9601–9621.

Schild, G. C., Oxford, J. S., de Jong, J. C., and Webster, R. G., 1983, Evidence for host-cell selection of influenza virus antigenic variants, *Nature* **303**:706–709.

Scholtissek, C., Rhode, W., von Hoyningen, V., and Rott, R., 1978, Genetic relatedness between the new 1977 epidemic strain (H1N1) of influenza and human influenza strains isolated between 1947 and 1957, *Virology* **87**:13–20.

Schulman, J. L., and Palese, P., 1977, Virulence factors of influenza A viruses: WSN virus neuraminidase required for plaque production in MDBK cells, *J. Virol.* **24**:170–176.

Schultze, B., Zimmer, G., and Herrler, G., 1993, Viral lectins for the detection of 9-O-acetylated sialic acid on glycoproteins and glycolipids, in: *Lectins and Glycobiology* (H. J. and S. Gabius, eds.), Springer-Verlag, Berlin, pp. 347–382.

Suzuki, Y., Suzuki, T., and Matsumoto, M., 1983, Isolation and characterization of receptor sialoglycoprotein for hemagglutinating virus of Japan (Sendai virus) from bovine erythrocyte membranes, *J. Biochem.* **93**:1621–1633.

Suzuki, Y., Suzuki, T., Matsunaga, M., and Matsumoto, M., 1985, Gangliosides as paramyxovirus receptor. Structural requirement of sialo-oligosaccharides in receptors for hemagglutinating virus of Japan (Sendai virus) and Newcastle disease virus, *J. Biochem.* **97**:1189–1199.

Suzuki, Y., Nagao, Y., Kato, H., Matsumoto, M., Nerome, K., Nakajima, K., and Nobusawa, E., 1986, Human influenza A virus hemagglutinin distinguishes sialyloligosaccharides in membrane-associated gangliosides as its receptor which mediates the adsorption and fusion processes of virus infection: Specificity for oligosaccharides and sialic acids and the sequence to which sialic acid is attached, *J. Biol. Chem.* **261**:17057–17061.

Szepanski, S., Gross, H. J., Brossmer, R., Klenk, H.-D., and Herrler, G., 1992, A single point mutation of the influenza C glycoprotein (HEF) changes the viral receptor binding activity, *Virology* **188**:85–92.

Szepanski, S., Veit, M., Pleschka, S., Klenk, H.-D., Schmidt, M.F.G., and Herrler, G., 1994, Posttranslational folding of the influenza C virus glycoprotein HEF: defective processing in cells expressing the cloned gene, *J. Gen. Virol.* **75**:1023–1030.

Tulip, W. R., Varghese, J. N., Baker, A. T., Van-Donkelaar, A., Laver, W. G., Webster, R. G., and Colman, P. M., 1991, Refined atomic structures of N9 subtype influenza virus neuraminidase and escape mutants, *J. Mol. Biol.* **221**:487–497.

Umetsu, Y., Sugawara, K., Nishimura, H., Hongo, S., Matsuzaki, M., Kitame, F., and Nakamura, K., 1992, Selection of antigenically distinct variants of influenza C virus by host cell, *Virology* **189**:740–744.

Varghese, J. N., and Colman, P. M., 1991, Three-dimensional structure of the neuraminidase of influenza virus A/Tokyo/3/67 at 2.2Å resolution, *J. Mol. Biol.* **221**:473–486.

Varghese, J. N., Laver, W. G., and Colman, P. M., 1983, Structure of the influenza virus glycoprotein antigen neuraminidase at 2.9Å resolution, *Nature* **393:**35–40.

Varghese, J. N., McKimm-Breschkin, J. L., Caldwell, J. B., Kortt, A. A., and Colman, P. M., 1992, The structure of the complex between influenza virus neuraminidase and sialic acid, the viral receptor, *Proteins* **14:**327–332.

Veit, M., Herrler, G., Schmidt, M.F.G., Rott, R., and Klenk, H.-D., 1990, The hemagglutinating glycoproteins of influenza B and C virus are acylated with different fatty acids, *Virology* **177:**807–811.

Vlasak, R., Muster, T., Lauro, A. M., Powers, J. C., and Palese, P., 1989, Influenza C esterase: Analysis of catalytic site, inhibition and possible function, *J. Virol.* **63:**2056–2062.

von Itzstein, M., Wu, W. Y., Kok, G. B., Pegg, M. S., Dyason, J. C., Jin, B., Van Phan, T., Smythe, M. L., White, H. F., Oliver, S. W., Colman, P. M., Varghese, J. N., Ryan, D. M., Woods, J. M., Bethell, R. C., Hotham, V. J., Cameron, J. M., and Penn, C. R., 1993, Rational design of potent sialidase-based inhibitors of influenza virus replication, *Nature* **363:**418–423.

Ward, C. W., Elleman, T. C., and Azad, A. A., 1982, Amino acid sequence of the pronase-released heads of neuraminidase subtype N2 from the asian strain A/Tokyo/3/67 of influenza virus, *Biochem. J.* **207:**91–95.

Webster, R. G., Air, G. M., Metzger, D. W., Colman, P. M., Varghese, J. N., Baker, A. T., and Laver, W. G., 1987, Antigenic structure and variation in an influenza virus N9 neuraminidase, *J. Virol.* **61:**2910–2916.

Weis, W., Brown, J. H., Cusack, S., Paulson, J. C., Skehel, J. J., and Wiley, D. C., 1988, Structure of the influenza virus haemagglutinin complexed with its receptor, *Nature* **333:**426–431.

Wharton, S. A., Weis, W., Skehel, J. J., and Wiley, D. C., 1989, Structure, function, and antigenicity of the hemagglutinin of influenza virus, in: *The Influenza Viruses* (R. M. Krug, ed.), Plenum Press, New York, pp. 153–173.

White, J., Helenius, A., and Gething, M.-J., 1982, Haemagglutinin of influenza virus expressed for a cloned gene promotes membrane fusion, *Nature* **300:**658–659.

Wilson, I. A., Skehel, J. J., and Wiley, D. C., 1981, Structure of the haemagglutinin membrane glycoprotein of influenza virus at 3 A resolution, *Nature* **289:**366–373.

Wilson, V. W., and Rafelson, M. E., Jr., 1967, Studies on the neuraminidases of influenza viruses. III. Stimulation of activity by bivalent cations, *Biochim. Biophys. Acta* **140:**160–166.

Chapter 10

Sialic Acid in Biochemical Pathology

Kunihiko Suzuki

1. INTRODUCTION

Free sialic acid constitutes a relatively small proportion of the total tissue sialic acid pool. The bulk of sialic acid in the body exists in sialoglycoconjugate form bound to either glycoproteins or glycolipids. Many proteins are glycosylated by complex carbohydrate chains terminated by a sialic acid residue (see Chapter 5). Glycosphingolipids are group of complex lipids that contain a long-chain base, sphingosine, as the basic building block (see Chapter 6). In almost all naturally occurring sphingolipids, sphingosine is acylated by a long-chain fatty acid, forming *N*-acylsphingosine, or ceramide. A complex hydrophilic side chain, consisting of carbohydrate, sialic acid, and other constituents, is attached to the terminal hydroxyl group of sphingosine. Glycoproteins and glycolipids are characteristic integral constituents of the plasma membrane in vertebrates. Their composition varies in different cell types and in different developmental stages. The compositional pattern can be further altered by viral transformation and oncogenesis. No genetic pathological condition related to abnormality in either synthesis or degradation of sialic acid itself is known. However, a series of genetic disorders do exist with the underlying abnormalities affecting either transport of sialic acid across the lysosomal membrane or intralysosomal degradation of sialic acid-containing glycoproteins and glycolipids. The latter category includes disorders in which removal of sialic acid is specifically impaired, and

---

Kunihiko Suzuki Brain and Development Research Center, Departments of Neurology and Psychiatry, University of North Carolina School of Medicine, Chapel Hill, North Carolina 27599.

Biology of the Sialic Acids, edited by Abraham Rosenberg. Plenum Press, New York, 1995.

those in which presence of sialic acid in the affected compounds is incidental. The scope of this chapter will be limited to genetically defined conditions in which alterations of sialic acid or sialic acid-containing compounds occur. Other aspects such as differentiation, infections, and oncogenesis are covered elsewhere in this volume. Table I lists the disorders that are included as the subjects of this chapter. For further detailed information on the biochemical aspects of these genetic disorders, the reader is referred to the recently published reference volume edited by Scriver *et al.* (1995).

2. SUMMARY

Salla disease, recognized relatively recently and named for the region of Finland where the disease is prevalent, is characterized by psychomotor retardation and ataxia with an abnormally high excretion of free sialic acid in the urine. The cause of the disease is an abnormal transport of sialic acid across the lysosomal membrane, possibly because of an abnormality in a specific sialic acid transporter protein. As a consequence, the tissue level of free sialic acid increases to 10–100 times the normal level. Among the disorders involving sialic acid-containing compounds, sialidosis exhibits several distinct clinical phenotypes and is caused by genetic deficiencies of the lysosomal sialidase (α-neuraminidase). Primary storage compounds are fragments of undigested glycoproteins with terminal sialic acid, which are also excreted excessively into urine. Removal of sialic acid from gangliosides does not appear affected in sialidosis patients. Two major categories of gangliosidoses are known, each affecting different gangliosides, namely, GM1 and GM2 gangliosides, respectively. GM1 gangliosidosis can manifest itself as infantile, juvenile, and adult forms with varying degrees of psychomotor and other neurological involvements, bone abnormalities, and visceromegaly. The underlying causes are genetic defects in lysosomal acid β-galactosidase which normally hydrolyzes the terminal β-galactose residue from GM1 ganglioside and its asialo derivative. Three genetically distinct forms of GM2 gangliosidosis exist, caused respectively by abnormalities in the genes coding for the β-hexosaminidase α-subunit (Tay–Sachs disease), β-hexosaminidase β-subunit (Sandhoff disease), and the GM2 activator protein (AB variant). All three gene products are essential for normal degradation of GM2 ganglioside *in vivo*. The clinical phenotype of GM2 gangliosidosis is diverse but abnormal accumulation of GM2 ganglioside primarily in neurons is the biochemical characteristic in all forms of GM2 gangliosidosis. Finally, galactosialidosis was considered earlier as an adult form of GM1 gangliosidosis on the basis of acid β-galactosidase deficiency but later recognized as a distinct disorder caused by a deficiency in a "protective protein" which is required to stabilize a β-galactosidase–sialidase complex. As a consequence, both β-galactosidase and

Table I
Genetic Disorders Involving Sialic Acid or Sialic Acid-Containing Compounds

Disease	Clinicopathological manifestations	Affected compounds	Enzymatic defects
Salla disease	Psychomotor retardation, ataxia, abnormally high excretion of free sialic acid in the urine, rare and severe infantile form	Free sialic acid	Sialic acid transporter?
Sialidosis	Two distinct clinical forms. The early onset form ("mucolipidosis I") more severe with bony abnormalities, organomegaly, and neurological manifestations. Late-onset form with macular cherry red spots and intractable myoclonus with intact intellect	Glycopeptides and oligosaccharides with terminal sialic acid	Lysosomal sialidase (lysosomal α-neuraminidase)
GM1 gangliosidosis	Slow growth, motor weakness, gray matter signs, infantile form with additional facial and skeletal abnormalities and organomegaly, swollen neurons	GM1 ganglioside Galactose-rich fragments of glycoproteins	GM1 ganglioside β-galactosidase
GM2 gangliosidosis			
Tay–Sachs disease	Severe gray matter signs, slow growth, motor weakness, hyperacusis, cherry red spot, head enlargement, swollen neurons A unique variant, "B1 variant," is commonly of later onset, slower progression, but otherwise similar to Tay–Sachs disease, milder pathology	GM2 ganglioside	β-Hexosaminidase A In "B1 variant," β-hexosaminidase A is normal against nonsulfated artificial substrates but deficient against GM2 ganglioside and sulfated artificial substrate
Sandhoff disease	Panethnic but otherwise virtually indistinguishable from Tay–Sachs disease	GM2 ganglioside, asialo-GM2 ganglioside, globoside Glycoprotein with terminal β-hexosamine	β-Hexosaminidase A and B
AB variant	Similar to Tay–Sachs disease	GM2 ganglioside	GM2 activator protein
Galactosialidosis	Similar to late-onset form of GM1 gangliosidosis, with varying degrees of additional mucopolysaccharidosis-like features	(Unknown)	"Protective protein" (secondary defect in GM1 ganglioside β-galactosidase and sialidase)

sialidase activities are secondarily deficient in the tissue. Specific genetic abnormalities on the gene level are not yet known for the sialic acid storage disease or for sialidosis. However, the genes coding for the lysosomal enzymes, GM2 activator protein, and the protective protein, responsible for all forms of GM1 and GM2 gangliosidosis and galactosialidosis have been cloned, characterized, and multiple disease-causing mutations have been identified.

3. DISEASES RELATED TO DEFECTIVE SIALIC ACID TRANSPORT

Salla Disease

When sialic acid-containing compounds are degraded, the liberated sialic acid cannot be further metabolized within the lysosome and must be transported out of the lysosome. A complex metabolic machinery appears to operate for this process, involving a transport protein specific for sialic acid. Free sialic acid is then metabolized by a cytoplasmic enzyme, *N*-acetylneuraminate lyase. If one of the essential components in this transport system is defective, free sialic acid cannot be removed from the lysosome and consequently accumulates abnormally since it cannot be metabolized intralysosomally. Since 1979, the unique genetic disorder Salla has been known in a delimited region of Finland. The disease is caused by defective transport of free sialic acid out of the lysosome (see appropriate section in Scriver *et al.*, 1995).

Clinical and Pathological Aspects. The disease is inherited as an autosomal recessive disorder and heterozygous individuals are entirely free of any phenotypic manifestations. The carrier frequency in northern Finland is estimated to be 1 in 40. The clinical onset of classical Salla disease is usually soon after birth with delayed motor development, ataxia of varying degrees with or without nystagmus. The delayed psychomotor development continues for many years into adulthood, ranging from total inability to walk to poor, spastic gait with extrapyramidal signs. Patients' IQ is usually severely limited. Patients are dysarthric and exhibit other basal ganglia signs. The disease takes a prolonged progressive course with slow deterioration of motor and mental capacities. Typical patients, however, may survive for several decades. In contrast to the typical Salla disease, a rarer infantile form of the disease, first recognized in 1982, takes a much more rapid, progressive course with complex clinical features. Patients start deteriorating soon after birth with ascites, organomegaly, and severely delayed psychomotor development. Bony abnormalities are often described. Multiple dysmorphic features are often noted. The disease is rapidly fatal within a few years. It is not known if the two clinical forms are allelic. However, patients with what appears to be an intermediate phenotype have been described more recently. Typical Salla disease patients excrete excess free sialic acid into urine at 5–10 times normal and infantile patients sometimes over 100 times normal. Intracellular vacuoles are prominent in all affected cells. That these vacuoles are altered lysosomes containing free sialic acid has been demonstrated by histochemistry as well as subcellular fractionation experiments.

Biochemistry and Enzymology. Free sialic acid is increased in all organs and tissues of the body, as well as in cultured fibroblasts and lymphoblasts from patients as the result of its accumulation in the lysosome. The serum level is also increased. Generally, the degree of accumulation is substantially more severe in infantile patients than in the typical older Salla disease patients. The accumulated material is *N*-acetylneuraminic acid. All enzymes directly associated with metabolism of sialic acids are known to be normal or sometimes elevated. They include *N*-acetylneuraminate lyase, CMP-NeuAc phosphodiesterase, NeuAc cytidyltransferase, and sialidase. When cultured cells were loaded with free sialic acid to achieve high intralysosomal concentrations equivalent to that in Salla disease, the rate of egress of free sialic acid from the lysosome was severely defective in Salla disease fibroblasts. In cystinosis, transport of free cysteine across the lysosomal membrane is defective. Egress of free cysteine was normal in Salla disease fibroblasts. This indicated strongly that the block in free sialic acid egress from the lysosome was specific and probably carrier-mediated, and that Salla disease is conceptually equivalent to cystinosis.

Molecular Genetics. The putative sialic acid transporter protein has not been isolated. Thus, no information is available regarding the molecular genetic aspects of Salla disease and its infantile variant.

4. DISEASES RELATED TO DEFECTIVE CATABOLISM OF SIALIC ACID-CONTAINING COMPOUNDS

4.1. Sialidosis

The term *sialidosis* refers to a series of disorders, probably allelic, caused by genetic abnormalities in the lysosomal sialidase which removes the terminal sialic acids from carbohydrate chains of glycoproteins. The activity of the sialidase which cleaves the sialic acid moieties from some of the gangliosides, such as GM3 and polysialogangliosides, is not defective (see appropriate section in Scriver *et al.*, 1995). Review of the earlier literature on sialidosis requires exceptional care. Some of the earlier cases reported as sialidosis may well have been cases of galactosialidosis, another distinct genetic disorder recognized much more recently (see below).

Clinical and Pathological Aspects. The disease is inherited as an autosomal recessive trait. Two major clinical forms are known. The more severe form, previously also known as "mucolipidosis I," primarily occurs in infants and juvenile age groups. The less severe form is known as the cherry-red spot–myoclonus syndrome and occurs primarily in juvenile and older patients. Patients with the most severe form of "mucolipidosis I" may exhibit congenital abnormalities, such as ascites or hydrops, and may die in the perinatal or immediately postnatal periods. Other patients develop rapidly progressive neurological manifestations, including psychomotor retardation, combined with mucopolysaccharidosis-like features, such as bone abnormalities, abnormal facial features, and hepatosplenomegaly. The course of the disease can be a few years to over 10 years. The milder "cherry-red spot–myoclonus" form of sialidosis usually has its onset in the juvenile age period. The presenting symptom may be impaired visual acuity, myoclonus, or difficulty walking. These manifestations, particularly the visual impairment, are progressive. There are typical bilateral cherry red spots at the macula and intractable myoclonic seizures. Characteristically, the mental function remains essentially intact. Patients' difficulty in communication is the result of the incessant myoclonus and occasionally other dystonic movements, but is not related to intellectual impairment. The course of the disease can be many years. All patients excrete excessive amounts of oligosaccharides and glycopeptides with terminal sialic acids, derived from degradation products of glycoproteins, into urine.

In mucolipidosis I patients, vacuolated lymphocytes and foamy bone marrow cells are common. Cellular vacuolation is often seen in the enlarged liver (more extensive in the Kupffer cells than in the hepatocytes) and spleen, CNS neurons, peripheral nerves, and the neurons in the myenteric plexus. Pathology is generally milder in the late-onset cherry-red spot-myoclonus form than in the more severe mucolipidosis I and is confined to the nervous system. CNS neurons prominently contain lipofuscin pigments.

Biochemistry and Enzymology. Detailed analyses of tissues from patients with sialidosis are lacking. However, extensive analytical information is available on the abnormal materials excreted into urine. Most of the urinary oligosaccharides have an *N*-acetylglucosamine residue at the reducing end and sialic acid at the nonreducing end. This is consistent with these oligosaccharides being derived from glycoproteins by action of endoglycosidase and undegradable further because of the genetic defect in the sialidase. The sialic acid residue is linked to the penultimate galactose by $\alpha 2 \rightarrow 6$ (80%) or $\alpha 2 \rightarrow 3$ (20%) linkage. Oligosaccharides of similar structures were found to accumulate in cultured fibroblasts fed with tritiated *N*-acetylmannosamine, which is known to be incorporated primarily into sialic acid. These analytical findings strongly suggest that the cellular vacuoles observed by histology and electron microscopy in organs of patients probably also contain excess amounts of these compounds. It is notable that no specific accumulation of either polysialogangliosides or GM3 ganglioside is known to occur in genetic sialidosis.

Deficient activity in lysosomal acid sialidase (α-neuraminidase) is the genetic defect underlying sialidosis. Sialic acid occurs at the nonreducing terminus of the carbohydrate chains of many glycoproteins and also as a characteristic component of gangliosides. However, there is no biochemical or enzymological evidence for specific abnormalities in ganglioside metabolism in sialidosis, even though sialic acid residues on the terminal galactose of many polysialogangliosides are readily susceptible to sialidase, and GM3 ganglioside is degraded to lactosylceramide by cleavage of the terminal sialic acid. This provides *a priori* evidence that there should be at least two lysosomal sialidases, specific for glycoproteins and gangliosides, respectively, and that it is the former and not the latter that is genetically defective in sialidosis. The defect is detectable in assay systems that utilizes artificial synthetic substrates. This indicates that the sialidase defective in sialidosis must be quantitatively predominant or that the ganglioside sialidase has no activity toward the water-soluble artificial substrates. While not directly pertinent, nonlysosomal sialidases are known, localized in the plasma membrane, the synaptic regions, and elsewhere, further complicating the picture (see Chapter 8).

Molecular Genetics. Little is known about the molecular genetics of sialidosis, since the responsible gene has not been cloned. The heterogeneity of sialidases in mammals, even within the lysosomal sialidases (see Chapter 8), is expected to add complexity when the gene becomes available for study, since the genetic relationship, if any, among them is not known. As indicated above, there should be at least two distinct lysosomal acid sialidases, each specific for glycoproteins and glycolipids. The glycoprotein sialidase defective in sialidosis has been mapped to human chromosome 10 pter→q23 (Mueller *et al.*, 1986). A few laboratories are actively purifying lysosomal sialidase to a sufficiently pure state for amino acid sequencing and eventual cloning of the gene. Its relative instability under many conditions commonly utilized for enzyme purification has hampered its successful purification.

4.2. Gangliosidoses

Two major categories of genetic disorders are the result of abnormal catabolism of different gangliosides: GM1 gangliosidosis and GM2 gangliosidosis. GM1 ganglioside is normally degraded to GM2 ganglioside by removal of the terminal β-galactoside residue by lysosomal acid β-galactosidase, and GM2 ganglioside is in turn further degraded to GM3 ganglioside by removal of its terminal *N*-acetylgalactosamine by the combined action of β-hexosaminidase A and the GM2 activator protein. Since β-hexosaminidase A isozyme is a heterodimer consisting of the α and β subunits, genetic defects in any of the three genes coding for the α or β subunits, or GM2 activator protein, respectively, result in genetically distinct GM2 gangliosidosis. Since gangliosides are highly enriched in neurons, gangliosidoses generally manifest themselves clinically as well as pathologically as primary diseases of the nervous system. Nevertheless, the catabolic systems involved in these disorders have substrate specificity beyond the gangliosides. These properties result in varying clinical and biochemical phenotypes. With the rapidly increasing information regarding mutations underlying these disorders, the genotype/phenotype correlation is attracting increasing interest.

4.2.1. GM1 Gangliosidosis

Genetic deficiency of lysosomal acid β-galactosidase in humans causes GM1 gangliosidosis, a neurological disorder (Sacrez *et al.*, 1967; Seringe *et al.*, 1968; Okada and O'Brien, 1968). Clinical manifestations of GM1 gangliosidosis are variable. The classical infantile form is a severe, rapidly progressive neurological as well as systemic disorder, and patients rarely survive for more than a few years. Patients with clinically, pathologically, and biochemically less severe juvenile and adult forms, however, have later onset, slower progression, and, in many adult patients, little mental involvement. Although it has been clear for many years that the underlying cause of these various clinical phenotypes is a genetic deficiency of acid β-galactosidase, numerous attempts at their enzymological delineation have failed. Recent availability of cDNA clones coding for the human acid β-galactosidase (Oshima *et al.*, 1988; Morreau *et al.*, 1989) made it feasible to examine the underlying genetic abnormalities on the gene level in different clinical phenotypes of GM1 gangliosidosis.

Clinical and Pathological Aspects. Until the mid-1960s, the only known "ganglioside storage disease" was the classical Tay–Sachs disease (a GM2 gangliosidosis). Existence of another entirely different category of gangliosidosis was firmly established by 1965 (Jatzkewitz and Sandhoff, 1963; Gonatas and Gonatas, 1965; O'Brien *et al.*, 1965). GM1 gangliosidosis is transmitted as a Mendelian autosomal recessive disorder. Although the most important clinical manifestations are for the most part neurological, the disease involves other systemic organs, such as the liver, spleen, and the skeletal structure. It is customary to divide the phenotype into "infantile," "juvenile," and "adult" forms according to the time of onset and the rate of progression. More recent advances in the molecular genetics of the disorder make this traditional classification conceptually obsolete because of the astronomical number of possible combinations of two mutant alleles, although its pragmatic usefulness remains.

Patients with the infantile form of GM1 gangliosidosis generally have their clinical onset by 6 months, although facial and bony abnormalities are often recognized at birth. In early stages, patients appear dull and hypotonic with retarded psychomotor development. Regression soon becomes evident leading to later manifestations, such as spasticity and seizures. Eventually, patients become deaf and blind, totally unresponsive to external stimuli. Macular cherry spots are common but not consistent. The entire clinical course rarely exceeds 2–3 years, and inevitable death occurs most commonly as a result of concurrent infections. Characteristically, neurological manifestations are primarily those of gray matter.

White matter and peripheral nerve involvements are relatively minor. Infantile patients show, in addition to the above neurological manifestations, clinical features that are similar to mucopolysaccharidoses, including facial deformity, macroglossia, radiologic bone abnormalities, and visceromegaly.

The late-infantile or juvenile form of the disease manifests itself later, usually after 1 year of age, with milder neurological signs and slower progression. The systemic involvements, prominent in infantile patients, are less so and may be absent. Patients with this phenotype often survive beyond 10 years. A typical case of this form was described early by Derry *et al.* (1968).

The adult or chronic form is very variable clinically. The clinical onset can be any time between a few to 30 years or older. Perhaps the first clearly delineated patients were described in 1977 by Suzuki *et al.* (1977). This phenotype appears to be fairly common in Japan, but cases are known in other ethnic groups. Slowly progressive dysarthria, gait difficulties, dystonic movements, and other extrapyramidal signs are the most prominent. Patients are often slightly or moderately impaired intellectually. Signs prominent in the younger forms, such as macular cherry-red spots, facial abnormalities, or visceromegaly, are not associated with the adult form. However, bony abnormalities are recorded, sometimes causing symptoms as a result of spinal compression. Clinical diagnosis of adult GM1 gangliosidosis is difficult, if not impossible, and patients with the above clinical picture should be screened for the possibility with appropriate enzymatic assays.

Pathology of the infantile and juvenile forms are qualitatively similar, with the infantile form being generally more severe. The gray matter is primarily affected, while the white matter and the peripheral nerves are relatively spared. Essentially all neurons of the body are enlarged and contain faintly granular material, which actually consists of discrete abnormal lamellar bodies ("membranous cytoplasmic bodies") under the electron microscope. The pathology of cortical neurons is for all practical purposes indistinguishable from that of classical Tay–Sachs disease (GM2 gangliosidosis). Many reactive astrocytes also contain abnormal cytoplasmic bodies. White matter often shows mild myelin loss with sudanophilia, which appears to be secondary to neuronal loss. Many reticuloendothelial organs, such as lung, liver, spleen, lymph nodes, and bone marrow, contain swollen foamy histiocytes with intracellular vacuoles containing strongly PAS-positive materials. This systemic pathology is present even in juvenile patients who show relatively few clinical signs of systemic organ involvement. The ultrastructural features of the systemic organs more resemble those of mucopolysaccharidoses and are different from neuropathology. Clear vacuoles in these cells, however, contain interwoven bundles of tubular structures approximately 200 Å wide.

The characteristic finding in the neuropathology of adult GM1 gangliosidosis is its selective distribution of lesions within the CNS (Goldman *et al.*, 1981;

Suzuki, 1991). Unlike in the younger phenotypes, cortical neurons tend to be relatively spared, and those in the basal ganglia are primarily affected. Even within the same region, severely swollen neurons can be found next to others that appear relatively normal or pyknotic. However, the ultrastructural features of affected neurons are qualitatively similar to those in the younger forms. This selective involvement of the basal ganglia can explain, at least on the phenomenological level, the predominant extrapyramidal manifestations in the adult disease. Again, systemic organs can show pathology similar to the younger forms even when there is no clinical evidence of their involvement.

Biochemistry and Enzymology. GM1 gangliosidosis was originally recognized as a disease entity by the massive accumulation of GM1 ganglioside in the brain. The total concentration of ganglioside in the brain at the terminal stage in the infantile disease can be 3–5 times normal. GM1 ganglioside, which constitutes approximately 22–25 mol% of total ganglioside in the normal gray matter, is present at 90–95 mol% in patients' gray matter (Suzuki *et al.*, 1969). Thus, the increase in GM1 ganglioside in the gray matter is actually up to 20 times normal. There is an associated moderate increase in asialo-GM1 ganglioside (Suzuki and Chen, 1967). There is also an abnormal increase in GM1 ganglioside in systemic organs, although the absolute amounts remain small (Suzuki *et al.*, 1971a). GM1 gangliosidosis was early on termed *generalized gangliosidosis* because of the abnormal accumulation of GM1 ganglioside in systemic organs. However, this terminology should be avoided since it is potentially misleading. Similar degrees of abnormal accumulation of the affected ganglioside, GM2, also occur in systemic organs of patients with Tay–Sachs disease. There is very limited information available for analytical biochemistry of the adult GM1 gangliosidosis because of the long survival of patients. In one well-characterized patient (Kobayashi and Suzuki, 1981), abnormal accumulation of GM1 ganglioside paralleled the distribution of pathological lesions in that the basal ganglia clearly contained a much larger quantity of GM1 ganglioside than cerebral cortex.

The main storage materials in the systemic organs are heterogeneous galactose-rich fragments of varying molecular weights derived from glycoproteins, keratan sulfate, and other carbohydrate-containing materials. The most insoluble of these was first described as "keratan sulfate-like" based on its behavior in preparative procedures, electrophoretic properties, and sugar composition (Suzuki, 1968). When all of these materials are combined, the amount in patients' tissues can be more than 20 times normal. Many compounds of smaller molecular weight of this group also appear in excess in patients' urine. Several extensive characterizations have been done on these materials, which appear to be for the most part derived from glycoproteins. These materials should, *a priori,* have a terminal β-galactose moiety at the nonreducing end.

The underlying cause of GM1 gangliosidosis is a genetic deficiency of the lysosomal acid β-galactosidase. The defect can be readily demonstrated either with conventional chromogenic or fluorogenic artificial substrates or with the natural substrate, GM1 ganglioside (Suzuki, 1977, 1987). Among several glycolipids that have the terminal β-galactosidase residue, GM1 ganglioside, asialo-GM1 ganglioside, and lactosylceramide are natural substrates for the enzyme. However, there is no prominent abnormal accumulation of lactosylceramide in GM1 gangliosidosis, because the other lysosomal β-galactosidase, galactosylceramidase, also hydrolyzes lactosylceramide (Tanaka and Suzuki, 1975, 1977). Genetic deficiency of galactosylceramidase is the cause of globoid cell leukodystrophy (Krabbe disease) and this enzyme is normal in GM1 gangliosidosis.

Molecular Genetics. Study of GM1 gangliosidosis entered the molecular biology phase in 1988–1989 when the cDNA coding for the normal human acid β-galactosidase was cloned and characterized. The human β-galactosidase gene has since been characterized (Morreau *et al.*, 1991). The human β-galactosidase gene is located on chromosome 3 (Shows *et al.*, 1979).

Initially eight mutations responsible for GM1 gangliosidosis were described, by coincidence, all among Japanese patients (Yoshida *et al.*, 1991; Nishimoto *et al.*, 1991). Within this population and within the relatively limited number of patients examined, one mutation each could be characterized as the cause of the juvenile and adult forms of the disease, respectively. Many of the adult and juvenile patients were homozygous for the respective mutations. While Nishimoto *et al.* (1991) did not detect catalytic activities of these mutant proteins in the conventional COS-1 cell expression system, Yoshida *et al.* (1991) found some residual activity of the juvenile mutant enzyme and even higher residual activity in the adult mutant enzyme in their ASVGM1-4 cell expression system. Their finding is consistent with the hypothesis that each of these mutations causes less severe, more slowly progressive clinical phenotypes. In contrast to the older forms, mutations responsible for the infantile form of GM1 gangliosidosis are highly heterogeneous even within this ethnic group. At least five different mutations were found among infantile patients. Additional mutations in other populations have since been described (Chakraborty *et al.*, 1991; Yoshida *et al.*, 1992; Mosna *et al., 1992; Boustany et al.*, 1993). The number of known disease-causing mutations will undoubtedly increase rapidly in the next few years.

4.2.2. GM2 Gangliosidosis

The genetic basis of GM2 gangliosidosis is more complex than that in GM1 gangliosidosis, because three gene products are essential for normal catabolism of GM2 ganglioside *in vivo*—β-hexosaminidase α and β subunits and the GM2 activator protein. They are coded by different genes. The gene coding for the β-hexosaminidase α subunit is located on human chromosome 15, while the other two genes are on chromosome 5. The two active isozymes of β-hexosaminidase both exist as dimers of these subunits: isozyme A is a heterodimer (αβ) and B is a homodimer (ββ). Hexosaminidase A is able to catabolize all known natural substrates with terminal β-hexosaminyl residues, in some instances, with help of the activator protein, while hexosaminidase B can degrade all natural substrates except for GM2 ganglioside. This isozyme composition and the substrate specificity result in two distinct disorders. When hexosaminidase α is genetically defective, only the A isozyme is inactivated, resulting primarily in defective GM2 ganglioside degradation (Tay–Sachs disease and variants), while genetic defects in the β subunit result in loss of both isozymes, since the β subunit is a constituent of both. Then, the biochemical abnormalities are much more complex because all natural substrates will be affected in addition to GM2 ganglioside. The primary function of the GM2 activator protein is to assist hexosaminidase A in the degradation of GM2 ganglioside *in vivo*. Therefore, genetic defects in the activator protein cause a condition similar to the hexosaminidase A deficiency and the resulting clinical disease mimics the classical Tay-Sachs disease.

4.2.2.a. β-Hexosaminidase α-Subunit Deficiency (Tay–Sachs Disease)

Clinical and Pathological Aspects. The classical infantile Jewish form of Tay–Sachs disease is the prototype of human ganglioside storage disorders. It has been known for over a century since the first descriptions in 1881 of the macular cherry-red spot by the British ophthalmologist Warren Tay (1881) and of detailed neurological manifestations by the U.S. neurologist Bernard Sachs several years later (Sachs, 1887). The disease is inherited as a Mendelian autosomal recessive trait and the classical infantile form of the disease occurs among the Ashkenazi Jewish population at an unusually high frequency. In that group, the gene frequency is approximately 1/25, i.e., 1 out of 25 individuals is a carrier of the mutant gene. The incidence in other ethnic groups is less than 1/100 of that in the high-risk Jewish population. The distribution of various phenotypic variants, however, appears to be relatively even among different ethnic groups. Patients with the classical infantile form of Tay–Sachs disease usually appear normal at birth but develop progressive psychomotor retardation from approximately 6 months after birth. The clinical manifestations are almost exclusively neurological with primary involvement of gray matter. Bilateral macular cherry-red spots are highly consistent. Regression and loss of already acquired mental and motor skills soon become obvious. By the age of 2–3 years, patients lose contact with the environment and reach the decerebrate stage. Hyperacusis and the resulting startle reactions are characteristically observed. In late stages, head enlargement is also common. Death usually occurs by 4 years.

During the period of intensive clinicopathological, analytical, and enzymological studies in the 1970s, it became clear that Tay–Sachs disease in fact consists of many clinical and enzymological variant forms that likely reflect the nature of different underlying gene abnormalities. The clinical phenotype varies widely among the late-onset, variant forms of hexosaminidase α-subunit deficiencies. The onset can be at any time from the late infantile period to the adult age. The macular cherry-red spots are less frequent, particularly in the adult patients. If a generalization can be made regarding the clinical manifestations of later-onset forms, involvement of the deeper brain structures is more prominent, compared to the overwhelming generalized gray matter involvement in the infantile form. The manifestations include dystonia, other extrapyramidal signs, such as ataxia, choreoathetotic movements, and the signs of spinocerebellar degeneration. In some older patients, mental capacity can be well preserved, although severe dysarthria often masks the preserved intelligence. Most of the older variants were recognized when the enzymatic diagnosis of the disease became feasible. The known phenotypic variations became so diverse that analyses on the gene level would seem to be the only way to sort out these complexities.

Reflecting the clinical picture, the pathology is also confined primarily to the CNS, particularly neurons. In the infantile form, essentially all neurons of the body are extremely swollen, being filled with finely granular material which pushes the normal subcellular organelles to the periphery of the perikarya. The classical work of Terry and Korey (1960) and Terry and Weiss (1963) established the most essential aspects of the pathology of infantile Tay–Sachs disease. The swollen neurons are filled with discrete multilamellar membranous bodies ("membranous cytoplasmic bodies"), approximately 1 μm in diameter. They contain histochemically demonstrable acid phosphatase activity and are recognized as altered secondary lysosomes. Another important pathological feature is abnormal neurite formation on the neuronal perikarya and the axonal meganeurite formation ("torpedoes") just beyond the axon hillock (Purpura and Suzuki, 1976). These abnormalities may well have an important bearing on the functional abnormalities of these neurons. The neuronal pathology in the older forms is generally less severe than that in the infantile form. While some neurons are swollen, others may by pyknotic. Cortical neurons are generally less affected than those in the basal ganglia, consistent with the clinical manifestations.

Biochemistry and Enzymology. Gangliosides as a new class of lipids were discovered by Klenk in the brains of patients with Tay–Sachs disease (Klenk, 1939). The total gray matter ganglioside content is usually increased to 4–5 times normal. GM2 ganglioside, normally a very minor constituent, comprises up to 90 mol% of the total ganglioside. Thus, the net increase of GM2 ganglioside can be several hundredfold over the normal level. It has been shown analytically that the membranous cytoplasmic bodies that fill swollen neurons are the repository of the increased GM2 ganglioside (Suzuki *et al.*, 1969). GM2 ganglioside is also increased in systemic organs but the absolute levels are very low, since these organs normally contain little GM2 ganglioside or higher gangliosides which are degradable to GM2.

The genetic defect in the β-hexosaminidase α subunit is the underlying cause of Tay–Sachs disease. This results in absent or catalytically defective β-hexosaminidase A, while the B isozyme remains normal, since the latter consists of only the β subunit. Because of the different substrate specificities of the two isozymes, β-hexosaminidase α deficiency causes a relatively specific catabolic defect of GM2 ganglioside, while most other β-hexosaminidase substrates are normally degraded. In many infantile patients, the α subunit may not be generated at all or rapidly degraded as a result of mutations. In these instances no hexosaminidase A isozyme is present. This can be readily detected by differential assays for the two isozymes (Okada and O'Brien, 1969). In the enzymologically unique B1 variant, however, a stable but catalytically inactive α subunit is generated which is still capable of associating with the β subunit to form a stable hexosaminidase A isozyme. Kytzia and Sandhoff (1985) clearly showed that each of the two subunits has its own catalytic site. The site on the α subunit is active toward the GM2 ganglioside with the help of the GM2 activator protein but inactive toward the artificial substrates, such as 4-methylumbelliferyl *N*-acetylglucosamine, which can be cleaved only by the catalytic site on the β subunit. Therefore, in B1 variant, both hexosaminidase A and B appear to be normal, if assays are done with the conventional artificial substrates but the catalytic activity toward GM2 ganglioside is defective, presenting a potential pitfall in the enzymatic diagnosis. This problem can be overcome with the use of 4-methylumbelliferyl *N*-acetylglucosamine 6-sulfate as the substrate, which is specifically cleaved by the active site on the α subunit.

Molecular Genetics. Tay–Sachs disease research entered its molecular biology era when a full-length cDNA coding for the human β-hexosaminidase α subunit was cloned and characterized (Myerowitz *et al.*, 1985; Korneluk *et al.*, 1986). Soon thereafter the genomic organization was characterized (Proia and Soravia, 1987). The gene is located on human chromosome 15. A major deletion at the 5′ end of the gene was soon identified as the underlying cause of the infantile Tay–Sachs disease prevalent among the French Canadian population (Myerowitz and Hogikyan, 1986, 1987) followed by the first point mutation in a patient with the B1 variant form of the disease (Ohno and Suzuki, 1988a). To everyone's surprise, the classical infantile Tay–Sachs disease among the Ashkenazi Jewish population turned out to be heterogeneous, consisting of two major mutations resulting from a splicing defect related to a base transversion at the donor site of intron 12 (Arpaia *et al.*, 1988; Myerowitz, 1988; Ohno and Suzuki, 1988b,c) and a four-base insertion in exon 11 (Myerowitz and Costigan, 1988). Another mutation underlying an adult phenotype prevalent in Israel and the surrounding area was also identified (Navon and Proia, 1989). These three mutations account for perhaps 90% of Tay–Sachs disease among the Ashkenazi Jewish population. A flood of mutation analyses followed and the most recent count showed nearly 60 known disease-causing abnormalities of the β-hexosaminidase α-subunit gene (Kaback *et al.*, 1993; Gravel *et al.*, 1994). While simplistic generalization is risky and can never be without exceptions, more severe, infantile forms tend to be caused by mutations that generate no detectable α-subunit mRNA while later-onset forms often have immunologically detectable but catalytically inactive enzyme protein.

A unique mutation that causes the B1 variant deserves special mention. The mutation is a base substitution G533→A in exon 5, which results in a change of the normal arginine to histidine. It was first identified in a Puerto Rican patient with late-infantile Tay–Sachs disease (Ohno and Suzuki, 1988a). This particular mutation, since termed DN-allele, has been found to be the causative mutation among many juvenile patients with the enzymological characteristics of the B1 variant and has more recently been traced to a region of northern Portugal as a possible site of origin (Tanaka *et al.*, 1988, 1990; Dos Santos *et al.*, 1991). The patients homozygous for this mutation are invariably of a juvenile phenotype, and a few late-infantile B1 variant patients have been found to have this mutation on one allele and an mRNA-negative infantile mutation on the other. A large number of juvenile Tay–Sachs patients described in the literature may well be accounted for by this mutation (Suzuki and Vanier, 1991).

4.2.2.b. β-Hexosaminidase β-Subunit Deficiency (Sandhoff Disease)

Clinical and Pathological Aspects. Patients with infantile Sandhoff disease are essentially indistinguishable clinically from infantile Tay–Sachs disease patients (Sandhoff, 1969). The clinical features of patients with late-onset/adult forms are also similar to those of Tay–Sachs disease patients of corresponding ages. There is no ethnic preponderance, unlike in infantile Tay–Sachs disease. Patients excrete a large amount of hexosamine-containing glycoprotein fragments, which can be of some diagnostic value (Suzuki *et al.*, 1971b). However, the differential diagnosis can easily be established by enzymatic assays.

Neuropathology is also generally similar to Tay–Sachs disease. Swollen neurons contain the membranous cytoplasmic bodies. However, the degree of secondary demyelination appears to be more severe than in Tay–Sachs disease. Because of additional accumulation of glycosphingolipids with a terminal hexosamine residue and of fragments of undigested glycoproteins both in the CNS and in systemic organs, strongly PAS-positive materials can be demonstrated when frozen specimens are stained with periodic acid–Schiff reagent. These materials are most prominent in hepatic cells, Kupffer cells, histiocytes in the spleen and lung, and in the renal tubular epithelium.

Biochemistry and Enzymology. Abnormal accumulation of GM2 ganglioside in the brain and systemic organs is similar to that in Tay–Sachs disease. However, unlike in Tay–Sachs disease, there are massive accumulations of other glycosphingolipids and glycoprotein fragments with terminal hexosamine residues. In the brain, the asialo derivative of GM2 ganglioside accumulates to a much higher level than seen in Tay–Sachs disease, almost to the same level as GM2 ganglioside. Globoside, which is not affected in Tay–Sachs disease, is the primary storage compound in the systemic organs. In addition, the systemic organs contain abnormal glycoprotein fragments.

These analytical findings can be explained on the bases of the enzymatic defect in Sandhoff disease. β-Hexosaminidase β subunit is genetically defective in this disorder. Therefore, both hexosaminidase A (αβ) and B (ββ) isozymes are equally deficient ("total hexosaminidase deficiency"). While hexosaminidase S (αα) does exist, it is catalytically inactive toward natural substrates. The end result is the inability to degrade all natural sphingolipid and glycoprotein substrates beyond the stage where the next step is removal of a hexosamine residue.

Molecular Genetics. Cloning and characterization of the β-hexosaminidase β cDNA and the gene provided an unexpected finding that the genes coding for the α and β subunits are homologous, even though they are located on different chromosomes (Korneluk *et al.*, 1986; Neote *et al.*, 1988). The gene coding for the β subunit is on human chromosome 5. Approximately two-thirds of the bases and amino acids are identical between the two subunits. Mutational analysis of patients with genetic hexosaminidase β-subunit deficiency has not been as extensive as for Tay–Sachs disease, although the number of identified disease-causing mutations is steadily increasing. A recent review lists a dozen such mutations (Gravel *et al.*, 1994). A very recently identified single base deletion near the 5′ end of the coding sequence in infantile patients in a large Maronite community in Cyprus should be added to this list (Hara *et al.*, 1994). A major deletion appears to be relatively common among infantile patients apparently of various ethnic origins (Neote *et al.*, 1990). A mutation within intron 12 which results in insertion of an extra 24 bases in the processed transcript causes either a mild juvenile phenotype or no apparent clinical symptoms. This mutation was initially found in a Canadian patient with a juvenile form of the disease (Nakano and Suzuki, 1989), was then found in another juvenile patient in Japan (Mitsuo *et al.*, 1990) and also in asymptomatic individuals in France ("hexosaminidase Paris") (Dlott *et al.*, 1990).

4.2.2.c. GM2 Activator Protein Deficiency ("AB Variant")

Clinical and Pathological Aspects. In 1969, Sandhoff described a patient with infantile "Tay–Sachs" disease who showed normal activities of β-hexosaminidase A and B, in contrast to others who showed defects in either β-hexosaminidase A or both A and B (Sandhoff, 1969). It was termed "AB variant" denoting the hexosaminidase isozymes that were normal. In retrospect, this article summarized the basic enzymology of the three distinct genetic forms of GM2 gangliosidosis involving three separate genes: β-hexosaminidase A deficiency (Tay–Sachs, "B variant"), A and B deficiency (Sandhoff, "0 variant"), and GM2 activator deficiency ("AB variant"). GM2 activator deficiency is exceedingly rare. Clinical and biochemical phenotype are very similar to those of the classical infantile Tay–Sachs disease (de Baecque *et al.*, 1975). There is an enormous accumulation of GM2 ganglioside in neurons reflected morphologically by the presence of the typical membranous cytoplasmic bodies. No older or late-onset forms have been described.

Biochemistry and Enzymology. Several years after the enzymological description of the first patient, it was recognized that a factor essential for degradation of GM2 ganglioside *in vivo* was missing in an AB variant patient (Conzelmann and Sandhoff, 1978). Inability of the patient's tissues to degrade GM2 ganglioside could be corrected by a protein factor extracted from normal kidney. Thus, existence of an activating factor essential for *in vivo* degradation of GM2 ganglioside was discovered based on the finding of a disease state caused by lack of such a factor. The GM2 activator protein in a small glycoprotein normally localized within the lysosome but is by itself catalytically inactive. Recent reviews containing detailed information on the general subject of the sphingolipid activator proteins have been published (Fürst and Sandhoff, 1992; Sandhoff *et al.*, 1994; Suzuki, 1994a,b).

Molecular Genetics. The functional gene coding for GM2 activator protein was mapped to human chromosome 5 by conventional somatic cell hybridization (Burg *et al.*, 1985) and later confirmed with DNA probes (Xie *et al.*, 1992a; Heng *et al.*, 1993; Swallow *et al.*, 1994). There is evidence for at least one pseudogene for the GM2 activator protein on chromosome 3 (Xie *et al.*, 1992a). The open frame codes for a precursor protein of 193 amino acids, including a signal sequence of 23 amino acids. The remainder of the N-terminus is presumably further processed proteolytically to a mature GM2 activator protein of 162 amino acids. The protein is glycosylated at the only site toward the N-terminus. Molecular cloning and characterization of the cDNA and the gene (Schröder *et al.*, 1989; Klima *et al.*, 1991) opened the way to characterize the genetic abnormalities in patients with GM2 activator deficiency. Two mutations have been described to date. The mutation in one of the earliest cases in the literature (de Baecque *et al.*, 1975) was first identified as T412C (C138R) (Schröder *et al.*, 1991; Xie *et al.*, 1992b). This substitution is at amino acid residue 107 in the mature activator protein. The mutation in another patient has recently been identified as G506C (R169P) (Schröder *et al.*, 1993). Although both of the identified mutations are point mutations within the mature GM2 activator protein, the activator protein was not detectable in tissues by immunological means.

4.3. Galactosialidosis (Protective Protein Deficiency)

The history of galactosialidosis is somewhat convoluted. Since patients show deficiencies of both β-galactosidase and sialidase (α-neuraminidase), they were often diagnosed erroneously as having either a variant of GM1 gangliosidosis or sialidosis. For example, an early somatic cell hybridization study showed that fibroblasts from a group of patients then classified as a GM1 gangliosidosis variant gave genetic complementation when fused with cells from patients with other phenotypes of GM1 gangliosidosis, indicating that a different gene is involved in these patients (Galjaard *et al.*, 1975). It was Wenger who first pointed out a combined deficiency of β-galactosidase and sialidase in many of these "variant" patients (Wenger *et al.*, 1978). A period of controversies followed regarding which of the two enzymatic deficiencies is the primary cause of this disease until van Diggelen *et al.* (1981, 1982) showed that β-galactosidase in fibroblasts from galactosialidosis patients had an abnormally short half life because of rapid proteolytic degradation *in vivo*. This led to the definitive delineation of this disorder as a genetic "protective protein" deficiency (d'Azzo *et al.*, 1982). See Chapter 8 for further details.

Clinical and Pathological Aspects. The mode of inheritance is autosomal recessive. The clinical phenotype of galactosialidosis is heterogeneous, ranging from an early onset, severe, and rapidly progressive infantile form to a late-onset, slowly progressive adult form. Patients with the infantile form can resemble closely those with infantile GM1 gangliosidosis with severe CNS involvement, macular cherry-red spots, visceromegaly, renal insufficiency, coarse facies, and skeletal abnormalities. The late-infantile form is essentially a later-onset, milder phenotype of the infantile disease. The juvenile/adult form of the disease appears to have a much higher incidence in Japan (Suzuki *et al.*, 1984, 1988). The main clinical manifestations are slowly progressive CNS symptoms, including motor disturbance and mental retardation, skeletal abnormalities, dysmorphism, macular cherry-red spots, and angiokeratoma. Patients survive well into adulthood. Galactosialidosis patients of all clinical types excrete an excess of a complex mixture of glycopeptide fragments which are rich in sialic acid. Not much is known about the pathology of patients with verified galactosialidosis.

Biochemistry and Enzymology. Very little is known about the analytical biochemistry of tissue constituents of patients with galactosialidosis. However, storage materials in cultured fibroblasts, fetal placenta, and those excreted into urine have been characterized in detail (van Pelt *et al.*, 1988a,b, 1989). They are predominantly sialylated glycopeptides, similar to those found in sialidosis patients.

The enzymatic characteristic of galactosialidosis is the simultaneous deficiency of both β-galactosidase and α-neuraminidase (Wenger *et al.*, 1978). It had been noted earlier that when β-galactosidase was purified, its molecular mass appeared to be approximately 54 kDa but that another protein component of approximately 32 kDa always co-purified with the 54-kDa component. d'Azzo *et al.* (1982) found that the 32-kDa component was lacking in galactosialidosis. On the basis of this finding and the earlier finding that the half-life of β-galactosidase in galactosialidosis patients was abnormally short but that it could be prolonged by addition of protease inhibitors, it was proposed that the function of the 32-kDa protein is to protect β-galactosidase and neuraminidase from proteolytic digestion within the lysosome by forming a complex with them and that genetic abnormality in this "protective protein" is the underlying cause of galactosialidosis. Furthermore, a partial reduction of the "protect protein" has been described in the juvenile form of the disease (Strisciuglio *et al.*, 1988).

Molecular Genetics. There has been some uncertainty regarding chromosomal localization of the "protective protein" gene. According to the most recent information, the human gene is on chromosome 20 (Wiegant *et al.*, 1991). Galjart *et al.* (1988, 1990) cloned human and murine cDNA coding for the "protective protein" of the respective species. Unexpectedly, the primary structure was homologous with carboxypeptidase. Consistent with this observation, simultaneous deficiency of carboxypeptidase has been described in galactosialidosis (Kase *et al.*, 1990; Tranchemontagne *et al.*, 1990). Given this apparent contradiction—the "protective protein" which is supposed to protect the two enzymes from proteolytic degradation is itself a peptidase—the exact mechanism by which the "protective protein" stabilizes β-galactosidase and sialidase is still not well understood.

To this author's knowledge, mutations responsible for human galactosialidosis have been described only in Japanese patients. One mutation was reported for the late-onset form of the disease prevalent in Japan (Shimmoto *et al.*, 1990; Suzuki *et al.*, 1991). A substitution from G (the third base from the 5′ donor site of intron 7) to A appears to cause skipping of exon 7 during processing of the transcript. All nine Japanese patients with the late-onset form had this mutation. Additional mutations have since been reported in the literature (Takano *et al.*, 1991; Shimmoto *et al.*, 1993).

5. CARBOHYDRATE-DEFICIENT GLYCOPROTEIN SYNDROME

A new genetic disorder, termed *carbohydrate-deficient glycoprotein syndrome*, has been attracting attention the past several years. The disease is characterized by retardation of psychomotor development and neurological manifestations. Glycoprotein fractions from patients have been noted to be less acidic than the normal counterparts. This is apparently related to lack of sialic acid on the carbohydrate chains. More severe type I and milder type II are recognized. In type I, some lysosomal enzymes are elevated moderately in the serum, whereas they are normal in type II. Genetic defects in glycosyltransferases at different steps of sequential addition of monosaccharides are suspected. Since it is not yet clear if this disorder is appropriate for inclusion in this chapter, only a few recent review articles are cited for interested readers (Hagberg *et al.*, 1993; Jaeken and Carchon, 1993; Jaeken *et al.*, 1993).

ACKNOWLEDGMENTS. The author thanks Professor Konrad Sandhoff and Dr. Roy Gravel for making copies of their review chapters available before publication (chapters in Scriver *et al.*, 1995). Work from the author's own laboratory has been supported over the years by a series of research grants from the NIH, in recent years by RO1-NS28997, RO1-NS24289, and a Mental Retardation Research Center Core grant, P30-HD03110.

REFERENCES

Arpaia, E., Dumbrille-Ross, A., Maler, T., Neote, K., Tropak, M., Troxel, C., Stirling, J. S., Pitts, J. S., Bapat, B., Lamhonwah, A. M., Mahuran, D. J., Schuster, S. M., Clarke, J.T.R., Lowden, J. A., and Gravel, R. A., 1988, Identification of an altered splice site in Ashkenazi Tay–Sachs disease, *Nature* **333:**85–86.

Boustany, R.-M., Qian, W.-H., and Suzuki, K., 1993, Mutations in acid β-galactosidase cause GM1-gangliosidosis in American patients, *Am. J. Hum. Genet.* **53:**881–888.

Burg, J., Conzelmann, E., Sandhoff, K., Solomon, E., and Swallow, D. M., 1985, Mapping of the gene coding for the human GM2 activator protein to chromosome 5, *Ann. Hum. Genet.* **49:**41–45.

Chakraborty, S., Rafi, M. A., and Wenger, D. A., 1991, A point mutation in the acid beta-galactosidase cDNA sequence of 2 adult patients with GM1 gangliosidosis, *Am. J. Hum. Genet.* **49(Suppl.):**95.

Conzelmann, E., and Sandhoff, K., 1978, AB variant of infantile GM2 gangliosidosis: Deficiency of a factor necessary for stimulation of hexosaminidase A-catalyzed degradation of ganglioside GM2 and glycolipid GA2, *Proc. Natl. Acad. Sci. USA* **75:**3979–3983.

d'Azzo, A., Hoogeveen, A., Reuser, A.J.J., Robinson, D., and Galjaard, H., 1982, Molecular defect in combined β-galactosidase and neuraminidase deficiency in man, *Proc. Natl. Acad. Sci. USA* **79:**4535–4539.

de Baecque, C. M., Suzuki, K., Rapin, I., Johnson, A. B., Wethers, D. L., and Suzuki, K., 1975, G_{M2}-gangliosidosis AB variant: Clinicopathological study of a case, *Acta Neuropathol.* **33:**207–226.

Derry, D. M., Fawcett, J. S., Andermann, F., and Wolfe, L. S., 1968, Late infantile systemic lipidosis: Major monosialo-gangliosidosis: Delineation of two types, *Neurology* **18:**340–348.

Dlott, B., d'Azzo, A., Quon, D.V.K., and Neufeld, E. F., 1990, Two mutations produce intron insertion in mRNA and elongated β-subunit of human β-hexosaminidase, *J. Biol. Chem.* **265:**17921–17927.

Dos Santos, M. R., Tanaka, A., Sá Miranda, M. C., Ribeiro, M. G., Maia, M., and Suzuki, K., 1991, GM2-gangliosidosis B1 variant: Analysis of β-hexosaminidase α gene mutations in eleven patients from a defined region in Portugal, *Am. J. Hum. Genet.* **49:**886–890.

Fürst, W., and Sandhoff, K., 1992, Activator proteins and topology of lysosomal sphingolipid catabolism, *Biochim. Biophys. Acta* **1126:**1–16.

Galjaard, H., Hoogeveen, A., Kleijer, W., de Wit-Verbeek, H. A., Reuser, A.J.J., Ho, M. W., and Robinson, D., 1975, Genetic heterogeneity in GM1-gangliosidosis, *Nature* **257:**60–62.

Galjart, N. J., Gillemans, N., Harris, A., van der Horst, G.T.J., Verheijen, F. W., Galjaard, H., and d'Azzo, A., 1988, Expression of cDNA encoding the human "protective protein" associated with lysosomal beta-galactosidase and neuraminidase: Homology to yeast proteases, *Cell* **54:**755–764.

Galjart, N. J., Gillemans, N., Meijer, D., and d'Azzo, A., 1990, Mouse "protective protein": cDNA cloning, sequence comparison and expression, *J. Biol. Chem.* **265:**4678–4684.

Goldman, J. E., Katz, D., Rapin, I., Purpura, D. P., and Suzuki, K., 1981, GM1-gangliosidosis presenting as dystonia. I. Clinical and pathological features, *Ann. Neurol.* **9:**465–475.

Gonatas, N. K., and Gonatas, J., 1965, Ultrastructural and biochemical observation on a case of systemic late infantile lipidosis and its relationship to Tay–Sachs disease and gargoylism, *J. Neuropathol. Exp. Neurol.* **24:**318–340.

Gravel, R. A., Clarke, J.T.R., Kaback, M. M., Mahuran, D., Sandhoff, K., and Suzuki, K., 1995, The GM2 gangliosidosis, in: *The Metabolic and Molecular Basis of Inherited Disease* (C. R. Scriver, A. L. Beaudet, W. S. Sly, and D. Valle, eds.), 7th ed., McGraw–Hill, New York, pp. 2839–2879.

Hagberg, B. A., Blennow, G., Kristiansson, B., and Stibler, H., 1993, Carbohydrate-deficient glycoprotein syndrome: Peculiear group of new disorders, *Pediatr. Neurol.* **9:**255–262.

Hara, Y., Ioannou, P., Drousiotou, A., Stylianidou, G., Anastasiadou, V., and Suzuki, K., 1994, Mutation analysis of a Sandhoff disease patient in the Maronite community in Cyprus, *Hum. Genet.* **94:**136–140.

Heng, H.H.Q., Xie, B., Shi, X.-M., Tsui, L.-C., and Mahuran, D. J., 1993, Refined mapping of the GM2 activator protein (GM2A) locus to 5q31.3-q33.1 distal to the spinal muscular atrophy locus, *Genomics* **18:**429–431.

Jaeken, J., and Carchon, H., 1993, The carbohydrate-deficient glycoprotein syndromes: An overview, *J. Inherit. Metab. Dis.* **16:**813–820.

Jaeken, J., Carchon, H., and Stibler, H., 1993, The carbohydrate-deficient glycoprotein syndromes: Pre-Golgi and Golgi disorders? *Glycobiology* **3:**423–428.

Jatzkewitz, H., and Sandhoff, K., 1963, On a biochemically special form of infantile amaurotic idiocy, *Biochim. Biophys. Acta* **70:**354–356.

Kaback, M., Lim-Steele, J., Dabholkar, D., Brown, D., Levy, N., and Zeiger, K., 1993, Tay–Sachs disease—Carrier screening, prenatal diagnosis, and the molecular era: An international perspective, *J. Am. Med. Assoc.* **270:**2307–2315.

Kase, R., Itoh, K., Takiyama, N., Oshima, A., Sakuraba, H., and Suzuki, Y., 1990, Galactosialidosis: Simultaneous deficiency of esterase, carboxy-terminal deamidase and acid carboxypeptidase activities, *Biochem. Biophys. Res. Commun.* **172:**1175–1179.

Klenk, E., 1939, Beiträge zur Chemie der Lipidosen. 3. Niemann-Picksche Krankheit und amaurotische Idiotie, *Z. Physiol. Chem.* **262:**128–143.

Klima, H., Tanaka, A., Schnabel, D., Nakano, T., Schröder, M., Suzuki, K., and Sandhoff, K., 1991, Characterization of full-length cDNAs and the gene coding for the human GM2 activator protein, *FEBS Lett.* **289:** 260–264.

Kobayashi, T., and Suzuki, K., 1981, Chronic GNM+-gangliosidosis presenting as dystonia. II. Biochemistry, *Ann. Neurol.* **9:**476–483.

Korneluk, R. G., Mahuran, D. J., Neote, K., Klavins, M. H., O'Dowd, B. F., Tropack, M., Willard, H. F., Anderson, M.-J., Lowden, J. A., and Gravel, R. A., 1986, Isolation of cDNA clones coding for the α subunit of human β-hexosaminidase: Extensive homology between the α- and β-subunits and studies on Tay–Sachs disease, *J. Biol. Chem.* **261:**8407–8413.

Kytzia, H.-J., and Sandhoff, K., 1985, Evidence for two different active sites on human hexosaminidase A—Interaction of GM2-activator protein with hexosaminidase A, *J. Biol. Chem.* **260:**7568–7572.

Mitsuo, K., Nakano, T., Kobayashi, T., Goto, I., Taniike, M., and Suzuki, K., 1990, Juvenile Sandhoff disease: A Japanese patient carrying a mutation identical to that found earlier in a Canadian patient, *J. Neurol. Sci.* **98:**277–286.

Morreau, H., Galjart, N. J., Gillemans, N., Willemsen, R., van der Horst, G.T.J., and d'Azzo, A., 1989, Alternate splicing of β-galactosidase messenger RNA generates the classic lysosomal enzyme and a β-galactosidase-related protein, *J. Biol. Chem.* **264:**20655–20663.

Morreau, H., Bonten, E., Zhou, X.-Y., and d'Azzo, A., 1991, Organization of the gene encoding human lysosomal β-galactosidase, *DNA Cell Biol.* **10:**495–504.

Mosna, G., Fattore, S., Tubiello, G., Brocca, S., Trubia, M., Gianazza, E., Gatti, R., Danesino, C.,

Minelli, A., and Piantanida, M., 1992, A homozygous missense arginine to histidine substitution at position 482 of the β-galactosidase in an Italian infantile GM1-gangliosidosis patient, *Hum. Genet.* **90:**247–250.

Mueller, O. T., Henry, W. M., Eddy, R. L., and Shows, T. B., 1986, Sialidosis and galactosialidosis: Chromosomal assignment of two genes associated with neuraminidase deficiency, *Proc. Natl. Acad. Sci. USA* **83:**1817–1821.

Myerowitz, R., 1988, Splice junction mutation in some Ashkenazi Jews with Tay–Sachs disease: Evidence against a single defect within this ethnic group, *Proc. Natl. Acad. Sci. USA* **85:**3955–3958.

Myerowitz, R., and Costigan, C., 1988, The major defect in Ashkenazi Jews with Tay–Sachs disease is an insertion in the gene for the α-chain of β-hexosaminidase, *J. Biol. Chem.* **263:**18587–18589.

Myerowitz, R., and Hogikyan, N. D., 1986, Different mutations in Ashkenazi Jewish and nonJewish French Canadians with Tay–Sachs disease, *Science* **232:**1646–1648.

Myerowitz, R., and Hogikyan, N. D., 1987, A deletion involving Alu sequences in the beta-hexosaminidase alpha chain gene of French Canadians with Tay–Sachs disease, *J. Biol. Chem.* **262:**15396–15399.

Myerowitz, R., Piekarz, R., Neufeld, E. F., Shows, T. B., and Suzuki, K., 1985, Human β-hexosaminidase α chain: Coding sequence and homology with the β-chain, *Proc. Natl. Acad. Sci. USA* **82:**7830–7834.

Nakano, T., and Suzuki, K., 1989, Genetic cause of a juvenile form of Sandhoff disease: Abnormal splicing of β-hexosaminidase β chain gene transcript due to a point mutation within intron 12, *J. Biol. Chem.* **264:**5155–5158.

Navon, R., and Proia, R. L., 1989, The mutations in Ashkenazi Jews with adult GM2 gangliosidosis, the adult form of Tay–Sachs disease, *Science* **243:**1471–1474.

Neote, K., Bapat, B., Dumbrille-Ross, A., Troxel, C., Schuster, S. M., Mahuran, D. J., and Gravel, R. A., 1988, Characterization of the human HEXB gene encoding lysosomal β-hexosaminidase, *Genomics* **3:**279–286.

Neote, K., McInnes, B., Mahuran, D. J., and Gravel, R., 1990, Structure and distribution of an Alu-type deletion mutation in Sandhoff disease, *J. Clin. Invest.* **86:**1524–1531.

Nishimoto, J., Nanba, E., Inui, K., Okada, S., and Suzuki, K., 1991, GM1-gangliosidosis (genetic β-galactosidase deficiency): Identification of mutations in different clinical phenotypes among Japanese patients, *Am. J. Hum. Genet.* **49:**566–574.

O'Brien, J. S., Stern, M. B., Landing, B. H., O'Brien, J. I., and Donnel, G. N., 1965, Generalized gangliosidosis: Another inborn error of metabolism? *Am. J. Dis. Child.* **109:**338–346.

Ohno, K., and Suzuki, K., 1988a, Mutation in GM2-gangliosidosis B1 variant, *J. Neurochem.* **50:**316–318.

Ohno, K., and Suzuki, K., 1988b, A splicing defect due to an exon–intron junctional mutation results in abnormal β-hexosaminidase α chain mRNAs in Ashkenazi Jewish patients with Tay–Sachs disease, *Biochem. Biophys. Res. Commun.* **153:**463–469.

Ohno, K., and Suzuki, K., 1988c, Multiple abnormal β-hexosaminidase α chain mRNAs in a compound-heterozygous Ashkenazi Jewish patient with Tay–Sachs disease, *J. Biol. Chem.* **263:**18563–18567.

Okada, S., and O'Brien, J. S., 1968, Generalized gangliosidosis: Beta-galactosidase deficiency, *Science* **160:**1002–1004.

Okada, S., and O'Brien, J. S., 1969, Tay–Sachs disease: Generalized absence of a β-D-N-acetyl-hexosaminidase component, *Science* **165:**698–700.

Oshima, A., Tsuji, A., Nagao, Y., Sakuraba, H., and Suzuki, Y., 1988, Cloning, sequencing, and expression of cDNA for human β-galactosidase, *Biochem. Biophys. Res. Commun.* **157:**238–244.

Proia, R. L., and Soravia, E., 1987, Organization of the gene encoding the human β-hexosaminidase α chain, *J. Biol. Chem.* **262:**5677–5681.

Purpura, D. P., and Suzuki, K., 1976, Distortion of neuronal geometry and formation of aberrant synapses in neuronal storage disease, *Brain Res.* **116:**1–21.

Sachs, B., 1887, On arrested cerebral development, with special reference to its cortical pathology, *J. Nerv. Ment. Dis.* **14:**541–553.

Sacrez, R., Juif, J. G., Gigonnet, J. M., and Gruner, J. E., 1967, Maladie de Landing, ou idiotie amaurotique infantile précose avec gangliosidose généralizéede type GM1, *Pediatrie* **22:**143–162.

Sandhoff, K., 1969, Variation of β-acetylhexosaminidase apttern in Tay–Sachs disease, *FEBS Lett.* **4:**351–354.

Sandhoff, K., Harzer, K., and Fürst, W., 1995, Sphingolipid activator proteins, in: *The Metabolic and Molecular Basis of Inherited Disease* (C. R. Scriver, A. L. Beaudet, W. S. Sly, and D. Valle, eds.), 7th ed., McGraw–Hill, New York, pp. 2727–2441.

Schröder, M., Klima, H., Nakano, T., Kwon, H., Quintern, L. E., Gärtner, S., Suzuki, K., and Sandhoff, K., 1989, Isolation of a cDNA encoding the human GM2-activator protein, *FEBS Lett.* **251:**197–200.

Schröder, M., Schnabel, D., Suzuki, K., and Sandhoff, K., 1991, A mutation in the gene of a glycolipid-binding protein (GM2 activator) that causes GM2-gangliosidosis variant AB, *FEBS Lett.* **290:**1–3.

Schröder, M., Schnabel, D., Young, E., Suzuki, K., and Sandhoff, K., 1993, Molecular genetics of GM2-gangliosidosis AB variant: A novel mutation and expression in BHK cells, *Hum. Genet.* **92:**437–440.

Scriver, C. R., Beaudet, A. L., Sly, W. S., and Valle, D. (eds.), 1995, *The Metabolic and Molecular Basis of Inherited Disease,* 7th ed., McGraw–Hill, New York.

Seringe, P., Plainfosse, B., Lautman, F., Lorilloux, J., Calamy, G., Berry, J.-P., and Watchi, J.-M., 1968, Gangliosidose généralizée du type Norman-Landing à GM1. Etude à propos d'un cas diagnostiqué du vivant du malade, *Ann. Pediat.* **15:**165–184.

Shimmoto, M., Takano, T., Fukuhara, Y., Oshima, A., Sakuraba, H., and Suzuki, Y., 1990, Japanese-type adult galactosialidosis: A unique and common splice junction mutation causing exon skipping in the protective protein/carboxypeptidase gene, *Proc. Jpn. Acad.* **66B:**217–222.

Shimmoto, M., Fukuhara, Y., Itoh, K., Oshima, A., Sakuraba, H., and Suzuki, Y., 1993, Protective protein gene mutations in galactosialidosis, *J. Clin. Invest.* **91:**2393–2398.

Shows, T. B., Scrafford-Wolff, L. R., Brown, J. A., and Meisler, M., 1979, GM1-gangliosidosis: Chromosome 3 assignment of a β-galactosidase A gene (β-GalA), *Somat. Cell Genet.* **5:**147–158.

Strisciuglio, P., Parenti, G., Giudice, C., Lijoi, S., Hoogeveen, A. T., and d'Azzo, A., 1988, The presence of a reduced amount of 32-kd "protective" protein is a distinct biochemical finding in late infantile galactosialidosis, *Hum. Genet.* **80:**304–306.

Suzuki, K., 1968, Cerebral GM1-gangliosidosis: Chemical pathology of visceral organs, *Science* **159:**1471–1472.

Suzuki, K., 1977, Globoid cell leukodystrophy (Krabbe disease) and GM1-gangliosidosis, in: *Practical Enzymology of the Sphingolipidoses* (R. H. Glew and S. P. Peters, eds.), Liss, New York, pp. 101–136.

Suzuki, K., 1987, Enzymatic diagnosis of sphingolipidoses, *Methods Enzymol.* **138:**727–762.

Suzuki, K., 1991, Neuropathology of late-onset gangliosidoses, *Dev. Neurosci.* **13:**205–210.

Suzuki, K., 1994a, Sphingolipid activator proteins, *Essays Biochem.* in press.

Suzuki, K., 1994b, Molecular basis of genetic sphingolipid activator protein deficiencies, *Trends Glycosci. Glycotechnol.* **6:**215–227.

Suzuki, K., and Chen, G. C., 1967, Brain ceramide hexosides in Tay–Sachs disease and generalized gangliosidosis (GM1-gangliosidosis), *J. Lipid Res.* **8:**105–113.

Suzuki, K., and Vanier, M. T., 1991, Biochemical and molecular aspects of late-onset GM2-gangliosidosis: B1 variant as the prototype, *Dev. Neurosci.* **13**:288–294.

Suzuki, K., Suzuki, K., and Kamoshita, S., 1969, Chemical pathology of GM1-gangliosidosis (generalized gangliosidosis), *J. Neuropathol. Exp. Neurol.* **28**:25–73.

Suzuki, Y., Crocker, A. C., and Suzuki, K., 1971a, GM1-gangliosidosis: Correlation of clinical and biochemical data, *Arch. Neurol.* **24**:58–64.

Suzuki, Y., Jacob, J. C., Suzuki, K., Kutty, K. M., and Suzuki, K., 1971b, G_{M2}-gangliosidosis with total hexosaminidase deficiency, *Neurology* **21**:313–328.

Suzuki, Y., Nakamura, N., Fukuoka, K., Shimada, Y., and Uono, M., 1977, β-Galactosidase deficiency in juvenile and adult patients. Report of six Japanese cases and review of literature, *Hum. Genet.* **36**:219–229.

Suzuki, Y., Sakuraba, H., Yamanaka, T., Ko, Y.-M., Iimori, Y., Okamura, Y., and Hoogeveen, A. T., 1984, Galactosialidosis: A comparative study of clinical and biochemical data on 22 patients, in: *The Developing Brain and Its Disorders* (M. Arima, Y. Suzuki, and H. Yabuuchi, eds.), University of Tokyo Press, Tokyo, pp. 161–175.

Suzuki, Y., Nanba, E., Tsuji, A., Yang, R.-C., Okamura-Oho, Y., and Yamanaka, T., 1988, Clinical and genetic heterogeneity in galactosialidosis, *Brain Dysfunct.* **1**:285–293.

Suzuki, Y., Sakuraba, H., Oshima, A., Yoshida, K., Shimmoto, M., Takano, T., and Fukuhara, Y., 1991, Clinical and molecular heterogeneity in hereditary β-galactosidase deficiency, *Dev. Neurosci.* **13**:299–303.

Swallow, D. M., Islam, I., Fox, M. F., Povey, S., Klima, H., Schepers, U., and Sandhoff, K., 1993, Regional localization of the gene coding for the GM2 activator protein (GM2A) to chromosome 5q32-33 and confirmation of the assignment of GM2AP to chromosome 3, *Ann. Hum. Genet.* **57**:187–193.

Takano, T., Shimmoto, M., Furuhara, Y., Itoh, K., Kase, R., Takiyama, N., Kobayashi, T., Oshima, A., Sakuraba, H., and Suzuki, Y., 1991, Galactosialidosis: Clinical and molecular analysis of 19 Japanese patients, *Brain Dysfunct.* **4**:271–280.

Tanaka, H., and Suzuki, K., 1975, Lactosylceramide β-galactosidase in human sphingolipidoses: Evidence for two genetically distinct enzymes, *J. Biol. Chem.* **250**:2324–2332.

Tanaka, H., and Suzuki, K., 1977, Substrate specificities of the two genetically distinct human brain β-galactosidases, *Brain Res.* **122**:325–335.

Tanaka, A., Ohno, K., and Suzuki, K., 1988, GM2-gangliosidosis B1 variant: A wide geographic and ethnic distribution of the specific β-hexosaminidase α chain mutation originally identified in a Puerto Rican patient, *Biochem. Biophys. Res. Commun.* **156**:1015–1019.

Tanaka, A., Ohno, K., Sandhoff, K., Maire, I., Kolodny, E. H., Brown, A., and Suzuki, K., 1990, GM2-gangliosidosis B1 variant: Analysis of β-hexosaminidase α gene abnormalities in seven patients, *Am. J. Hum. Genet.* **46**:329–339.

Tay, W., 1881, Symmetrical changes in the region of the yellow spot in each eye of an infant, *Trans. Ophthalmol. Soc. UK* **1**:155.

Terry, R. D., and Korey, S. R., 1960, Membranous cytoplasmic granules in infantile amaurotic idiocy, *Nature* **188**:1000–1002.

Terry, R. D., and Weiss, M., 1963, Studies in Tay–Sachs disease: II. Ultrastructure of cerebrum, *J. Neuropathol. Exp. Neurol.* **22**:18–55.

Tranchemontagne, J., Michaud, L., and Potier, M., 1990, Deficient lysosomal carboxypeptidase activity in galactosialidosis, *Biochem. Biophys. Res. Commun.* **168**:22–29.

van Diggelen, O. P., Schram, A. W., Sinnot, M. L., Smith, P. J., Robinson, D., and Galjaard, H., 1981, Turnover of β-galactosidase in fibroblasts from patients with genetically different types of β-galactosidase deficiency, *Biochem. J.* **200**:143–151.

van Diggelen, O. P., Hoogeveen, A. T., Smith, P. J., Reuser, A.J.J., and Galjaard, H., 1982,

Enhanced proteolytic degradation of normal β-galactosidase in the lysosomal disease with combined β-galactosidase and neuraminidase deficiency, *Biochim. Biophys. Acta* **703:**69–76.

van Pelt, J., Kamerling, J. P., Vliegenthart, J.F.G., Hoogeveen, A. T., and Galjaard, H. A., 1988a, Comparative study of the accumulated sialic acid-containing oligosaccharides from human galactosialidosis and sialidosis fibroblasts, *Clin. Chim. Acta* **174:**325–335.

van Pelt, J., van Kuik, J. A., Kamerling, J. P., Vliegenthart, J.F.G., van Diggelen, O. P., and Galjaard, H., 1988b, Storage of sialic acid-containing carbohydrates in the placenta of a human galactosialidosis fetus—Isolation and structural characterization of 16 sialyloligosaccharides, *Eur. J. Biochem.* **177:**327–338.

van Pelt, J., Hard, K., Kamerling, J. P., Vliegenthart, J.F.G., Reuser, A.J.J., and Galjaard, H., 1989, Isolation and structural characterization of twenty-one sialyloligosaccharides from galactosialidosis urine. An intact N,N′-diacetylchitobiose unit at the reducing end of a diantennary structure, *Biol. Chem. Hoppe-Seyler* **370:**191–203.

Wenger, D. A., Tarby, T. J., and Wharton, C., 1978, Macular cherry-red spots and myoclonus with dementia: Coexistent neuraminidase and β-galactosidase deficiencies, *Biochem. Biophys. Res. Commun.* **82:**589–595.

Wiegant, J., Galjart, N. J., Raap, A. K., and d'Azzo, A., 1991, The gene encoding human protective protein (PPGB) is on chromosome 20, *Genomics* **10:**345–349.

Xie, B., Kennedy, J. L., McInnes, B., Auger, D., and Mahuran, D., 1992a, Identification of a processed pseudogene related to the functional gene encoding the GM2 activator protein: Localization of the pseudogene to human chromosome 3 and the functional gene to human chromosome 5, *Genomics* **14:**796–798.

Xie, B., Wang, W., and Mahuran, D. J., 1992b, A Cys138-to-Arg substitution in the GM2 activator protein is associated with the AB variant form of GM2 gangliosidosis, *Am. J. Hum. Genet.* **50:**1046–1052.

Yoshida, K., Oshima, A., Shimmoto, M., Fukuhara, Y., Sakuraba, H., Yanagisawa, N., and Suzuki, Y., 1991, β-Galactosidase gene mutations in GM1-gangliosidosis: A common point mutation among Japanese adult-chronic cases, *Am. J. Hum. Genet.* **49:**435–442.

Yoshida, K., Oshima, A., Sakuraba, H., Nakano, T., Yanagisawa, H., Inui, K., Okada, S., Uyama, E., Namba, R., Kondo, K., Iwasaki, S., Takamiya, K., and Suzuki, Y., 1992, GM1 gangliosidosis in adults: Clinical and molecular analysis of 16 Japanese patients, *Ann. Neurol.* **31:**328–332.

Chapter 11

Neurite Loss Caused by Ganglioside Undersialylation

Abraham Rosenberg

1. INTRODUCTION

An emerging thought in sialobiology holds that lessened sialylation of gangliosides by whatever means leads in the growing neuron to a quick breakdown of neurital processes followed by neuronal loss of function. This chapter will take a brief look at findings stemming from work around that idea now in full swing in the author's laboratory.

Simpler neuronal gangliosides, i.e., those with a shorter neutral sugar backbone, have their sialyl oligosaccharide chains built in an endoplasmic reticulum–*cis*-Golgi intermediate organellar compartment, and they make up most of the glycocalyx of the outer lipid bilayer in the perikaryal plasma membrane of the developing neuron. However, higher gangliosides built with a longer neutral sugar backbone arise biosynthetically from simpler gangliosides as they go on their membrane-generating way from endoplasmic reticulum through the Golgi to the plasma membrane of the outgrowing neurite.

Dangerously raising calcium ion levels within the neuron, divalent cationophores have in part a blocking effect on the buildup of higher neuritogenic gangliosides, causing neurites quickly to fall apart. At the same time, simpler gangliosides remain stable or increase in the face of divalent cationophore treatment, and the neuronal perikaryon endures.

Abraham Rosenberg Neuropsychiatric Institute and Brain Research Institute, University of California, Los Angeles, California 90024-1759.

Biology of the Sialic Acids, edited by Abraham Rosenberg. Plenum Press, New York, 1995.

Chronic ethanol, even at rather low levels, lessens sialylation of all neuronal gangliosides, thereby slowing the morphogenetic pace in the neuron, and leading to the central nervous system impairment that is a key feature of the fetal alcohol syndrome. As the evidence to be presented in this chapter suggests, biosynthetic sialylation of gangliosides may prove to be a quite vulnerable point of assault in the etiology of a variety of important developmental neuronodystrophies.

2. GANGLIOSIDE SIALYLATION AS A KEY TO NEURITOGENESIS

Gangliosides (sialoglycosphingolipids) quantitatively are the major structural complex lipid elements in the exofacial molecular bilayer of the phenotypically morphodifferentiating central nervous system neuron (Hogan *et al.*, 1988; Van Echten and Sandhof, 1989; Rosenberg *et al.*, 1992; Igarishi *et al.*, 1994). In addition to playing a major role in concert with the other structural complex glyco- and phospho-sphingolipids which are organized as specific components in the outer lipid bilayer of the neuronal plasma membrane, gangliosides can act as positive or negative modifiers of biologically active plasma membrane proteins, particularly of membrane-bound protein kinases (Chan, 1987; see also Chapters 6 and 7). An intensive analysis of the developmental patterns of ganglioside sialylation correlating with neurite outgrowth in cultured embryonic chick brain cortical neurons (Rosenberg *et al.*, 1992) has shown that neurite outgrowth depends closely on production by the developing neuron of neuritogenic, growth cone-forming, plasma membrane elements endowed with a quite massive content of gangliosides, in large part those of the more complex species (Rosenberg *et al.*, 1992; Igarishi *et al.*, 1994). Hence, the functionality of specific sialylation mechanisms (see Chapter 3) that govern initial biosynthesis as well as replacement turnover of those gangliosides has emerged as a key determinant impinging directly on the phenomenon of neuritogenesis. Figure 1 gives a quantitative picture of ganglioside sialylation in cultured embryonic chick cortical neurons during their progression from preneuritogenic precursor form, through early neuritogenesis, then full neurite outgrowth, and finally during turnover maintenance of developed neurital structure.

In the earliest stage when the neurons are perikaryoform cells without real neurites, only gangliosides of the short-chain lactosyl and gangliotriosyl species (GM3, GM2, GD3, and GD2) are sialylated. Hence, at this stage there is a productive expression of activity only of sialyl transferases 1 and 2 (see Chapter 3) which respectively synthesize GM3 and GD3. GM2 (not shown) and GD2 are the products of *N*-acetyl galactosaminylation of GM3 and GD3 (see Chapter 6).

As shown in Figure 1, with the onset of vigorous neurite outgrowth, sialyl-labeled GM1 and GD1b, the respective terminal-galactosylation products of GM2 and GD2, make an appearance in the neuron. So also does the product of

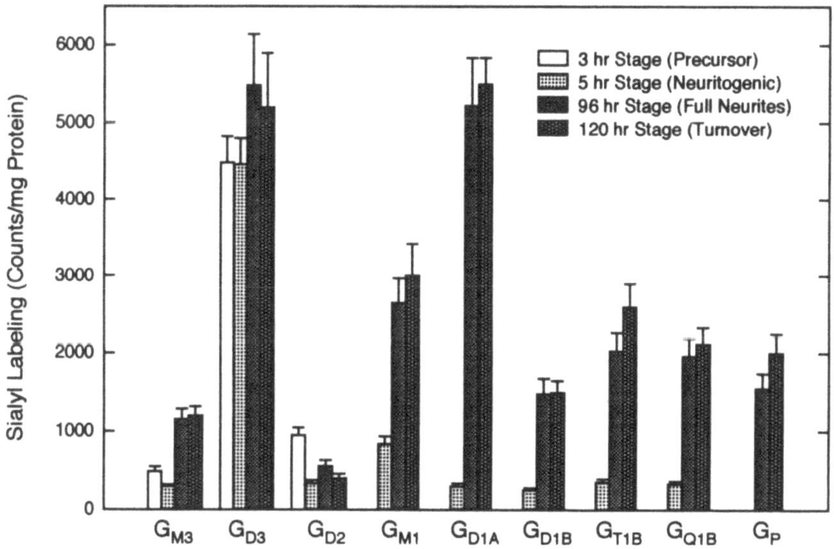

FIGURE 1. Sialyl-labeling of the gangliosides in cultured embryonic chick brain cortical neurons from the preneuritogenic precursor neuron stage through full neuritogenesis. The precursor neurons transferred from the telencephalon of 8-day chick embryo brain were cultured in serum-free medium with EGF as neurotrophic factor. Ganglioside sialic acid was labeled metabolically with the obligate sialic acid precursor [^{3}H]-ManNAc.

sialylation of the terminal galactose of the newly formed gangliotetraosyl back bone in GM1; thus, sialyl-labeled GD1a, which is produced by sialylation of GM1 through the action of sialyl transferase 4, makes an appearance concomitantly with neuritogenesis. Likewise, sialyl-labeled GT1b and GQ1b make an appearance along with the onset of neurite outgrowth. Thus, neuritogenesis corresponds to expression of the products of activity of sialyl transferase 4, which sialylates GD1b yielding GT1b, as well as sialyl transferase 5, which further sialylates the terminal sialyl group on GT1b, yielding GQ1b. The data suggest that sialyl transferase 4 and 5 may promote neuritogenesis.

While the sialyl-labeling of the simpler gangliosides remains relatively stable, an upswing in labeling of the terminally sialylated higher gangliotetraosyl ceramide ganglioside species GD1a, GT1b, and GQ1b paces full neurite outgrowth, following which sialyl-labeling of gangliosides continues with further turnover maintenance of the embryonic neuronal plasma membrane. During this latter turnover stage, substantial sialyl-labeling of GP is observable, presumably mediated by sialyl transferase 5. This is the same enzyme that produces GQ1b, and eventually further sialylates it, yielding GP.

Regulation of Ganglioside Sialylation

The regulatory elements in the mechanisms for ganglioside sialylation necessarily include the developmental expression of ganglioside-specific sialyl transferases and, moreover, the expression and biological processing of vital Ca^{2+}-dependent proteins that control pro-plasma membrane vesicle formation and trafficking (Feizi, 1994; Bergeron *et al.*, 1994) as well as the functional organization of the cytoskeletal pathway over which vesicles travel through the Golgi loci holding the specific glycosyl and sialyl transferases which build the sialyl oligosaccharide components of the various species of neuronal gangliosides on their way to residence in the plasma membrane. Impairment of any of these elements regulating ganglioside sialylation singly or in combination may lead to neuron damage stemming from a loss of ganglioside-dependent neuritogenesis (Rosenberg *et al.*, 1992).

3. IONIC BLOCKING OF GANGLIOSIDE-DEPENDENT NEURITOGENESIS

Treatment with the monovalent cationophore monensin decreases biosynthesis of the gangliosides in the neurite-bearing embryonic chick brain cortical neuron (Hogan *et al.*, 1988). The biosynthetic block induced by monensin coincides with an almost total loss of neurites. The effects of monensin stem from import of excessive Na^{+} which may also heighten intraneuronal Ca^{2+} in homeostatic exchange for Na^{+}.

Neuritogenesis appears to be particularly closely regulated by intraneuronal Ca^{2+} levels (Mattson, 1990). Excitatory amino acid neuronotoxicity may result from prolonged abnormally heightened intraneuronal Ca^{2+} (Choi, 1987), and this phenomenon is considered a causative factor in neuron degeneration attendant on neuronal anoxia (Rothman and Olney, 1986), Huntington's disease (Coyle and Schwarcz, 1976), and Alzheimer's disease (Greenamyre and Young, 1989).

The resting concentration of intraneuronal cytoplasmic free Ca^{2+} in cultured embryonic neurons ranges between 70 and 80 nM (Mattson *et al.*, 1991). Exposure of the neurons to 1 μM A23187 (calcium ionophore) causes an import of Ca^{2+} which is only partially offset by neuronal organellar sequestration of Ca^{2+} and Ca^{2+}-export mechanisms (Carafoli, 1987) so that A23187 induces a resultant prolonged five-fold heightening of intraneuronal Ca^{2+}. When cultured chick brain cortical neurons at the 72-hr fully neurited stage (see Figure 2) were exposed for 24 hr to 1 μM A23187 (Rosenberg and Noble, 1993), the neurital field disintegrated, but the neuronal perikarya did not, as shown in Figure 2.

The effects of A23187 exposure on ganglioside biosynthesis are shown in Figure 3. There was no difference in biosynthetic labeling of GM3 in A23187-treated neurons as compared with control untreated neurons, but labeling in

GD3, the product of sialylation of GM3 through the action of sialyl transferase 2, increased by 100%, and GM2, the product of *N*-acetyl galactosaminylation of GM3 through the action of *N*-acetyl galactosaminyl transferase 1, increased over 600%. Both GD3 and GM2 are the immediate glycosylation products of GM3. (See Chapter 3 for a discussion of these enzymes.) Perhaps these transferases, sialyl transferase 2 and *N*-acetyl galactosaminyl transferase 1, were activated by heightened intraneuronal Ca^{2+}. However, a more likely (but not the only) possibility is that heightened Ca^{2+} blocked transport of the product gangliosides GD3 and GM2 from the endoplasmic reticulum/*cis*-Golgi intermediate compartment where sialyltransferase 2 and *N*-acetyl galactosaminyl transferase 1 reside, resulting in metabolic backup of those gangliosides. Failure of their entry into the medial/*trans*-Golgi compartment where the enzymes are found that build those precursor gangliosides into the higher gangliosides GM1, GD1a, GD1b, GT1b, and GQ1b—namely, galactosyl transferase 2 and sialyl transferases 4 and 5 (see Chapter 3)—would occasion a drastic reduction in biosynthetic labeling of the higher neuritogenic gangliosides as shown in Figure 3. An expected result of this kind of biosynthetic distortion by heightened Ca^{2+} would be continued production of perikaryal plasma membrane which is rich in GD3 and GD2, a dissipation of neurital plasma membrane which needs to be rich in the higher gangliosides, and a reversal to preneuritogenic morphology as seen in Figure 2. Treatment of

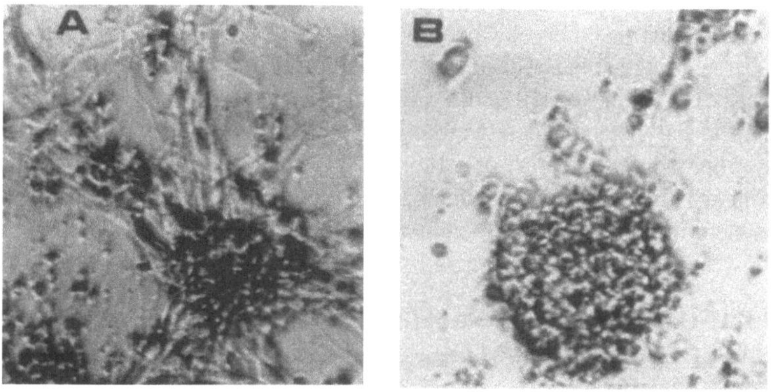

FIGURE 2. The effect of exposure to 1 μM A23187 (CA^{2+} ionophore) on the neurites of cultured embryonic chick brain cortical neurons. Precursor neurons were cultured in serum-free neurotrophic medium for 96 hr. (A) Control neurons. A cluster of granule-form neurons with small spheroid perikarya and extensive neurital field has developed. Their neurites interlace with those of peripheral Purkinje-form neurons with large ovoid perikarya and multipolar Golgi-form neurons. (B) Cortical neurons developed in culture for 96 hr as in A, but with 1 μM A23187 in the culture medium for the last 24 hr of culture. Practically all neurites are lost, giving the neurons the appearance of having a precursor neuron morphology. A comparison of the sialyl-labeling in the gangliosides of the control and A23187-exposed neurons is presented in Figure 3.

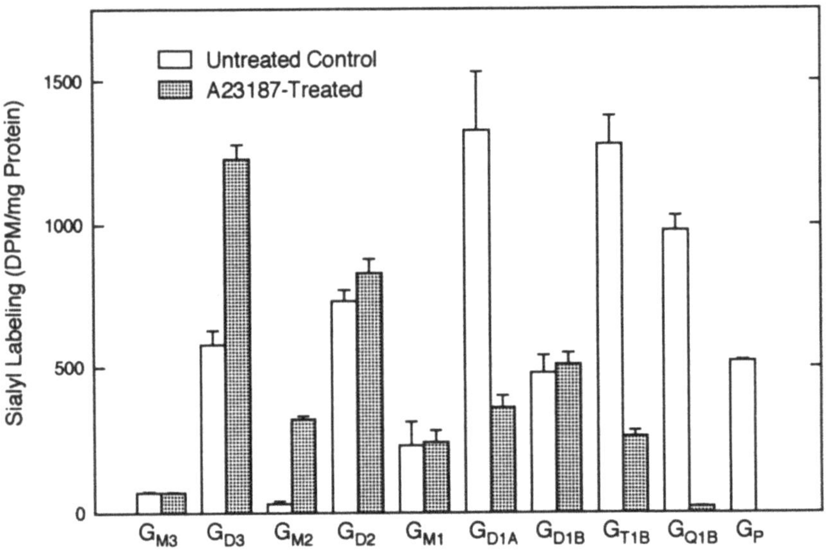

FIGURE 3. Sialyl-labeling in the ganglioside species of control neurite-bearing 96-hr chick embryo cortical neurons in culture compared with the sialyl-labeling in neurons exposed to Ca^{2+} ionophore A23187 for the last 24 hr of their culture span.

the neurite-bearing embryonic neurons with A23187, by heightening intraneuronal Ca^{2+}, brings about a decreased sialylation of the major neuritogenic gangliosides GT1b and GQ1b and a correlative loss of neurites. By implication, decreased sialylation of these neuritogenic gangliosides may underlie the neuronodystrophy that is a feature of pathological conditions such as neuronal anoxia, as well as Huntington's and Alzheimer's disease, since such pathological conditions stem in part from Ca^{2+}-heightening effects of glutamate excitotoxicity.

4. GANGLIOSIDE UNDERSIALYLATION IN THE FETAL ALCOHOL SYNDROME

Chronic exposure to ethanol impairs neuritogenesis-dependent development of embryonic cortical neurons (Miller, 1993) thereby leading to the symptoms of central nervous system dysfunction that are a major component of the fetal alcohol syndrome. Dystrophic consequences of exposure of the early neuron to ethanol have long been recognized to include impaired neuritogenesis (Dow and Riopelle, 1985). As indicated in Section 2, sialylation of the higher neuritogenic gangliosides paces neuritogenesis in a model system for study of cortical neuron evolvement.

Exposure of cultured embryonic chick neurons to 25 mM ethanol, a level easily achievable *in utero* during the first gestational trimester in the imbibing pregnant human, results in a drastic lessening of sialylation of all major gangliosides, as shown in Figure 4, thereby producing a correlated widespread loss of neurites (Rosenberg and Noble, 1994). An analysis of the mode of attenuated ganglioside sialylation brought about by ethanol exposure of the neuron reveals a major difference between the effect of ethanol and that of heightened intraneuronal Ca^{2+}.

With heightened intraneuronal Ca^{2+}, sialylation of GM3 is not impeded, and sialylation of the simpler gangliosides GD3, GM2, and GD2, which are derived from GM3 are substantially increased, while sialylation of the higher neuritogenic gangliosides GD1a, GT1b, GQ1b, and GP is lowered a great deal, so that inhibition of biosynthetic transport is evoked as the impairment mechanism as explained in Section 3. With ethanol, however, all of the major neuronal gangliosides show greatly decreased sialylation, as seen in Figure 4. It is of interest that ethanol induces severe attenuation of sialylation of GM3 (the initial precursor for eventual buildup of the neuritogenic ganglioside GD1a), and ethanol also induces severe attenuation of sialylation of GD3 (the initial precursor for buildup of the neuritogenic gangliosides GT1b and GQ1b). The diminished sialylation of the higher neuritogenic gangliosides may be accounted for in part

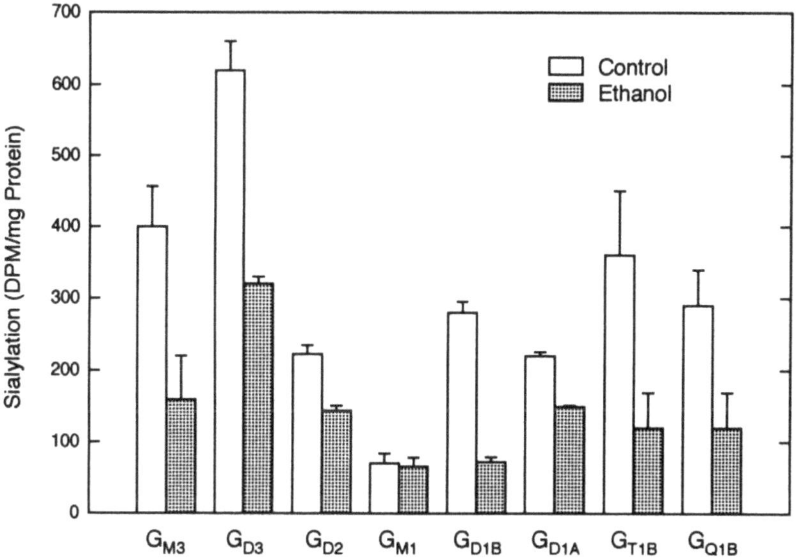

FIGURE 4. Effects on the sialyl-labeling of ganglioside species on exposure of 96-hr chick brain cortical neurons in culture to 25 mM ethanol in the culture medium for the last 24 hr of their culture span.

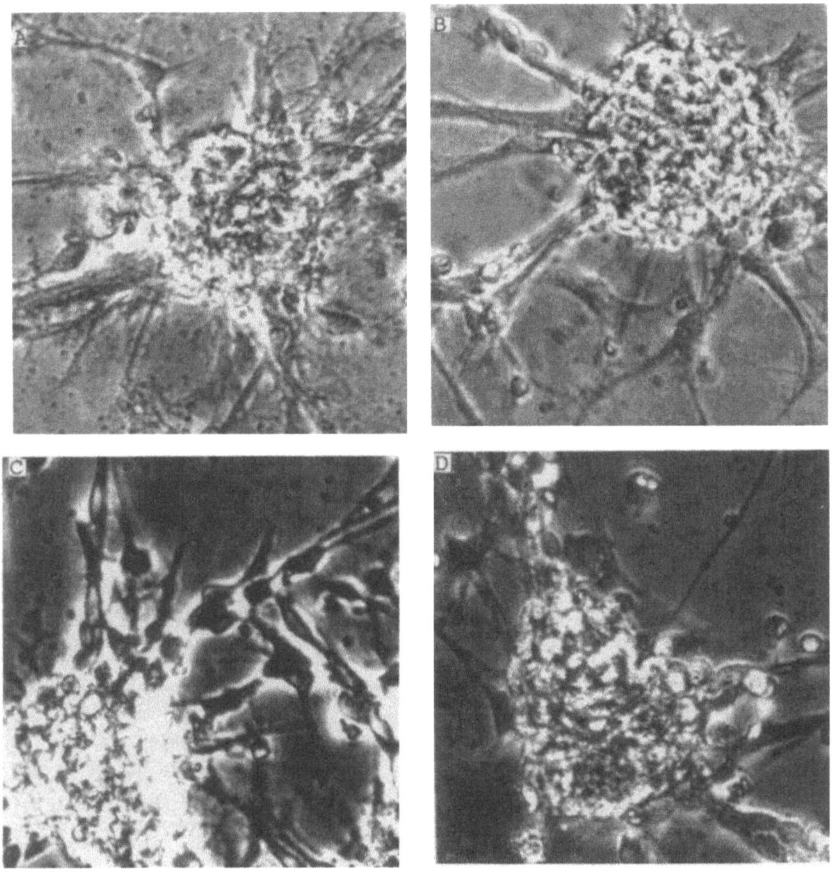

FIGURE 5. Neurite loss induced by exposure of 96-hr neurite-bearing cultured embryonic chick brain cortical neurons to graded moderate levels of ethanol in their serum-free neurotrophic culture medium for the last 24 hr of their culture span. (A) Control neurons, no ethanol; (B) 2.5 mM ethanol; (C) 10 mM ethanol; (D) 25 mM ethanol.

by the attenuated level of production of the initial precursor gangliosides GM3 and GD3, which are the obligate substrates for the buildup of the higher gangliosides. While undersialylation of these latter simpler gangliosides may be related in some way to impaired activity of GM3- and GD3-synthesizing ectosialyl transferases 1 and 2 (Rosenberg and Noble, 1992) which operate at a decreased level during ethanol-impaired cytoskeleton-regulated endosomal recycling of plasma membrane elements (Rosenberg and Noble, 1992), or whether it is caused by ethanol-impaired developmentally regulated genetic expression of sialyl transferases 1 or 2, or inhibition of expression of enzymatic activity at the transcriptional or cotranslational activation levels of production of these enzymes is a new question in sialobiology now under careful scrutiny.

As shown in Figure 5A, embryonic chick cortical precursor neurons at the 96-hr (neurite turnover/maintenance) developmental stage in serum-free culture (Rosenberg and Noble, 1992) have developed into full neurite-bearing cortical neuronal forms. In greatest number are granule neuronlike forms that have aggregated into clusters of small perikarya from which massive neurite bundles emanate, and peripheral to the perikaryal clusters of the granule neuron forms lie Purkinje-like neurons with large ovoid perikarya and finally there are multipolar Golgi neuronlike forms, with neurites from all of these cortical neuronlike subtypes interlacing to form a robust neuritial field. Exposure of the neurons to 25 mM ethanol for the last 24 hr of their 96-hr developmental period in culture leads to a drastic evolvement of neurite loss in all of the cortical neuronal forms as seen in Figure 5D. However, intact neuronal perikarya still are in evidence.

With only 10 mM ethanol in the culture medium of the neurons for the last 24 hr of their 96-hr culture span, an ethanol level that would extrapolate to the induction of a state of mild inebriation in the pregnant human, neuritial destruction is almost as severe as that produced by 25 mM ethanol, Figure 5C, and so is the attenuation of ganglioside sialylation (data not shown). Only at a very low ethanol level, i.e., 2.5 mM, does attenuation of the neuritial field appear not to be significant, as seen in Figure 5B, and attenuation of ganglioside sialylation is not statistically significant at this ethanol level (data not shown). These findings lead to the caveat that chronic exposure of the developing cortical neuron to moderate levels of ethanol may induce serious impairment of ganglioside-dependent neuritogenesis by inhibiting ganglioside sialylation, leading inexorably to the central nervous system aspect of the fetal alcohol syndrome.

REFERENCES

Bergeron, J.J.M., Brenner, M. B., Thomas, D. Y., and Williams, D. B., 1994, Calnexin: A membrane-bound chaperone of the endoplasmic reticulum, *Trends Biochem. Sci.* **19**:124–128.

Carafoli, E., 1987, Intracellular calcium homeostasis, *Annu. Rev. Biochem.* **56**:395–443.

Chan, K.-F.J., 1987, Ganglioside-modulated protein phosphorylation, *J. Biol. Chem.* **262**:5248–5255.

Choi, D. W., 1987, Ionic dependence of glutamate neurotoxicity, *J. Neurosci.* **7**:369–379.

Coyle, J. T., and Schwarcz, R., 1976, Lesion of striatal neurons with kainic acid provides a model for Huntington's chorea, *Nature* **263**:244–246.

Dow, K., and Riopelle, R. J., 1985, Ethanol neurotoxicity: Effects on neurite formation and neurotrophic factor production in vitro, *Science* **228**:591–593.

Feizi, T., 1994, Evidence for carbohydrate-mediated interactions between the neural cell-adhesion-molecules NCAM and L1, *Trends Biochem. Sci.* **19**:233–234.

Greenamyre, J. T., and Young, A. B., 1989, Excitatory amino acids and Alzheimer's disease, *Neurobiol. Aging* **10**:593–602.

Hogan, M. V., Saito, M., and Rosenberg, A., 1988, Influence of monensin on ganglioside anabolism and neurite stability in cultured chick neurons, *J. Neurosci. Res.* **20**:390–394.

Igarishi, M., Waki, H., Saito, S., Komiya, Y., and Ando, S., 1994, Characteristics of gangliosides including O-acetylated species in growth cone membranes at several developmental stages in rat forebrain, *Dev. Brain Res.* **78**:17–24.

Mattson, M. P., 1990, Antigenic changes similar to those seen in neurofibrillary tangles are elicited by glutamate and Ca^{2+} influx in cultured hippocampal neurons, *Neuron* **4**:105–117.

Mattson, M. P., Rychlik, B., Chu, C., and Chrystakos, S., 1991, Evidence for calcium-reducing and excitoprotective roles for the calcium-binding protein calbindin-D_{28K} in cultured hippocampal neurons, *Neuron* **6**:41–51.

Miller, M. W., 1993, Migration of cortical neurons is altered by gestational exposure to ethanol, *Alcohol Clin. Exp. Res.* **17**:304–314.

Rosenberg, A., and Noble, E. P., 1992, Complex higher gangliosides made in Golgi spur outgrowth of neurites in chick brain neurons, *Brain Dysfunct.* **5**:184–191.

Rosenberg, A., and Noble, E. P., 1993, Inhibition of neuritogenic ganglioside biosynthesis in chick brain neurons by calcium ionophore A23187, *Biochem. Biophys. Res. Commun.* **190**:522–528.

Rosenberg, A., and Noble, E. P., 1994, Ethanol attenuation of ganglioside sialylation and neuritogenesis, *Alcohol* **11**:565–569.

Rosenberg, A., Sauer, A., Noble, E. P., Gross, H.-J., Chang, R., and Brossmer, R., 1992, Developmental patterns of ganglioside sialylation coincident with neuritogenesis in cultured embryonic chick brain neurons, *J. Biol. Chem.* **267**:10607–10612.

Rothman, S. M., and Olney, J. W., 1986, Glutamate and the pathology of hypoxic-ischemic brain damage, *Ann. Neurol.* **19**:105–111.

Van Echten, G., and Sandhof, K., 1989, Modulation of ganglioside biosynthesis in primary cultured neurons, *J. Neurochem.* **52**:207–214.

Index